Intelligent Computing Techniques in Biomedical Imaging

Methods, Case Studies, and Applications

Intelligent Computing Techniques in Biomedical Imaging

Methods, Case Studies, and Applications

Edited by

Bikesh Kumar Singh
Department of Biomedical Engineering, National Institute of Technology Raipur, Raipur, Chhattisgarh, India

G. R. Sinha
GSFC University, Vadodara, Gujarat, India

ELSEVIER

ACADEMIC PRESS
An imprint of Elsevier

Academic Press is an imprint of Elsevier
125 London Wall, London EC2Y 5AS, United Kingdom
525 B Street, Suite 1650, San Diego, CA 92101, United States
50 Hampshire Street, 5th Floor, Cambridge, MA 02139, United States

ISBN: 978-0-443-15999-2

For Information on all Academic Press publications
visit our website at https://www.elsevier.com/books-and-journals

Publisher: Mara Conner
Acquisitions Editor: Chris Katsaropoulos
Editorial Project Manager: Tessa Kathryn
Production Project Manager: Maria Bernard
Cover Designer: Matthew Limbert

Typeset by MPS Limited, Chennai, India

Contents

Section IV
Intelligent techniques for medical image analysis

Section V
Modern applications (Case studies)

List of contributors

K. Adalarasu School of Electrical and Electronics Engineering, SASTRA Deemed to be University, Thanjavur, Tamil Nadu, India

K. Aravindh School of Electrical and Electronics Engineering, SASTRA Deemed to be University, Thanjavur, Tamil Nadu, India

K. Ghousiya Begum School of Electrical and Electronics Engineering, SASTRA Deemed to be University, Thanjavur, Tamil Nadu, India

Ajoy K. Behera Department of Pulmonary Medicine and TB, AIIMS, Raipur, Chhattisgarh, India

Kanmani Buddhi Department of Electronics and Telecommunication Engineering, BMS College of Engineering, Bengaluru, Karnataka, India

Trilok Chand Department of Computer Science Engineering, Punjab Engineering College (Deemed to be University), Chandigarh, Punjab, India

D.K. Chandraker Reactor Engineering Division, Bhabha Atomic Research Centre, Mumbai, Maharashtra, India

Devendra K. Chouhan Department of Orthopedics, Post Graduate Institute of Medical Education & Research, Chandigarh, Punjab, India

R. Keerthik Dhivya School of Electronics Engineering, Vellore Institute of Technology (VIT), Chennai, Tamil Nadu, India

Arun Kumar Dhuraisamy Department of Robotics and Automation, PSG College of Technology, Coimbatore, Tamil Nadu, India

Padma Ganasala Electronics and Communications Engineering, Gayatri Vidya Parishad College of Engineering, Visakhapatnam, Andhra Pradesh, India

Ranganath T. Ganga Department of Pulmonary Medicine and TB, AIIMS, Raipur, Chhattisgarh, India

Saurabh Gupta Department of Biomedical Engineering, National Institute of Technology Raipur, Raipur, Chhattisgarh, India

Subhanshu Gupta Department of Computer Science and Engineering, National Institute of Technology Raipur, Raipur, Chhattisgarh, India

M. Jagannath School of Electronics Engineering, Vellore Institute of Technology (VIT), Chennai, Tamil Nadu, India

Pankaj Jain Department of Biomedical Engineering, National Institute of Technology Raipur, Raipur, Chhattisgarh, India

Pankaj Kumar Jain Artificial Intelligence and Data Science, Jio Institute, Navi Mumbai, Maharashtra, India; School of Biomedical Engineering, Indian Institute of Technology (BHU), Varanasi, Uttar Pradesh, India

Sonam Jain Department of Product Management, Medicare Imaging, Navi Mumbai, Maharashtra, India

Moushmi Kar Department of Electronics, PTSNS University, Shahdol, Madhya Pradesh, India

Shivansh Kaushik Department of Chemistry, University of California Riverside, Riverside, CA, United States

Ashima Khosla Department of Computer Science Engineering, Punjab Engineering College (Deemed to be University), Chandigarh, Punjab, India

Vaibhav Koshta Department of Biomedical Engineering, National Institute of Technology Raipur, Raipur, Chhattisgarh, India

Roopa R. Kulkarni Department of Electronics and Communication Engineering, Dayananda Sagar Academy of Technology & Management, Bengaluru, Karnataka, India

Danyal Mahmood Department of Electronic Engineering, Faculty of Engineering and Green Technology, Universiti Tunku Abdul Rahman, Kampar, Malaysia

Susmita Ghosh Mazumadar Department of Electronics, PTSNS University, Shahdol, Madhya Pradesh, India

N. Nathiya School of Advanced Sciences, Vellore Institute of Technology (VIT), Chennai, Tamil Nadu, India

Humaira Nisar Department of Electronic Engineering, Faculty of Engineering and Green Technology, Universiti Tunku Abdul Rahman, Kampar, Malaysia

Ching Wei Ong Department of Electronic Engineering, Faculty of Engineering and Green Technology, Universiti Tunku Abdul Rahman, Kampar, Malaysia

Prashant P. Patavardhan Department of Electronics and Communication Engineering, RV Institute of Technology & Management, Bengaluru, Karnataka, India

Mahesh Prakash Department of Radiology, Post Graduate Institute of Medical Education & Research, Chandigarh, Punjab, India

A.D. Prasad Civil Engineering Department, National Institute of Technology Raipur, Raipur, Chhattisgarh, India

Prateek Pratyasha Department of Biomedical Engineering, National Institute of Technology Raipur, Raipur, Chhattisgarh, India

Deepak Saini Department of Information System, LM Thapar School of Management, Derabassi Campus, TIET, Patiala, Punjab, India

Padma Selvaraj Department of Computer Science & Technology, Madanapalle Institute of Technology & Science, Madanapalle, Andhra Pradesh, India

Neeraj Sharma School of Biomedical Engineering, Indian Institute of Technology (BHU), Varanasi, Uttar Pradesh, India

S.J. Shivasundar School of Electrical and Electronics Engineering, SASTRA Deemed to be University, Thanjavur, Tamil Nadu, India

Bikesh Kumar Singh Department of Biomedical Engineering, National Institute of Technology Raipur, Raipur, Chhattisgarh, India

Nivedita Singh Department of Electronics and Telecommunication Engineering, Government Engineering College, Raipur, Chhattisgarh, India

Rohit Raj Singh Department of Biomedical Engineering, National Institute of Technology Raipur, Raipur, Chhattisgarh, India

G.R. Sinha GSFC University, Vadodara, Gujarat, India

S.L. Sinha Department of Mechanical Engineering, National Institute of Technology Raipur, Raipur, Chhattisgarh, India

Dilip Singh Sisodia Department of Computer Science and Engineering, National Institute of Technology Raipur, Raipur, Chhattisgarh, India

S. Sivaprakash Department of Software Systems, School of Computer Science and Engineering, Vellore Institute of Technology, Vellore, Tamil Nadu, India

G. Subashini Department of Robotics and Automation, PSG College of Technology, Coimbatore, Tamil Nadu, India

Kalyan Tadepalli Artificial Intelligence and Data Science, Jio Institute, Navi Mumbai, Maharashtra, India

R.H. Talwekar Department of Electronics and Telecommunication Engineering, Government Engineering College, Raipur, Chhattisgarh, India

Hardik Thakkar Department of Biomedical Engineering, National Institute of Technology Raipur, Raipur, Chhattisgarh, India

Priyanka Bibay Thakkar Department of Electronics and Telecommunication Engineering, Government Engineering College, Raipur, Chhattisgarh, India

Harish Velingkar Department of Electronics and Communication Engineering, Agnel Institute of Technology & Design, Assagao, Goa, India

Shashi Kant Verma Department of Mechanical Engineering, National Institute of Technology Durgapur, Durgapur, West Bengal, India

P. Vijayakumar School of Electronics Engineering, Vellore Institute of Technology (VIT), Chennai, Tamil Nadu, India

Preface

In the constantly changing field of biomedical imaging, the combination of advanced computing methods with diagnostic approaches has become a powerful driving factor. This book, *Intelligent Computing Techniques in Biomedical Imaging*, showcases the impressive collaboration between computational progress and the complex field of medical imaging. Biomedical imaging is crucial for uncovering the complexities of the human body, providing a noninvasive glimpse into its detailed structures and functions. By incorporating advanced computing methods such as machine learning, deep learning, and image processing algorithms, the field of biomedical imaging has reached new levels of sophistication.

In this thorough investigation, we unite a group of respected researchers, practitioners, and visionaries who have explored the merging of advanced computing and biomedical imaging. Their work covers a wide range of applications, such as disease diagnosis, image segmentation, feature extraction, and image reconstruction. The combination of these advanced methodologies not only improves the precision and effectiveness of diagnostic processes but also opens the way for innovative research. This book has been carefully constructed to be a valuable tool for professionals, researchers, and students exploring the ever-changing field of biomedical imaging. Every chapter explores deeply into the latest techniques, methodologies, and case studies that highlight the power of advanced computing in overcoming challenges and unleashing possibilities in the field of medical imaging.

The book is divided into 5 sections which includes 20 chapters. The organization of the book is as follows:

Section I: Prerequisites: This section provides an overview of medical image analysis, intelligent techniques, and their practical applications. The section covers the fundamentals of probability and statistics. Altogether, this section is segmented into three chapters.

Section II: Biomedical Imaging: This section covers the fundamentals of data acquisition and preprocessing. The section contains a chapter that outlines the steps involved in computer-aided diagnosis and the evaluation of the developed model's performance.

Section III: Medical Image Description and Representation: This portion of the book contains chapters that are concerned with the extraction of features and the selection of features in conjunction with computer vision and medical pictures.

Section IV: Intelligent Techniques for Medical Image Analysis: This section of this book investigates chapters discussing advanced algorithms such as machine learning and deep learning techniques applied in various medical imaging scenarios. The chapter explores into image segmentation and template-matching algorithms for classifying knee osteoarthritis grades. Aside from pulmonary imaging, the research also covers digital breast tomosynthesis versus 2D mammography and Alzheimer's detection.

Section V: Modern Applications: This section includes 10 case studies: (1) detecting pediatric renal disease at an early stage, (2) imaging in neuro-developmental disorders (3) identifying drowsiness in sleep apnea patients using multiple biosignals, (4) identifying obstructive sleep apnea using a single-channel electrocardiogram, (5) topographic maps of anxiety in the brain, (6) intelligent computing in brain–computer interface, (7) identifying diabetic retinopathy from greyscale fundus images, (8) developing a framework for identifying leg postures based on 1D force data through machine learning models, (9) exploring artificial intelligence for biomedical applications related to fluid-material dynamics, and (10) MRI and SPECT image fusion based on an adaptive fusion rule in the transform domain.

As editors, we want to express our sincere appreciation to the contributors who have generously shared their knowledge and perspectives to bring this compilation to life. Their dedication to pushing the boundaries has certainly enhanced the overall understanding of the biomedical imaging community. We trust this book will act as an enlightening manual, nurturing a more profound comprehension of the interdependent connection between advanced computing and biomedical imaging. May it inspire further innovation, collaboration, and discoveries that push the boundaries of what is possible in the pursuit of better healthcare through advanced imaging technologies.

Bikesh Kumar Singh
G.R. Sinha

Section I

Prerequisites

Chapter 1

Introduction to intelligent medical image analysis

Rohit Raj Singh[1], Bikesh Kumar Singh[1] and G.R. Sinha[2]

[1]Department of Biomedical Engineering, National Institute of Technology Raipur, Raipur, Chhattisgarh, India, [2]GSFC University, Vadodara, Gujarat, India

1.1 Brief history of medical imaging

First, X-rays were classified in 1895. In a remarkable span of less than a decade, X-ray imaging was widely implemented as a diagnostic procedure in major institutions across Western countries.

During that period, exposure periods for a single image were extremely lengthy and sensitivity was low. Radiation injuries were prevalent during the early years of diagnostic and recreational X-ray use when research on the biological effects of X-rays was limited [1].

The next big breakthrough was magnetic resonance imaging (MRI), which was discovered in the late 1970s. Due to the utilization of the Fourier transform in the generation of the cross-sectional image, MRI also depends on digital data processing. Subsequent to that juncture, advancements have been more gradual in nature, albeit marking significant improvements in both image quality and acquisition speed [1].

In 1972, Geoffrey Hounsfield unveiled an innovative device that utilized X-rays to produce cross-sectional images as opposed to planar ones. The technique was termed tomography. Computed tomography (CT) or computerized axial tomography (CAT) is the imaging modality that pioneered image creation using digital computers [2].

1.2 Why medical imaging

The primary advantage of medical image processing is its capacity to facilitate comprehensive, noninvasive anatomy exploration. To improve patient outcomes, develop more precise diagnoses, and construct superior medical apparatus and drug delivery systems, it is possible to create and assess three-dimensional models of the pertinent anatomy. It has become one of the most significant instruments for the advancement of medicine in recent years.

The continuous enhancement of image quality in conjunction with advanced software tools enables precise digital reproduction of anatomical structures, including those of varying dimensions and with distinct properties like soft tissues and bone.

By integrating actual anatomical geometries into simulation models, statistical analysis, and measurement, it is possible to attain a more comprehensive understanding of phenomena such as the interplay between medical devices and patient anatomy.

1.3 How medical imaging works

Medical image processing begins with the acquisition and reconstruction of raw data from CT or MRI images into a format compatible with the software involved. Typical image processing input consists of a voxel (3D pixel) grid encompassing a three-dimensional (3D) bitmap of greyscale intensities. The greyscale intensity of a CT scan is determined by X-ray absorption, whereas in MRI it is based on the signal strength emitted by proton particles both during relaxation and subsequent to the application of extremely strong magnetic fields.

Typically, the reconstructed image volume is segmented and edited to isolate and represent distinct anatomical regions, including bone and tissue, for the benefit of medical users.

Intelligent Computing Techniques in Biomedical Imaging. DOI: https://doi.org/10.1016/B978-0-443-15999-2.00015-3

This chapter's objective is to help scientists choose a suitable AI technique and then mentor them in figuring out the optimal plan of action using medical imaging. In addition, the purpose of this article is to familiarize readers with the fundamentals of artificial intelligence (AI) before getting into real problems that need to be solved using AI techniques. Machine learning and deep learning are two examples of essential concepts that are explained through demonstrations. Discourse will be devoted to subjects concerning the diverse AI methodologies, including supervised and unsupervised learning.

1.4 Types of medical imaging

Medical imaging techniques (MITs) are autonomous approaches that do not require surgical incisions to visualize the interior of the body. It was implemented to aid in the diagnosis and treatment of various medical syndromes. Diverse MITs exist, each of which entails unique advantages and disadvantages [3].

1.4.1 X-ray projection imaging

The earliest medical imaging modality was X-ray imaging, which was implemented in practice shortly after its discovery in 1895. As an alternative to the conventional method of image formation on photosensitive film, which is utilized for X-ray imaging projection, direct digital X-ray imaging is gaining popularity. X-rays are classified as high-energy photons, and their generation, as well as the generation of X-ray contrast, are both contingent on atomic interaction with electrons within the inner shell. "X-rays" are produced through the deceleration of kinetic electrons upon interaction with a solid target. A fraction of the kinetic energy of electrons is released as X-ray particles. Element ions are accelerated in the direction of the anode by means of a high-voltage electrostatic field after being emitted into the vacuum at the cathode of an X-ray tube. A partial loss of energy occurs when electrons are deflected in the dense metallic anode material as a result of the coulombic field generated by the nucleus (Fig. 1.1) [1].

The nucleus' coulombic field deflects high-energy electrons approaching the anode material. Under perfect conditions (X), the electron collides directly with the nucleus, emitting one X-ray photon with its full kinetic energy. A near "fly-by" (Y) is more likely, but the electron loses less energy, producing a longer-wavelength, lower-energy X-ray photon.

X-ray generation is a straightforward process wherein electrons undergo acceleration due to a potential difference, resulting in their conversion into electromagnetic radiation [4]. A generator and an X-ray tube, each of which is contained within a vacuum, are the fundamental elements of X-ray generation. In addition to a vacuum enclosure and high voltage cables that connect the cathode and anode components to the X-ray generator, an X-ray tube comprises a cathode and an anode that are separated by a brief distance [5].

1.5 Computed tomography

CT, alternatively referred to as CT scanning or CAT scanning, is a medical imaging modality that generates intricate cross-sectional images of the body through the utilization of X-rays and computer processing. By offering a noninvasive method of visualizing the intricate internal structures of bones, organs, and tissues, it enables medical practitioners to diagnose and monitor an extensive array of medical conditions [6].

CT, as opposed to projection X-ray imaging, permits the generation of an absorber map in two or three dimensions. CT contrast in soft tissue is considerably superior to that of projection X-ray imaging; however, the spatial resolution of a clinical whole-body CT scanner is considerably diminished in comparison to conventional X-ray imaging [2].

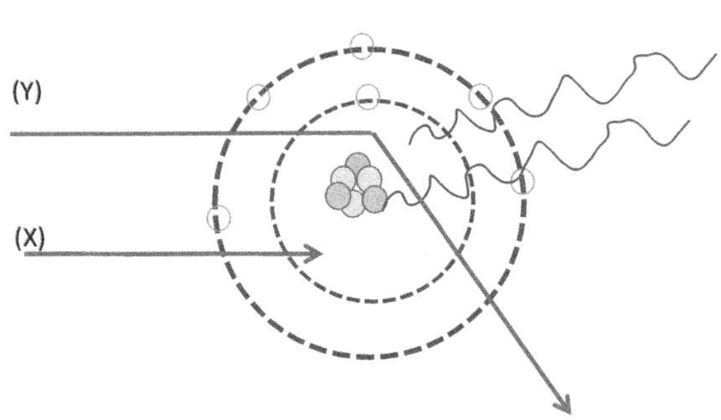

FIGURE 1.1 Generation of X-rays via electron deceleration.

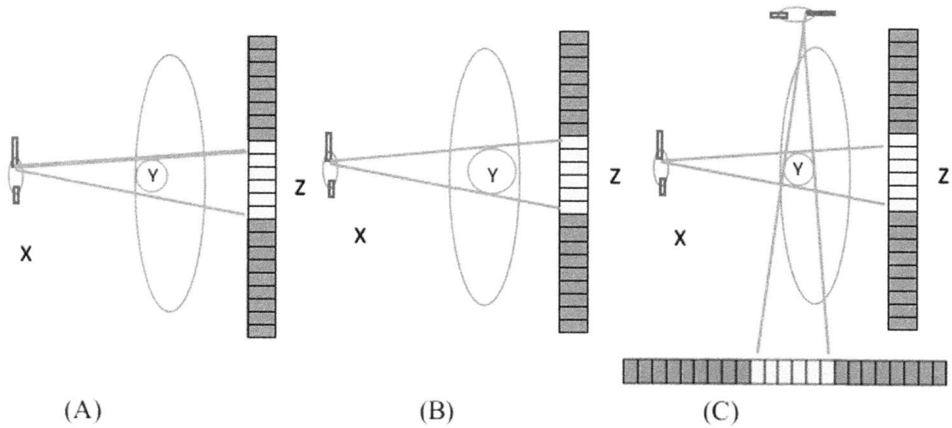

FIGURE 1.2 Image uncertainty in radiography.When an object (B), such as a tumor, attenuates the X-rays emitted by the source (A), a silhouette (C) is produced on the detector. Despite the considerable discrepancy in the location of the tumor, examples X and Y exhibit a substantial degree of similarity in the projected image. Z is the user-entered text. Two X-ray images that are perpendicular to one another, such as an anterior-posterior projection and a lateral projection, can be obtained to reduce the ambiguity.

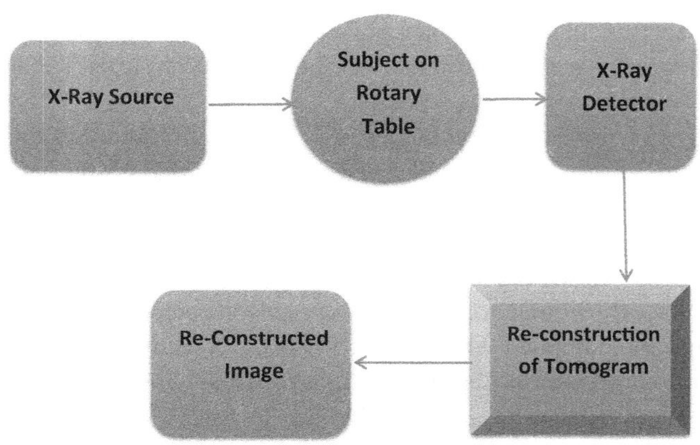

FIGURE 1.3 Computed tomography block-diagram.

Cross-sectional imaging (CT) was first implemented in 1971 and is the first imaging modality to require the assistance of a computer for image reconstruction: the cross-sectional image is generated by transforming a series of X-ray projections. Contrast, image quality, spatial resolution, and acquisition time have all undergone significant development since the initial CT scanners were introduced. Contemporary clinical CT scanners operate at breakneck speeds, generating two-dimensional cross-sectional images in under one second. Sophisticated CT microscopes yield voxels with a size smaller than 10 μm in the in-plane direction.

1.5.1 Principle of computed tomography imaging

The precise spatial coordinates of areas of interest (e.g., tumors or lesions) are not discernible in projection images, including standard X-ray projection images. As an illustration, the X-ray projection images produced by the two objects (B) in Fig. 1.2A and B are comparable.

Radiologists often obtain two perpendicular projections, denoted as lateral and AP (anterior and posterior, respectively), as depicted in Fig. 1.2C.

The principal aim of CT is to acquire precise and cross-sectional images of the bodily internal structures to facilitate diagnosis and treatment. As per our definition, a map can be classified as an image if it is sampled at a finite resolution. Tomography, an imaging technique originating from the Greek verbs "to cut" and "to graph," eliminates the requirement for a physical scalpel to visualize a cross-sectional incision (Fig. 1.3) [7].

1.6 Nuclear imaging

Nuclear imaging, alternatively referred to as nuclear medicine, is a medical imaging modality that visualizes and evaluates the functionality of bodily organs and tissues through the use of radiopharmaceuticals, which are minute quantities of radioactive substances. In contrast to alternative imaging modalities such as CT scans and X-rays, which

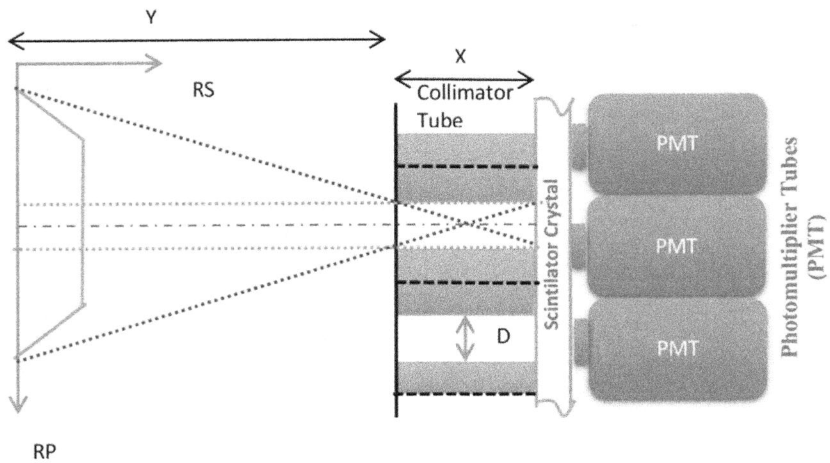

FIGURE 1.4 An illustrative depiction that highlights the spatial resolution of an assemblage of collimated photomultiplier tubes.

predominantly convey anatomical details, nuclear imaging directs its attention toward physiological processes. In general, these radioactive compounds are coupled to pharmacologically active substances ("radiopharmaceuticals") that accumulate specifically in tumors and other anatomical sites.

By employing volumetric computerized image reconstruction or projection techniques, it is possible to ascertain the spatial distribution of the radiopharmaceutical [8]. Metabolic processes can be imaged and utilized for diagnostic purposes in this manner.

The function of tissues and organs is investigated via nuclear imaging. During the procedure, a minute quantity of a radioactive substance is employed to aid in the examination. Body tissue absorbs the radioactive substance, which is known as a radionuclide (radiopharmaceutical or radioactive tracer). Similar to CT, three-dimensional reconstructions are acquired, resulting in the development of single-photon emission computed tomography (SPECT). Positron emission tomography (PET), an alternative technology, operates by utilizing positron emitters to generate coincident pairs of gamma photons.

The manner in which the radiotracer is administered during a nuclear medicine examination varies; it may be subcutaneously injected, orally ingested, or inhaled as a gas. Ultimately, it accumulates within the targeted organ or anatomical region and emits gamma radiation as a means of energy transmission. This energy is detected by a gamma camera, an instrument, or a PET scanner. These devices collaborate with a computer to quantify the quantity of radiotracer assimilated by the body and generate specialized images that provide intricate information regarding the composition and operation of tissues and organs (Fig. 1.4).

The conversion layer facilitates the conversion of high-energy γ-photons to visible light by employing a scintillation crystal. Angle-emitted photons are attenuated through the use of a collimator layer. The collimators, which are contained within a lead layer, are cylindrical apertures measuring X in length and D in diameter. The relative sensitivity RS exhibits a trapezoidal shape as the radial position RP from the center of the collimator tube increases in relation to the distance Y from the detector array. Furthermore, the RS curve gradually intersects with the sensitivity curve of the adjacent photomultiplier tubes, which is denoted in gray.

1.7 Magnetic resonance imaging

MRI, or magnetic resonance imaging, is a volumetric imaging technique that shares some similarities with CT. Nevertheless, the fundamental physical principles that underlie this differ significantly from CT. MRI relies on the spatial arrangement of protons within a robust magnetic field, while CT produces contrast by means of the interaction between electrons in the atomic shell and high-energy radiation.

1.7.1 Working of magnetic resonance imaging

MRI utilizes robust magnets to generate a magnetic field of such strength that compels protons within the body to align with it. Following this, a radiofrequency current is pulsed through the patient, which induces protons to spin erratically in opposition to the magnetic field's attraction. MRI sensors can detect the energy released when the radiofrequency field is deactivated, resulting from the realignment of protons with the magnetic field. The quantity of energy released and the time required for the protons to realign with the magnetic field are variables that are influenced by both the surrounding environment and the chemical composition of the molecules. Medical professionals possess the capability to differentiate between different varieties of tissues by leveraging these magnetic characteristics.

Types of MRI:

- T1-weighted images are frequently employed in structural imaging due to their substantial anatomical detail.
- T2-weighted images are frequently employed to identify abnormalities such as tumors or inflammation due to their sensitivity to changes in tissue water content.
- Functional MRI: by detecting alterations in blood flow and oxygenation, this variant maps and measures brain activity.
- Diffusion-weighted imaging is a diagnostic technique that assists in the identification of conditions such as stroke by analyzing the motion of water molecules within tissues.
- Magnetic resonance angiography is a diagnostic technique that analyses vascular conditions by means of visualizing blood vessels.

1.8 Ultrasound imaging

Ultrasound imaging, also known as sonography, is an MIT that generates dynamic, real-time images of the internal structures of the body using high-frequency sound pulses [9]. Pressure waves in the low megahertz range undergo refraction and partial reflection at boundaries due to their speed-of-sound propagation through tissue. The presence of tissue echogenic heterogeneities is therefore correlated with ultrasound contrast. The determination of the depth of an echogenic object is possible by analyzing the travel duration of the echo. Two-dimensional surveys are practicable through the emission of focused sound waves in distinct directions. An exceptional level of image quality is the result of a complex interplay between inhomogeneous tissue and echoes, variations in the speed of sound across tissues, and a significant noise component caused by weak signal and high amplification.

The operation of ultrasound imaging is predicated on the propagation of sound waves. High-frequency sound waves are transmitted into the body by a transducer; the resulting echoes, which are generated through processing of the waves' reflection off of tissues, constitute the images. Determining the depth and composition of the structures under examination is the time required for the reverberation to return. In contrast to alternative imaging modalities like X-rays or CT scans, ultrasound operates without the use of ionizing radiation, rendering it a risk-free procedure suitable for a wide range of patient populations, including expectant women.

1.8.1 Area of application

- Ultrasound is extensively employed within the field of obstetric care to assess fetal health and monitor fetal development. Moreover, it serves a pivotal function in gynecological examinations by providing valuable assistance in the identification of anomalies within the uterus and ovary cysts.
- Ultrasound aids in the visualization of the heart's structure and function in cardiology. Echocardiography, an advanced variant of ultrasound, plays a critical role in the evaluation of cardiac function, detection of abnormalities, and assessment of cardiac anatomy.
- Imaging abdominal organs such as the pancreas, liver, kidneys, and gallbladder is facilitated by ultrasound. It is frequently employed in the diagnosis of liver disease, kidney abnormalities, and gallstones.
- Ultrasound is a valuable tool in the assessment of joint health, soft tissue, and muscle health. By means of guiding injections, evaluating injuries, and visualizing structures in close proximity to the body's surface, it is commonly utilized in the field of sports medicine.
- Doppler ultrasound is a diagnostic modality employed to evaluate vascular conditions and blood flow. It is essential for the detection of blood clotting, the evaluation of blood vessels, and the diagnosis of conditions like deep vein thrombosis.

1.8.2 Modern advancement in ultrasound imaging

- The acquisition of real-time four-dimensional (4D) and 3D images has become feasible due to technological advancements in ultrasound. This facilitates a clearer perception of intricate structures and insights into dynamic phenomena, including the motion of the fetus.

- The introduction of contrast agents into the circulation has the potential to augment the visibility of specific structures in the context of ultrasound imaging. CEUS is especially beneficial for detecting abnormalities in organs like the liver and evaluating blood flow.
- The proliferation of portable ultrasound devices has enabled the implementation of point-of-care imaging across a diverse range of clinical environments. Handheld devices that are compact in size enable healthcare professionals to conduct rapid assessments in emergency situations, at the bedside, or in environments with limited resources.

1.8.3 Future of ultrasound imaging

- Automation of routine tasks, enhancement of precision, and early detection of abnormalities are all potential benefits of incorporating AI into ultrasound image analysis.
- Therapeutic ultrasound is increasingly being recognized for its potential applications beyond diagnostics, including targeted drug delivery, tissue ablation, and noninvasive therapies for conditions like tumors.

Ultrasound imaging is an extremely versatile and essential instrument utilized in the domain of medical diagnostics. Ultrasound's pervasive influence across various medical specialties is attributed to its noninvasive characteristics, real-time functionality, and ongoing technological progress. These attributes provide clinicians with invaluable insights that contribute to precise diagnoses and improved patient care. With the continuous advancement and integration of technology, including AI, ultrasound imaging is poised for a brighter future characterized by increased accuracy and a wider range of medical imaging applications.

1.9 Artificial intelligence in medical imaging

Throughout history, the field of medical image analysis has predominantly been dependent on the proficiency of clinicians and radiologists, who engaged in the manual interpretation of images to formulate diagnoses and strategies for treatment. AI has arisen as a game-changer due to the exponential growth of digital imaging and the availability of large datasets; it is capable of processing vast quantities of data to detect patterns, anomalies, and correlations that may elude the human eye. The integration of AI into medical imaging has emerged as a transformative force, offering the potential to improve diagnostic accuracy, efficiency, and overall patient care, as technology continues to advance.

AI in medical imaging entails the analysis and interpretation of medical images, including X-rays, CT scans, MRIs, and ultrasounds, through the implementation of machine learning algorithms. By being trained on enormous datasets of annotated medical images, these algorithms can discover patterns and anomalies that are difficult for human observers to consistently identify. In a number of crucial domains, the convergence of AI and medical imaging contains immense potential:

- Diagnostics and image interpretation: medical image anomalies and subtle patterns are a specialty of AI algorithms. They can facilitate the precise and rapid interpretation of images by radiologists, thereby promoting the early detection of various ailments including cancer, cardiovascular disease, and neurological disorders. In the end, this may enhance patient outcomes through the facilitation of timely interventions.
- Automation of routine tasks: in medical imaging analysis, AI can automate tedious and repetitive duties, freeing up healthcare professionals to concentrate on more intricate facets of patient care. An example of how AI algorithms can optimize the diagnostic process is by providing support in organ localization, segmentation, and measurement.
- Personalized medicine: by integrating medical images with additional patient data, including genetic information and electronic health records, AI algorithms have the capability to customize treatment strategies for each individual patient. The implementation of this individualized approach holds promise for enhancing treatment outcomes and optimizing therapeutic strategies.
- Enhanced work-flow mechanism: the incorporation of AI into medical imaging workflows has the potential to increase diagnostic turnaround times and decrease operational inefficiency. In hectic healthcare environments, automated image analysis can priorities critical cases, thereby optimizing the overall workflow.
- Quality improvement: AI has the potential to enhance image quality and promote standardization through the detection and rectification of artefacts and inconsistencies. By adhering to these standards, diagnostic images are produced, thereby enhancing the dependability and reproducibility of the findings.

Although there are significant potential advantages, the implementation of AI in medical imaging also poses certain difficulties. Care must be taken to address concerns such as data privacy, regulatory compliance, and the necessity for rigorous validation of AI algorithms. Furthermore, it is imperative to maintain the confidence of healthcare professionals and ensure a smooth integration with preexisting healthcare systems to achieve effective implementation.

1.10 Conclusion

The incorporation of AI into the analysis of medical imaging signifies a paradigm shift within the National Health Service of the United Kingdom. Through the integration of healthcare professionals' knowledge and the capabilities of machine learning algorithms, AI possesses the capacity to significantly transform diagnostics, enhance patient outcomes, and propel the progress of personalized medicine. The ongoing advancements and progress in this domain suggest that the integration of artificial intelligence and medical imaging will significantly transform the way in which healthcare is provided.

References

[1] J.T. Bushberg, J.A. Seibert, E.M. Leidholdt, J.M. Boone, The Essential Physics of Medical Imaging, Lippincott Williams & Wilkins, Philadelphia, 2002.

[2] W.A. Kalender, Computed Tomography: Fundamentals, System Technology, Image Quality, Applications, Publicis;, Erlangen, 2011.

[3] H. Kasban, M.A.M. El-Bendary, D.H. Salama, A comparative study of medical imaging techniques, Int. J. Inf. Sci. Intell. Syst. 4 (2) (2015) 37–58.

[4] J.A. Seibert, The AAPM/RSNA physics tutorial for residents. X-ray generators, Radiographics. 17 (6) (1997) 1533–1557.

[5] J.A. Seibert, X-ray imaging physics for nuclear medicine technologists. Part 1: basic principles of x-ray production, J. Nucl. Med. Technol. 32 (3) (2004) 139–147.

[6] C.P. Ho, R.W. Kim, M.B. Schaffler, D.J. Sartoris, Accuracy of dual-energy radiographic absorptiometry of the lumbar spine," cadaver study, Radiology 176 (1) (1990) 171.

[7] Hedrick, D.L. Hykes, D.E. Starchman, Ultrasound physics and instrumentation, Elsevier Mosby, St. Louis, 2005.

[8] A.C. Kak, M. Slaney, Principles of Computerized Tomographic Imaging, IEEE Press, New York, 1998. http://www.slaney.org/pct/pct-toc.html Accessed 08.12.

[9] Hounsfield, Computerized transverse axial scanning (tomography), Part 1. Description of system, Br. J. Radiol. 46 (552) (1973) 1016–1022.

Chapter 2

Introduction to intelligent techniques and applications

Pankaj Jain and Saurabh Gupta
Department of Biomedical Engineering, National Institute of Technology Raipur, Raipur, Chhattisgarh, India

Biomedical imaging (BI) is a branch of medical science that involves the application of various imaging methods to visualize the internal structures of the human anatomy. It has revolutionized the way we diagnose and treat diseases, providing physicians with noninvasive methods to study and monitor the anatomy and functioning of the human body [1]. Also to diagnose, treat, and monitor diseases, BI is essential. Imaging techniques such as X-rays, computed tomography (CT) scans, and magnetic resonance imaging (MRI) play a crucial role in identifying and diagnosing a wide range of conditions, including cancers, cardiovascular diseases, and neurological disorders [2]. Positron emission tomography (PET) scans is needed to evaluate cancer treatments and monitor Alzheimer's condition [3]. BI also plays a significant role in biomedical research. Imaging techniques like PET and MRI are useful for studying the impact of novel medications on the body and identify biomarkers that can be used to predict treatment outcomes.

2.1 History and evolution of biomedical imaging

BI has a rich history, starting with the invention of X-rays in 1895. Since then, imaging techniques have evolved rapidly, providing noninvasive ways to visualize the internal structures of the body [4]. CT scans, MRI, ultrasound imaging, and single photon emission computerized tomography (SPECT) are some commonly used imaging techniques that have become an integral part of healthcare and biomedical research [5].

Recent advancements in BI are focused on improving the resolution, accuracy, and safety of imaging techniques. Techniques like optical coherence tomography [6] and photoacoustic imaging [7] offer higher resolution and safety compared to traditional imaging techniques. Artificial intelligent (AI) algorithms are also playing an increasingly important role in BI, by assessing huge databases to detect abnormalities and patterns that might be challenging for human experts to acknowledge [8] (Fig. 2.1).

BI has completely transformed the way for diagnose, to treat, and monitor diseases. With the continual advancement of technology, it is reasonable to anticipate BI to become even more sophisticated, further improving patient outcomes and advancing medical research.

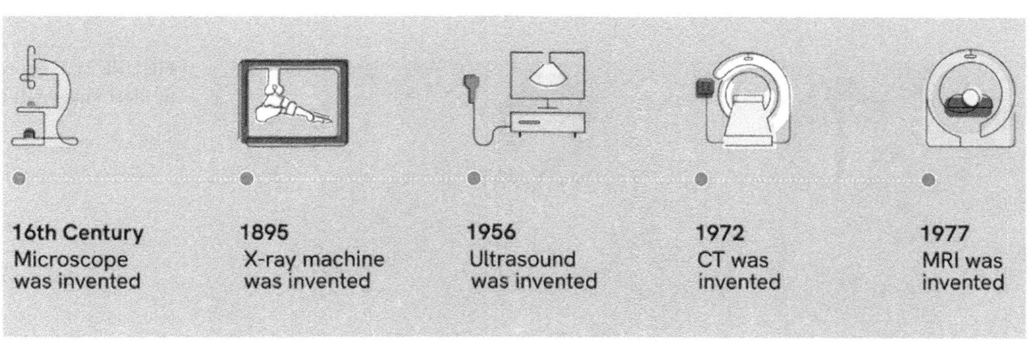

FIGURE 2.1 Evolution of biomedical imaging [https://chanzuckerberg.com/blog/the-past-present-and-future-of-medical-imaging/].

Intelligent Computing Techniques in Biomedical Imaging. DOI: https://doi.org/10.1016/B978-0-443-15999-2.00012-8

2.2 Importance of biomedical imaging

BI is a crucial tool in healthcare and research, allowing physicians and researchers to visualize the internal structures and functions of the body without invasive procedures. Its importance lies in its ability to improve patient outcomes, reduce healthcare costs, and advance medical research.

BI plays a vital role in healthcare by actively contributing to the early identification and diagnosis of diseases, tracking the effectiveness of treatments, and tailoring personalized treatment strategies to meet the individual requirements of patients. It also plays a critical role in drug development by enabling researchers to study the impact of novel medications and evaluate their efficacy [9] (Fig. 2.2).

Personalized medicine is an emerging field that uses BI to tailor healthcare to individual patient characteristics. This allows physicians to develop customized treatment plans that are specifically designed to address the individualized needs of each patient, improving treatment outcomes and reducing healthcare costs [10].

BI has also revolutionized surgical procedures by allowing surgeons to operate with greater accuracy and precision. It reduces the risk of surgical complications, improves surgical outcomes, and reduces recovery time [11].

Although there have been notable advancements in imaging technology that have greatly enhanced diagnostic precision and patient care, the field still faces certain challenges. Now let us discuss the various challenges in BI.

2.3 Challenges in biomedical imaging

One of the most significant challenges in BI is the interpretation of images. Medical images can be complex and difficult to read, requiring specialized training and expertise to accurately diagnose and treat patients. Additionally, there can be significant variability in the appearance of normal and abnormal tissues, leading to difficulties in distinguishing between the two. Moreover, the interpretation of images can be a time-consuming and subjective process, with different clinicians potentially coming to different conclusions. Efforts are underway to improve image analysis techniques through advanced computational techniques, like machine learning (ML) and AI [12].

Another challenge in BI is the cost of imaging procedures. Medical imaging (MI) is often expensive, with some imaging modalities costing thousands of dollars per scan. This can be a significant financial burden on patients, especially those without insurance or with high out-of-pocket costs. As a result, some patients may delay or avoid imaging studies, leading to delays in diagnosis and treatment. Efforts are underway to develop more affordable imaging modalities and to reduce the cost of existing imaging technologies [13–15].

A related challenge in BI is the potential for harm to patients. Some imaging studies, such as CT scans, involve the use of ionizing radiation, which can potentially elevate the chances of cancer and other diseases. Additionally, some patients may experience adverse reactions to contrast agents used in imaging studies, such as allergic reactions or kidney damage. Efforts are underway to develop imaging technologies that minimize the potential for harm to patients, such as low-dose CT scans and noninvasive imaging modalities [16–18].

Another challenge in BI is the standardization of imaging protocols. Different imaging centers may use different imaging protocols, leading to variability in image quality and potential errors in diagnosis. Efforts are underway to develop standardized imaging protocols that can be used across different centers, allowing for more consistent image quality and better diagnostic accuracy [19,20].

Finally, data management and privacy are significant challenges in BI. Medical images can be large and complex, requiring specialized hardware and software to store and manage. Additionally, there are significant concerns around patient privacy and the potential for data breaches. Efforts are underway to develop more secure and efficient methods

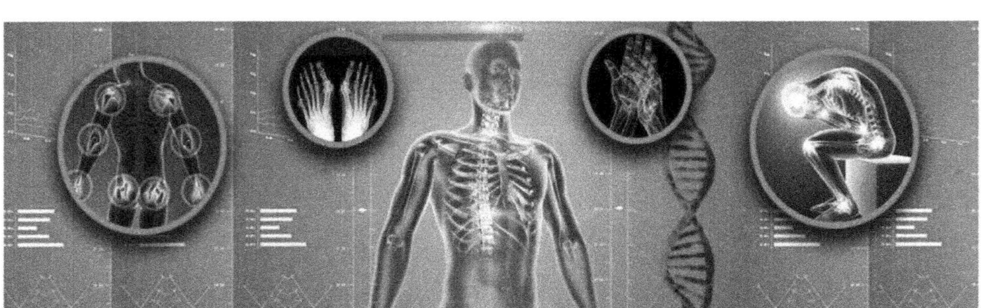

FIGURE 2.2 Importance of biomedical imaging [9].

of data storage and management, such as cloud-based imaging systems and block chain-based data management systems [21,22] (Fig. 2.3).

BI is a critical component of modern medicine, but it is not without its challenges. Efforts are underway to improve image analysis techniques, reduce the cost and potential harm of imaging procedures, standardize imaging protocols, and improve data management and privacy. With ongoing research and development, we can expect even more innovative solutions to the challenges facing BI.

2.3.1 Innovative solutions to the challenges in biomedical imaging

BI plays a critical role in diagnosing and treating various medical conditions, but it poses several challenges, including complexity, subjectivity, and cost. Innovative solutions are needed to overcome these challenges and improve the accuracy, consistency, and objectivity of interpretation, ultimately improving patient outcomes.

One solution is utilizing AI and ML algorithms, computer-aided diagnosis (CAD) systems, standardized reporting systems, imaging biomarkers, 3D imaging technology, collaborative networks, and virtual reality/augmented reality technologies, which assist radiologists in interpretation, facilitate communication between healthcare providers, provide

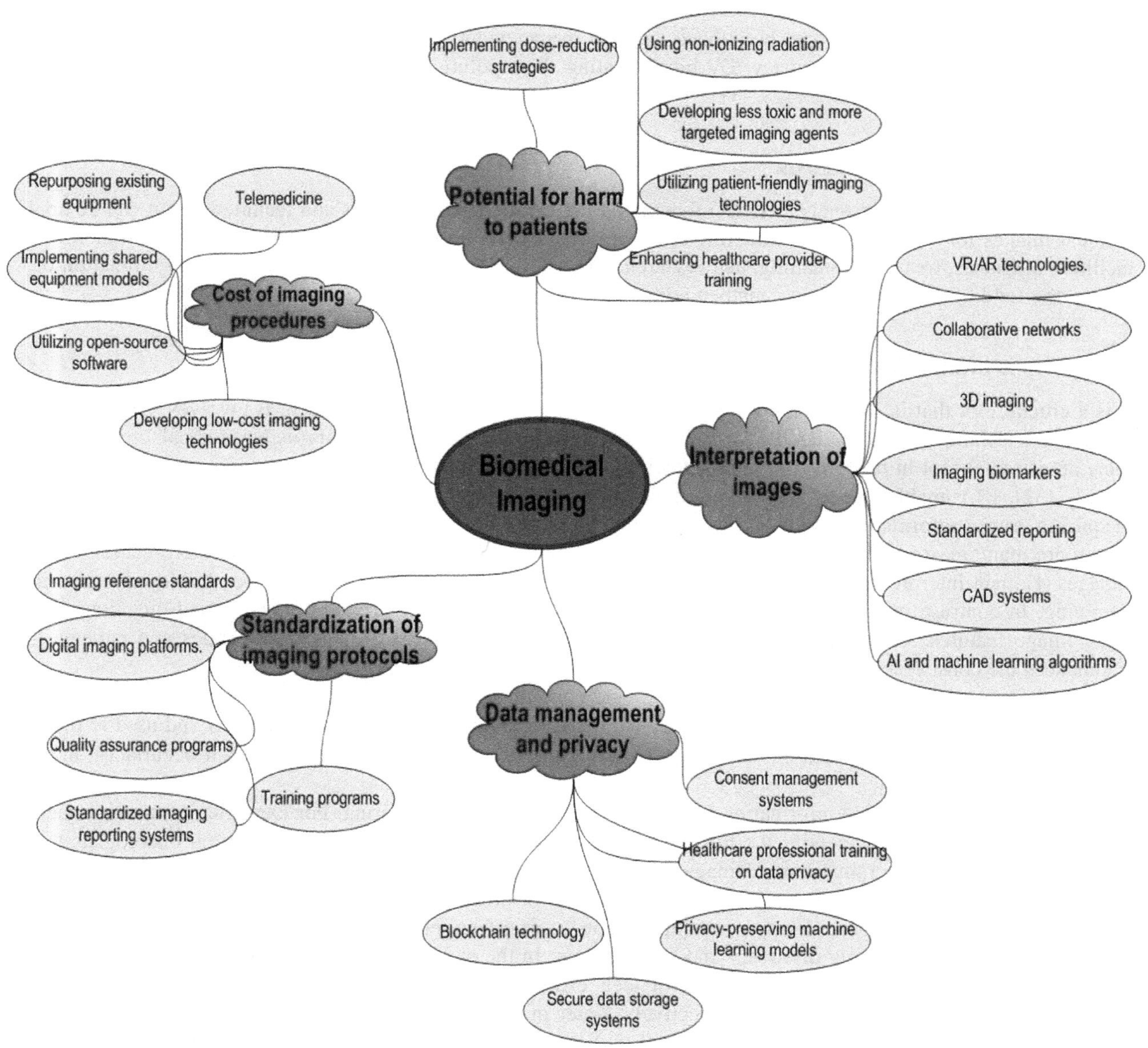

FIGURE 2.3 Biomedical imaging challenges and innovative solutions.

quantitative measures of biological processes, offer comprehensive views of organs and tissues, connect healthcare providers, and provide an immersive visualization of images, respectively [23,24].

Cost is a significant barrier to accessing high-quality BI care. To address this challenge, existing equipment can be repurposed, shared equipment models can be implemented, open-source software can be utilized, telemedicine can be utilized to diagnose and treat patients remotely, and low-cost imaging technologies can be developed, significantly reducing the cost of imaging procedures [25].

New imaging technologies have revolutionized the field of medicine, but they also pose potential risks to patients. To minimize potential harm, nonionizing radiation can be utilized, new imaging agents can be developed, dose-reduction strategies can be implemented, patient-friendly imaging technologies can be utilized, and healthcare provider training can be enhanced [26–28].

Standardized imaging protocols are essential to ensure diagnostic imaging is performed consistently and accurately across different centers. To overcome the challenges of developing such protocols, quality assurance programs can be implemented, imaging reference standards can be developed, training programs can be implemented, standardized imaging reporting systems can be developed, and digital imaging platforms can be developed, facilitating secure sharing of imaging studies between different centers.

Finally, BI generates vast amounts of data, which can be challenging to manage and pose significant privacy risks. To address these challenges, secure data storage systems can be implemented, privacy-preserving ML models can be developed, block chain technology can be utilized, consent management systems can be developed, and healthcare professionals can be trained on data privacy. By implementing these solutions, patient data can be managed safely and securely, improving patient outcomes [29–31].

2.3.2 Biomedical image analysis

The study of biomedical image analysis (BIA) focuses on the creation of algorithms and techniques that can be applied to medical images for the purpose of extracting relevant information and insights. This critical component of MI helps to facilitate diagnosis, treatment planning, and research. BIA also involves effective interpretation and utilization of MI data from the field of computer science, mathematics, and engineering.

2.3.3 Why is biomedical image analysis important?

BIA is a crucial tool that is used in many fields of medical research and clinical practice. It involves the utilization of computational techniques and methodologies to derive objective and quantitative data using medical images. MI has become an essential tool in modern medicine, allowing doctors to visualize internal organs, tissues, and structures noninvasively [32]. BIA enables researchers and clinicians to analyze these images in a systematic and efficient manner, for acquiring crucial information on the anatomy and functioning of the body.

There are many reasons why BIA is important. First, it allows for more accurate and reliable interpretation of medical images. Human interpretation of images is often subjective and can vary depending on the individual's experience and training. In contrast, computational methods and algorithms used in BIA provide objective and quantitative measures of image features, which can be reproduced consistently across different imaging modalities and patients [33]. This enhances the consistency and precision of diagnostic and medical planning.

Second, BIA can help to identify specific diseases and conditions by analyzing the image features. For example, in cancer diagnosis, features such as tumor size, shape, and density can be extracted from the images and used to differentiate between benign and malignant tumors [34]. This helps doctors to diagnose the condition accurately and plan appropriate treatment [35,36].

Third, BIA provides valuable information for treatment planning and monitoring. For example, in radiation therapy, image analysis can be used to identify the tumor and its location accurately. This allows radiation oncologists to precisely target the tumor while minimizing damage to healthy tissues [37]. Similarly, in neurology, image analysis can be used to identify brain regions that are affected by neurological disorders, such as Parkinson's disease or epilepsy. This helps doctors to plan surgical procedures or other treatments more effectively [38].

Fourth, BIA is crucial for drug discovery and development. In the early stages of drug development, image analysis can be used to screen potential drug candidates for their efficacy in treating specific diseases. This helps to identify promising drug candidates quickly and cost-effectively. Later in the drug development process, image analysis can be used to monitor the effects of the drug on the target tissues or organs. This helps to ensure that the drug is safe and effective before it is approved for clinical use [39,40].

Finally, BIA can help to advance medical research by providing insights into the underlying biological mechanisms of diseases and conditions. By analyzing the image features, researchers can identify new biomarkers, which can be used to develop new diagnostic tests and treatments. This can lead to better outcomes for patients and improve overall healthcare [41,42].

2.3.4 How is biomedical image analysis done?

BIA employs diverse computational methods and algorithms, by enabling the extraction of meaningful data from medical images, aiding in accurate interpretation and diagnosis. The process typically involves several steps, including image acquisition, image preprocessing, feature extraction, and classification [43,44].

The process begins with image acquisition, which involves the use of various MI modalities to capture medical images of the body. The quality of the images obtained during the acquisition phase plays a crucial role in the accuracy of the subsequent analysis [45,46] (Fig. 2.4).

The next step in BIA is image preprocessing. This involves several techniques such as noise reduction, image enhancement, and image registration (IMR) to improve the quality of the images obtained during the acquisition phase. This is done to remove any artifacts or distortions in the images and to normalize the images for further analysis [47,48].

Feature extraction is the next step in BIA. It involves the identification and extraction of relevant features from the images. These features may include the size, shape, texture, and intensity of various structures and tissues within the images. Feature extraction can be done manually or through automated methods such as ML and deep learning (DL) algorithms [49,50]. Classification is the final step in BIA. It involves the use of various classification algorithms to

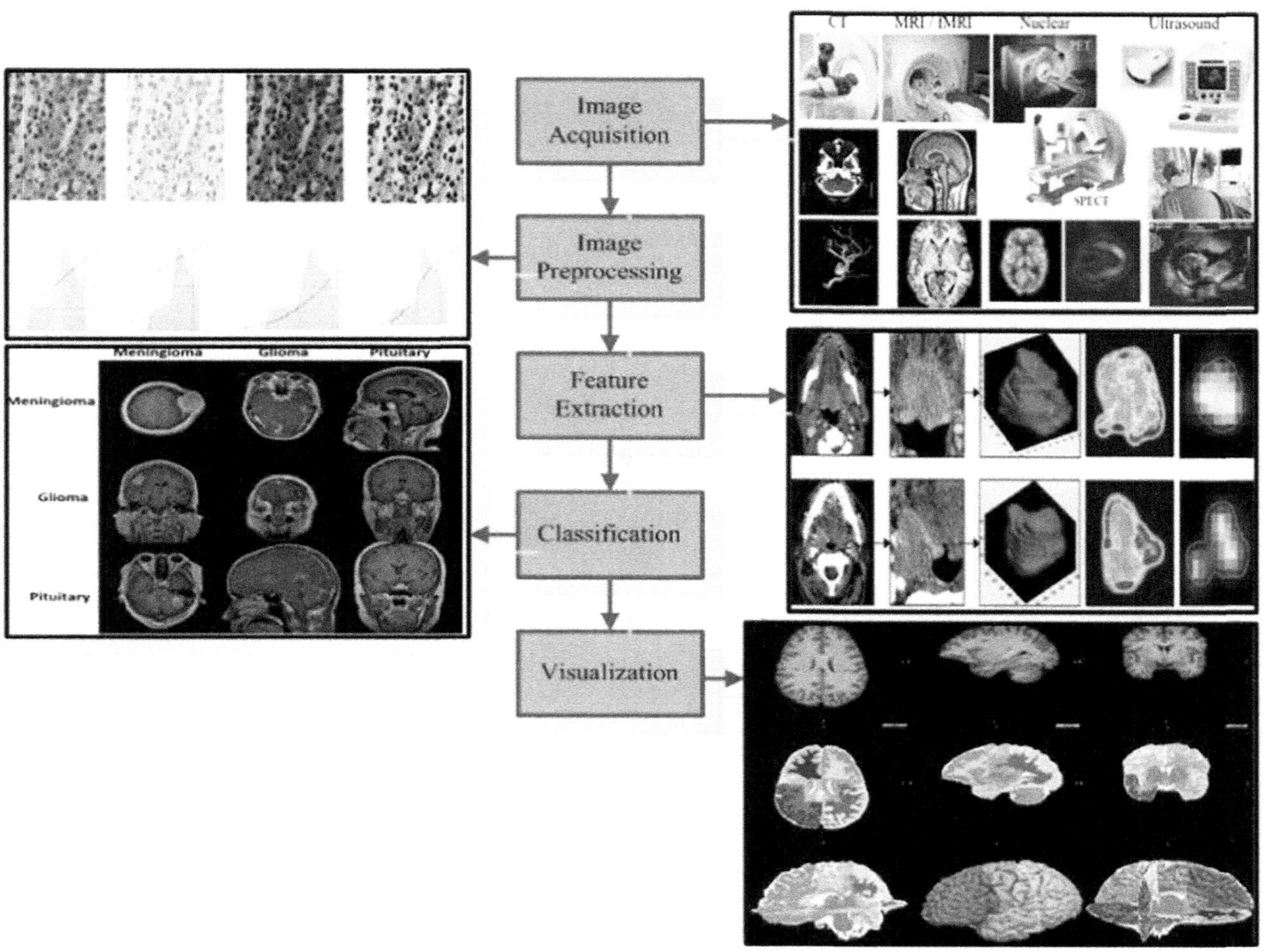

FIGURE 2.4 Biomedical imaging analysis [38].

categorize the extracted features and to identify specific diseases or conditions. The classification process may involve the use of supervised or unsupervised learning algorithms, depending on the availability of labeled data. The output of the classification process can be used to aid diagnosis, treatment planning, and monitoring of diseases and conditions [51,52].

Visualization: Visualization is the technique of clearly and intuitively presenting the findings of image analysis.

The use of ML and DL algorithms in BIA has experienced a BI evolution in recent years. These algorithms allow for the development of automated image analysis pipelines that can analyze large volumes of data quickly and accurately. ML algorithms [53,54] can learn from labeled data to predict outcomes for new and unseen data, while DL algorithms [55,56] can extract complex features from the images using artificial neural networks (Fig. 2.5).

However, there are several challenges associated with BIA. These challenges include the accessibility of effective labeled data, complexity in the image features, need for computational resources, and interpretability of the algorithms. Addressing these challenges is essential to ensure the continued advancement of BIA and its widespread implementation in clinical practice and research.

2.3.5 Intelligent techniques for biomedical image analysis

BIA has experienced substantial growth, driven by technological advancements and the abundance of extensive datasets. It encompasses the preprocessing, interpretation, and visualization of medical images from modalities like MRI

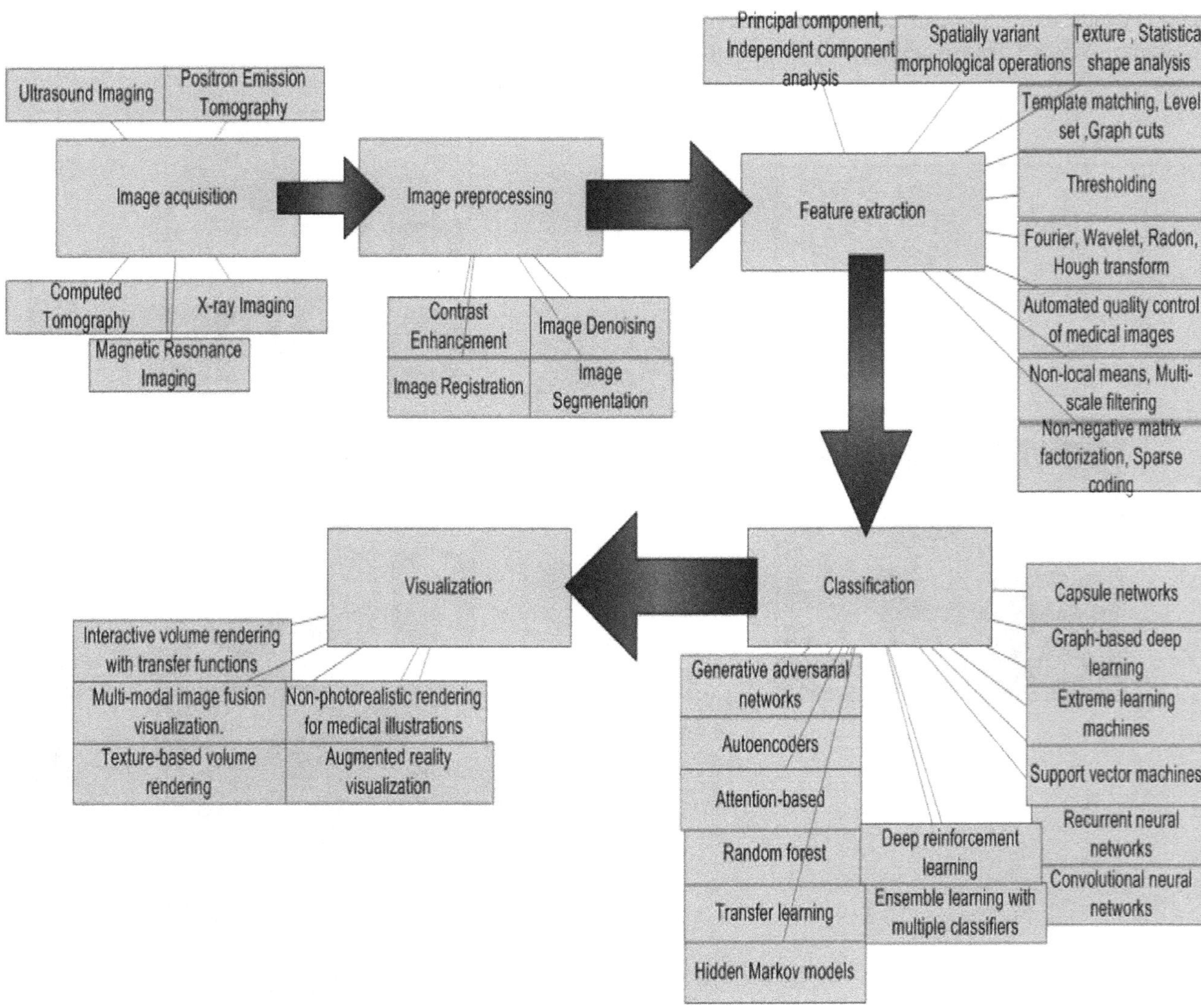

FIGURE 2.5 Biomedical imaging analysis workflow.

and ultrasound. Intelligent techniques such as AI, ML, DL, CAD, IMR, image segmentation, image fusion and image visualization have significantly contributed in enhancing the precision and effectiveness of BIA tasks.

AI technique has been extensively used in BIA due to its capability to extract information from data and generate prediction accordingly. AI algorithms have been used to classify images, identify abnormalities, and aid in disease detection. ML belongs to subfield of AI concentrating on creating techniques that are capable of learning from data and get better over time. In BIA, ML algorithms are used to segment images, detect tumors, and classify images [57,58]. DL belongs to subfield of ML that employs neural networks to perform complex tasks. DL have achieved remarkable success in BIA, detecting abnormalities like tumors, lesions, and other pathologies. Convolutional neural networks (CNNs), a popular DL algorithm, have been extensively employed in BIA due to their proficiency in learning and identifying features present in medical images [59,60].

CAD is another intelligent technique used in BIA. CAD systems use algorithms to analyze medical images and provide diagnostic information to radiologists and other healthcare professionals. These systems have been used to identify tumors, evaluate the severity of diseases, and assist in the diagnosis of various conditions [61,62].

IMR is a crucial process in BIA that involves aligning multiple images for comparison and analysis. In BIA, several images of an identical individual are frequently acquired at various instances or utilizing various imaging modalities. IMR algorithms use various techniques such as feature detection and optimization to align images and enable accurate comparison. Image segmentation is another important technique used in BIA. Segmentation entails splitting an image into multiple section or segments to enable analysis and measurement. This technique has been utilized in multiple conditions, such as identifying tumors, measuring organ volumes, and quantifying tissue characteristics [63,64].

Image fusion involves combining multiple images of the same patient or tissue to create a composite image with improved resolution, contrast, or other characteristics. Image fusion algorithms use various techniques such as wavelet transforms and fuzzy logic to merge images and create a more detailed and informative representation. Image visualization is an essential part of BIA, as it enables healthcare professionals to interpret and communicate findings to patients and other stakeholders. Visualization techniques such as volume rendering and surface rendering are used to create three-dimensional representations of medical images, which can be used to visualize complex structures and abnormalities [65].

All these techniques empower professionals to diagnose quickly and precisely and improve patient outcomes by enabling earlier detection and treatment of diseases. With continued technological advancements, we can anticipate further breakthroughs to see further innovations in this field that will continue to drive progress in MI and analysis.

2.3.6 Intelligent machine learning techniques

ML has become an increasingly important tool in BI, with applications ranging from image segmentation, registration, and classification to personalized medicine and drug discovery. ML has the potential to revolutionize BI, enabling more accurate diagnosis, more personalized treatment, and more efficient drug discovery. Some intelligent ML techniques used in BIA are given below.

2.3.7 Support vector machines

Support vector machine (SVM) is a popular ML technique in BIA for classification and regression tasks. SVM operates by identifying an optimal hyperplane that separates input data into different classes or predicts continuous output variables [66]. In SVM, data are transformed into a higher-dimensional space, and the optimal hyperplane is determined by maximizing the margin between the closest data points of different classes. The goal of the SVM algorithm is to find the hyperplane that achieves maximum margin separation and minimizes classification error [67] (Fig. 2.6).

SVM offers several advantages in BIA. They can effectively handle high-dimensional feature spaces, exhibit resistance to overfitting, and demonstrate good generalization capabilities for new data. Furthermore, SVM can address nonlinear relationships between input variables using kernel functions that transform the data into higher-dimensional spaces where linear separation becomes possible [68]. BIA tasks benefit from SVM in various ways, such as the classification of medical images, tissue segmentation, and detection of abnormal regions. SVM have exhibited high accuracy and robustness, even with noisy or incomplete data.

However, SVM have potential limitations, particularly in terms of computational complexity, especially when dealing with large datasets or high-dimensional feature spaces [69]. Nonetheless, this challenge can be addressed by parallelizing the training process across multiple processors or employing feature selection techniques to reduce dimensionality. Selecting an appropriate kernel function can also be demanding, necessitating domain expertise.

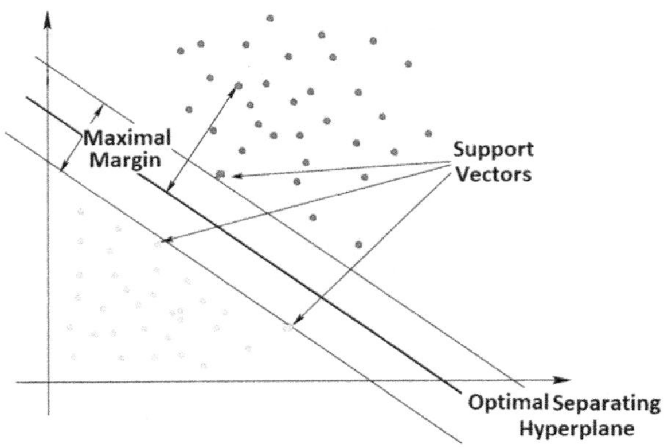

FIGURE 2.6 Support vector machines in biomedical image analysis [66].

2.3.8 Random forests

Random forests (RF) are widely employed in BIA as a ML algorithm for classification and regression tasks. They belong to the ensemble learning family, combining multiple decision trees (DT) to make predictions [70].

The RF algorithm trains multiple DT on different subclass of the training data, with each tree providing an independent prediction. The outcome is then obtained by adding the predicted results of all the forest's trees. In classification tasks, this aggregation is typically achieved through majority voting, while in regression tasks, the predictions are averaged [71] (Fig. 2.7).

RF offer several advantages in BIA. They effectively handle high-dimensional feature spaces and identify significant features contributing to classification or regression tasks. Furthermore, they exhibit resistance to overfitting, as the aggregation of multiple DT reduces model variance. RF have been successfully applied to various BIA tasks, including image segmentation, classification of medical images, and disease diagnosis. They demonstrate high accuracy and robustness, even with noisy or incomplete data [72].

However, one potential limitation of RF lies in their computational complexity, especially for large datasets or high-dimensional feature spaces. Nonetheless, this limitation can be addressed by parallelizing the training process across multiple processors or utilizing feature selection techniques to reduce data dimensionality.

2.3.9 K-nearest neighbors

K-nearest neighbors (KNN) is a ML algorithm extensively used in both classification and regression tasks, including BIA. In the context of image classification, KNN is employed to assign an image to a specific predefined class. Notably, KNN is a nonparametric algorithm that does not rely on assumptions about the data distribution. Instead, it follows a straightforward and intuitive approach to classification [73,74] (Fig. 2.8).

The KNN algorithm stores all available training data and, when a new image is presented, calculates the distances between the new image and all the training images. Subsequently, it selects the K-nearest training images and assigns the new image to the class with the highest number of nearest neighbors. This process enables KNN to classify biomedical images based on various criteria, such as distinguishing between normal and abnormal images or assessing disease severity. For instance, KNN can be utilized to classify mammograms as benign or malignant by evaluating specific features [75].

One limitation of the KNN algorithm is its computational expense, particularly when dealing with large datasets. The algorithm necessitates calculating the distance between the new image and all training images for each prediction, resulting in increased computational demands. Additionally, the performance of the algorithm can be influenced by the choice of the hyperparameter K, which determines the number of nearest neighbors considered for classification.

2.3.10 Logistic regression

Logistic regression (LR) is a widely employed statistical method in BIA, particularly for binary classification tasks. As a supervised learning algorithm, it enables the prediction of the probability that an image belongs to a specific class or category. BIA benefits from LR in various tasks, including image classification, segmentation, and detection. The algorithm operates by modeling the relationship between input image features and the probability of class membership [76] (Fig. 2.9).

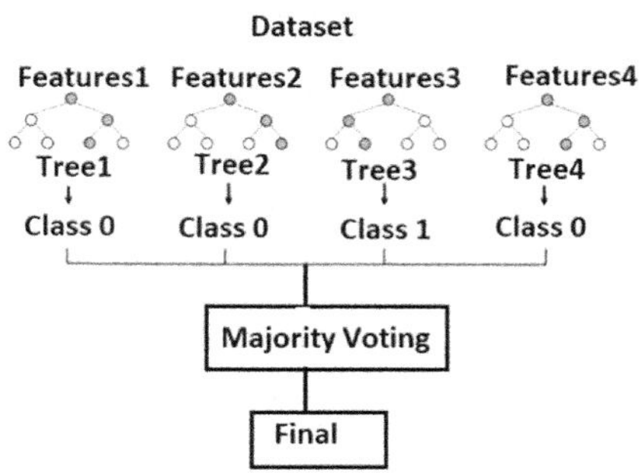

FIGURE 2.7 Random forest in biomedical image analysis [71].

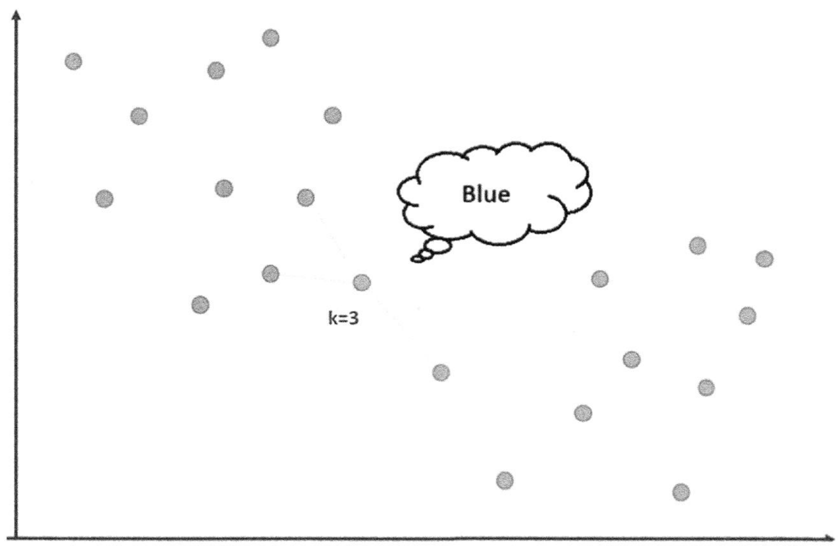

FIGURE 2.8 K-nearest neighbors in biomedical image analysis [74].

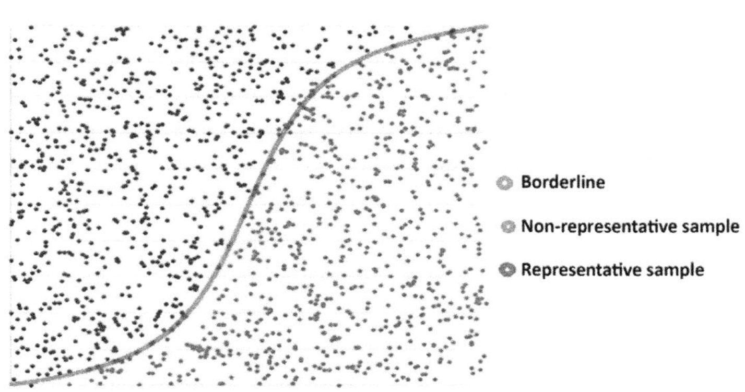

FIGURE 2.9 Logistic regression in biomedical image analysis.

LR assumes a linear relationship between the input features and the probability of belonging to a particular class. To capture this relationship, it utilizes a logistic function, also known as a sigmoid function, which maps the input features to a probability value ranging from 0 to 1.

Training of the LR model involves using a labeled dataset, where each image possesses a binary label representing its class. By leveraging this data, the model learns the connection between the input features and the probability of belonging to the positive or negative class. Following model training, it becomes possible to classify new images by predicting the probability of class membership. The predicted probability can subsequently be thresholded to assign the image to the positive or negative class [77].

2.3.11 Decision trees

DT are widely utilized in BIA as a popular ML technique. As a supervised learning algorithm, DT can handle both classification and regression tasks [78]. The fundamental principle behind DT involves constructing a tree-like model that represents decisions and their consequences based on input image features. Each node in the tree corresponds to a decision based on one or multiple features, while each branch represents the outcome of that decision. The leaf nodes denote the final decision or prediction [79].

In the realm of BIA, DT finds application in various tasks such as image classification, segmentation, and detection. For example, a DT model can be trained to predict the presence of a specific disease in an image based on its visual characteristics [80].

DT models are capable of accommodating both continuous and categorical input features, as well as handling missing data. They offer interpretability, shedding light on the significant features influencing predictions. Moreover, DT models can be trained with limited labeled data, making them suitable for tasks with restricted labeled datasets (Fig. 2.10).

However, one potential challenge with DT models is their susceptibility to overfitting, where the model becomes excessively complex and starts fitting noise instead of capturing the underlying patterns in the training data. To mitigate overfitting, techniques like pruning and ensemble learning can be employed.

Ensemble learning involves combining multiple DT models to enhance overall performance and alleviate overfitting. Techniques like bagging and boosting are commonly employed, where different subsets of the data are used to train multiple DT models or training examples are adaptively weighted [81].

2.3.12 Ensemble learning

Ensemble learning (EL) is a valuable technique in ML that leverages the power of combining multiple models to enhance prediction accuracy and robustness. In the domain of BIA, EL plays a crucial role in improving the performance of ML models, particularly when dealing with complex and diverse datasets. One commonly used EL is bagging, also known as bootstrap aggregating, where multiple models are trained on different subsets of the data. The individual models' outputs are then combined to make a final prediction, aiding in reducing overfitting and enhancing generalization capabilities [82].

Boosting is another effective EL approach that focuses on refining weaker models iteratively. During each iteration, the misclassified samples' weights are increased, prioritizing them in subsequent iterations. This process continues until a predefined number of iterations or until the error rate stabilizes, resulting in a stronger model [83] (Fig. 2.11).

EL finds application in various BIA tasks, such as image classification, segmentation, and detection. For instance, multiple models can be trained on different image features like texture, color, and shape. By combining their outputs, the overall accuracy of the predictions can be improved. EL also enables the integration of multiple data sources, such as imaging and genomic data, to enhance disease diagnosis or prognosis accuracy. Moreover, it proves beneficial in addressing the data imbalance issue often encountered in biomedical image datasets. By training multiple models on different data subsets and merging their outputs, EL helps in mitigating the impact of data imbalance [84].

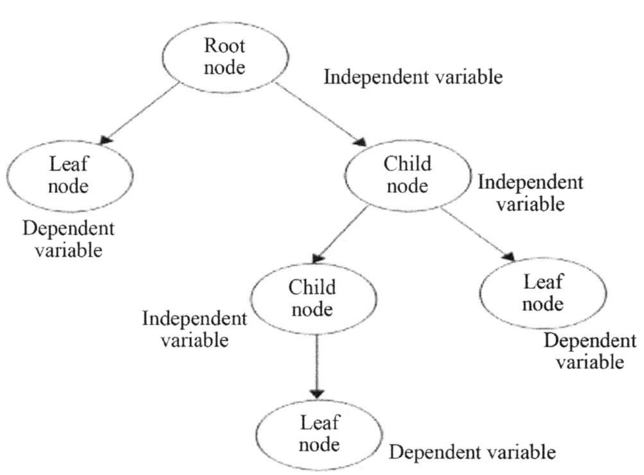

FIGURE 2.10 Decision trees in biomedical image analysis [79].

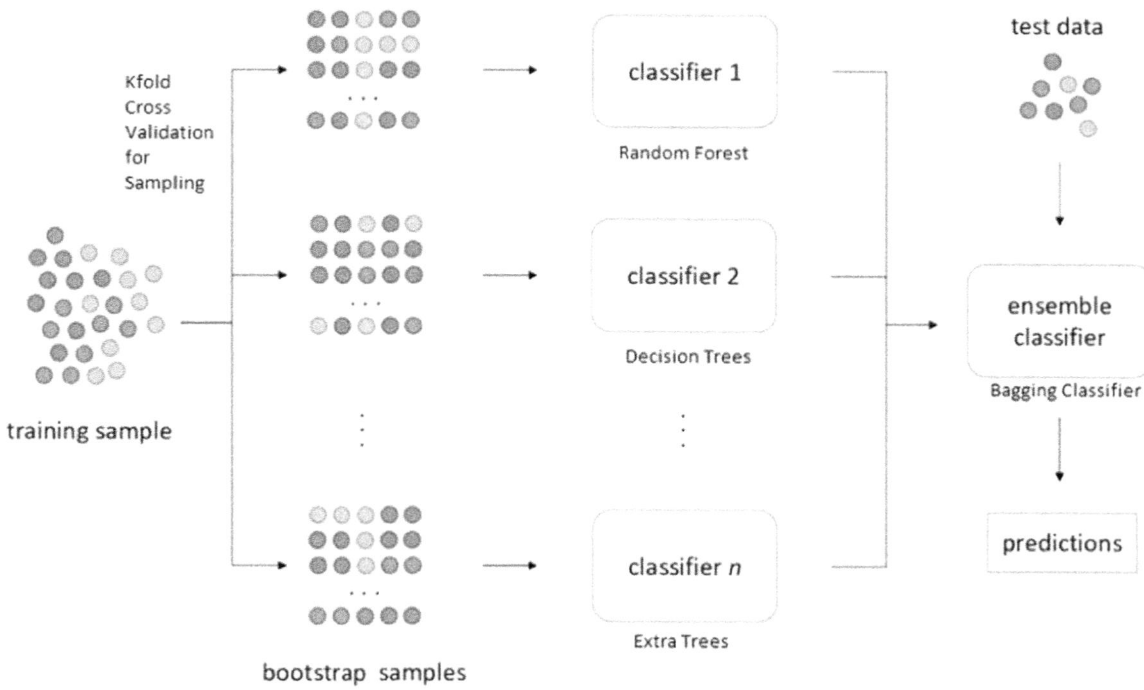

FIGURE 2.11 Ensemble learning in biomedical image analysis [83].

2.3.13 Adaptive boosting

Adaptive boosting (AdaBoost), an EL algorithm, is extensively utilized in BIA. It combines multiple weak classifiers to create a strong classifier. The algorithm assigns weights to each data point and sequentially trains the classifier, with subsequent weak classifiers focusing more on misclassified samples from previous iterations [85].

In BIA, AdaBoost finds application in various tasks, such as cancerous cell detection in microscopy images, lung nodule detection in CT scans, and disease outcome prediction. It has also been employed for brain tumor segmentation and retinal image classification (Fig. 2.12).

AdaBoost excels in enhancing classification accuracy by leveraging the combination of weak classifiers. It effectively handles high-dimensional data and copes with imbalanced datasets. The algorithm is relatively straightforward to implement and has demonstrated successful outcomes in diverse MI tasks [86].

One potential limitation of AdaBoost is its sensitivity to dataset noise. If the dataset contains a substantial amount of noisy data, the accuracy of the final classifier may be compromised. Furthermore, AdaBoost may require significant computational resources due to its multiple iterations of training on the same dataset. However, with the advancements in computing resources, this limitation is becoming less significant.

2.3.14 Hidden Markov models

Hidden Markov models (HMMs) have found extensive application in BIA, especially for analyzing sequential and spatially ordered data. HMMs are well-suited for tasks involving the segmentation, detection of abnormalities, and classification of biomedical images. They are particularly valuable when the underlying structure of the data is influenced by hidden variables or states that cannot be directly observed [87] (Fig. 2.13).

HMMs function by modeling the probability distribution of an observed sequence based on a sequence of hidden states. While the hidden states cannot be directly observed, their presence and transitions can be inferred from the observed sequence. HMMs assume that the probability of each observation is solely dependent on the current hidden state, and state transitions adhere to a Markov process [88].

One key advantage of HMMs is their ability to capture complex dependencies between observed variables. In BIA, this capability is particularly relevant for tasks like medical image segmentation, where the location of one structure may be contingent on the position of another structure. However, it should be noted that HMMs can be computationally

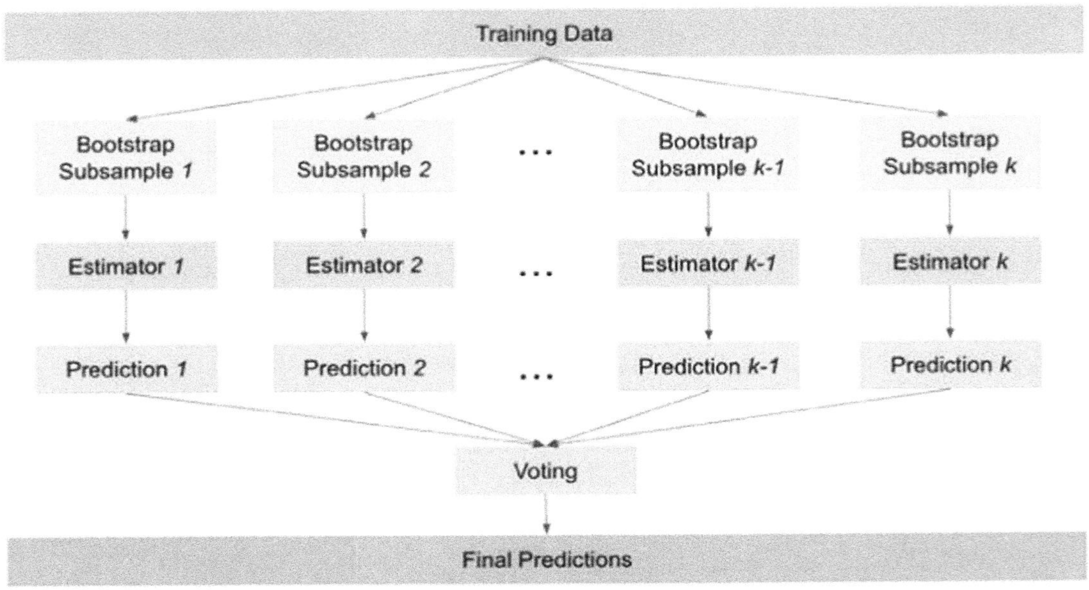

FIGURE 2.12 Adaptive boosting in biomedical image analysis [85].

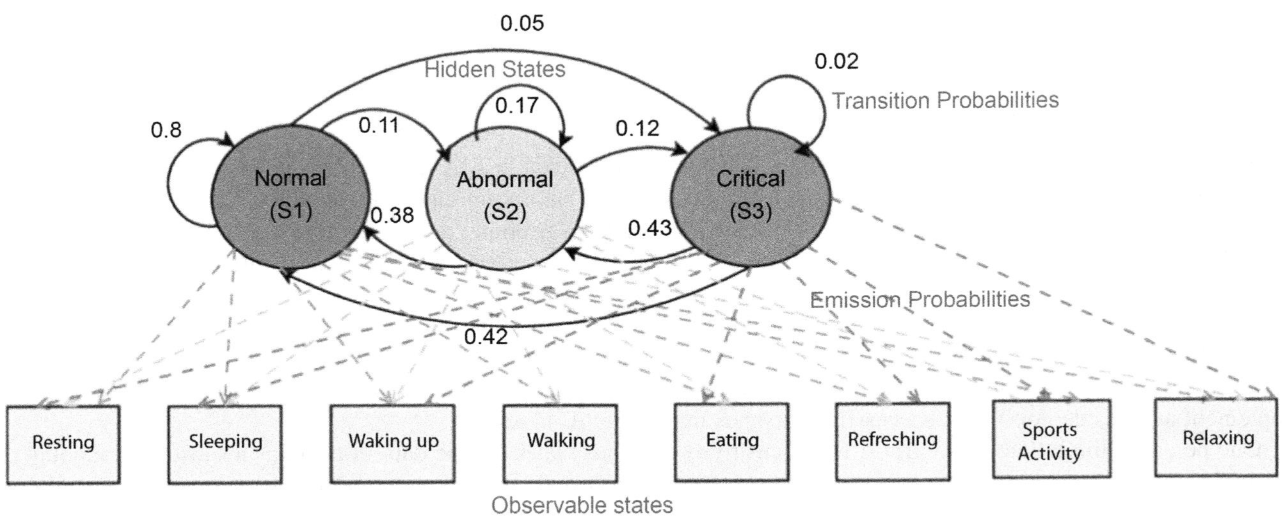

FIGURE 2.13 Hidden Markov model in biomedical image analysis [88].

demanding, particularly when dealing with high-dimensional image data. Moreover, HMMs may not be optimal for tasks where the hidden states are ill-defined or challenging to estimate accurately [89].

2.3.15 Gaussian mixture models

Gaussian mixture models (GMMs) are widely used in BIA for clustering and classification tasks. They assume that the observed data are generated by a combination of Gaussian distributions, with each Gaussian representing a distinct cluster in the data [90]. GMMs have found application in various BIA tasks, such as the segmentation and classification of medical images, and the identification of different tissue types within images. They are particularly valuable when dealing with data that are not easily separable or linearly separable, and when the underlying data distribution is unknown.

GMMs operate by estimating the parameters of the underlying Gaussian distributions and the weights assigned to each distribution in the mixture model. These parameters are typically estimated using maximum likelihood or Bayesian approaches. Once the parameters are estimated, GMMs can be utilized to cluster or classify new data points based on their likelihood of belonging to each Gaussian distribution [91].

2.3.16 Principal component analysis

Principal component analysis (PCA) is a widely utilized method in BIA for dimensionality reduction and feature extraction. Its primary goal is to find the principal components of a dataset, which are the directions in the feature space that account for the most variance in the data [92].

In the field of BIA, PCA finds applications in various tasks such as image compression, segmentation, and classification. By reducing the dimensionality of high-dimensional image data, PCA simplifies the analysis process and decreases computational requirements. Moreover, PCA allows for feature extraction from images that are most relevant to a specific task, such as detecting abnormal regions in medical images. The methodology of PCA involves projecting the original image data onto a lower-dimensional subspace that captures the highest variance in the data. The principal components are obtained through an eigenvalue decomposition of the covariance matrix of the original data, with the eigenvectors representing the principal components and the corresponding eigenvalues indicating the amount of variance captured by each principal component [93] (Fig. 2.14).

One significant advantage of PCA is its ability to reduce the dimensionality of the data while retaining the most important information. This facilitates efficient computation and analysis. However, it is worth noting that PCA may not be suitable for all types of data and might not yield optimal results when the data exhibit complex structures.

2.3.17 Multitask learning

Multitask learning (MtL) is a valuable ML technique employed in BIA, where a single model is trained to simultaneously perform multiple related tasks. This approach encompasses training a model to handle tasks such as image classification, segmentation, and detection. The key advantage of MtL in this domain is the ability to enhance the performance of each individual task by leveraging shared learned representations [94]. For instance, the model can learn to detect tumors and segment organs within the same image, leading to improved accuracy in both tasks (Fig. 2.15).

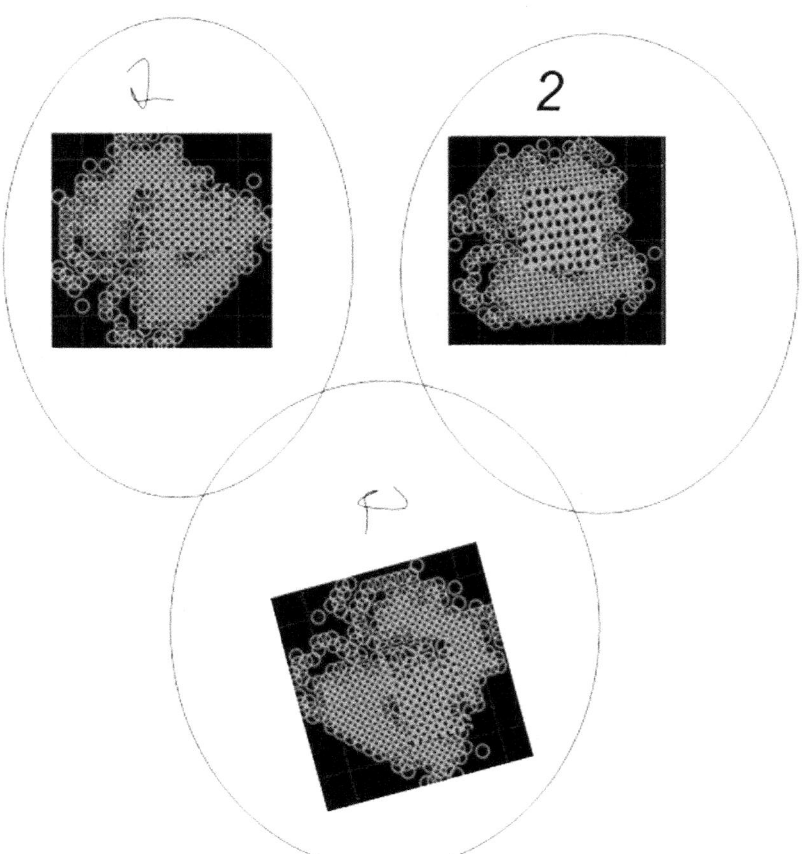

FIGURE 2.14 Principal component analysis in biomedical image analysis [92].

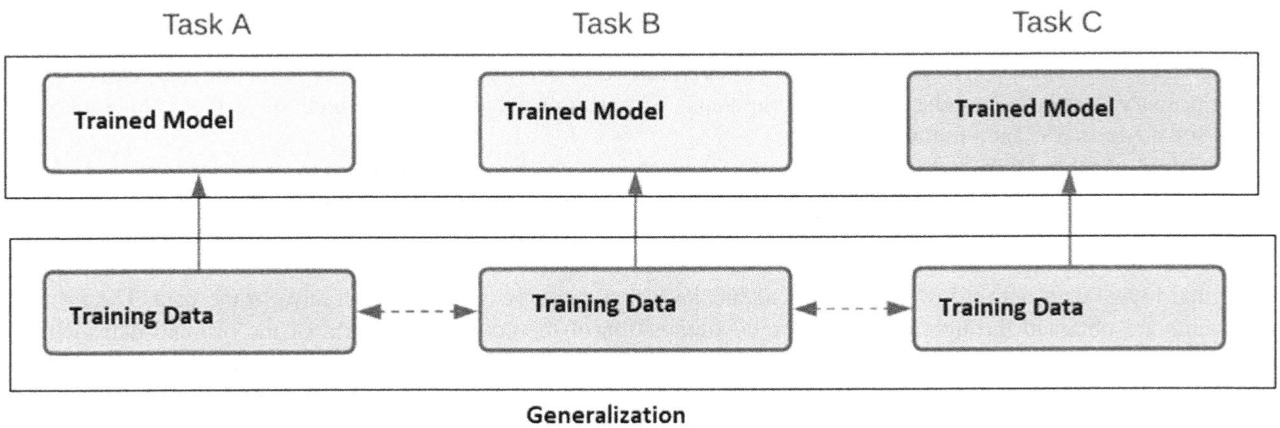

FIGURE 2.15 Multitask learning in biomedical image analysis [94].

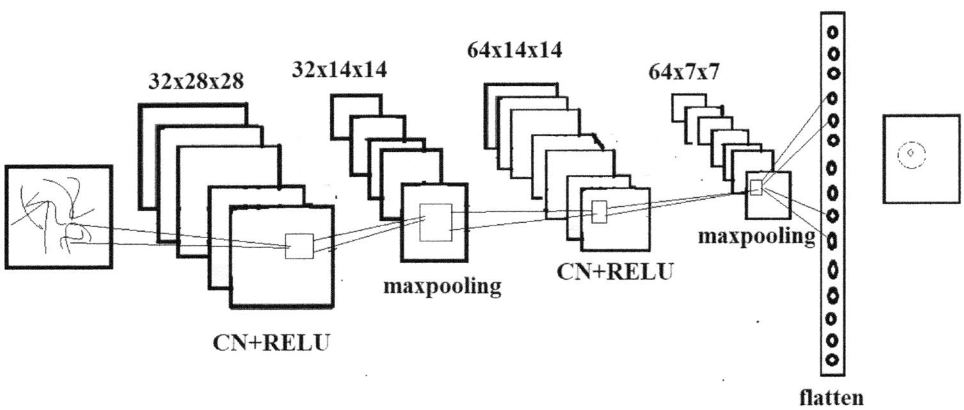

FIGURE 2.16 Convolutional neural network in biomedical image analysis [98].

Another notable benefit of MtL in BIA is its potential to reduce the amount of required training data. By jointly learning shared features across tasks, the model can leverage a larger dataset, contributing to more robust representations compared to training each task individually [95].

2.3.18 Intelligent deep learning techniques

Intelligent DL techniques are being increasingly utilized in biomedical applications to analyze complex data and images. These techniques use neural networks to recognize patterns and learn from data, allowing for more accurate and personalized diagnoses, treatment plans, and drug development. DL algorithms can be trained on large datasets to identify subtle features and predict outcomes, making them useful for tasks such as tumor detection, genomics, and drug discovery. As technology advances, intelligent DL techniques are expected to play an even more significant role in advancing biomedical research and improving patient outcomes.

2.3.19 Convolutional neural networks

CNNs have emerged as a powerful DL architecture for tackling image analysis tasks, including biomedical applications. In the field of BIA, CNNs have exhibited remarkable success in tasks such as image classification, segmentation, and detection [96]. CNNs are constructed with multiple layers, including convolutional layers, pooling layers, and fully connected layers. Convolutional layers are responsible for learning filters that detect various features like edges, corners, and textures within the images. Pooling layers subsequently downsample the generated feature maps, retaining the most relevant information. Finally, fully connected layers map the learned features to the corresponding output class probabilities [97] (Fig. 2.16).

BIA greatly benefits from CNNs as they have been successfully employed in various tasks, including identifying cancerous cells in histopathological images, detecting tumors in MI scans and segmenting organs in radiological images. CNNs excel in these applications due to their ability to automatically learn intricate patterns and complex relationships among input pixels, eliminating the need for explicit feature engineering [98,99].

One notable advantage of CNNs in BIA is their capacity to handle large datasets. By training on extensive datasets, CNNs can learn a wide range of features that prove beneficial for diverse tasks. Moreover, the integration of CNNs in BIA can streamline the process, saving both time and resources by automating image analysis and reducing the reliance on manual intervention [100].

2.3.20 Recurrent neural networks

Recurrent neural networks (RNNs) are a powerful class of DL algorithms that have found extensive applications in various domains, including BIA. RNNs excel at capturing sequential patterns in data by maintaining a memory state that can be updated at each time step, enabling them to model complex temporal dependencies [101].

In the field of BIA, RNNs have demonstrated significant potential in disease progression prediction. For instance, in the context of Alzheimer's disease, RNNs have been employed to analyze temporal patterns in MI data, such as MRI scans. By examining the evolution of brain structures over time, RNNs can accurately forecast disease progression, enabling personalized treatment planning for patients [102] (Fig. 2.17).

Additionally, RNNs have proven valuable in disease detection and classification tasks within BIA. By analyzing microscopy images, RNNs can detect cancer cells and predict prognosis. Through the examination of spatial and temporal patterns in cell behavior, RNNs offer accurate cancer detection and prognostic insights [104].

Furthermore, RNNs have been leveraged to analyze time-series data from wearable sensors, including electrocardiogram (ECG) signals, for heart disease detection and prediction. RNNs can effectively capture temporal patterns in ECG data, enabling the accurate identification of abnormalities and prediction of heart disease risks [103].

2.3.21 Generative adversarial networks

Generative adversarial networks (GANs) are a powerful class of DL algorithms that have made significant contributions to various fields, including BIA. GANs consist of two neural networks, namely the generator and the discriminator, engaged in a competitive process that enhances the quality of generated images. The ability of GANs to generate realistic images resembling real data has rendered them invaluable in BIA [105].

A prominent application of GANs in this domain is image synthesis. GANs have been employed to generate synthetic medical images resembling real medical images. By training the GAN on a dataset of authentic medical images, the generator learns to produce synthetic images that closely resemble real images. These synthetic images can augment the training dataset, enhancing the performance of ML algorithms in tasks like segmentation and classification [106] (Fig. 2.18).

GANs have also demonstrated their effectiveness in image enhancement, particularly for low-quality medical images such as ultrasound images. By training the GAN on a dataset comprising low-quality images and their corresponding high-quality counterparts, the generator learns to enhance the low-quality images and generate high-quality versions. This capability enables the enhancement of image quality, aiding in more accurate analysis and diagnosis.

Furthermore, GANs have been utilized in BIA for image segmentation tasks. For instance, GANs have been employed to segment brain tumors from MRI images. Training the GAN on a dataset comprising MRI images and

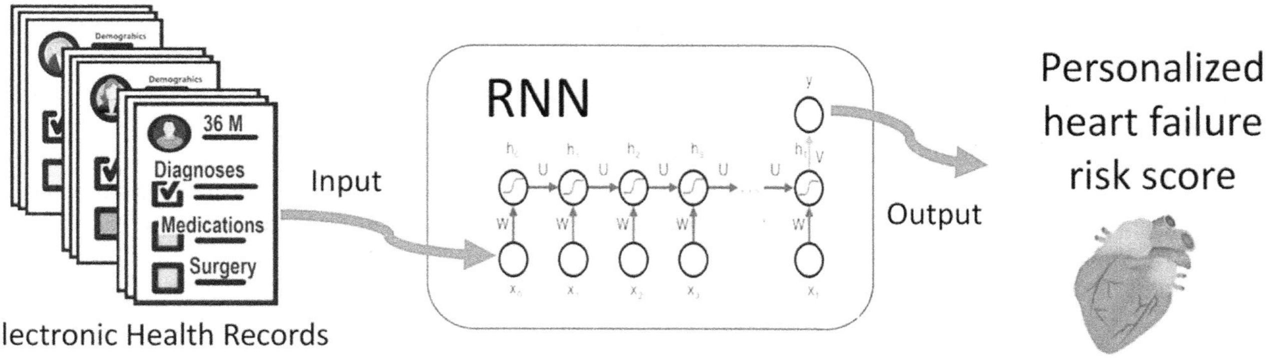

FIGURE 2.17 Recurrent neural networks in biomedical image analysis [103].

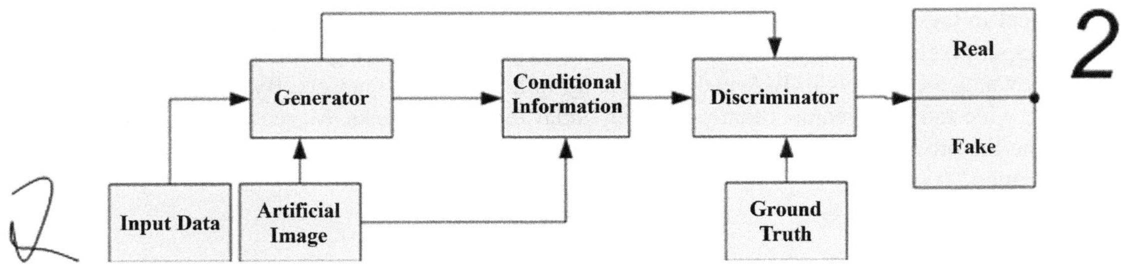

FIGURE 2.18 Generative adversarial networks in biomedical image analysis [106].

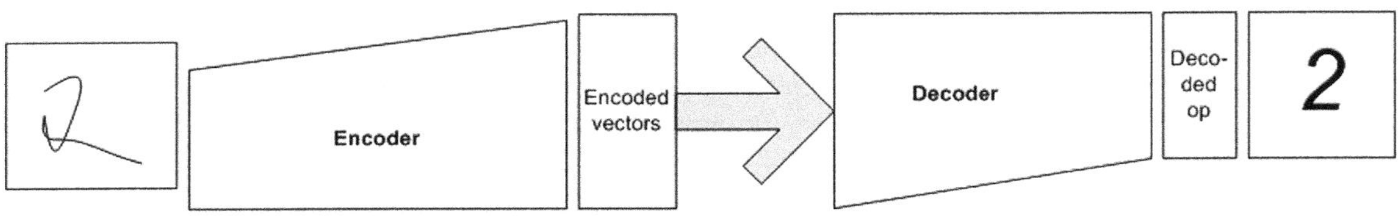

FIGURE 2.19 Autoencoder in biomedical image analysis [108].

their corresponding tumor segmentation masks allows the generator to learn to segment tumors in new MRI images. This facilitates more precise tumor identification and assists medical professionals in diagnosis and treatment planning [107].

2.3.22 Autoencoders

Autoencoders (AEs) are neural networks that excel in learning to encode and reconstruct data by compressing it into a lower-dimensional representation. In the context of BIA, AEs have proven to be valuable for several tasks, including image denoising, image compression, and feature extraction [108] (Fig. 2.19).

The architecture of AEs consists of an encoder network responsible for compressing the input image into a latent code, and a decoder network that reconstructs the image from the latent code. During training, the encoder and decoder networks work collaboratively to minimize the discrepancy between the input image and its reconstructed counterpart [109].

In BIA, AEs offer the advantage of learning meaningful representations from medical images. By training on a large dataset of medical images, the encoder network can extract crucial features from the images, which can subsequently be utilized for various downstream tasks, such as disease diagnosis and prognosis.

Moreover, AEs can effectively address the challenge of image denoising. They can learn to eliminate noise from noisy medical images by reconstructing a clean image from the noisy input. This capability is particularly beneficial in MI, where images often exhibit noise due to factors such as low-dose radiation [110].

2.3.23 Transfer learning

Transfer learning (TL) is a powerful technique that is widely used in BIA to address the challenges of limited data availability. In this approach, pretrained models are leveraged to solve new tasks by reusing the knowledge learned from large-scale datasets. TL offers several advantages, including improved accuracy, enhanced generalization, and reduced training time, by minimizing the need for extensive labeled data [111].

Fine-tuning is one of the most popular TL techniques in BIA. It involves retraining a pretrained model on a new dataset specific to the task at hand. For instance, in medical image classification, a pretrained model can be fine-tuned using a new dataset to accurately classify images as healthy or diseased. Similarly, in medical image segmentation, a pretrained model can be fine-tuned to precisely delineate anatomical structures like the heart or lungs. Furthermore, in medical image detection, fine-tuning a pretrained model enables the identification of abnormalities such as tumors or lesions [112] (Fig. 2.20).

Another TL technique is feature extraction, where the pretrained model serves as a feature extractor for a new task. The model is used to extract relevant features from input images, which are then fed into a separate classifier or regressor. Feature extraction has proven useful in medical image classification and diagnosis, where pretrained models extract informative features from medical images to predict the likelihood of a disease or condition. TL has demonstrated its

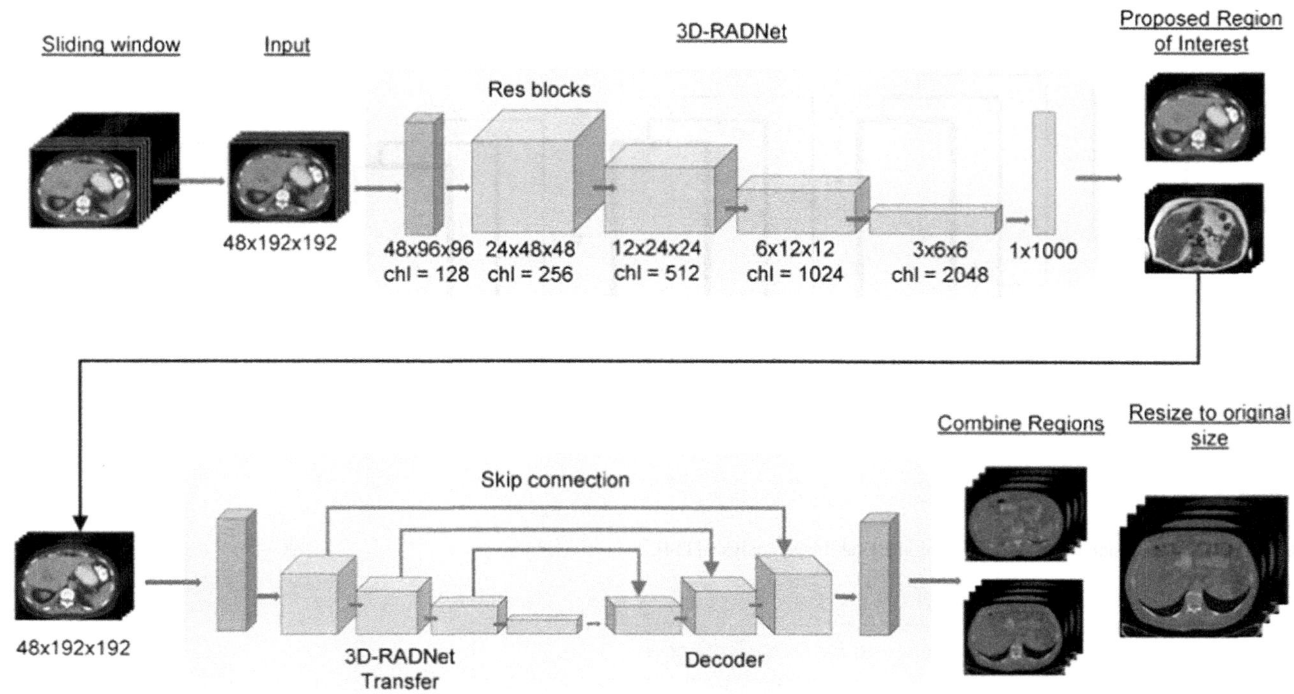

FIGURE 2.20 Transfer learning in biomedical image analysis [112].

effectiveness across various domains of BIA. By harnessing the knowledge encoded in pretrained models, TL improves the accuracy and efficiency of image analysis tasks, even with limited labeled data. It has emerged as a valuable tool in the field, offering significant advancements in MI research and clinical applications [113].

2.3.24 Deep Bayesian learning

Deep Bayesian learning (DBL) is a powerful approach that combines the strengths of Bayesian methods and DL to enhance predictions and reduce uncertainty in data analysis. It has gained significant attention in the field of BIA, where it can be applied to improve the performance of deep neural networks in various tasks such as medical image classification, segmentation, and reconstruction. By incorporating probability distributions over model parameters, DBL models are able to effectively quantify uncertainty and make predictions even in situations where training data is limited or noisy [114] (Fig. 2.21).

One of the key advantages of DBL is its ability to mitigate overfitting in deep neural networks by utilizing Bayesian priors to regularize the model. This regularization helps to prevent the model from becoming overly complex and provides a more robust and reliable estimation of parameters. Moreover, DBL allows for model selection, where different model architectures and hyperparameters can be compared and evaluated in a principled Bayesian framework. This enables the selection of the most appropriate model configuration for a specific BIA task, leading to improved performance and generalization. Furthermore, DBL offers a valuable tool for uncertainty quantification in medical image analysis. In clinical decision-making, uncertainty plays a critical role, as inaccurate predictions can have significant implications for patient care. DBL models provide a means to assess and quantify uncertainty in their predictions, allowing practitioners to better understand the reliability and confidence of the results. This information can guide clinical decisions and support more informed and personalized patient management [115].

2.3.25 Capsule and Siamese networks

Capsule networks (CNs) are DL architectures that aim to enhance image recognition performance by capturing hierarchical relationships among different parts of an image. They provide a robust representation of image components that is invariant to geometric transformations, making them well-suited for BIA [116]. For instance, CNs have been successfully applied to detect and segment lung nodules in CT images, achieving state-of-the-art outcomes. By modeling the interdependencies among different components of the nodules, CNs can significantly improve the accuracy of detection and segmentation (Fig. 2.22).

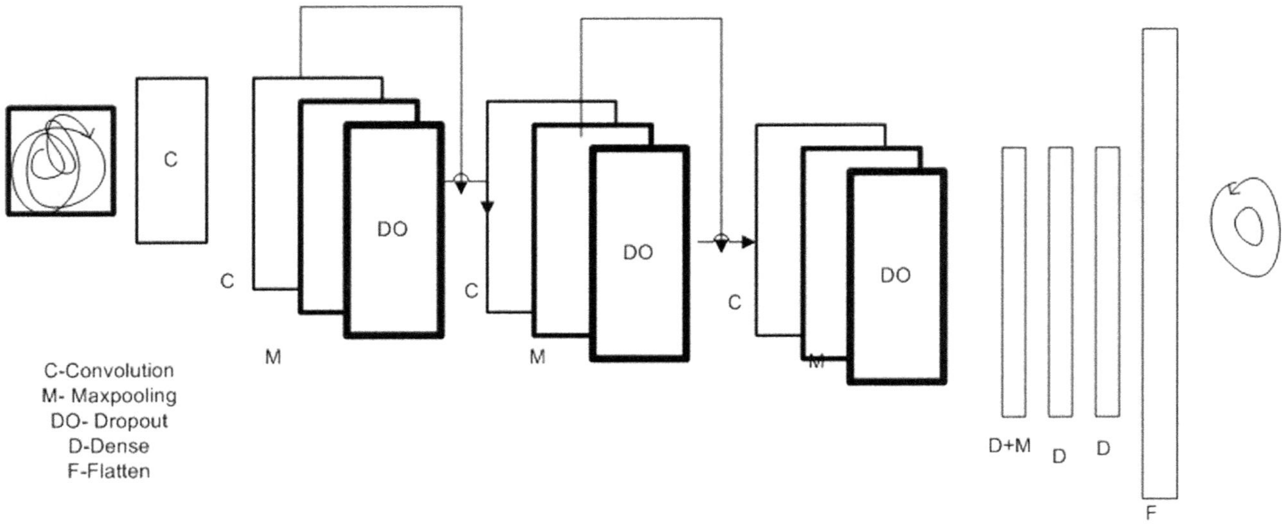

FIGURE 2.21 Bayesian deep learning in biomedical image analysis [114].

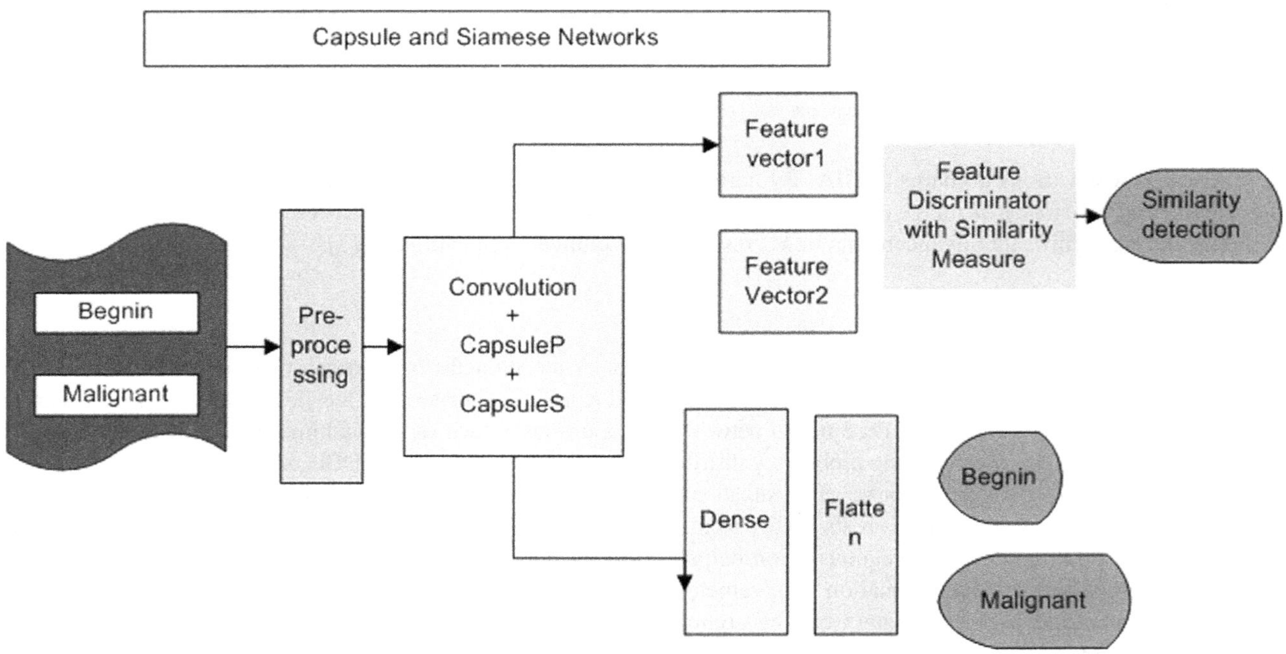

FIGURE 2.22 Capsule and Siamese networks in biomedical image analysis [117].

On the other hand, Siamese networks (SNs) are deep neural networks specifically designed for image comparison and similarity determination. In the context of BIA, SNs find utility in tasks such as IMR, where aligning two images is required. Additionally, SNs can be employed for image classification and retrieval, where the goal is to identify similar images either with respect to a reference image or a database of images. A notable application of SNs is in the classification of skin lesions, where they have demonstrated state-of-the-art performance [117].

2.3.26 Attention mechanisms

Attention mechanisms (AMs) are a type of DL technique that enables the network to focus on the most relevant parts of an image. AMs can be used in BIA for tasks such as object detection and segmentation. By focusing on the most important parts of an image, AMs can improve the accuracy of detection and segmentation [118]. For example, AMs have been used for the segmentation of retinal vessels in fundus images, achieving state-of-the-art results. By focusing on the most relevant parts of the image, AMs can improve the accuracy of segmentation and reduce false positives [119] (Fig. 2.23).

2.3.27 Deep reinforcement learning

Deep reinforcement learning (DRL) is a DL approach that involves training an agent to make decisions based on feedback received from its environment. In the field of BIA, DRL has shown promise in tasks like medical image segmentation and diagnosis. For instance, in the segmentation of brain tumors from MRI images, DRL has achieved state-of-the-art results. By training an agent to learn the optimal tumor segmentation strategy, DRL can enhance the accuracy of the segmentation process while minimizing the reliance on manual annotation [120]. Furthermore, DRL has been applied to medical diagnosis, such as the detection of diabetic retinopathy from fundus images. By training an agent to classify retinal images and make diagnostic predictions, DRL can improve the accuracy of diagnosis and potentially reduce the need for extensive expert knowledge. This approach leverages the agent's ability to learn complex patterns and make decisions based on learned experiences, contributing to more efficient and accurate medical diagnoses [121] (Fig. 2.24).

2.3.28 Deep belief and convex networks

Deep belief networks (DBNs) are a class of deep neural networks that excel at extracting meaningful features from raw data. In the realm of BIA, DBNs have been successfully employed for tasks like image classification and segmentation [122]. For instance, in cancer classification from pathology images, DBNs have demonstrated state-of-the-art performance. By autonomously learning to extract crucial image features, DBNs enhance classification accuracy while minimizing the requirement for manual feature engineering. DBNs have also been leveraged for image segmentation, including liver segmentation from CT images. By learning to identify vital liver features from the images, DBNs can enhance segmentation accuracy and diminish the necessity for laborious manual annotation [123].

Deep convex networks (DCNs), on the other hand, are a type of deep neural network specifically designed for solving convex optimization problems. In BIA, DCNs have found applications in tasks such as medical image

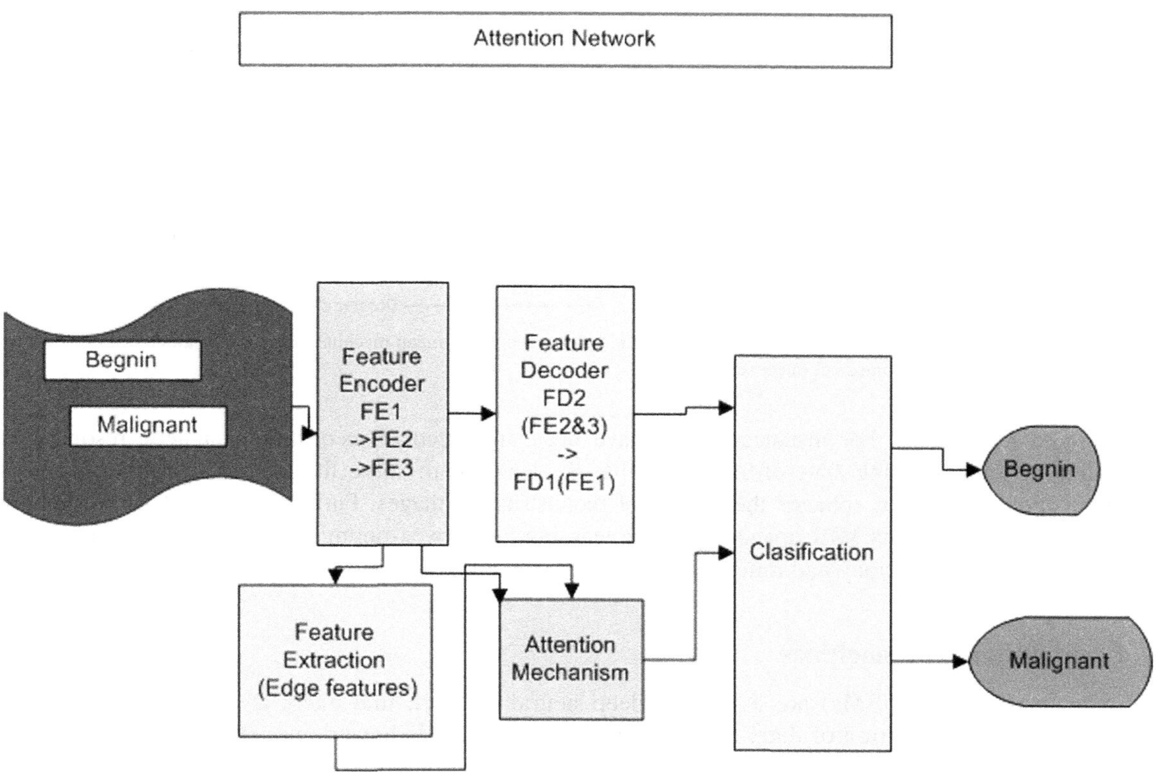

FIGURE 2.23 Attention mechanisms in biomedical image analysis [118].

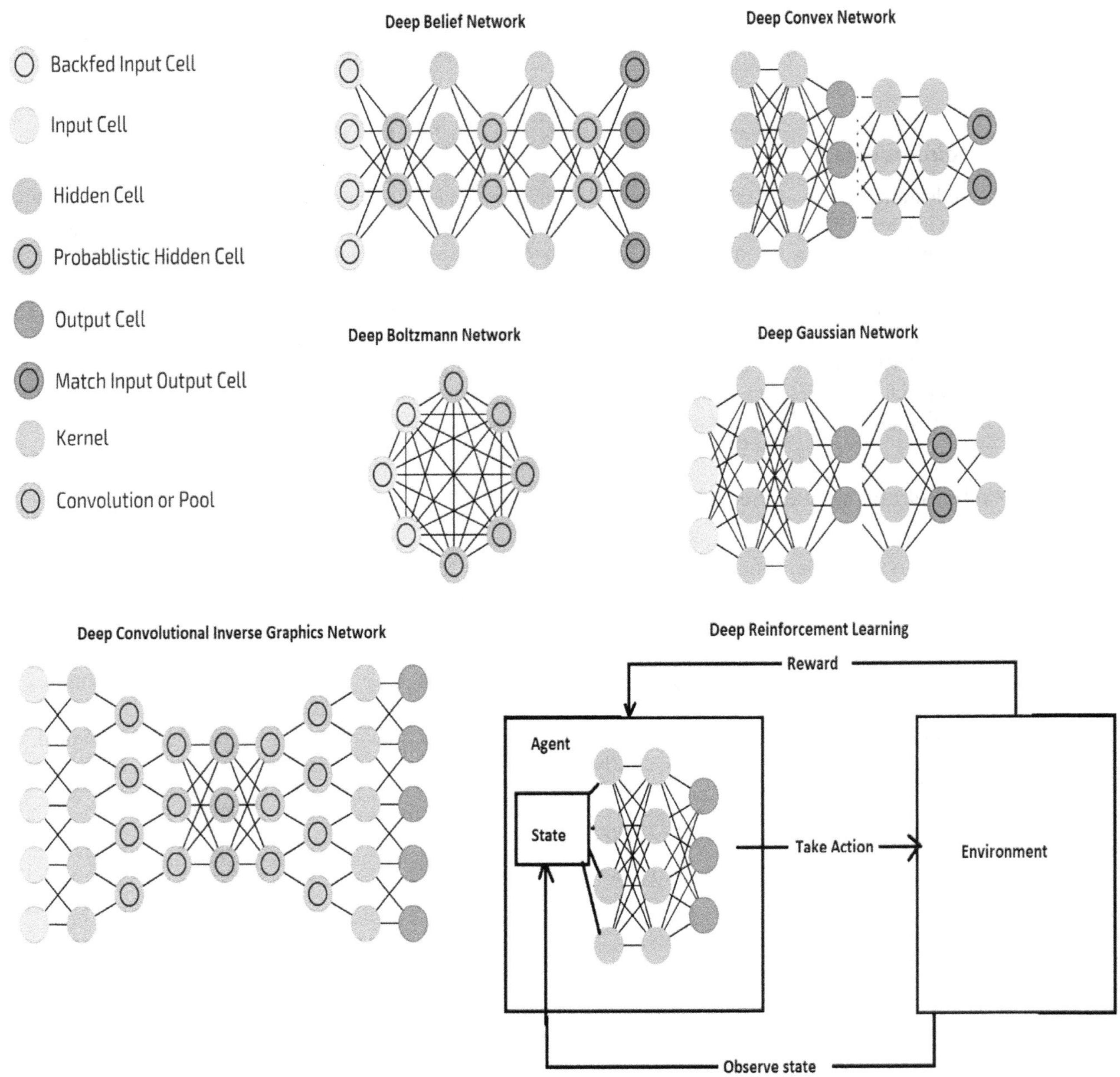

FIGURE 2.24 Deep reinforcement learning, deep belief and convex networks, deep Boltzmann machines, deep Gaussian processes, and deep convolutional inverse graphics networks in biomedical image analysis.

reconstruction and registration. For instance, DCNs have been employed to reconstruct magnetic resonance images from undersampled data, yielding state-of-the-art results. By learning to solve the convex optimization problem of image reconstruction, DCNs can enhance the quality of reconstructed images. Furthermore, DCNs have been utilized for IMR, including aligning brain MRI images. Through learning the convex optimization problem of image alignment, DCNs improve registration accuracy and mitigate the need for manual intervention [124].

2.3.29 Deep Boltzmann machines

Deep Boltzmann machines (DBMs) are a type of deep neural network that excel at learning and modeling the underlying probability distribution of data. In the domain of BIA, DBMs have demonstrated their efficacy in tasks like medical image segmentation and classification [125]. For instance, DBMs have been utilized for heart segmentation from MRI images, achieving remarkable results. By learning the probability distribution that characterizes the heart from the images, DBMs enhance the accuracy of segmentation while mitigating the reliance on manual annotation [126]. Additionally, DBMs have found applications in image classification, such as the diagnosis of diabetic retinopathy through the classification of retinal images. By learning the probability distribution that represents the features of the retinal images, DBMs improve the accuracy of classification and reduce the dependence on expert knowledge [127].

2.3.30 Deep Gaussian processes

Deep Gaussian processes (DGPs) are a sophisticated extension of Gaussian processes (GPs) that introduce multiple layers of hidden variables into the model. In the field of BIA, DGPs have proven valuable in tasks such as medical image segmentation and registration [128]. Specifically, DGPs have been employed to segment lungs from CT images, achieving state-of-the-art outcomes. By employing multiple layers of hidden variables to model the distribution of lung images, DGPs enhance the accuracy of segmentation while reducing the reliance on manual annotation [129]. Furthermore, DGPs have demonstrated their effectiveness in IMR, particularly in aligning brain images. By leveraging multiple layers of hidden variables to model the distribution of brain images, DGPs improve the accuracy of registration and diminish the need for manual intervention [130].

2.3.31 Deep convolutional inverse graphics networks

Deep convolutional inverse graphics networks (DCIGNs) are a powerful class of deep generative models that excel at inferring the inherent 3D structure of objects from 2D images. In the realm of BIA, DCIGNs have found application in critical tasks such as medical image segmentation and diagnosis. One notable application of DCIGNs is in the segmentation of the heart from MRI images, where they have achieved remarkable results. By leveraging their ability to infer the underlying 3D structure of the heart, DCIGNs enhance the accuracy of segmentation, leading to improved diagnostic outcomes. DCIGNs have also demonstrated efficacy in IMR, particularly in aligning brain images. Through their capacity to infer the underlying 3D structure of the brain, DCIGNs enhance the accuracy of registration while reducing the dependence on manual intervention [131].

2.3.32 Convolutional recurrent neural networks

Convolutional recurrent neural networks (CRNNs) are a powerful class of deep neural networks that integrate the spatial processing capabilities of CNNs with the sequential processing capabilities of recurrent neural networks (RNNs). In the domain of BIA, CRNNs have proven to be highly effective for tasks such as medical image segmentation and diagnosis [101]. An illustrative application of CRNNs is in the segmentation of the colon from endoscopic images, where they have achieved state-of-the-art performance. By combining the spatial processing of CNNs with the sequential processing of RNNs, CRNNs can capture both local and contextual information, leading to improved accuracy in segmentation and diagnosis. Furthermore, CRNNs have been successfully employed for image classification tasks, specifically in the classification of ECG signals for the diagnosis of heart disease. By leveraging the joint capabilities of spatial and sequential information processing within CRNNs, these models can effectively analyze ECG signals, enhancing the accuracy of diagnosis while reducing reliance on expert knowledge [132] (Fig. 2.25).

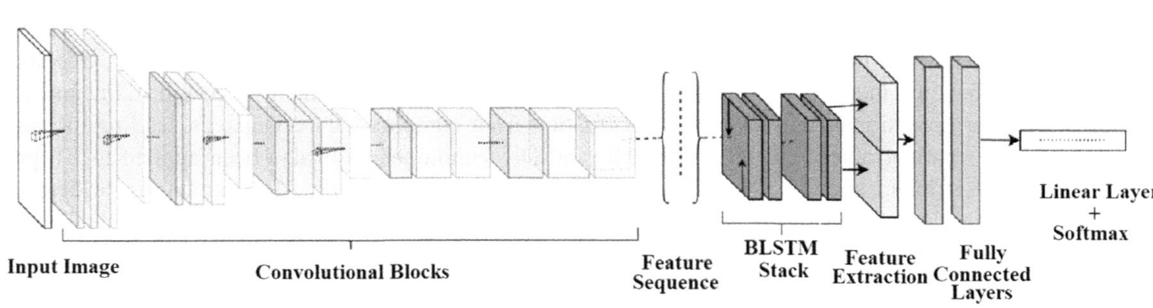

FIGURE 2.25 Convolutional recurrent neural networks in biomedical image analysis [101].

2.4 Application of biomedical image analysis

BIA is a rapidly evolving field with numerous applications in healthcare. It involves the application of advanced computational techniques to medical images acquired from various imaging modalities, such as MRI, CT, ultrasound, and PET. BIA plays a crucial role in enabling automated and quantitative analysis of medical images, facilitating accurate diagnosis, treatment planning, and disease monitoring. Let us explore some specific examples of the application of BIA in different areas of healthcare.

1. Cancer Detection and Diagnosis: One of the primary applications of BIA is in the detection and diagnosis of cancer. Various imaging modalities can provide detailed information about the structure and function of organs and tissues, aiding in the identification of abnormal tissues such as tumors. BIA algorithms can analyze medical images to automatically detect and quantify these abnormalities. For instance, CAD systems have been developed for breast cancer detection using mammography images. These systems analyze mammograms to identify suspicious regions that can be further evaluated by radiologists, improving the accuracy of breast cancer detection.
2. Cardiovascular Disease Analysis: BIA is widely employed in the analysis of cardiovascular diseases. Imaging techniques like echocardiography, MRI, and CT provide valuable insights into the heart and blood vessels. BIA algorithms can extract quantitative measures of heart function, such as ejection fraction, wall motion abnormalities, and blood flow velocity. In the analysis of coronary artery disease, CT images can visualize the coronary arteries and detect stenosis (narrowing). BIA algorithms can quantify the degree of stenosis, guiding treatment decisions such as the need for angioplasty or bypass surgery.
3. Neurological Disorder Analysis: BIA is extensively used in the analysis of neurological disorders. Imaging modalities like MRI and PET offer detailed information about brain structure and function. BIA algorithms can extract quantitative measures of brain function, such as cerebral blood flow, oxygen consumption, and metabolism. In Alzheimer's disease analysis, MRI can detect brain atrophy, a characteristic feature of the disease. BIA algorithms quantify the degree of atrophy and track disease progression, enabling monitoring of treatment effectiveness and the development of new therapies.
4. Surgical Planning and Guidance: BIA plays a crucial role in assisting surgeons in surgical planning and guidance. MI techniques, including CT, MRI, and ultrasound, provide preoperative images of the body. BIA algorithms analyze these images to identify critical structures such as blood vessels and nerves, aiding in planning the safest and most effective surgical approach. During surgery, real-time imaging can guide the surgeon, allowing for more precise and less invasive procedures.
5. Personalized Medicine: BIA contributes to the development of personalized treatment plans for patients. By analyzing medical images, physicians can determine the most appropriate treatment based on individual patient characteristics. For example, MRI-based image analysis has been used to predict the response to neoadjuvant chemotherapy in patients with breast cancer. This information helps tailor treatment strategies, improving patient outcomes and reducing unnecessary interventions.
6. Drug Development: BIA is utilized in evaluating the efficacy of new drugs in clinical trials. By analyzing medical images, researchers can assess whether a drug has the desired effect on the target tissue or organ. BIA enables the objective measurement of treatment response, potentially accelerating the drug development process and reducing the cost of clinical trials.
7. Public Health Research: BIA contributes to public health research and policymaking. By analyzing medical images from large populations, researchers can identify patterns and trends in disease prevalence and risk factors. This information helps in the development of public health interventions to improve population health. For instance, BIA has been used to analyze imaging data for the early detection and monitoring of infectious diseases like COVID-19.

In conclusion, BI and analysis have revolutionized the field of healthcare by enabling noninvasive and accurate diagnosis, monitoring, and treatment of diseases. While there are challenges in BI, innovative solutions and advanced analysis techniques have made it possible to overcome these challenges and improve patient outcomes. One such solution is BIA, which has become increasingly important in recent years. By using sophisticated algorithms and software, BIA can help healthcare professionals extract useful information from complex medical images, leading to more accurate diagnoses and better patient outcomes. Intelligent ML and DL techniques have also been applied to BI, providing even more accurate and efficient analysis of medical images. These techniques allow for the automatic detection and segmentation of abnormalities in medical images, providing valuable insights to healthcare professionals. The application of BIA has been used in various medical fields, including radiology, cardiology, and neurology. With continued advancements in technology and analysis techniques, BI and analysis will undoubtedly continue to have a significant impact on the field of healthcare.

References

[1] S. Hussain, I. Mubeen, N. Ullah, S.S.U.D. Shah, B.A. Khan, M. Zahoor, et al., Modern diagnostic imaging technique applications and risk factors in the medical field: a review, BioMed. Res. Int 2022 (2022).

[2] A. Umar, S. Atabo, A review of imaging techniques in scientific research/clinical diagnosis, MOJ Anat. Physiol. 6 (5) (2019) 175–183.

[3] I.R. Duffy, A.J. Boyle, N. Vasdev, Improving PET imaging acquisition and analysis with machine learning: a narrative review with focus on Alzheimer's disease and oncology, Mol. Imaging 18 (2019).

[4] Y.M.Y. Abdallah, History of medical imaging, Arch. Med. Health Sci. 5 (2) (2017) 275.

[5] H. Kasban, M.A.M. El-Bendary, D.H. Salama, A comparative study of medical imaging techniques, Int. J. Inf. Sci. Intell. Syst. 4 (2) (2015) 37–58.

[6] M.J. Gounis, G.J. Ughi, M. Marosfoi, D.K. Lopes, D. Fiorella, H.G. Bezerra, et al., Intravascular optical coherence tomography for neurointerventional surgery, Stroke 50 (1) (2019) 218–223.

[7] Z. Hosseinaee, M. Le, K. Bell, P.H. Reza, Towards non-contact photoacoustic imaging, Photoacoustics 20 (2020) 100207.

[8] G. Rong, A. Mendez, E.B. Assi, B. Zhao, M. Sawan, Artificial intelligence in healthcare: review and prediction case studies, Engineering 6 (3) (2020) 291–301.

[9] M. Javaid, A. Haleem, R.P. Singh, R. Suman, S. Rab, Significance of machine learning in healthcare: features, pillars and applications, Int. J. Intell. Netw. 3 (2022) 58–73.

[10] L.H. Goetz, N.J. Schork, Personalized medicine: motivation, challenges, and progress, Fertil. Steril. 109 (6) (2018) 952–963.

[11] D.A. Hashimoto, G. Rosman, D. Rus, O.R. Meireles, Artificial intelligence in surgery: promises and perils, Ann. Surg. 268 (1) (2018) 70–76.

[12] D. Shen, G. Wu, H.I. Suk, Deep learning in medical image analysis, Annu. Rev. Biomed. Eng. 19 (2017) 221–248. A.S. Lundervold, A. Lundervold, An overview of deep learning in medical imaging focusing on MRI. Zeitschrift für Medizinische Physik 29 (2) (2019) 102–127.

[13] M. Hadian, A. Jabbari, E. Mazaheri, M. Norouzi, What is the impact of clinical guidelines on imaging costs? J. Educ. Health Promotion 10 (2021).

[14] H.Y. Jung, J.R. Vest, M.A. Unruh, L.M. Kern, R. Kaushal, HITEC Investigators, Use of health information exchange and repeat imaging costs, J. Am. Coll. Radiol. 12 (12) (2015) 1364–1370.

[15] R. Sarwal, U. Prasad, K.M. Gopal, S. Kalal, D. Kaur, A. Kumar, et al., Investment Opportunities in India's Healthcare Sector, OSF Preprints, 2021.

[16] S.P. Power, F. Moloney, M. Twomey, K. James, O.J. O'Connor, M.M. Maher, Computed tomography and patient risk: facts, perceptions and uncertainties, World J. Radiol. 8 (12) (2016) 902.

[17] P.K. Nguyen, J.C. Wu, Radiation exposure from imaging tests: is there an increased cancer risk? Expert. Rev. Cardiovasc. Ther. 9 (2) (2011) 177–183.

[18] J.R. Parikh, R.A. Geise, E.I. Bluth, C.E. Bender, G. Sze, A.K. Jones, et al., Potential radiation-related effects on radiologists, Am. J. Roentgenol. 208 (3) (2017) 595–602.

[19] V. Venkataraman, T. Browning, I. Pedrosa, S. Abbara, D. Fetzer, S. Toomay, et al., Implementing shared, standardized imaging protocols to improve cross-enterprise workflow and quality, J. Digital Imaging 32 (2019) 880–887.

[20] X.T. Li, R.Y. Huang, Standardization of imaging methods for machine learning in neuro-oncology, Neuro-Oncol. Adv 2 (Supplement_4) (2020) iv49–iv55.

[21] S. Dash, S.K. Shakyawar, M. Sharma, S. Kaushik, Big data in healthcare: management, analysis and future prospects, J. Big Data 6 (1) (2019) 1–25.

[22] L. Ismail, H. Materwala, A.P. Karduck, A. Adem, Requirements of health data management systems for biomedical care and research: scoping review, J. Med. Internet Res. 22 (7) (2020) e17508.

[23] E. von Ende, S. Ryan, M.A. Crain, M.S. Makary, Artificial intelligence, augmented reality, and virtual reality advances and applications in interventional radiology, Diagnostics 13 (5) (2023) 892.

[24] Z. Ahmed, K. Mohamed, S. Zeeshan, X. Dong, Artificial intelligence with multi-functional machine learning platform development for better healthcare and precision medicine, Database (2020) 2020.

[25] P. Rao, S.H. Fischer, M.E. Vaiana, E.A. Taylor, Barriers to price and quality transparency in health care markets, Rand Health Q. 9 (3) (2022).

[26] A.S. Ahuja, The impact of artificial intelligence in medicine on the future role of the physician, PeerJ 7 (2019) e7702.

[27] A.A. Seyhan, C. Carini, Are innovation and new technologies in precision medicine paving a new era in patients centric care? J. Transl. Med. 17 (2019) 1–28.

[28] A. Haleem, M. Javaid, R.P. Singh, R. Suman, Medical 4.0 technologies for healthcare: features, capabilities, and applications, Internet Things Cyber-Phys. Syst. 2 (9) (2022).

[29] G.A. Kaissis, M.R. Makowski, D. Rückert, R.F. Braren, Secure, privacy-preserving and federated machine learning in medical imaging, Nat. Mach. Intell. 2 (6) (2020) 305–311.

[30] P. Tagde, S. Tagde, T. Bhattacharya, P. Tagde, H. Chopra, R. Akter, et al., Blockchain and artificial intelligence technology in e-health, Environ. Sci. Pollut. Res. 28 (2021) 52810–52831.

[31] R. Kumar, D. Singh, K. Srinivasan, Y.C. Hu, AI-powered blockchain technology for public health: a contemporary review, open challenges, and future research directions, Healthcare, 11, MDPI, 2022, p. 81. 1.

[32] S. Hussain, I. Mubeen, N. Ullah, S.S.U.D. Shah, B.A. Khan, M. Zahoor, et al., Modern diagnostic imaging technique applications and risk factors in the medical field: a review, BioMed. Res. Int. 2022 (2022).

[33] J.W. Prescott, Quantitative imaging biomarkers: the application of advanced image processing and analysis to clinical and preclinical decision making, J. Digital Imaging 26 (2013) 97–108.

[34] R. García-Figueiras, S. Baleato-González, A.R. Padhani, A. Luna-Alcalá, J.A. Vallejo-Casas, E. Sala, et al., How clinical imaging can assess cancer biology, Insights Imaging 10 (2019) 1–35.

[35] R.M. Rangayyan, Biomedical Image Analysis, CRC Press, 2004.

[36] M. Haidekker, Advanced Biomedical Image Analysis, John Wiley & Sons, 2010. H.K. Huang, Biomedical image processing. Crit. Rev. Bioeng. 5(3) (1981) 185–271.

[37] S. Goyal, T. Kataria, Image guidance in radiation therapy: techniques and applications, Radiol. Res. Pract. (2014) 2014.

[38] A.M. Lozano, N. Lipsman, H. Bergman, P. Brown, S. Chabardes, J.W. Chang, et al., Deep brain stimulation: current challenges and future directions, Nat. Rev. Neurol. 15 (3) (2019) 148–1160.

[39] Y.X. Wang, M. Deng, Medical imaging in new drug clinical development, J. Thorac. Dis. 2 (4) (2010) 245.

[40] J.R. Lindner, J. Link, Molecular imaging in drug discovery and development, Circ. Cardiovasc. Imaging 11 (2) (2018) e005355.

[41] R.M. Califf, Biomarker definitions and their applications, Exp. Biol. Med. 243 (3) (2018) 213–221.

[42] F. Aeffner, M.D. Zarella, N. Buchbinder, M.M. Bui, M.R. Goodman, D.J. Hartman, et al., Introduction to digital image analysis in whole-slide imaging: a white paper from the digital pathology association, J. Pathol. Inform. 10 (1) (2019) 9.

[43] G.J. Awcock, R. Thomas, Applied Image Processing, Macmillan, Basingstoke, UK, 1995, pp. 111–118.

[44] M.M. Petrou, C. Petrou, Image Processing: The Fundamentals, John Wiley & Sons, 2010.

[45] V.K. Mishra, S. Kumar, N. Shukla, Image acquisition and techniques to perform image acquisition, SAMRIDDHI: A J. Phys. Sci. Eng. Technol. 9 (01) (2017) 21–24.

[46] M.B. Williams, E.A. Krupinski, K.J. Strauss, W.K. Breeden III, M.S. Rzeszotarski, K. Applegate, et al., Digital radiography image quality: image acquisition, J. Am. Coll. Radiol. 4 (6) (2007) 371–388.

[47] J. Chaki, N. Dey, A Beginner's Guide to Image Preprocessing Techniques, CRC Press, 2018.

[48] V. Mayya, U. Kulkarni, D.K. Surya, U.R. Acharya, An empirical study of preprocessing techniques with convolutional neural networks for accurate detection of chronic ocular diseases using fundus images, Appl. Intell. 53 (2) (2023) 1548–1566.

[49] J. Jing, T. Gao, W. Zhang, Y. Gao, C. Sun, Image feature information extraction for interest point detection: a comprehensive review, IEEE Trans. Pattern Anal. Mach. Intell. (2022).

[50] F. Al-Areqi, M.Z. Konyar, Effectiveness evaluation of different feature extraction methods for classification of Covid-19 from computed tomography images: a high accuracy classification study, Biomed. Signal. Process. Control. 76 (2022) 103662.

[51] Y. Guan, Y. Han, S. Liu, Deep learning approaches for image classification techniques. In 2022 IEEE International Conference on Electrical Engineering, Big Data and Algorithms (EEBDA) (pp. 1132–1136). IEEE, 2022.

[52] R.S. Hawezi, F.S. Khoshaba, S.W. Kareem, A comparison of automated classification techniques for image processing in video internet of things, Comput. Electr. Eng. 101 (2022) 108074.

[53] Y. Zhang, J.M. Gorriz, D.R. Nayak, Optimization algorithms and machine learning techniques in medical image analysis, Math. Biosci. Eng. 20 (3) (2023) 5917–5920.

[54] K. Hammernik, T. Küstner, B. Yaman, Z. Huang, D. Rueckert, F. Knoll, et al., Physics-driven deep learning for computational magnetic resonance imaging: combining physics and machine learning for improved medical imaging, IEEE Signal. Process. Mag. 40 (1) (2023) 98–114.

[55] S. Chakraborty, K. Mali, An overview of biomedical image analysis from the deep learning perspective, Res. Anthol. Improv. Med. Imaging Tech. Anal. Interven. (2023) 43–59.

[56] K. Aggarwal, M.M. Jimeno, K.S. Ravi, G. Gonzalez, S. Geethanath, Developing and deploying deep learning models in brain MRI: a review, arXiv Prepr. arXiv 2301 (2023) 01241.

[57] Y. Kumar, A. Koul, R. Singla, M.F. Ijaz, Artificial intelligence in disease diagnosis: a systematic literature review, synthesizing framework and future research agenda, J. Ambient. Intell. Humanized Comput. (2022) 1–28.

[58] S. Quazi, Artificial intelligence and machine learning in precision and genomic medicine, Med. Oncol. 39 (8) (2022) 120.

[59] C. Malamateniou, S. McFadden, Y. McQuinlan, A. England, N. Woznitza, S. Goldsworthy, et al., Artificial intelligence: guidance for clinical imaging and therapeutic radiography professionals, a summary by the Society of Radiographers AI working group, Radiography 27 (4) (2021) 1192–1202.

[60] L. Alzubaidi, J. Zhang, A.J. Humaidi, A. Al-Dujaili, Y. Duan, O. Al-Shamma, et al., Review of deep learning: concepts, CNN architectures, challenges, applications, future directions, J. Big Data 8 (2021) 1–74.

[61] R.A. Castellino, Computer aided detection (CAD): an overview, Cancer Imaging 5 (1) (2005) 17.

[62] H.J. Park, S.M. Kim, B. La Yun, M. Jang, B. Kim, J.Y. Jang, et al., A computer-aided diagnosis system using artificial intelligence for the diagnosis and characterization of breast masses on ultrasound: added value for the inexperienced breast radiologist, Medicine 98 (3) (2019).

[63] I.R.I. Haque, J. Neubert, Deep learning approaches to biomedical image segmentation, Inform. Med. Unlocked 18 (2020) 100297.

[64] N. Sharma, L.M. Aggarwal, Automated medical image segmentation techniques, J. Med. Phys. 35 (1) (2010) 3.

[65] L. Zhou, M. Fan, C. Hansen, C.R. Johnson, D. Weiskopf, A review of three-dimensional medical image visualization, Health Data Science, 2022, 2022. J.P. Agrawal, B.J. Erickson, C.E. Kahn Jr, Imaging informatics: 25 years of progress. Yearb. Med. Inform. 25 (S 01) (2016) S23–S31.

[66] D.P. Kaucha, P.W.C. Prasad, A. Alsadoon, A. Elchouemi, S. Sreedharan, Early detection of lung cancer using SVM classifier in biomedical image processing. In 2017 IEEE International Conference on Power, Control, Signals and Instrumentation Engineering (ICPCSI) (pp. 3143–3148). IEEE, 2017.

[67] J. Eali, S. Neal, D. Bhattacharyya, T.R. Nakka, S.P. Hong, A novel approach in bio-medical image segmentation for analyzing brain cancer images with U-NET semantic segmentation and TPLD models using SVM, Traitement Du. Signal. 39 (2) (2022).

[68] N.B. Bahadure, A.K. Ray, H.P. Thethi, Image analysis for MRI based brain tumor detection and feature extraction using biologically inspired BWT and SVM, Int. J. Biomed. Imaging (2017) 2017.

[69] M.M. Rahman, S.K. Antani, G.R. Thoma, A learning-based similarity fusion and filtering approach for biomedical image retrieval using SVM classification and relevance feedback, IEEE Trans. Inf. Technol. Biomed. 15 (4) (2011) 640–646.

[70] S.K. Mohapatra, M.N. Mohanty, Big data analysis and classification of biomedical signal using random forest algorithm, New Paradigm in Decision Science and Management: Proceedings of ICDSM 2018, Springer, Singapore, 2020, pp. 217–224.

[71] T.M. Oshiro, P.S. Perez, J.A. Baranauskas, How many trees in a random forest? In Machine Learning and Data Mining in Pattern Recognition: 8th International Conference, MLDM 2012, Berlin, Germany, July 13–20, 2012. Proceedings 8 (pp. 154–168). Springer, Berlin Heidelberg, 2012.

[72] H. Seo, M. Badiei Khuzani, V. Vasudevan, C. Huang, H. Ren, R. Xiao, et al., Machine learning techniques for biomedical image segmentation: an overview of technical aspects and introduction to state-of-art applications, Med. Phys. 47 (5) (2020) e148–e167.

[73] N.E.A. Khalid, S. Ibrahim, P.N.M.M. Haniff, MRI brain abnormalities segmentation using K-nearest neighbors (k-NN), Int. J. Comput. Sci. Eng. 3 (2) (2011) 980–990.

[74] T.T. Htay, S.S. Maung, Early stage breast cancer detection system using glcm feature extraction and k-nearest neighbor (k-NN) on mammography image, 2018 18th International Symposium on Communications and Information Technologies (ISCIT), IEEE, 2018, pp. 171–175.

[75] A. Alarabeyyat, M. Alhanahnah, Breast cancer detection using k-nearest neighbor machine learning algorithm. In 2016 9th International Conference on Developments in eSystems Engineering (DeSE) (pp. 35–39). IEEE, 2016.

[76] X.H. Cao, I. Stojkovic, Z. Obradovic, A robust data scaling algorithm to improve classification accuracies in biomedical data, BMC Bioinform. 17 (1) (2016) 1–10.

[77] D. Zhang, D. Shen, Alzheimer's Disease Neuroimaging Initiative, Multi-modal multi-task learning for joint prediction of multiple regression and classification variables in Alzheimer's disease, NeuroImage 59 (2) (2012) 895–907.

[78] X. Chen, M. Wang, H. Zhang, The use of classification trees for bioinformatics, Wiley Interdiscip. Rev. Data Min. Knowl. Discov. 1 (1) (2011) 55–63.

[79] B.T. Jijo, A.M. Abdulazeez, Classification based on decision tree algorithms for machine learning, Evaluation 6 (2021) 7.

[80] E. Gokgoz, A. Subasi, Comparison of decision tree algorithms for EMG signal classification using DWT, Biomed. Signal. Process. Control. 18 (2015) 138–144.

[81] C. Grana, D. Borghesani, R. Cucchiara, Optimized block-based connected components labeling with decision trees, IEEE Trans. Image Process. 19 (6) (2010) 1596–1609.

[82] R.S. Akinbo, O.A. Daramola, Ensemble machine learning algorithms for prediction and classification of medical images, Machine Learning-Algorithms Model. Appl. (2021).

[83] I.D. Mienye, Y. Sun, A survey of ensemble learning: concepts, algorithms, applications, and prospects, IEEE Access. 10 (2022) 99129–99149.

[84] B. Zolfaghari, L. Mirsadeghi, K. Bibak, K. Kavousi, Cancer prognosis and diagnosis methods based on ensemble learning, ACM Comput. Surv. (2023).

[85] T.A. Assegie, R.L. Tulasi, N.K. Kumar, Breast cancer prediction model with decision tree and adaptive boosting, IAES Int. J. Artif. Intell. 10 (1) (2021) 184.

[86] J. Wang, X. Chen, H. Lu, L. Zhang, J. Pan, Y. Bao, et al., Feature-shared adaptive-boost deep learning for invasiveness classification of pulmonary subsolid nodules in CT images, Med. Phys. 47 (4) (2020) 1738–1749.

[87] S.N.N. Htun, T.T. Zin, P. Tin, Image processing technique and hidden Markov model for an elderly care monitoring system, J. Imaging 6 (6) (2020) 49.

[88] M. Rahul, N. Kohli, R. Agarwal, Facial expression recognition using local binary pattern and modified hidden Markov model, Int. J. Adv. Intell. Paradig. 17 (3–4) (2020) 367–378.

[89] K. Aas, L. Eikvil, R.B. Huseby, Applications of hidden Markov chains in image analysis, Pattern Recognit. 32 (4) (1999) 703–713.

[90] A. Chaddad, Automated feature extraction in brain tumor by magnetic resonance imaging using Gaussian mixture models, J. Biomed. Imaging 2015 (2015) 8.

[91] R. Alroobaea, S. Rubaiee, S. Bourouis, N. Bouguila, A. Alsufyani, Bayesian inference framework for bounded generalized Gaussian-based mixture model and its application to biomedical images classification, Int. J. Imaging Syst. Technol. 30 (1) (2020) 18–30.

[92] D. Nandi, A.S. Ashour, S. Samanta, S. Chakraborty, M.A. Salem, N. Dey, Principal component analysis in medical image processing: a study, Int. J. Image Min. 1 (1) (2015) 65–86.

[93] M. Mudrova, A. Procházka, Principal component analysis in image processing. In Proceedings of the MATLAB Technical Computing Conference, Prague, 2005.

[94] K.H. Thung, C.Y. Wee, A brief review on multi-task learning, Multimed. Tools Appl. 77 (2018) 29705–29725.

[95] G. Crichton, S. Pyysalo, B. Chiu, A. Korhonen, A neural network multi-task learning approach to biomedical named entity recognition, BMC Bioinform. 18 (1) (2017) 1–14.

[96] H. Barzekar, Z. Yu, C-Net: a reliable convolutional neural network for biomedical image classification, Expert. Syst. Appl. 187 (2022) 116003.

[97] Z. Zhou, J. Shin, L. Zhang, S. Gurudu, M. Gotway, J. Liang, Fine-tuning convolutional neural networks for biomedical image analysis: actively and incrementally. In Proceedings of the IEEE Conference on Computer Vision and Pattern Recognition (pp. 7340–7351), 2017.

[98] M.A. Abdou, Literature review: efficient deep neural networks techniques for medical image analysis, Neural. Comput. Appl. 34 (8) (2022) 5791–5812.

[99] D.R. Sarvamangala, R.V. Kulkarni, Convolutional neural networks in medical image understanding: a survey, Evolut. Intell. 15 (1) (2022) 1–22.

[100] H. Yu, L.T. Yang, Q. Zhang, D. Armstrong, M.J. Deen, Convolutional neural networks for medical image analysis: state-of-the-art, comparisons, improvement and perspectives, Neurocomputing 444 (2021) 92–110.

[101] A. Chakravarty, J. Sivaswamy, RACE-net: a recurrent neural network for biomedical image segmentation, IEEE J. Biomed. Health Inform. 23 (3) (2018) 1151–1162.

[102] M. Nguyen, T. He, L. An, D.C. Alexander, J. Feng, B.T. Yeo, et al., Predicting Alzheimer's disease progression using deep recurrent neural networks, NeuroImage 222 (2020) 117203.

[103] A. Carfì, C. Motolese, B. Bruno, F. Mastrogiovanni, Online human gesture recognition using recurrent neural networks and wearable sensors, 2018 27th IEEE International Symposium on Robot and Human Interactive Communication (RO-MAN), IEEE, 2018, pp. 188–195.

[104] T.C. Hollon, B. Pandian, A.R. Adapa, E. Urias, A.V. Save, S.S.S. Khalsa, et al., Near real-time intraoperative brain tumor diagnosis using stimulated Raman histology and deep neural networks, Nat. Med. 26 (1) (2020) 52–58.

[105] X. Yi, E. Walia, P. Babyn, Generative adversarial network in medical imaging: a review, Med. Image Anal. 58 (2019) 101552.

[106] T. Iqbal, H. Ali, Generative adversarial network for medical images (MI-GAN), J. Med. Syst. 42 (11) (2018) 231.

[107] L. Lan, L. You, Z. Zhang, Z. Fan, W. Zhao, N. Zeng, et al., Generative adversarial networks and its applications in biomedical informatics, Front. Public. Health 8 (2020) 164.

[108] M. Astaraki, Ö. Smedby, C. Wang, Prior-aware autoencoders for lung pathology segmentation, Med. Image Anal. 80 (2022) 102491.

[109] X. Zhang, H. Dou, T. Ju, J. Xu, S. Zhang, Fusing heterogeneous features from stacked sparse autoencoder for histopathological image analysis, IEEE J. Biomed. Health Inform. 20 (5) (2015) 1377–1383.

[110] J. Ehrhardt, M. Wilms, Autoencoders and variational autoencoders in medical image analysis, Biomedical Image Synthesis and Simulation, Academic Press, 2022, pp. 129–162.

[111] P. Kora, C.P. Ooi, O. Faust, U. Raghavendra, A. Gudigar, W.Y. Chan, et al., Transfer learning techniques for medical image analysis: a review, Biocybern. Biomed. Eng. 42 (1) (2022) 79–107.

[112] D. Anand, D. Tank, H. Tibrewal, A. Sethi, Self-supervision vs. transfer learning: robust biomedical image analysis against adversarial attacks, 2020 IEEE 17th International. Symposium on Biomedical Imaging (ISBI), IEEE, 2020, pp. 1159–1163.

[113] S. Niu, M. Liu, Y. Liu, J. Wang, H. Song, Distant domain transfer learning for medical imaging, IEEE J. Biomed. Health Inform. 25 (10) (2021) 3784–3793.

[114] S. Ahmed, D. Dera, S.U. Hassan, N. Bouaynaya, G. Rasool, Failure detection in deep neural networks for medical imaging, Front. Med. Technol. 4 (2022) 919046.

[115] A. Filos, S. Farquhar, A.N. Gomez, T.G. Rudner, Z. Kenton, L. Smith, et al., A systematic comparison of Bayesian deep learning robustness in diabetic retinopathy tasks, arXiv Prepr. arXiv 1912 (2019) 10481.

[116] J. Chen, Y. Zou, Y. Wang, Wireless capsule endoscopy video summarization: a learning approach based on Siamese neural network and support vector machine. In 2016 23rd International Conference on Pattern Recognition (ICPR) (pp. 1303–1308). IEEE, 2016.

[117] G. Madhu, A. Govardhan, V. Ravi, S. Kautish, B.S. Srinivas, T. Chaudhary, et al., DSCN-net: a deep Siamese capsule neural network model for automatic diagnosis of malaria parasites detection, Multimed. Tools Appl. 81 (23) (2022) 34105–34127.

[118] M. Shaikh, V.A. Kollerathu, G. Krishnamurthi, Recurrent attention mechanism networks for enhanced classification of biomedical images. In 2019 IEEE 16th International Symposium on Biomedical Imaging (ISBI 2019) (pp. 1260–1264). IEEE, 2019.

[119] X. Tang, B. Zhong, J. Peng, B. Hao, J. Li, Multi-scale channel importance sorting and spatial attention mechanism for retinal vessels segmentation, Appl. Soft Comput. 93 (2020) 106353.

[120] S.K. Zhou, H.N. Le, K. Luu, H.V. Nguyen, N. Ayache, Deep reinforcement learning in medical imaging: a literature review, Med. Image Anal. 73 (2021) 102193.

[121] A. Panwar, G. Semwal, S. Goel, S. Gupta, Stratification of the lesions in color fundus images of diabetic retinopathy patients using deep learning models and machine learning classifiers. In Edge Analytics: Select Proceedings of 26th International Conference—ADCOM 2020 (pp. 653–666). Springer Singapore, Singapore, 2022.

[122] M. Kim, C. Yan, D. Yang, Q. Wang, J. Ma, G. Wu, Deep learning in biomedical image analysis, Biomedical Information Technology, Academic Press, 2020, pp. 239–263.

[123] N. Coudray, P.S. Ocampo, T. Sakellaropoulos, N. Narula, M. Snuderl, D. Fenyö, et al., Classification and mutation prediction from non–small cell lung cancer histopathology images using deep learning, Nat. Med. 24 (10) (2018) 1559–1567.

[124] A. Jiménez-Sánchez, S. Albarqouni, D. Mateus, Capsule networks against medical imaging data challenges. In Intravascular Imaging and Computer Assisted Stenting and Large-Scale Annotation of Biomedical Data and Expert Label Synthesis: 7th Joint International Workshop, CVII-STENT 2018 and Third International Workshop, LABELS 2018, Held in Conjunction with MICCAI 2018, Granada, Spain, September 16, 2018, Proceedings 3 (pp. 150–160). Springer International Publishing, 2018.

[125] P.R. Jeyaraj, E.R.S. Nadar, Deep Boltzmann machine algorithm for accurate medical image analysis for classification of cancerous region, Cogn. Comput. Syst. 1 (3) (2019) 85–90.

[126] O. Oktay, E. Ferrante, K. Kamnitsas, M. Heinrich, W. Bai, J. Caballero, et al., Anatomically constrained neural networks (ACNNs): application to cardiac image enhancement and segmentation, IEEE Trans. Med. Imaging 37 (2) (2017) 384–395.

[127] R.D. Hjelm, V.D. Calhoun, R. Salakhutdinov, E.A. Allen, T. Adali, S.M. Plis, Restricted Boltzmann machines for neuroimaging: an application in identifying intrinsic networks, NeuroImage 96 (2014) 245–260.

[128] Á.E. Esteban, M. López-Pérez, A. Colomer, M.A. Sales, R. Molina, V. Naranjo, A new optical density granulometry-based descriptor for the classification of prostate histological images using shallow and deep Gaussian processes, Comput. Methods Prog. Biomed. 178 (2019) 303–317.

[129] W. Dong, S. Sun, M. Yin, A multi-view deep learning model for pathology image diagnosis, Appl. Intell. 53 (6) (2023) 7186–7200.

[130] A. Jones, F.W. Townes, D. Li, B.E. Engelhardt, Alignment of spatial genomics and histology data using deep Gaussian processes, BioRxiv 2022 (2022).

[131] T.D. Kulkarni, W.F. Whitney, P. Kohli, J. Tenenbaum, Deep convolutional inverse graphics network, Adv. Neural Inf. Process. Syst. 28 (2015).

[132] Z. Xiong, M.P. Nash, E. Cheng, V.V. Fedorov, M.K. Stiles, J. Zhao, ECG signal classification for the detection of cardiac arrhythmias using a convolutional recurrent neural network, Physiol. Meas. 39 (9) (2018) 094006.

Chapter 3

Basics of probability and statistics using Python

Kanmani Buddhi
Department of Electronics and Telecommunication Engineering, BMS College of Engineering, Bengaluru, Karnataka, India

3.1 Introduction

Data is a value that represents a physical parameter. Data can be the marks scored by students in a class; the height of a person; the number of hours a person spends with his/her mobile phone; the number of times the heart beats in a second; the human blood pressure; the number of vehicles crossing an identified location in an hour; the height of the aircraft; or any other data to represent a real physical quantity. These data may be independent and sometimes dependent. For example, there can exist dependency between the height of the person, the age, and the blood pressure; similarly, there can be a dependency between the average grade secured by the student and the annual salary he/she secures through campus placements; similarly one can attempt to measure the happiness/health quotient and the annual salary. We recognize that data exist in all engineering domains, including medical and finance. There exists a need to capture the data, process the data, arrive at useful statistical parameters, and eventually arrive at suitable conclusions. The data can be discrete or continuous. In data science, we obtain the data and attempt to develop a model based on the available data. The model can then be used to predict/obtain the value for new input data. In the field of medical imaging, data analytics (image analytics) plays an important role as the diagnosis will be based on the inference arrived at. For example, the initial medical image can be focused on identifying the tumor, and further classifying the tumor as benign or malignant [1–4]. Hence, the methods and algorithms used for analysis and modeling the system need to be accurate. Given any dataset, that represents any physical parameter, we can obtain useful statistical parameters like the mean, the deviation from the mean, the standard deviation, the variance, the skewness, and the kurtosis. To ensure consistency in the comparison of the statistical averages, it becomes useful to normalize the initial set of values.

The probability of occurrence of an event is the probability of favorable outcomes to the total number of outcomes. It is interesting to note that the probability is closely related to the statistical averages. Since data can be discrete or continuous, we have the probability of discrete events and the probability of continuous events. Hence, in this Chapter, we have three broad sections: the statistical parameters of a given dataset; the discrete probability distribution; and the continuous probability distribution [5,6].

In Section 3.2, we commence with the statistical averages of a given dataset. We then introduce various parameters used to describe a discrete probability distribution. This is then followed by typical parameters to describe the continuous probability distribution. Here, the emphasis is on the mathematical equations that describe various parameters. In Section 3.3, we have numerical examples for each of the concepts introduced in Section 3.2. The examples are carefully selected so that there is an introduction to concepts essential for data science. The Appendix has the Python code for Python Functions, while access to the Python code developed for various examples is provided [7].

3.2 The basics of probability and statistics

3.2.1 Statistical analysis of data

Let us consider a set of values x_1, x_2, x_3,.... x_N, which represent a real physical parameter. The physical parameters can be marks scored in an examination; the temperature in a room; the number of people at a location; the blood pressure of a human being; the speed of a vehicle; the number of products sold in a month; the age of people in a hall; or any

Intelligent Computing Techniques in Biomedical Imaging. DOI: https://doi.org/10.1016/B978-0-443-15999-2.00017-7

other physical parameter. While the physical parameter can actually represent the variation in an identified domain, irrespective of the nature of the type of the physical parameter, the dataset is eventually a finite sequence of numbers. Given a dataset, we can obtain statistical parameters like the mean, the mode, the median, the variance, the standard deviation, the skewness, and the kurtosis. These parameters, help in describing the given dataset. The mean of a dataset is the average value and is given by Eq. (3.1).

$$\text{Mean} = \mu_x = E[X] = \frac{1}{N}\sum_{i=1}^{N} x_i \tag{3.1}$$

The mode is the value which has the maximum occurrence, while the median is that value of the dataset, which divides the data into two equal groups when the data is arranged either in ascending or descending order.

The variance of a dataset, given by Eq. (3.2), is the measure of the deviation of the data from the mean. A high variance implies large deviation from the mean. The square root of the variance is known as the standard deviation, and is given by Eq. (3.3). The mean and standard deviation are usually the two important statistical parameters of any given dataset.

$$\begin{aligned} \text{Variance} = \sigma^2 &= E\left[(X-\mu_x)^2\right] \\ &= \frac{1}{N}\sum_{i=1}^{N}(x_i-\mu_x)^2 \\ &= \left(\frac{1}{N}\sum_{i=1}^{N}{x_i}^2\right) - {\mu_x}^2 \end{aligned} \tag{3.2}$$

$$\text{Standard deviation} = \sigma = \sqrt{\text{Variance}} \tag{3.3}$$

The skewness of the distribution, given by Eq. (3.4), is another statistical parameter, and it gives the measure of symmetry of the dataset. The skewness for symmetric data is zero, while a dataset with a tail in the right half, implies positive skewness, while a tail in the left half of the distribution, implies negative skewness.

$$\text{Skewness} = \frac{1}{\sigma^3}E\left[(X-\mu_x)^3\right] \tag{3.4}$$

The kurtosis of the distribution, given by Eq. (3.5), is another statistical parameter, and it gives the measure of the distribution in comparison to the normal distribution, which has a kurtosis of 3.

$$\text{Kurtosis} = \frac{1}{\sigma^4}E\left[(X-\mu_x)^4\right] \tag{3.5}$$

Any given dataset can be described by the above statistical parameters. Given two different datasets, these parameters can be different and may not serve as a measure for comparison. Hence, it is convenient to represent any dataset by its equivalent normalized distribution, given by Eq. (3.6), which is derived by subtracting the mean and dividing by the standard deviation. This normalized dataset now has zero mean and unit standard deviation.

$$\begin{aligned} &\text{Normalized distribution } X_N\text{:} \frac{(X-\mu_x)}{\sigma} \\ &\mu_{x_N} = E[X_N] = 0 \ \sigma_{x_N} = 1 \end{aligned} \tag{3.6}$$

Often we have two datasets that are related to each other, for example: the performance of a student in the continuous internal evaluation (CIE) and performance in the semester-end examination have a correlation; similarly the weight of an individual and his/her body mass index (BMI) are related. The number of values in each dataset is identical and represents different physical parameters. There is a need to have a metric for comparing two such datasets. One of the parameters is the Covariance, as given by Eq. (3.7), and represents the variation of each around the mean. Another parameter is the correlation coefficient, given by Eq. (3.8). The correlation coefficient for any two datasets, takes a value in the range -1 to $+1$. The correlation is positive when the two distributions have a similar variation, while the correlation is negative when they have an opposite variation. Similarly, the correlation is zero, when there is no dependence of data in the two distributions.

$$\text{Covariance} = \text{Cov}(X, Y) = E\left[\left(X - \mu_x\right)\left(Y - \mu_y\right)\right] = \frac{1}{N}\sum_{i=1}^{N}\left(x_i - \mu_x\right)\left(y_i - \mu_y\right) \tag{3.7}$$

$$\text{Correlation} = \rho_{X,Y} = \frac{\text{Cov}(X, Y)}{\sigma_x \sigma_y} = \frac{E\left[\left(X - \mu_x\right)\left(Y - \mu_y\right)\right]}{\sigma_x \sigma_y} \tag{3.8}$$

Sometimes, the given dataset may have a linear relationship between input parameter x, and the corresponding output parameter y, as given by Eq. (3.9). The line is represented by two parameters: the slope m, and the intercept/bias b.

$$y = mx + b \tag{3.9}$$

In such cases, we have the dataset and there is a need to obtain the two parameters of the linear model, that form the best fit for the given data so that the mean square error is minimized. The parameters of a given dataset can be obtained from Eq. (3.10), and the corresponding mean-square-error (mse), is given by Eq. (3.11).

$$\begin{aligned} &\text{Linear model: } \hat{y} = \hat{m}x + \hat{b} \\ &\text{where,} \quad \hat{m} = \frac{\text{Cov}(X, Y)}{\sigma_x^2} \\ &\hat{b} = \mu_y - \hat{m}\mu_x \end{aligned} \tag{3.10}$$

The mean square error, is given by Eq. (3.11).

$$\text{mse} = \frac{1}{N}\sum \left(\hat{y} - \hat{y}_{hat}\right)^2 \tag{3.11}$$

Some of the nonlinear models to fit a dataset are the hyperbolic tangent, tanh(x) given by Eq. (3.12); the Sigmoid(x) given by Eq. (3.13), and the rectifier linear unit, ReLU(x) given by Eq. (3.14).

$$\tanh(x) = \frac{(e^x - e^{-x})}{(e^x + e^{-x})} \tag{3.12}$$

$$\text{Sigmoid}(x) = \frac{1}{(1 + e^{-x})} \tag{3.13}$$

$$\text{ReLU}(x) = \begin{cases} x & x \geq 0 \\ 0 & \text{otherwise} \end{cases} \tag{3.14}$$

During simulation studies, we add noise to a generated dataset and try to fit a model. Here, we add the noise with uniform distribution or the Gaussian noise with zero mean and unit standard deviation. In any programming language, there exist functions for generating noise, and every simulation run results in a different sequence. To ensure consistency and enable comparison of simulation results, we set the seed of the random function, so that repeated runs of the code result in identical random sequence.

3.2.2 Discrete probability distributions

Probability of an event is defined as the ratio of the number of desired events, to the total number of events. Some examples of discrete events are: the toss of a coin (with heads/tails as possible outcomes; the roll of a dice (with numbers 1–6 as possible outcomes); the number of people arriving at a bank; the number of mails arriving in a post office; the number of students who are COVID positive. In these discrete events, we cannot exactly indicate the numbers. However, we can predict the probability of the event. Let us consider a discrete sample set X, with N, distinct possible events, and then the probability of the event is given by Eq. (3.15).

$$\begin{aligned} P(k) = P(X = k) \quad &\text{with} \quad k = 1, 2, \ldots, N \\ = \frac{\text{Number of desirable events}}{\text{Total number of events}} \end{aligned} \tag{3.15}$$

The sum of all individual probabilities is equal to 1, as indicated by Eq. (3.16).

$$\sum_{k=1}^{N} P(k) = 1 \tag{3.16}$$

For any given discrete probability distribution function, we can compute the parameters like: mean given by Eq. (3.17); the variance given by Eq. (3.18); the standard deviation given by Eq. (3.19); the skewness given by Eq. (3.20), and the kurtosis given by Eq. (3.21).

$$\text{Mean} = \mu = \sum_{k=1}^{N} kP(k) \tag{3.17}$$

$$\begin{aligned}\text{Variance} = \sigma^2 &= E\left[(X-\mu)^2\right] \\ &= \sum_{k=1}^{N}(X-\mu)^2 P(k)\end{aligned} \tag{3.18}$$

$$\text{Standard deviation} = \sigma = \sqrt{\text{Variance}} \tag{3.19}$$

$$\begin{aligned}\text{Skewness} &= \frac{1}{\sigma^3}\text{E}\left[(X-\mu)^3\right] \\ &= \frac{1}{\sigma^3}\sum_{k=1}^{N}(X-\mu)^3 P(k)\end{aligned} \tag{3.20}$$

$$\begin{aligned}\text{Kurtosis} &= \frac{1}{\sigma^4}\text{E}\left[(X-\mu)^4\right] \\ &= \frac{1}{\sigma^4}\sum_{k=1}^{N}(X-\mu)^4 P(k)\end{aligned} \tag{3.21}$$

The probability of an event is closely related to the information of the event and is given by Eq. (3.22). It can be observed that the probability and information are inversely related.

$$\text{Information} = \log_2 \frac{1}{P(k)} = -\log_2 P(k) \tag{3.22}$$

There are two more parameters that help describe a given probability distribution: the Entropy given by Eq. (3.23) and the Gini-index given by Eq. (3.24).

$$\text{Entropy} = H(X) = -\sum_{k=1}^{N} P(k)\log_2 P(k) \quad \text{with} \quad k = 1, 2, \ldots, N \tag{3.23}$$

$$\text{Gini-index}(X) = 1 - \sum_{k=1}^{N} P^2(k) \tag{3.24}$$

The Bayes' theorem helps compute the conditional probability of two events and is given by Eq. (3.25). Here, P

$$\begin{aligned}&P(A/B) = \frac{P(B/A)P(A)}{P(B)} \\ &\text{Here, } P(B/A), \text{ is the likelihood} \\ &\quad P(A), \text{ is the prior} \\ &\quad P(B), \text{ is the marginal} \\ &\quad P(A/B), \text{ is the posterior}\end{aligned} \tag{3.25}$$

Eq. (3.26) gives the probability of the union and the intersection of two events. Mutually exclusive events refer to the situation, where the occurrence of one rules out the occurrence of the other. Similarly, two events are independent, if the occurrence of one does not affect the occurrence of the other.

$$\begin{aligned} P(A \cup B) &= P(A) + P(B) - P(A \cap B) \\ &= P(A) + P(B) \text{ if A and B are mutually exclusive} \\ P(A \cap B) &= P(A/B)P(B) = P(B/A)P(A) \\ &= P(A)P(B) \text{ if A and B are independent} \end{aligned} \tag{3.26}$$

3.2.2.1 Poisson distribution

The Poisson distribution is a discrete probability distribution which can be used to represent most of the arrival processes. The Poisson arrival process is described by Eq. (3.27) and is completely characterized by the parameter λ. Typical arrival processes are: the number of people arriving in a bank; the number of people arriving in a restaurant; the number of parcels arriving in a post office; the number of phone calls made by the people in a city; the number of packets arriving in a node. It is interesting to observe that while the number of arrivals is discrete, the time between arrivals is continuous. The interarrival time can be represented by the continuous exponential distribution function (to be discussed in Section 3.2.3). Students arriving for a lecture; or people arriving to watch a movie, do not belong to the Poisson arrival process.

$$P(k) = \frac{\lambda e^{-\lambda}}{k!} \quad \text{for} \quad k = 0, 1, 2, \ldots., N \tag{3.27}$$

3.2.2.2 Binomial distribution

The binomial distribution is a discrete probability distribution, which can be used to represent a number of favorable outcomes, in the occurrence of N, independent binary events. For example, the number of *Heads*, in the simultaneous toss of N coins; the number of 1s in a binary digit of N bits; the binomial distribution is given by Eq. (3.28), with p, being the probability of the favorable event.

$$\begin{aligned} P(k) &= C_k^N (p)^k (1-p)^{N-k} \quad \text{for} \quad k = 0, 1, 2, \ldots, N \\ \text{with} \quad C_k^N &= \frac{N!}{k!(N-k)!} \end{aligned} \tag{3.28}$$

3.2.3 Continuous probability distributions

The probability density function (PDF), $f_X(x)$, of a continuous random variable X, is defined by Eq. (3.29) and satisfies the basic rules of probability. Some examples of continuous random variables are: temperature in the classroom; the height of students in the class; the time between arrival of two consecutive people in a bank; the phase difference between two independent signal generators in the laboratory; and the speed of a vehicle. Most of the parameters of the discrete probability distribution function and the continuous probability distribution function can be related. However, it is to be noted that in continuous probability distribution, the probability that the random variable X takes a particular value is zero (as indicated in Eq. 3.29), and it is nonzero only when X lies in finite range.

$$\begin{aligned} &\text{Probability density function: PDF} = f_X(x) \\ &\text{with} \quad \int_{-\infty}^{\infty} f_X(x)dx = 1 \\ &P(X = a) = 0 \\ &P(a \le X \le b) = \int_a^b f_X(x)dx \end{aligned} \tag{3.29}$$

The cumulative distribution function (CDF), $F_X(a)$, of the random variable X, is given by Eq. (3.30). It can be observed that $F_X(-\infty) = 0$, while $F_X(+\infty) = 1$.

$$\text{Cumulative distribution function: CDF} = F_X(a) = P(X \leq a)$$

$$\text{Hence,} \quad F_X(a) = \int_{-\infty}^{a} f_X(x)\,dx$$

$$\text{and,} \quad P(X \geq a) = \int_{a}^{\infty} f_X(x)\,dx = 1 - F_X(a) \tag{3.30}$$

$$F_X(-\infty) = 0$$

$$F_X(\infty) = 1$$

The mean, the variance, the standard deviation, the skewness, and the kurtosis of the continuous random variable X, are given by Eqs. (3.31 through 3.35).

$$\text{Mean} = \mu = \int_{-\infty}^{\infty} x f_X(x)\,dx \tag{3.31}$$

$$\begin{aligned} \text{Variance} = \sigma^2 &= E\left[(x-\mu)^2\right] \\ &= \textstyle\int_{-\infty}^{\infty} (x-\mu)^2 f_X(x)\,dx \\ &= E\left[x^2\right] - \mu^2 \end{aligned} \tag{3.32}$$

$$\text{Standard deviation} = \sigma = \sqrt{\text{Variance}} \tag{3.33}$$

$$\begin{aligned} \text{Skewness} &= \frac{1}{\sigma^3} E\left[(x-\mu)^3\right] \\ &= \frac{1}{\sigma^3} \int_{-\infty}^{\infty} (x-\mu)^3 f_X(x)\,dx \end{aligned} \tag{3.34}$$

$$\begin{aligned} \text{Kurtosis} &= \frac{1}{\sigma^4} E\left[(x-\mu)^4\right] \\ &= \frac{1}{\sigma^4} \int_{-\infty}^{\infty} (x-\mu)^4 f_X(x)\,dx \end{aligned} \tag{3.35}$$

3.2.3.1 *Uniform distribution*

The uniform distribution function is an example of a continuous random variable and is given by Eq. (3.36a). The mean and standard deviation of the Uniform distribution is given by Eq. (3.36b). Phase difference between two independent signal generators is an example of a random variable with uniform distribution function, and the phase angle varies uniformly in the range 0–2π.

$$U(a,b) = \begin{cases} 1/(b-a) & a \leq x \leq b \\ 0 & \text{otherwise} \end{cases} \tag{3.36a}$$

$$\mu = \frac{(a+b)}{2} \qquad \sigma = \frac{(b-a)}{\sqrt{12}} \tag{3.36b}$$

3.2.3.2 *Exponential distribution*

The exponential distribution function is an example of a continuous random variable and is given by Eq. (3.37a). The mean and standard deviation of the exponential distribution are given by Eq. (3.37b). The interval time between two consecutive arrivals of a Poisson arrival process follows the exponential distribution.

$$E(a) = \begin{cases} ae^{-ax} & x \geq 0 \\ 0 & \text{otherwise} \end{cases} \tag{3.37a}$$

$$\mu = \frac{1}{a} \quad \sigma = \frac{1}{a} \tag{3.37b}$$

Using Eq. (3.38), we can compute the probability of the random variable being less than or greater than a given value.

$$\begin{aligned} &P(X \leq \beta) = F_X(\beta) = \int_0^\beta ae^{-ax}dx = \left(1 - e^{-a\beta}\right) \\ &P(X \geq \beta) = (1 - F_X(\beta)) = e^{-a\beta} \end{aligned} \tag{3.38}$$

Eq. (3.39) gives the conditional probability of the random variable X, being greater than $X > (\beta + \alpha)$, given that is greater than $X > \beta$. We see that this is a memory-less process, as the amount of time already lapsed has no impact on the amount of time we need to wait. Let us consider a person waiting in the bus stop for the bus to college since ten minutes. Let us assume another friend joins him in the bus stop for the same bus to college. Both of them have equal probability of getting the bus, and this is an example of the memory-less process.

$$\begin{aligned} &P\left((X \geq (\beta + \alpha))/(X \geq \beta)\right) = P(X \geq \alpha) \\ &\text{Since,} \quad e^{-a(\alpha+\beta)}/e^{-a\beta} = e^{-a\alpha} \end{aligned} \tag{3.39}$$

3.2.3.3 Gaussian distribution

The Gaussian distribution function, also known as the Normal distribution, is an example of a continuous random variable, given by Eq. (3.40), and is specified by the two parameters: the mean μ, and the standard deviation σ. Some examples of the Gaussian distribution are: the marks scored by students in a course; the height of students in a class; and the weight of students in a class. Most of the variables usually have the normal distribution, with the mean and the variance being different.

$$f_X(x) = N(\mu, \sigma) = \frac{1}{\sigma\sqrt{2\pi}} e^{-(x-\mu)^2/(2\sigma^2)} \tag{3.40}$$

The Normal distribution with zero-mean and unit standard deviation is given by Eq. (3.41), and represents the distribution when any Gaussian distribution is normalized.

$$N(0, 1) = \frac{1}{\sqrt{2\pi}} e^{-\mu^2/2} \tag{3.41}$$

A unique feature of the Gaussian distribution is that, irrespective of the mean and standard deviation, the probability of X going from $(\mu - \sigma)$ to $(\mu + \sigma)$, is always 0.6827. Similarly the area of the curve of X going from $(\mu - 2\sigma)$ to $(\mu - 2\sigma)$, is always 0.9545, while the area under 3σ limits is 0.9973, as given by Eq. (3.42).

$$\begin{aligned} &\int_{(\mu-\sigma)}^{(\mu+\sigma)} f_X(x)dx = \frac{1}{\sigma\sqrt{2\pi}} \int_{(\mu-\sigma)}^{(\mu+\sigma)} e^{-(x-\mu)^2/(2\sigma^2)}dx = 0.6827 \\ &\text{Similarly,} \quad \int_{(\mu-2\sigma)}^{(\mu+2\sigma)} f_X(x)dx = 0.9545 \\ &\text{and,} \quad \int_{(\mu-3\sigma)}^{(\mu+3\sigma)} f_X(x)dx = 0.9973 \end{aligned} \tag{3.42}$$

3.3 Numerical examples

In this section, we consider examples, related to the concepts of: statistics, discrete probability, and continuous probability. The examples help comprehend the concepts introduced in Section 3.2. Every concept of probability and statistics

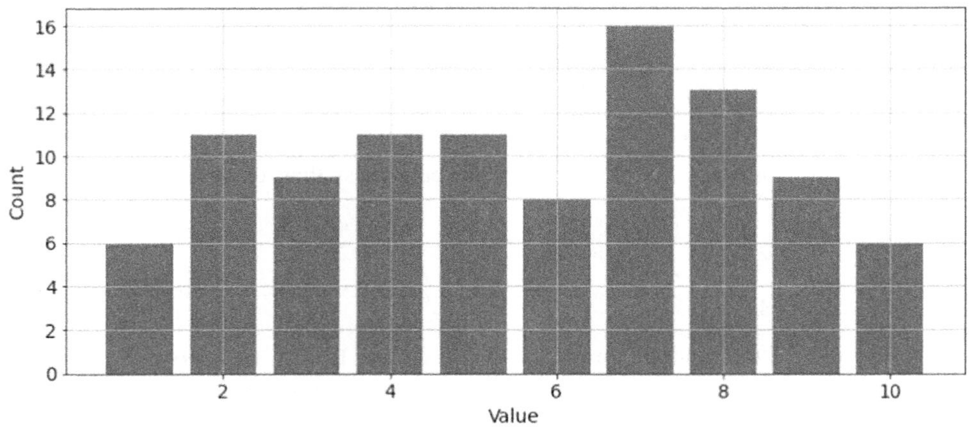

FIGURE 3.1 The random sequence of 100 integers generated in the range of 1–10, with count of each integer.

has the Python subfunction included in the Appendix. The examples are accompanied by the essential block of the Python code *(the code for print, plot of the results is not included).* Here, the use of the programming tool, Python, is purely to help comprehend the concepts, and not to learn Python. Hence, the code may not be the most efficient, and the interested reader may arrive at alternate and efficient code.

3.3.1 Examples in statistics

Here, we begin with examples in statistics and then consider examples of discrete probability followed by examples in continuous probability. There is an attempt to have relevant examples for every concept introduced in Section 3.2.

3.3.1.1 Example on mean, mode, and median

In this example, we generate a sequence of 100 random numbers, in the range 1–10, using the Python code given below. We then compute the mean, mode and median of the generated dataset. To ensure repeatability of the results, the seed of the random generator is set to a constant.

```
N = 100
np.random.seed(108)
X = np.random.randint(1, 11, N)
mean = my_mean(X)
mode = my_mode(X)
median = my_median(X)
```

Fig. 3.1 has the details of the data sequence generated. It can be observed that this dataset has a mean of 5.59, the mode is 7, and the median is 6.

3.3.1.2 Example on mean and standard deviation of related datasets

In this example, we generate a sequence X, of 90 random numbers, in the range 30–60, using the Python code given below, and shown in Fig. 3.2. We then generate a new random sequence Y given by $X + 10$; the random $Z = X \times 10$; and random sequence, W being normalized X. We compute the mean, and standard deviation of each of these data sequences, derived from the initial sequence X.

```
N = 90
np.random.seed(108)
X = np.random.randint(30, 61, N)
value, count=np.unique(X,return_counts=True)
Y = X+10
Z = X*10
W = my_Normalized_Distribution(X)
```

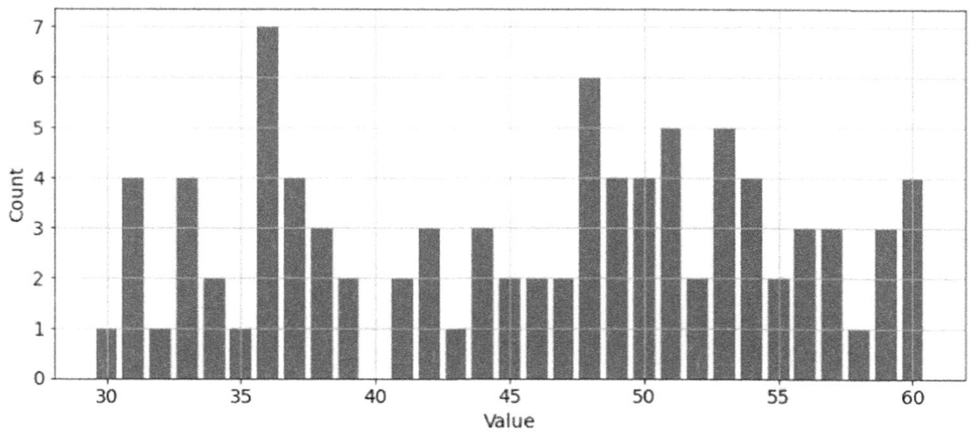

FIGURE 3.2 The random sequence of 90 integers generated in the range of 30–60, with count of each integer.

TABLE 3.1 The mean and standard deviation of the datasets *X, Y, Z*, and *W*.

Parameter/dataset	X	Y	Z	W
Mean	45.7778	55.7778	457.7778	0.0
Standard deviation	8.7886	8.7886	87.886	1.0

From Table 3.1 and comparing the results of the distribution X and Y, it can be observed that addition of a constant, changes the mean, but does not change the standard deviation. Similarly, comparing the results of the distribution X and Z, it can be observed that multiplication by a constant, changes both the mean and the standard deviation. The normalized dataset has zero mean and unit standard deviation.

3.3.1.3 Example on mean and standard deviation of the normalized distribution

In this example, we generate a sequence X, of 30 random numbers, in the range 1–10, using the Python code given below. We then obtain the corresponding normalized distribution Y.

```
N = 30
np.random.seed(108)
X = np.random.randint(1, 11, N)
Y = my_Normalized_Distribution(X)
```

The mean and standard deviation of the random sequence X is given by 5.8 and 2.5482, while that of the sequence Y is 0 and 1. The sequence X and Y are shown in Fig. 3.3.

3.3.1.4 Example on generating a random sequence and its count

In this example, we generate a sequence X, of 1000 random numbers, in the range 1–10, using the Python code given below.

```
N = 1000
np.random.seed(9)
X = np.random.randint(1,11,N)
value, count=np.unique(X,return_counts=True)
```

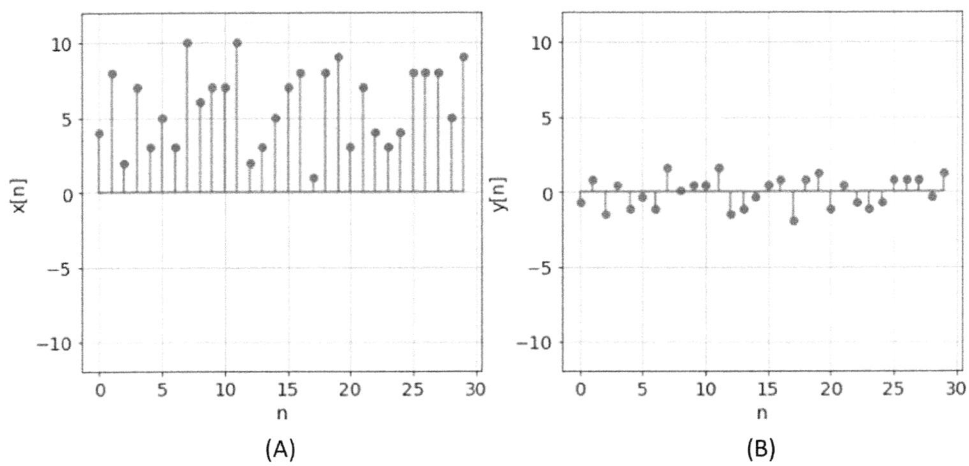

FIGURE 3.3 (A) The random sequence of *X*; and (B) the corresponding normalized sequence *Y*.

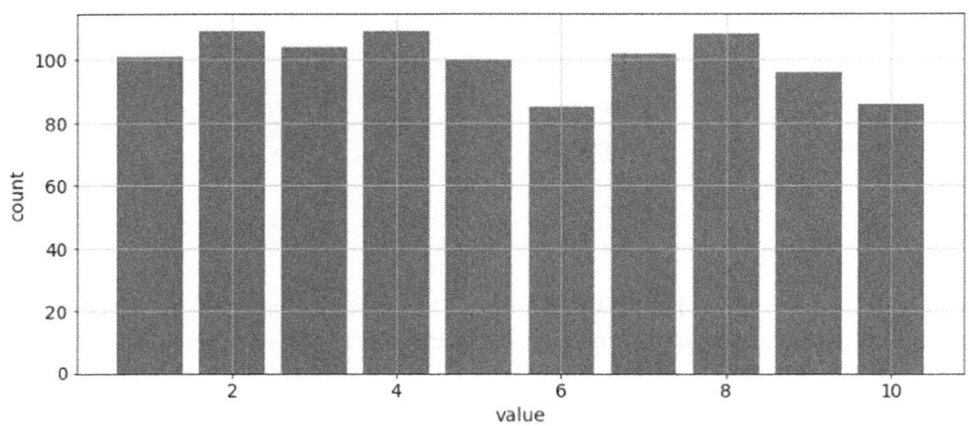

FIGURE 3.4 The bar graph of the random sequence of 1000 digits generated in the range of 1–10.

We count the number of distinct values and then compute the mean and standard deviation of the dataset. Fig. 3.4 has details of the bar graph of the generated dataset. It is observed that the mean is 5.379, while the standard deviation is 2.8512.

3.3.1.5 Example on mean and standard deviation of the whole and a random sample

In this example, we generate a set of 1000 integers in the range of 0–100. We then obtain a smaller set of this initial sequence, as shown in Fig. 3.5. We compute the mean and standard derivation of the original sequence, and the sample of the original.

```
N = 1000
np.random.seed(9)
X = np.random.randint(0,101,N)
value, count=np.unique(X,return_counts=True)
Sample_of_X = np.random.choice(X, 200)
```

It can be observed that the mean of the original dataset is 49.42, while that of the sample is 48.87. Hence, there is little difference in the mean. However, the standard deviation of the original is 28.9236, while that of the sample is 12.7831. Hence, there is a significant reduction in the standard deviation of the sample of a dataset.

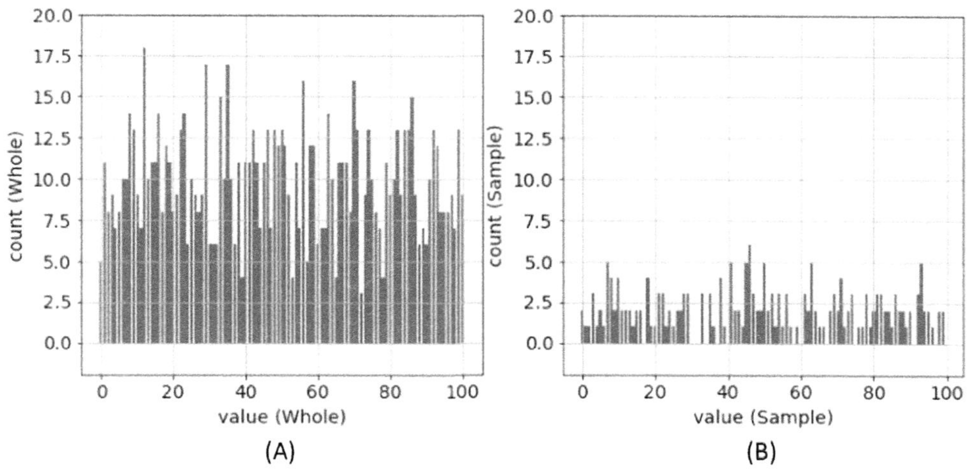

FIGURE 3.5 (A) The bar graph of the original dataset with 1000 values; and (B) the bar graph of the sample of the original (a random selection of 200 samples from the original).

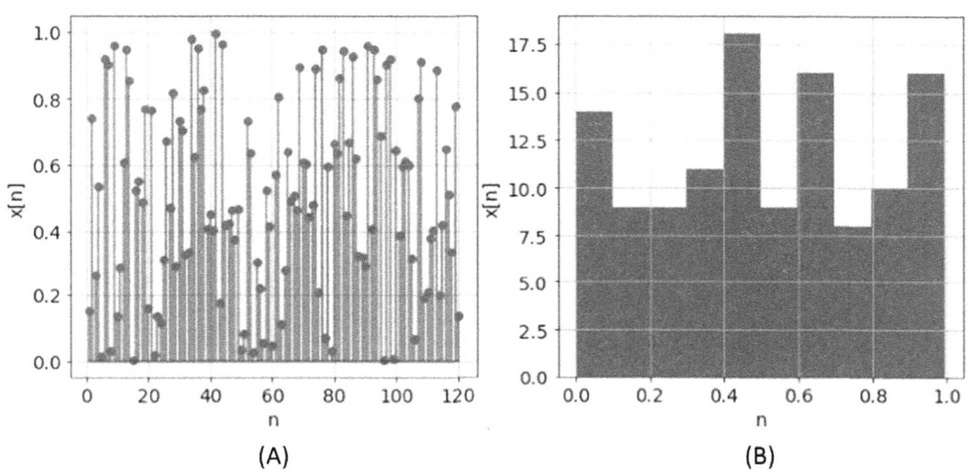

FIGURE 3.6 (A) The random sequence of X with uniform distribution; and (B) the corresponding bar graph.

3.3.1.6 Example on generating noise with uniform distribution

In this example, we generate a random sequence X, of 120 real numbers, in the range 0–1, with uniform distribution, using the Python code given below. Fig. 3.6 has details of the generated sequence, along with the bar graph. It is observed that the mean is 0.4715, while the standard deviation is 0.3035.

```
N = 120
np.random.seed(12)
X = np.random.random(N)
```

3.3.1.7 Example on generating noise with Gaussian distribution

In this example, we generate a random sequence X, of 300 real numbers, in the range 0–1, with normal distribution, using the Python code given below. Fig. 3.7 has details of the generated sequence, along with the bar graph. It is observed that the mean is 0.0482, while the standard deviation is 1.0043.

```
N = 300
np.random.seed(18)
X = np.random.normal(0,1,N)
```

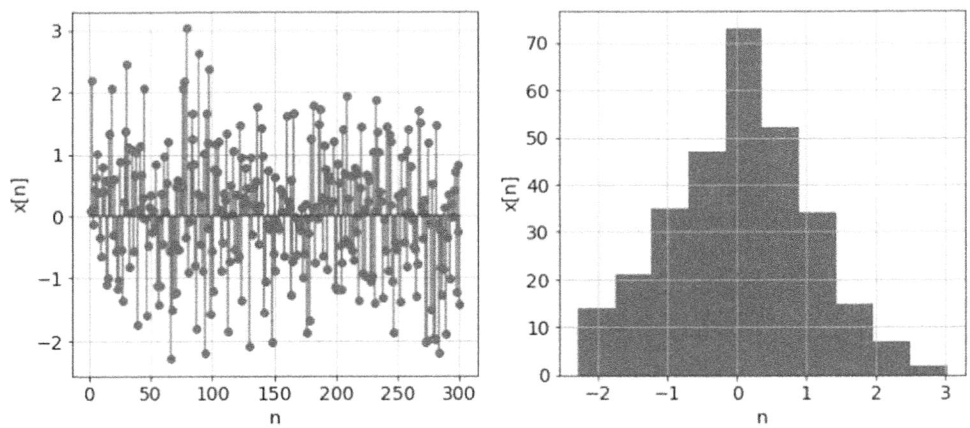

FIGURE 3.7 (A) The random sequence of X with Gaussian distribution; and (B) the corresponding bar graph.

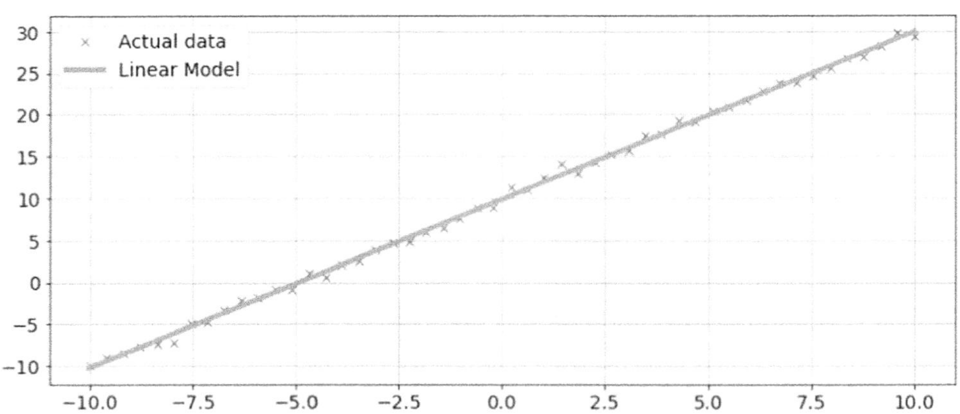

FIGURE 3.8 The generation of linearly related dataset with added Gaussian noise $N(0,0.5)$, together with the linear-fit for the generated dataset.

3.3.1.8 Example on generating linear dataset with added Gaussian noise

In this example, we generate a set of linearly related sequence Y, of 50 real numbers, in the range -10 to 10. We then add Gaussian noise of zero mean and standard deviation 0.5, using the Python code given below.

```
N, m , b  = 50, 2, 10
X = np.linspace(-10,10,N)
Y = my_line(m,b,X)
np.random.seed(45)
noise = np.random.normal(0,0.5,N)
Y = Y + noise
mew_x = my_mean(X)
mew_y = my_mean(Y)
sigma_x = my_STD(X)
# Predicted values
m_hat = my_Covariance(X,Y)/sigma_x/sigma_x
b_hat = mew_y - m_hat*mew_x
Y_hat = m_hat*X + b_hat
# Mean square Error
mse = np.sum((Y-Y_hat)*(Y-Y_hat))/N
```

We then predict the parameters of the linear dataset. Fig. 3.8 has details of the generated linear dataset, together with the linear model. It can be observed that the predicted slope is 2.0059 (while the actual slope is 2), and the predicted intercept is 9.8568 (while the actual intercept is 10). The mean-square-error of the predicted linear model is 0.2228.

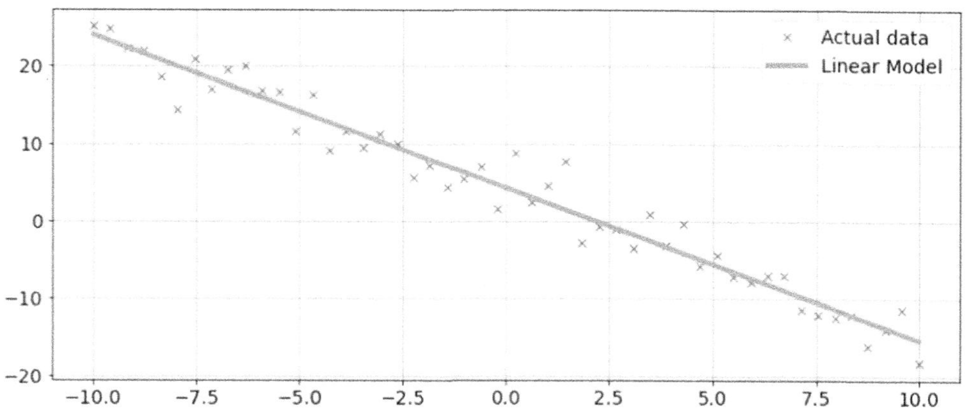

FIGURE 3.9 The generation of linearly related dataset with added Gaussian noise $N(0,2.5)$, together with the linear-fit for the generated dataset.

We now repeat the example, with small changes in the linear model and the noise added. We now add the Gaussian noise of zero mean and standard deviation 2.5. Fig. 3.9 has details of the generated linear dataset, together with the linear model. We predict the parameters of the linear dataset. It can be observed that the predicted slope is -1.9706 (while the actual slope is -2), and the predicted intercept is 4.2842 (while the actual intercept is 5). The mean-square-error of the predicted linear model is 5.5706. The error is higher in this example, as the noise added is high.

3.3.1.9 Example on covariance and correlation

In this example, we generate a set of linearly related sequence X, and Y of 90 real numbers, in the range -5 to 5, with added Gaussian noise, using the Python code given below.

```
N = 90
x = np.linspace(-5,5,N)
np.random.seed(45)
noise = np.random.normal(0,3,N)

m1, b1 = 8, 2
X = my_line(m1,b1,x)
X = X + noise


m2, b2  = -5, 5
Y = my_line(m2,b2,x)
Y = Y + 2*noise
covariance = my_Covariance(X,Y)
correlation = my_Correlation(X,Y)
```

Fig. 3.10 has the scatter plot of the two datasets X and Y. The covariance of the dataset is -306.6244, while the correlation is -0.8797. The correlation in this example is negative. Sometimes in Eq. (3.7), to account for the sample distribution instead of the whole distribution, N is replaced by (N-1) while computing the covariance.

In this example, we generate a set of linearly related sequence X, and Y of 50 real numbers, in the range -5 to 5, with added Gaussian noise, using the Python code given below. Fig. 3.11 has the scatter plot of the two datasets X and Y. The covariance of the dataset is 99.3969, while the correlation is 0.98. The correlation in this example is positive.

```
N = 50
x = np.linspace(-5,5,N)
noise = np.random.normal(0,1,N)
m1, b1 = 5, 30
X = my_line(m1,b1,x)
X = X + 2*noise
m2, b2  = 2, -20
Y = my_line(m2,b2,x)
Y = Y + 2*noise
```

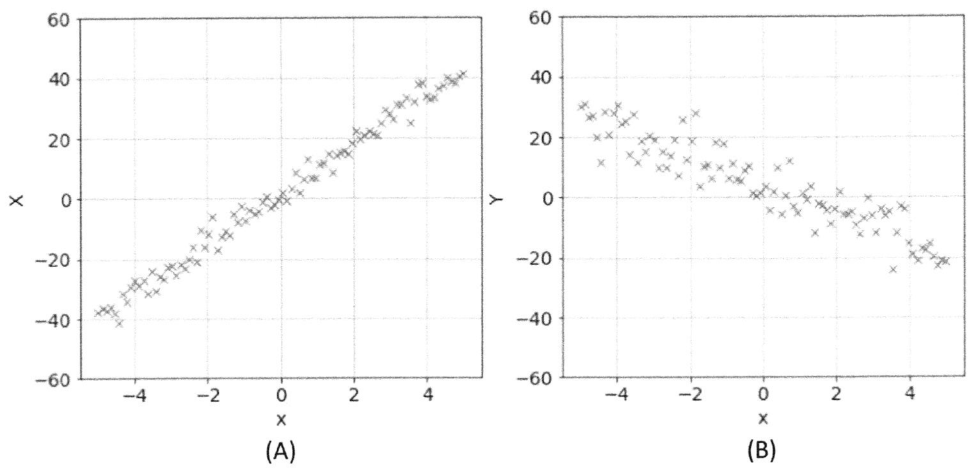

FIGURE 3.10 (A) The linearly related dataset X; (B) The linearly related dataset Y, having negative correlation.

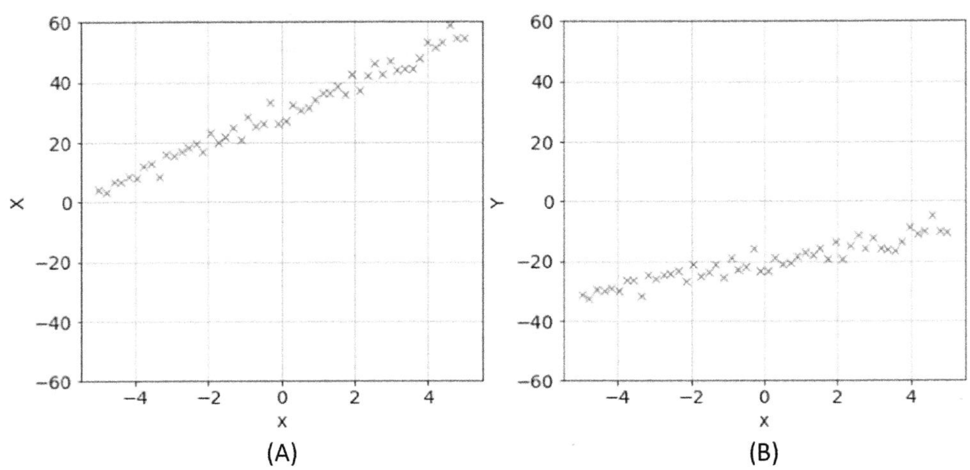

FIGURE 3.11 (A) The linearly related dataset X; (B) the linearly related dataset Y, having positive correlation.

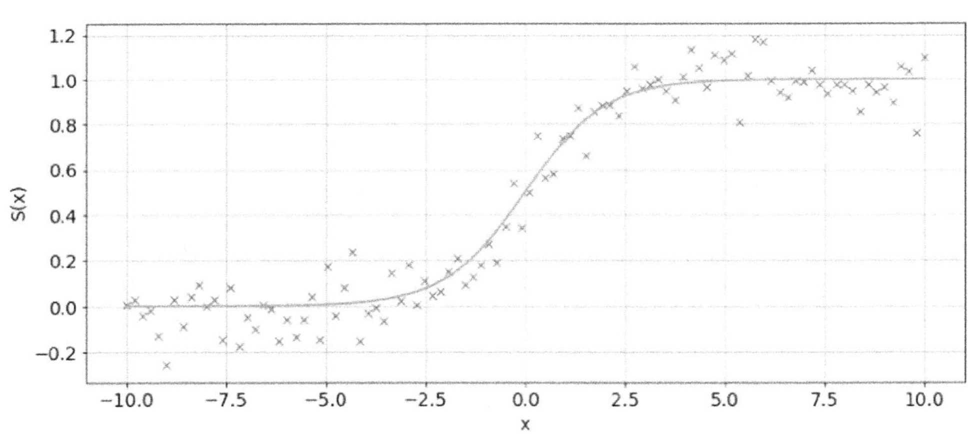

FIGURE 3.12 The plot of the sigmoid function, and with added Gaussian noise.

3.3.1.10 Example on sigmoid with added noise

In this example, we generate a set of data points related through the sigmoid function, for X in the range -10 to 10, with added Gaussian noise, using the Python code given below. Fig. 3.12 has the plot of the data points and the sigmoid function.

```
N = 100
x = np.linspace(-10,10, N)
Sx = my_Sigmoid(x)
np.random.seed(45)
noise = np.random.normal(0,0.1,N)
Sx_noisy = Sx + noise
```

3.3.1.11 Example on tanh(x) with added noise

In this example, we generate a set of data points related through the tan hyperbolic, for X in the range -10 to 10, with added Gaussian noise, using the Python code given below. Fig. 3.13 has the plot of the data points and the tan hyperbolic function.

```
N = 100
x = np.linspace(-10,10, N)
tanhx = my_tanhx(x)
np.random.seed(45)
noise = np.random.normal(0,0.1,N)
tanhx_noisy = tanhx + noise
```

3.3.1.12 Example on ReLU(x) with added noise

In this example, we generate a set of data points related through the ReLU (rectifier linear unit), for X in the range -10 to 10, with added Gaussian noise, using the Python code given below. Fig. 3.14 has the plot of the data points and the ReLU function.

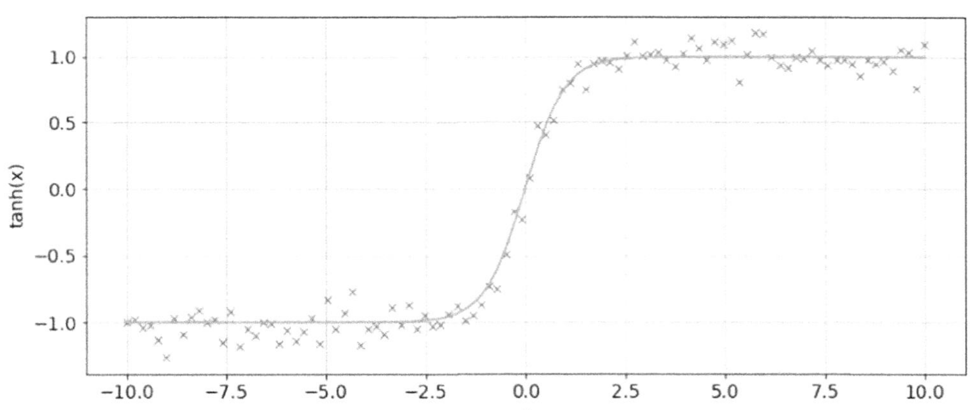

FIGURE 3.13 The plot of the tan hyperbolic function, and with added Gaussian noise.

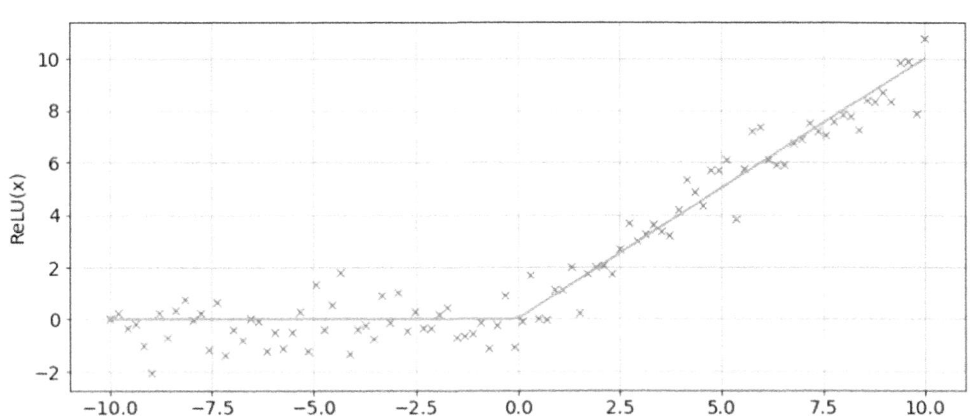

FIGURE 3.14 The plot of the ReLU (rectified linear unit) function and with added Gaussian noise.

```
N = 100
x = np.linspace(-10,10, N)
ReLU = my_ReLU(x)
np.random.seed(45)
noise = np.random.normal(0,0.8,N)
ReLU_noisy = ReLU + noise
```

3.3.2 Examples in discrete probability

Having seen examples in statistical averages, we shall now consider examples in discrete probability distribution, like tossing a coin, rolling a dice, the Poisson distribution, the binomial distribution, and making a decision based on computing the probability. We also have an example on an introduction to the confusion matrix, and the Type-I and Type-II errors.

3.3.2.1 Example of tossing a coin N *times*

In this example, we attempt to simulate the toss of a coin 12 times and assume heads can be represented by 0, while tails can be represented by 1, using the Python code given below.

```
N = 12
np.random.seed(99)
A = np.random.randint(0,2, N)
value, count = np.unique(A, return_counts=True)
P0=count[0]/N
P1=count[1]/N
P = count/N
Mew = my_Mew_Discrete(value,P)
STD = my_STD_Discrete(value,P)
Entropy = my_Entropy(P)
Gini_Index = my_Gini_Index(P)
```

The generated binary sequence is given by: [1, 1, 1, 0, 1, 0, 0, 0, 1, 0, 1, 1]. In this example, we have five heads and seven tails, and hence the probability of the head is .4167, while the probability of the tail is .5833. Hence, this distribution has a mean of 0.5833, and standard deviation of 0.493. Also, the entropy of this probability distribution is .9799, and Gini-index is 0.4861.

3.3.2.2 Example of rolling a dice

In this example, we attempt to simulate the roll of a dice 60 times, using the Python code given below.

```
N = 60
np.random.seed(108)
Dice  = np.random.randint(1,7, N)
```

Fig. 3.15 has the count of the various outcomes of the roll of the dice, and hence the probability distribution is given by: [.15, .1667, .15, .2, .1667, .1667]. We then have the mean of the distribution as 3.5667, and standard deviation of 1.677. The entropy of this probability distribution is 2.578, and Gini-index is 2.585.

3.3.2.3 Example of Poisson distribution

In this example, we compute the probability distribution function of the Poisson distribution with parameters, as given by Python code below.

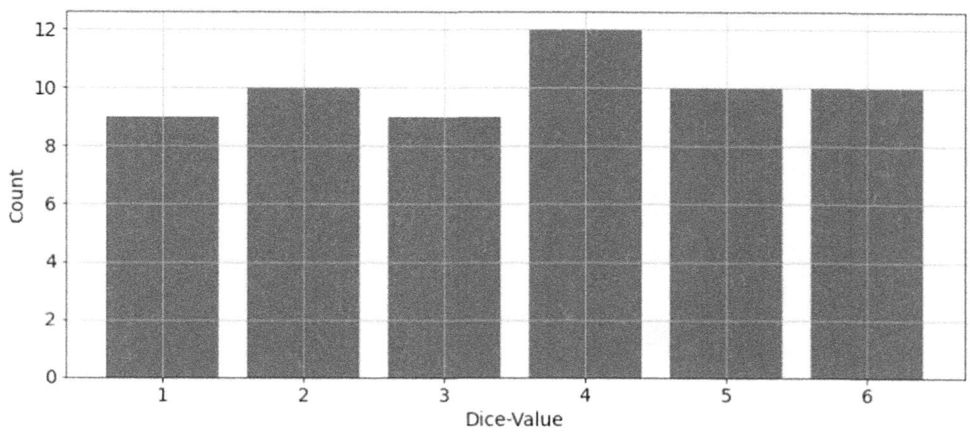

FIGURE 3.15 The attempt to simulate the roll of a dice 60 times.

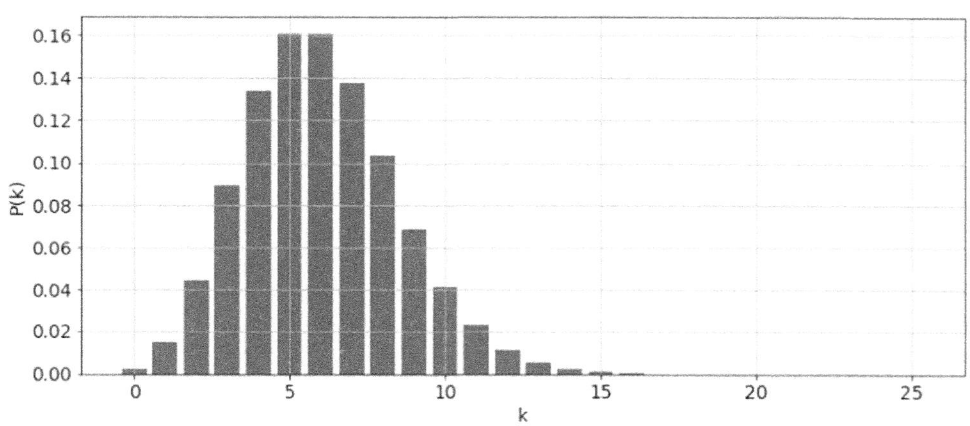

FIGURE 3.16 The probability distribution function for Poisson distribution function, with $\lambda = 4$ and $N = 25$.

```
lamda = 4
N = 25
k=np.arange(N+1)
PK =my_Poisson(lamda, N)
```

Fig. 3.16 has the probability distribution function. The mean of the distribution is 6, the variance is 5.8, standard deviation is 2.4494, the skewness is 0.4082, and the kurtosis is 3.1667. The entropy of this probability distribution is 3.3172, and Gini-index is 0.8836.

We now repeat the simulation, with $\lambda = 10.5$, and the corresponding probability density function is given by Fig. 3.17. The mean of the distribution is 10.499, the variance is 10.4898, standard deviation is 3.2388, the skewness is 0.3052, and the kurtosis is 3.0772. The entropy of this probability distribution is 3.7305, and Gini-index is 0.9124.

3.3.2.4 Example of binomial distribution

In this example, we compute the probability distribution function of the binomial distribution with the probability of the success being .18 using Python code given below. Here, the biased coin is tossed 30 times, and we compute the probability, $P(k)$, of getting k, heads.

```
p = 0.18      # probability of Success
N = 30        # number of times coin tossed
PK = my_Binomial(p, N)
```

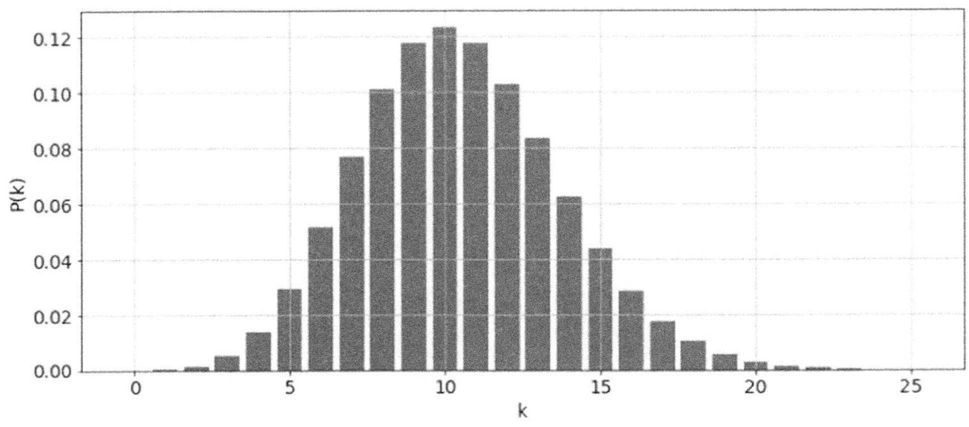

FIGURE 3.17 The probability distribution function for Poisson distribution function, with $\lambda = 10.5$ and $N = 25$.

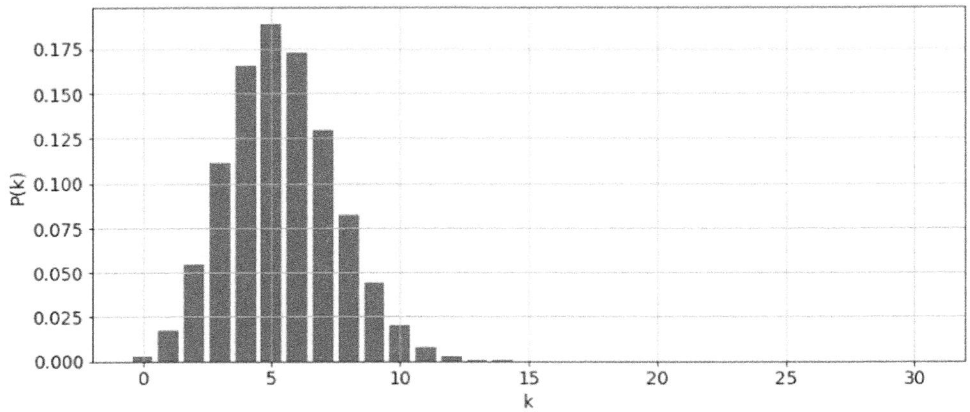

FIGURE 3.18 The binomial probability distribution function for probability of success. 18, and $N = 30$.

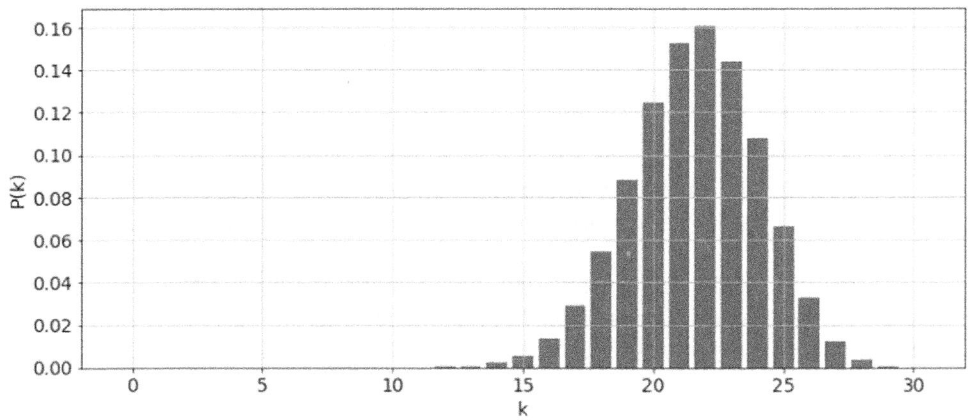

FIGURE 3.19 The binomial probability distribution function for probability of success. 72, and $N = 30$.

Fig. 3.18 has the probability distribution function. The mean of the distribution is 5.4, the variance is 4.428, standard deviation is 2.1043, the skewness is 0.3041, and the kurtosis is 3.0258. The entropy of this probability distribution is 3.1073, and Gini-index is 0.8657.

We now repeat the simulation, with $P = .72$, and the corresponding probability density function is given by Fig. 3.19. The mean of the distribution is 21.6, the variance is 6.048, standard deviation is 2.4593, the skewness is -0.1789, and the kurtosis is 2.9653. The entropy of this probability distribution is 3.3409, and Gini-index is 0.8855.

3.3.2.5 Example of conditional probability

In this example, we consider the details of registration for an elective course on, probability and statistics, with 300 students, as given by Table 3.2.

From Table 3.2, we can compute the various probabilities, as given by Eq. (3.43a), and the conditional probabilities are given by Eq. (3.43b). Eq. (3.43c), shows the Bayes' conditional probability.

$$\begin{aligned} P(\text{Registered}) &= 180/300 = 0.6 \\ P(\text{Not registered}) &= 120/300 = 0.4 \\ P(\text{Know}) &= 135/300 = 0.45 \\ P(\text{Do not know}) &= 165/300 = 0.55 \end{aligned} \tag{3.43a}$$

$$\begin{aligned} P(\text{Know/registered}) &= 45/180 = 0.25 \\ P(\text{Do not know/registered}) &= 135/180 = 0.75 \\ P(\text{Know/not registered}) &= 90/120 = 0.75 \\ P(\text{Do not know/not registered}) &= 30/120 = 0.75 \end{aligned} \tag{3.43b}$$

$$\begin{aligned} P(\text{Know/registered})P(\text{Registered}) &= P(\text{Registered/know})P(\text{Know}) \\ (45/180)(180/300) &= (45/135)(135/300) \end{aligned} \tag{3.43c}$$

From Table 3.2, we can compute all required probabilities. However, it is easier to comprehend the same by the probability tree, shown in Fig. 3.20, as given by Eqs. (3.44a)–(3.44c).

$$\begin{aligned} P(\text{Know}) &= P(\text{Know/registered})P(\text{Registered}) \\ &\quad + P(\text{Know/not registered})P(\text{Not registered}) \\ &= (45/180)(180/300) + (90/120)(120/300) = 0.45 \end{aligned} \tag{3.44a}$$

$$\begin{aligned} P(\text{Do not know}) &= P(\text{Do not know/registered})P(\text{Registered}) \\ &\quad + P(\text{Do not know/not registered})P(\text{Not registered}) \\ &= (135/180)(180/300) + (30/120)(120/300) = 0.55 \end{aligned} \tag{3.44b}$$

$$\begin{aligned} P(\text{Know/registered})P(\text{Registered}) &= P(\text{Registered/know})P(\text{Know}) \\ (45/180)(180/300) &= (45/135)(135/300) \end{aligned} \tag{3.44c}$$

3.3.2.6 Example of making a decision through conditional probability

In this example, we consider the dataset of Table 3.3, where the decision is on going to play, or not going to play. This decision is based on four input parameters: Outlook, Temperature, Humidity, and Wind. Using conditional probability, we can arrive at the decision, for any other input parameter set. Here, it is assumed, the four input parameters are independent.

In this example, we have the final decision, for going to play being Yes or No, and is dependent on the four input parameters. Let us consider the situation, of the need to arrive at the decision, when the input parameters are: Outlook-Sunny; Temperature – Cool; Humidity – High and Wind – weak, a combination not included in the above dataset. We use the Bayes' conditional probability to arrive at the decision. Eq. (3.45), has the probability of the decision being YES, or NO.

$$\begin{aligned} P(NO) &= (5/14) \\ P(YES) &= (9/14) \end{aligned} \tag{3.45}$$

TABLE 3.2 Elective course registration details of 300 students.

	Know (probability and statistics)	Do not know (probability and statistics)
Registered (for the elective)	45	135
Not registered (for the elective)	90	30

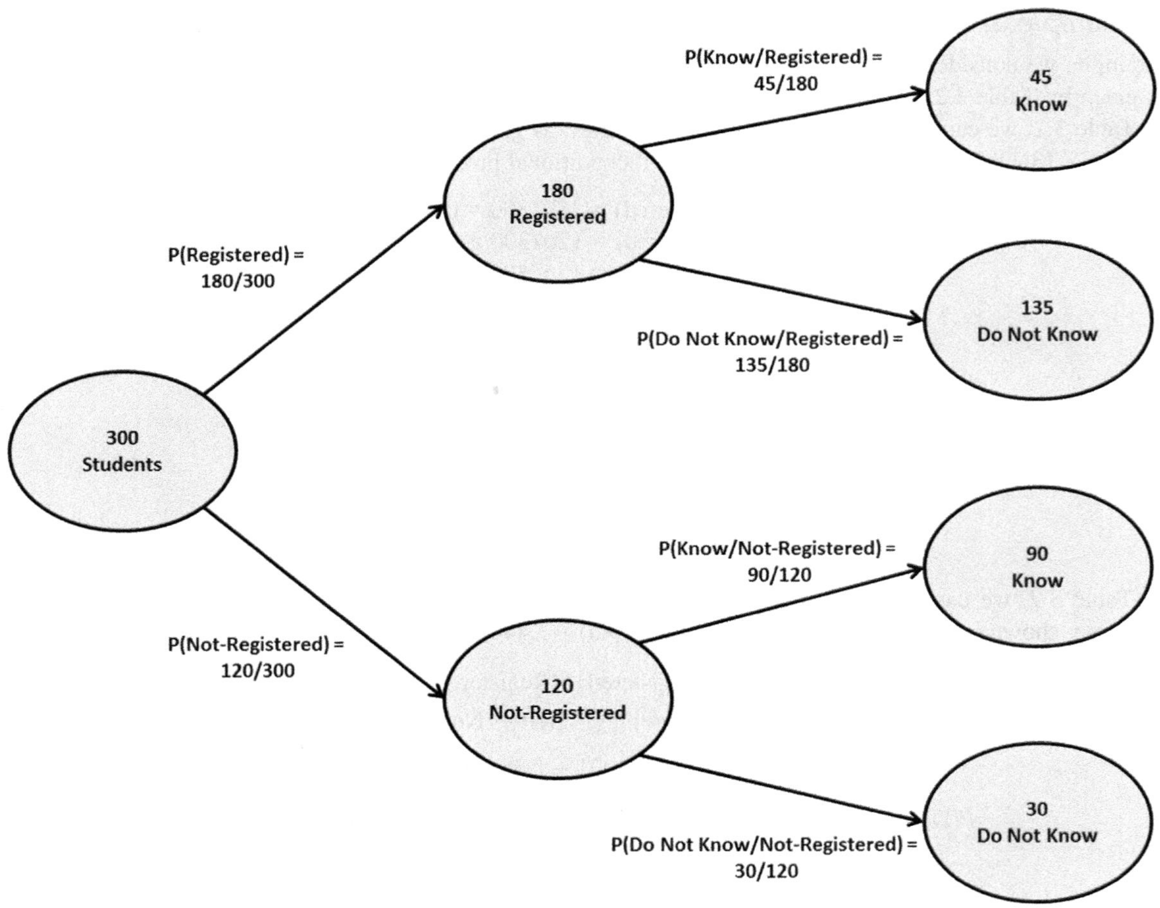

FIGURE 3.20 The Probability tree of the given Elective registration details of Table 3.2.

TABLE 3.3 The dataset contributing to the decision on playing (Yes/No).

Outlook (x_1)	Temperature (x_2)	Humidity (x_3)	Wind (x_4)	Decision (y)
Sunny	Hot	High	Weak	NO
Sunny	Hot	High	Strong	NO
Rain	Cool	Normal	Strong	NO
Sunny	Mild	High	Weak	NO
Rain	Mild	High	Strong	NO
Overcast	Hot	High	Weak	YES
Rain	Mild	High	Weak	YES
Rain	Cool	Normal	Weak	YES
Overcast	Cool	Normal	Strong	YES
Sunny	Cool	Normal	Weak	YES
Rain	Mild	Normal	Weak	YES
Sunny	Mild	Normal	Strong	YES
Overcast	Mild	High	Strong	YES
Overcast	Hot	Normal	Weak	YES

Eq. (3.46a) gives the conditional probability of the decision being a NO, for the given input variables. Similarly, (3.46b) gives the conditional probability of the decision being a YES, for the given input variables.

$$\begin{aligned} &P(\text{NO}/(\text{Sunny, Cool, High, Weak})) \\ &= \{P((\text{Sunny, Cool, High, Weak})/\text{NO})\}P(\text{NO}) \\ &= \{P(\text{Sunny}/\text{NO})P(\text{Cool}/\text{NO})P(\text{High}/\text{NO})P(\text{Weak}/\text{NO})\}P(\text{NO}) \\ &= \{(3/5)(1/5)(4/5)(2/5)\}(5/14) = 0.0137 \end{aligned} \tag{3.46a}$$

$$\begin{aligned} &P(\text{YES}/(\text{Sunny, Cool, High, Weak})) \\ &= \{P((\text{Sunny, Cool, High, Weak})/\text{YES})\}P(\text{YES}) \\ &= \{P(\text{Sunny}/\text{YES})P(\text{Cool}/\text{YES})P(\text{High}/\text{YES})P(\text{Weak}/\text{YES})\}P(\text{YES}) \\ &= \{(2/9)(3/9)(3/9)(6/9)\}(9/14) = 0.0103 \end{aligned} \tag{3.46b}$$

In this example, the conditional probability of NO is greater than the conditional probability of YES, the recommended decision is NO. This is one method of arriving at the decision. One can convert the given dataset into a decision tree based on the Entropy or the Gini-index, and then for any input combination, one has to just walk down the tree to arrive at the decision.

3.3.2.7 Example of probability and the confusion matrix

In this example, we consider details of 3000 students, of a college, who had to take up the mandatory initial COVID test on campus, which had 45 students being COVID-positive. The initial tests resulted in few False negatives and few False positives, as shown in Table 3.4. One can compute all probabilities and conditional probabilities either directly from the Table.

In Table 3.4, it can be observed that the diagonal elements are correct predictions. The off-diagonal needs to be zero, however, due to errors in algorithms, it is likely that the off-diagonal elements are not zero. In this example, we have, 55 false positives, recognized as Type-I error. This error is not that dangerous, as on further tests in the actual situation being a false positive will be identified and the person can go home with happiness. On the other hand, we have 15 students who are actually positive but have been incorrectly identified as negative. This case is of greater concern than the Type-I error. Hence, all focus on prediction algorithms is towards reducing false negatives or the Type-II error. Table 3.3 is also known as the confusion matrix, and the accuracy and precision are parameters that help measure the quality of the prediction algorithm and is given by Eqs. (3.47a) and (3.47b). Other parameters which help measure the quality of the prediction algorithm are: sensitivity or the true positive rate; selectivity or the true negative rate; the balanced accuracy; the F1 score.

$$\begin{aligned} \text{Accuracy} &= \frac{(\text{TP} + \text{TN})}{(\text{TP} + \text{TN}) + (\text{FP} + \text{FN})} \\ &= \frac{\text{Correct predictions}}{\text{Correct predictions} + \text{Incorrect predictions}} \end{aligned} \tag{3.47a}$$

TABLE 3.4 The confusion matrix of the actual and predicted COVID tests.

Actual/predicted details	Predicted: COVID positive	Predicted: COVID negative
Actual: COVID positive	30 Correct prediction: True Positive (TN)	15 Type-II error: False Negative (FN)
Actual: COVID negative	55 Type-I error: False Positive (FN)	2900 Correct prediction: True Negative (TN)

$$\begin{aligned}\text{Precision} &= \frac{\text{TP}}{(\text{TP}+\text{FP})}\\ &= \frac{\text{True positive}}{\text{True positive}+\text{False positive}}\end{aligned} \tag{3.47b}$$

3.3.3 Examples in continuous probability

Having seen examples in discrete probability distribution, we shall now consider examples in continuous probability distribution, the uniform distribution, the exponential distribution, and the Gaussian distribution.

3.3.3.1 Example of uniform distribution function

In this example, we consider the uniform distribution function described by Eq. (3.6a), for U(−1, 1), using Python code given below.

```
a, b = -1, 1
x = np.linspace(-10,10,400)
PDF = my_Uniform_Distribution(a,b, x)
CDF = my_integral(x, PDF)
```

The plot of the probability density function, U(−1,1) and the corresponding cumulative distribution function are given in Fig. 3.21. This distribution has mean = 0, standard deviation = 0.5793, skewness = 0, and kurtosis = 1.794.

We now consider the uniform distribution U(−5,5). The plot of the probability density function and the cumulative distribution function are given in Fig. 3.22. This distribution has mean = 0, standard deviation = 2.894, skewness = 0, and kurtosis = 1.797.

3.3.3.2 Example of exponential distribution function

In this example, we consider the Exponential distribution function described by Eq. (3.7a), for E(2), using Python code given below.

```
a = 2
x = np.linspace(-2,15,500)
PDF = my_Exponential_Distribution(a, x)
CDF = my_integral(x, PDF)
```

The plot of the probability density function, E(2) and the cumulative distribution function are given in Fig. 3.23. This distribution has mean = 0.5, standard deviation = 0.5034, skewness = 1.9459, and kurtosis = 8.7702.

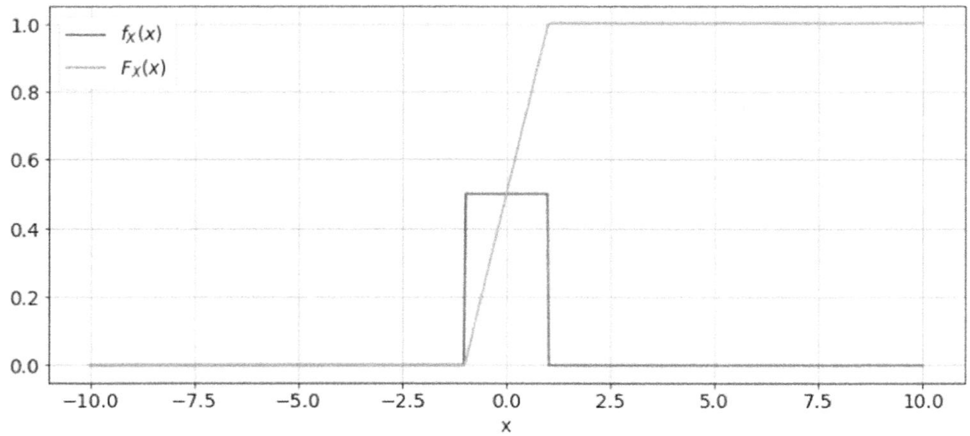

FIGURE 3.21 The probability density function and the cumulative distribution function for the uniform distribution U(−1,1).

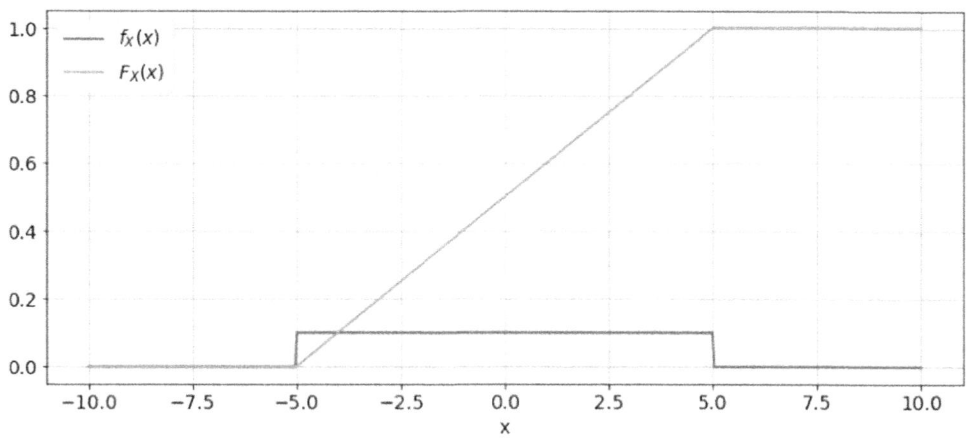

FIGURE 3.22 The probability density function and the cumulative distribution function for the uniform distribution U(−5,5).

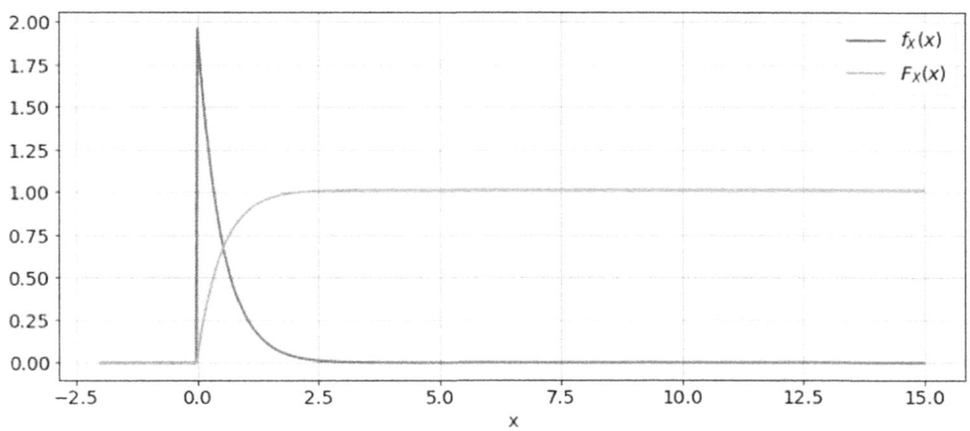

FIGURE 3.23 The probability density function and the cumulative distribution function for the exponential distribution function, E(2.0).

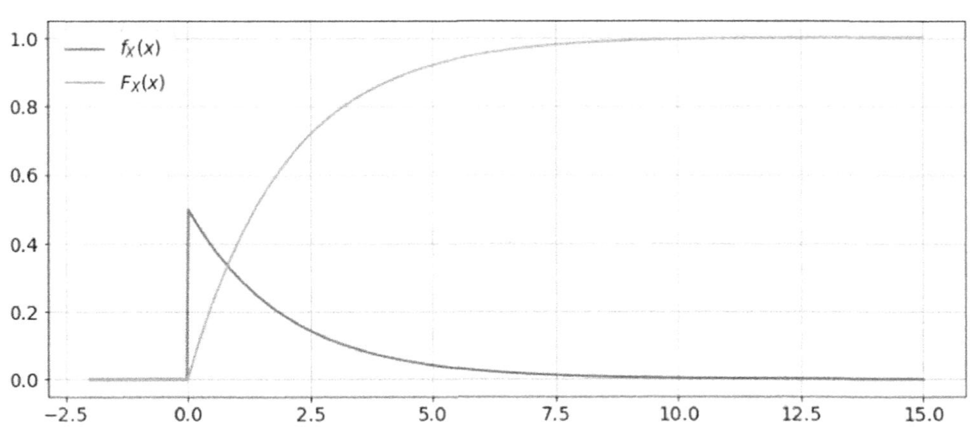

FIGURE 3.24 The probability density function and the cumulative distribution function for the exponential distribution function, E(0.5).

We now consider the exponential distribution E(0.5). The plot of the probability density function and the cumulative distribution function are given in Fig. 3.24. This distribution has mean = 1.9906, standard deviation = 1.9716, skewness = 1.8407, and kurtosis = 7.4778.

3.3.3.3 *Example of Gaussian distribution function*

In this example, we consider the Gaussian distribution function described by Eq. (3.40), for E(2), using Python code given below. Fig. 3.25, has the three Gaussian distributions.

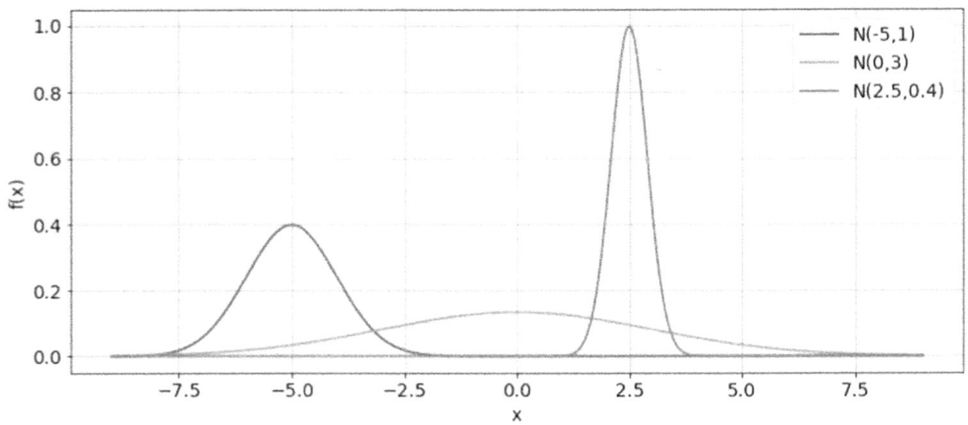

FIGURE 3.25 The plot of three Gaussian density functions: *N*(-5,1), *N*(0,3), and *N*(2.5,0.4).

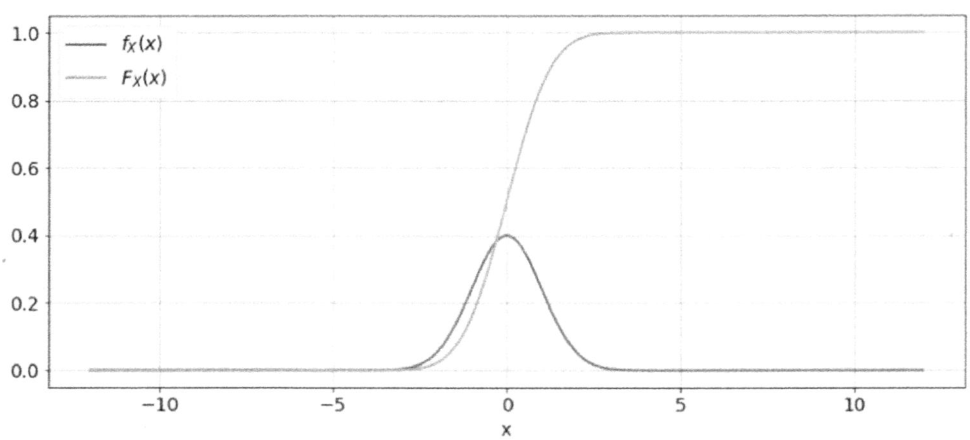

FIGURE 3.26 The probability density function and the cumulative distribution function for the Gaussian distribution function, *N*(0,1).

```
x=np.linspace(-9,9,300)
gauss1 = my_gaussian(-5,1,x)
gauss2 = my_gaussian(0,3,x)
gauss3 = my_gaussian(2.5,0.4,x)
```

We now consider the Gaussian distribution *N*(0,1), generated using the Python code given below. The plot of the probability density function and the cumulative distribution function are given in Fig. 3.26. All Gaussian distributions have, skewness = 0, kurtosis = 3; and the area within one sigma limits = 0.6827, the area within-two-sigma limits = 0.9545; the area within three sigma limits = 0.9973. These parameters do not depend on the mean and standard deviation of the Gaussian distribution.

```
mew, sigma = 0, 1
x = np.linspace(-10,10,300)
PDF = my_gaussian(mew, sigma, x)
CDF = my_integral(x, PDF)
```

Fig. 3.27 has Gaussian distribution *N*(0,3), the plot of the probability density function, and the corresponding cumulative distribution function.

Fig. 3.28 has Gaussian distribution *N*(2.5, 0.4), the plot of the probability density function, and the corresponding cumulative distribution function.

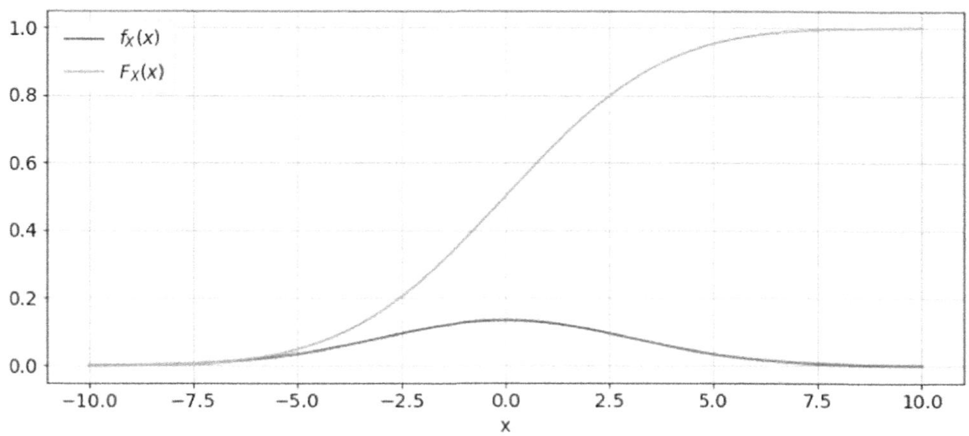

FIGURE 3.27 The probability density function and the cumulative distribution function for the Gaussian distribution function, *N*(0,3).

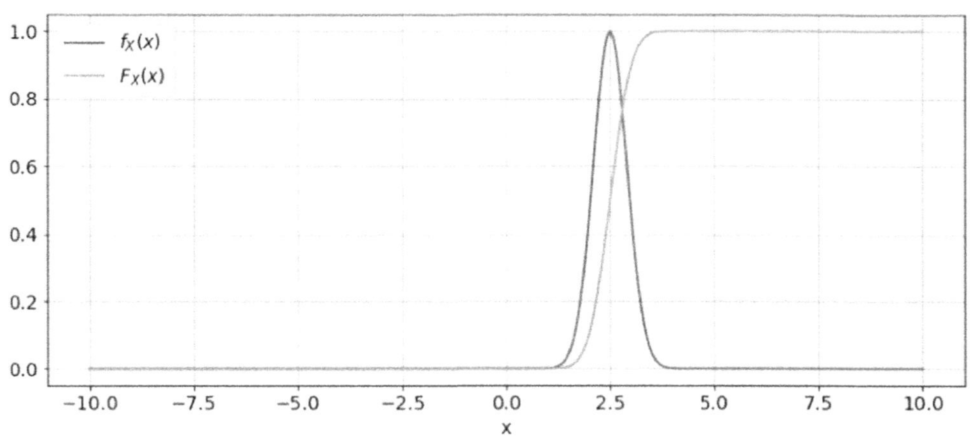

FIGURE 3.28 The probability density function and the cumulative distribution function for the Gaussian distribution function, *N*(2.5,0.4).

3.4 Conclusions

In this Chapter, we have attempted to discuss various concepts of statistics and discrete and continuous probability distributions. Every concept is accompanied by suitable examples, together with the Python code. This is purely to provide a quick review of frequently used concepts. An interested reader can explore further, and also apply the concept to any desired situation.

Acknowledgments

Sincere acknowledgments to the professional working environment provided by the college management towards pursuing individual passion. Acknowledgments are to the entire team from Knowledge Network of Indian Institute of Technology, Ghandhinagar, for having organized the Two week online Faculty Development Program on, "Data Science and Analytics," under the TEQIP –III Initiative (MHRD and Government of Gujarat), during January 2021, for introducing participants, to the essential concepts of Data Science Concepts, and the Python programming language.

Appendix

Python functions for concepts in statistics

```
def my_mean(x):
    mean = np.sum(x)/len(x)
    return mean

def my_mode(x):
    value, count=np.unique(x,return_counts=True)
    mode = value[np.argmax(count)]
    return mode

def my_median(x):
    A = np.sort(x)
    m=int(N/2)
    median = A[m]
    return median

def my_STD(x):
    mean = my_mean(x)
    A = np.power((x-mean),2)
    variance = np.sum(A)/N
    STD = np.sqrt(variance)
    return STD

def my_skewness(x):
    mean = my_mean(x)
    STD = my_STD(x)
    A = np.power((x-mean),3)
    skewness = np.sum(A)/N/np.power(STD,3)
    return skewness

def my_kurtosis(x):
    mean = my_mean(x)
    STD = my_STD(x)
    A = np.power((x-mean),4)
    kurtosis = np.sum(A)/N/np.power(STD,4)
    return kurtosis
```

```
def my_Covariance(x,y):
    N = len(x)
    mean_x = my_mean(x)
    mean_y = my_mean(y)
    XY = np.sum((x-mean_x)*(y-mean_y))
    Covariance = XY/N
    return Covariance

def my_Correlation(x,y):
    COV = my_Covariance(x,y)
    STD_x = my_STD(x)
    STD_y = my_STD(y)
    Correlation = COV/STD_x/STD_y
    return Correlation

def my_Normalized_Distribution(x):
    mean = my_mean(x)
    STD = my_STD(x)
    X_N = (x-mean)/STD
    return X_N

def my_line(m, b, x):
    y = m*x + b
    return y

def my_tanhx(x):
    A = np.exp(x)
    B = np.exp(-x)
    y = (A-B)/(A+B)
    return y

def my_Sigmoid(x):
    y = 1/(1+np.exp(-x))
    return y

def my_ReLU(x):
    y = []
    for n in range(N):
        val = max(0,x[n])
        y.append(val)
    return y
```

Python functions for concepts in discrete probability distribution

```
def my_Entropy(P):
    N = len(P)
    Entropy =0
    for n in range(N):
        Entropy = Entropy - P[n]*np.log2(P[n])
    return Entropy

def my_Gini_Index(P):
    N = len(P)
    Gini_Index = 1
    for n in range(N):
        Gini_Index = Gini_Index - P[n]*P[n]
    return Gini_Index

def my_Mew_Discrete(k,P):
    Mew = np.sum(k*P)
    return Mew

def my_STD_Discrete(k,P):
    N = len(P)
    Mew = np.sum(k*P)
    A = (k-Mew)*(k-Mew)
    variance = np.sum(A*P)
    STD = np.sqrt(variance)
    return STD

def my_Skewness_Discrete(k,P):
    Mew =my_Mew_Discrete(k,P)
    STD = my_STD_Discrete(k,P)
    A = np.power((k-Mew),3)
    Skewness = np.sum(A*P)/np.power(STD,3)
    return Skewness

def my_Kurtosis_Discrete(k,P):
    Mew =my_Mew_Discrete(k,P)
    STD = my_STD_Discrete(k,P)
    A = np.power((k-Mew),4)
    Kurtosis = np.sum(A*P)/np.power(STD,4)
    return Kurtosis
```

```
def my_factorial(k):
    val = 1
    for i in range(1,k+1):
        val = val*i
    return val

def my_NCK(N,K):
    m = min(K,(N-K))
    num, den, val = 1,1,1
    for i in range(N,m,-1):
        num = num *i
        den = den *val
        val = val +1
    NCK = num/den
    return NCK

def my_Poisson(lmd, N):
    PK =[]
    for i in range(N+1):
        temp = (lmd**i) * (np.exp(-lmd))/my_factorial(i)
        PK.append(temp)
    return PK

def my_Binomial(p, N):
    q = (1-p)
    PK = []
    for k in range(N+1):
        N_C_K = my_NCK(N,k)
        prob = N_C_K * np.power(p,k) * np.power(q,(N-k))
        PK.append(prob)
    return PK
```

Python functions for concepts in continuous probability distribution

```
def my_integral(x,y):
    N = len(x)
    integral = []
    integral.append(0)
    delta_x = x[1] - x[0]
    for i in range(N-1):
        temp1 = delta_x * (y[i+1] + y[i]) * 0.5
        temp2 = integral[i] + temp1
        integral.append(temp2)
    return integral

def my_derivative(x,y):
    N = len(x)
    dy_by_dx = []
    delta_x = x[1] - x[0]
    for i in range(N-1):
        temp = (y[i+1]-y[i])/delta_x
        dy_by_dx.append(temp)
    dy_by_dx.append(temp)#repeating the last value, for array size
    return dy_by_dx

def my_mean_continuous(x, PDF):
    mean = my_integral(x, x*PDF)
    return mean[-1]

def my_STD_continuous(x, PDF):
    mean = my_mean_continuous(x, PDF)
    A = np.power((x-mean),2)
    B = my_integral(x, A*PDF)
    sigma = np.sqrt(B[-1])
    return sigma
```

```
def my_Skewness_continuous(x, PDF):
    mean = my_mean_continuous(x, PDF)
    sigma = my_STD_continuous(x,PDF)
    A = np.power((x-mean),3)
    B = my_integral(x, A*PDF)
    Skewness = B[-1]/(sigma**3)
    return Skewness

def my_Kurtosis_continuous(x, PDF):
    mean = my_mean_continuous(x, PDF)
    sigma = my_STD_continuous(x,PDF)
    A = np.power((x-mean),4)
    B = my_integral(x, A*PDF)
    Kurtosis = B[-1]/(sigma**4)
    return Kurtosis

def my_Uniform_Distribution(a,b, x):
    N  = len(x)
    fx = []
    for n in range(N):
        if x[n]<a or x[n]>b:
            val = 0
        else:
            val = 1/(b-a)
        fx.append(val)
    return fx

def my_Exponential_Distribution(a,x):
    N = len(x)
    fx = []
    for n in range(N):
        if x[n]<0:
            val = 0
        else:
            val= a*np.exp(-a*x[n])
        fx.append(val)
    return fx
def my_gaussian(mew,sigma,x):
    A = 1/sigma/np.sqrt(2*np.pi)
    B = np.exp(-0.5*(x-mew)*(x-mew)/sigma/sigma)
    return A*B
```

References

[1] B. Kanmani, R.M. Vasu, Diffuse optical tomography using intensity measurements and the a priori acquired regions of interest: theory and simulations, J. Phys. Med. Biol. 50 (2005) 247–264 (stacks.iop.org/pb/50/247).

[2] B. Kanmani, R.M. Vasu, Diffuse optical tomography through solving a system of quadratic equations: theory and simulations, J. Phys. Med. Biol. 51 (2006) 981–998 (stacks.iop.org/PMB/51/981).

[3] B. Kanmani, R.M. Vasu, Contrast-detail and noise-tolerance analysis for detection and reconstruction of absorbing inhomogeneities with diffuse optical tomography using single- and phase-correlated dual source schemes, J. Phys. Med. Biol. 52 (2007) 1409–1429 (stacks.iop.org/PMB/52/1409).

[4] B. Kanmani, R.M. Vasu, Diffuse tomographic imaging, after estimation of location and approximate value of the object inhomogeneities through backprojection of the data, Internet Report, Saratov Fall Meet: Optical Technologies in Biophotonics and Medicine V, October 7–10, Russia, 2003. Available as Proceedings SPIE Volume 5475, pp108-14, 2004.

[5] M.R. Spiegel, J.J Schiller, R.A. Srinivasan, Schaum's outline series, Probability and Statistics, Third ed., McGraw Hill Education (India) Pvt Ltd., 2010.

[6] A. Papoulis, Random Probability, Variables and Stochastic Processes, McGraw-Hill Companies, 1984.

[7] Link to the Python Code: https://colab.research.google.com/drive/1li8R_evRKlOb-EuYh_pyDK1rioUu5gDS?usp = sharing.

Section II

Biomedical imaging (data acquisition and pre-processing)

Chapter 4

Computer-aided diagnosis and performance evaluation

Hardik Thakkar, Bikesh Kumar Singh and Saurabh Gupta
Department of Biomedical Engineering, National Institute of Technology Raipur, Raipur, Chhattisgarh, India

4.1 Introduction

Computer-aided diagnosis (CAD) has now become an integral part of clinical procedure in hospitals and medical centers across the globe. It is a rapidly growing field that has revolutionized the way medical professionals diagnose and treat various diseases. It involves the use of computer algorithms and machine learning techniques to assist doctors and other healthcare professionals in analyzing medical images, such as X-rays, MRI scans, CT scans, and detecting abnormalities that might be difficult for the human eye to identify.

CAD systems has several significant benefits, which includes (1) Increased Accuracy: CAD systems can accurately detect abnormalities and provide a diagnosis, reducing the risk of human error. (2) Early Detection: CAD can detect diseases at an early stage, allowing for timely intervention and treatment, which can increase the chances of successful recovery. (3) Improved Patient Outcomes: CAD can help identify the most appropriate treatment plan for the patient, leading to improved health outcomes and quality of life. (4) Reduced Healthcare Costs: Early detection and accurate diagnosis can lead to a reduction in healthcare costs by avoiding unnecessary diagnostic procedures and treatments. (5) Accessibility: CAD can be used in remote areas where access to medical experts may be limited, allowing for a wider range of patients to receive accurate diagnosis and treatment. Overall, CAD has the potential to revolutionize the healthcare industry by improving accuracy, efficiency, and accessibility of medical diagnosis and treatment. However, the performance of these CAD systems is critical to their usefulness in clinical practice.

The performance evaluation is an essential aspect of CAD development, allowing researchers to assess the system's accuracy, sensitivity, specificity, and overall effectiveness. Performance evaluation metrics provide quantitative measures to evaluate CAD system performance, and rigorous testing can identify areas for improvement and increase the system's reliability and usefulness in clinical settings. In this context, this area of research continues to evolve, with ongoing efforts to develop and improve performance evaluation methods for CAD systems.

4.2 Background

CAD is currently the most studied topics in medical physics, diagnostic radiology, and imaging nowadays. In the 1960s, the first initiatives toward computerized medical picture analysis were made. In the 1980s, a significant shift occurred in the use of computer output for medical diagnosis. Specifically, researchers began focusing on CAD rather than solely relying on automated computer diagnosis. This shift marked a turning point in the field, prompting increased and more rigorous research efforts.

During the early 1980s, the Kurt Rossmann Laboratories, which was part of the Department of Radiology (University of Chicago), initiated comprehensive and organized investigations into various CAD approaches. Prior to this, they had been primarily focused on the research regarding effects of images (digital) on radio diagnosis, while other researchers were exploring the development of a picture archiving and communication system (PACS). Although PACS was believed to have potential for managing radiologic images, and benefiting hospitals economically, it did not

Intelligent Computing Techniques in Biomedical Imaging. DOI: https://doi.org/10.1016/B978-0-443-15999-2.00006-2

appear to offer significant clinical advantages for radiologists at that time. Consequently, the research groups arrived at the idea of CAD, as they believed that the full potential of digital images could be harnessed in the day-to-day tasks of radiologists, such as image interpretation and diagnosis.

4.3 Literature survey

Researchers started exploring and contributing toward CAD systems from 1980 to 1990. In one of the early research articles [1], authors had discussed the changes of that era, including data storage problems, poor trial design, bias, selective reporting of results, and statistical naivety. In another article [2], researchers used CAD systems for patients with acute abdominal pain. The diagnostic and decision-making capabilities were evaluated for two distinct time periods: a baseline period before a computer system was installed, and a test period after the system was provided to assist with diagnosis. Both periods were comparable in terms of case types and the rate at which cases were accumulated. One of the noted article published in that era include image feature analysis and CAD for automatic detection of micro classifications in mammography [3].

In the next decade (1990–2000), many research articles were published. In one of the research articles, authors [4] gave a concise summary of different CAD methods for identifying and distinguishing pulmonary nodules, interstitial opacities, clustered micro-calcifications, and masses in chest radiographs and mammograms. An ROC analysis revealed that radiologists' diagnostic accuracy was significantly enhanced when they had access to the computer output. In another article [5], authors provide a concise overview of the reasoning and techniques involved in computer-based analysis of breast images for the diagnosis of breast lesions. Specifically, they explore how computers can be employed to evaluate the probability of malignancy in mammograms, ultrasound, or magnetic resonance imaging (MRI) scans of the breast, leading to improved diagnostic precision and a reduction in the number of benign lesions requiring biopsy. In another article [6], authors introduced a computer-aided automatic diagnostic framework designed for the detection of early-stage lung cancer nodules from thoracic helical CT images. They applied diagnostic approach encompassing both analytical and diagnostic processes. They first employ the fuzzy clustering algorithm to separate the lung and pulmonary blood vessel sections. Then, they use image processing techniques to evaluate the properties of these regions. Next, they formulated diagnostic rules based on the extracted features to assist in identifying potential nodule locations. In another article [7], authors demonstrated the role of texture analysis in image processing for CAD system.

Another research article [8] demonstrated the use of CAD system for diagnosis in chest radiography. In this article, they reviewed 150 papers on CAD system for chest images. In this paper, they provide a summary of their findings by enumerating a range of resolved and unresolved issues, as well as potential clinical applications, by identifying some emerging patterns in the area and by offering suggestions for future investigations. Based on limitations and scope of enhancement in CAD system, another article [9] was published. The article explored how the limitations of the human eye and brain's visual system, as well as the occurrence of overlapping structures in images and the prevalence of typical cases in screening programs, inspire the use of computer-based methods. These techniques have the potential to improve detection and diagnostic accuracy, ultimately leading to better patient outcomes. The effectiveness of computerized image analysis for interpreting medical images depends on two critical factors. First, the computer systems must be capable of accurately identifying and analyzing potentially concerning lesions. Second, radiologists must be able to integrate the information provided by the computer output into their decision-making process. Another article [10] demonstrated a historical review, status of research at that point of time, and the direction about the future. The paper discussed the efficiency of CAD system for diagnosis of breast cancer and at the same time the inefficiency of for many different types of lesions.

In the next decade (2010–2020), many notable contributions were found exploring CAD system application for different diseases diagnosis. In [11], author demonstrated a road map for CAD system to be incorporated in routine clinical practice from laboratory. Authors discussed about the reasons that why CAD systems not integrated part of each of clinic like the availability of large high quality database, etc. In a recent article [12], authors had discussed about the use of deep learning models in CAD system. The use of deep learning technology has demonstrated its potential to enhance the performance of CAD, which presents opportunities for CAD to be utilized as a concurrent reader or even a primary screener, ultimately leading to improved accuracy and workflow. In another recent article [13], the authors have discussed the development and effective application of several CAD techniques, including temporal subtraction, in various radiological modalities such as plain radiography, conventional scintigraphy, ultrasound, PET, CT, and MRI. Some of these techniques are already in use in clinical settings and efforts are underway to integrate CAD tool into routine clinical practice across all radiological subspecialties. The CAD system is expected to become an integrated part of PACS and a standard for diagnostic examinations in daily clinical practice. A systematic survey was presented [14] on CAD system developments. In this article, authors highlighted the opportunities and scope of research in CAD system.

The article provides a comprehensive survey of almost 250 research articles on CAD systems. A recent review article [15] on CAD retinopathy presented that how CAD system can considerably mitigate the burden on the ophthalmologists. The article had discussed about the existing methods for retinal feature extraction and automatic analysis. In another review article on CAD system for lung cancer, the authors had discussed about the challenges and the methodologies. The focus of the paper was on lung cancer diagnosis, which is regarded as one of the most significant, interesting, and difficult challenges in the field of oncology. In another paper [16], standard online dataset was used containing nine features. The study utilized a dominance-based feature selection approach and evaluated the performance of various combinations of dominant feature sets. Results showed that using the top four or top five dominant features produced classifier performances that were nearly as good as those using all features. The best performing classifier was found to be the artificial neural network (ANN), which achieved best accuracy of 99.6%.

In recent years, CAD system is tested with deep learning model and completely automated AI model for disease management. Recent research articles demonstrated the performance of these models. In a recent review article [17], the discussion was on diagnostic imaging and application of AI in CAD systems. The overall review was based on how AI can be used for medical image enhancement process and consequently useful for CAD systems. In another article for skin cancer detection using CAD system [18], the study's initial steps involved reducing image noise through the use of a median filter. Subsequently, a novel approach to image segmentation, which relied on a convolutional neural network (CNN) optimized via satin bowerbird optimization, was presented and evaluated using a confusion matrix. A recent article based on development of CAD system for breast ultrasound images using CNN model was discussed. The classification of breast tumors into benign or malignant categories has been accomplished with breast ultrasound and CAD. However, traditional CAD software has encountered difficulties, such as the challenge of designing handcrafted features and the difficulty of confirming overfitting problems. The research article describes the use of a CAD system for tumor diagnosis. They employed a technique based on image fusion on ultrasound images that merges various image content representations and ensembles different CNN architectures. The study incorporated VGGNet, ResNet, and DenseNet in their CNN-based approach. The proposed method achieved an accuracy of 91.10%. The novel machine learning technique was introduced by the authors for CAD system in a recent article [19]. The aim was to optimize the dendritic neuron model using a heuristic optimization algorithm and design a tidy visual classifier, which can be implemented on a hardware system for CAD to mitigate the sensitivity of the backpropagation algorithm to initial conditions and the potential issue of getting trapped in local minima. A new evolutionary dendritic neuron model has been proposed. This model was optimized using the algorithm named gbest-guided artificial bee colony (GABC).

Overall, the research on the performance evaluation of CAD systems over the few decades has shown significant progress in developing and validating methods for assessing the accuracy and clinical impact of these systems. These methods have the potential to improve the clinical utility of CAD systems and their ability to improve patient outcomes. However, further research is needed to refine and validate these methods and to address the challenges associated with the integration of CAD systems into clinical practice.

4.4 Architecture of computer-aided diagnosis systems

CAD systems are designed to assist physicians and other healthcare professionals in the interpretation of medical images or other diagnostic data. These systems use various techniques from machine learning, artificial intelligence, and image processing to analyze the data and provide diagnostic support. Fig. 4.1 shows the architecture of CAD systems.

Data acquisition: This component involves the acquisition of medical images or other diagnostic data from various sources, such as X-ray machines, MRI scanners, CT scanners, or ultrasound machines.

Preprocessing: The acquired data are preprocessed to correct for any artifacts or noise that may be present in the images. This step involves techniques such as filtering, segmentation, and feature extraction.

Feature extraction: This component involves extracting relevant features from the preprocessed data that are likely to be useful in making a diagnosis. This step can involve various techniques such as texture analysis, edge detection, and shape analysis.

Machine learning: These algorithms are used to investigate the extracted features and to learn patterns in the data that can be used for classification and diagnosis. Various machine learning techniques such as decision trees, support vector machines, and neural networks can be used in this step.

Decision support: The machine learning component outputs a diagnostic prediction or a probability score that can be used as decision support for the healthcare professional. The output can be in the form of a binary classification, a probability score, or a heat map highlighting regions of interest in the image.

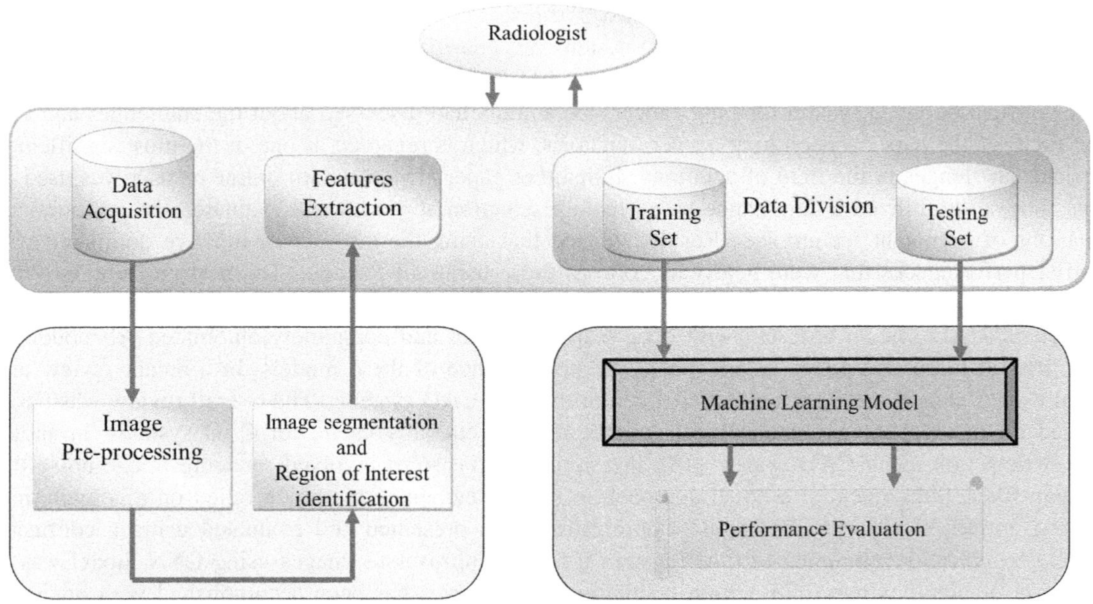

FIGURE 4.1 Architecture of computer-aided diagnosis system.

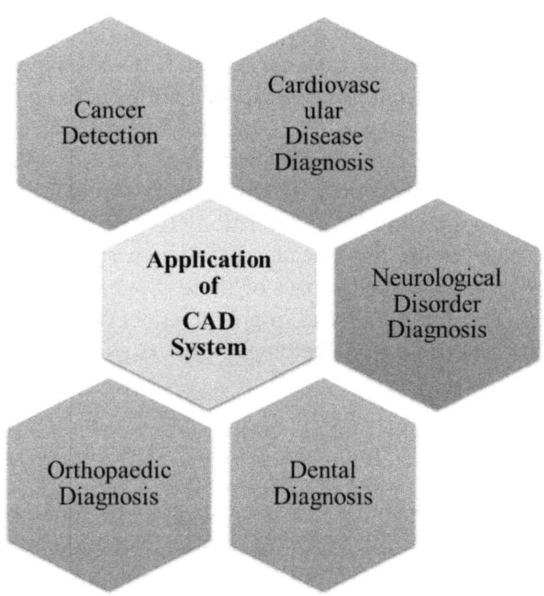

FIGURE 4.2 Applications of computer-aided diagnosis system.

User interface: The output from the CAD system is typically presented to the healthcare professional through a user interface that allows them to review the diagnostic support provided by the system and make their own diagnosis based on the information provided.

Data management: The CAD system also includes a data management component that is responsible for storing and managing the large amounts of data generated by the system, including the acquired images, preprocessed data, and machine learning models. Overall, the architecture of a CAD system is designed to provide accurate and reliable diagnostic support to healthcare professionals while also ensuring that the system is easy to use and manage.

4.5 Applications of computer-aided diagnosis systems

CAD systems have numerous applications in the field of medical diagnostics. Some of the key applications of CAD systems is shown in Fig. 4.2.

Cancer detection: CAD systems could be used to detect various types of cancers. These systems can help radiologists identify suspicious lesions and tumors on medical images, improving the accuracy and efficiency of cancer diagnosis. CAD systems are used to detect cancer by analyzing medical images, such as mammograms, CT scans, and MRIs. These systems use advanced algorithms and machine learning techniques to assist radiologists and other healthcare professionals in identifying suspicious lesions or tumors that may be indicative of cancer. The CAD system starts by preprocessing the medical image to remove any noise or artifacts that may affect the accuracy of the analysis. The image is then segmented to identify regions of interest, such as masses or calcifications. Relevant features are extracted from the segmented regions and these features are used to categorize the region as either benign or malignant.

The classification stage involves the use of machine learning algorithms, or ANN networks to analyze the extracted features and make a diagnosis. The CAD system may compare the extracted features to a database of known cancer cases, allowing it to make a more accurate diagnosis. The output of the CAD system is typically a visual representation of the analyzed region, with suspicious areas highlighted for further examination by the radiologist or other healthcare professional. The CAD system may also provide a numerical score or probability of malignancy to assist in making the final diagnosis. CAD systems have been shown to improve the accuracy and efficiency of cancer diagnosis, particularly for mammography and lung cancer screening. However, they are not without their limitation, and false positives and false negatives can occur. Therefore the final diagnosis should always be made by a trained healthcare professional, and CAD systems should be used as a tool to assist in the diagnosis, not as a replacement for human expertise.

Cardiovascular disease diagnosis: CAD systems can be used for cardiovascular disease diagnosis by analyzing medical images of the heart and blood vessels, such as CT scans, MRIs, or echocardiograms. These systems use advanced algorithms and machine learning techniques to assist healthcare professionals in identifying blockages, aneurysms, and other abnormalities that may be indicative of cardiovascular disease. The initial step in the CAD system is to preprocess the medical image to eliminate any noise or artifacts that might affect the accuracy of the analysis. Next, the image is segmented to identify areas of interest, such as the coronary arteries or heart chambers. Relevant features are then extracted from the segmented regions, such as the size and shape of the vessels or the flow of blood. Similar to the process in cancer detection, these extracted features are then used to classify the region as either normal or abnormal. In the classification stage, machine learning algorithms are used to analyze the extracted features and make a diagnosis. To improve the accuracy of the diagnosis, the CAD system may compare the extracted features to a database of known cardiovascular disease cases.

In addition, the CAD system may provide a numerical score or probability of disease to assist in making the final diagnosis. CAD systems have been shown to enhance the accuracy and efficiency of cardiovascular disease diagnosis, especially in detecting early-stage disease or asymptomatic patients.

Neurological disorder diagnosis: CAD systems can assist in the diagnosis of neurological disorders. These systems can analyze medical images of the brain to identify patterns of abnormality, providing valuable insights into the progression and severity of the disease. CAD systems can be used for neurological disorder diagnosis by analyzing medical images of the brain, such as MRI or CT scans. These systems use advanced algorithms and machine learning techniques to assist healthcare professionals in identifying patterns of abnormality that may be indicative of neurological disorders, such as Parkinson's disease, multiple sclerosis, schizophrenia, or Alzheimer's disease. The CAD system first processes the medical image to eliminate any noise or artifacts that may affect the analysis's accuracy. The image is then segmented to identify regions of interest, such as gray matter, white matter, or cerebrospinal fluid. Relevant features, such as the thickness of cortical layers, volume of brain structures, or degree of connectivity between different regions, are extracted from the segmented areas. These extracted features are then used to classify the region as normal or abnormal using machine learning algorithms such as linear discriminant, decision tree, or random forests. To increase the diagnosis's accuracy, the CAD system may compare the extracted features to a database of known neurological disorder cases.

The CAD system outputs a visual representation of the analyzed brain regions with abnormal areas highlighted for further examination by healthcare professionals. Additionally, the system may provide a numerical score or probability of disease to assist in the final diagnosis. CAD systems have demonstrated improved accuracy and efficiency in neurological disorder diagnosis, particularly for identifying early-stage diseases or subtle changes in brain structure or function. However, false positives and false negatives can occur, and thus trained healthcare professionals should always make the final diagnosis.

Dental diagnosis: CAD systems can be used in dental diagnosis by analyzing medical images of the teeth and jaw, such as panoramic radiographs, cone-beam CT scans, or intraoral images. These systems use advanced algorithms and machine learning techniques to assist dental professionals in identifying dental caries, periodontal disease, or other dental abnormalities. The CAD system starts by preprocessing the medical image to remove any noise or artifacts that may affect the accuracy of the analysis. The image is then segmented to identify regions of interest, such as the teeth, bone, or soft tissue. Relevant features are extracted from the segmented regions, such as the size and shape of the teeth, the presence of cavities, or the degree of bone loss. The extracted features are used to classify the region as either normal or abnormal. The CAD system can utilize a database of known dental disease cases to compare the extracted features and make a more precise diagnosis. The system generates a visual representation of the analyzed dental regions with abnormal areas highlighted for further examination by the dental professional. CAD systems have been demonstrated to improve the efficiency of dental diagnosis, particularly for identifying early-stage disease or subtle changes in tooth or bone structure. But, as discussed earlier, these systems have limitations. Thus a trained dental professional must always make the final diagnosis, and CAD systems should be used just as an aid in the diagnosis.

Orthopedic diagnosis: CAD systems can assist in the diagnosis of musculoskeletal conditions, such as arthritis and bone fractures. These systems can analyze medical images of bones and joints to identify areas of damage or abnormality, allowing doctors to provide appropriate treatment. CAD systems can be used for orthopedic diagnosis by analyzing medical images of bones and joints, such as X-rays, CT scans, or MRI scans. These systems use advanced algorithms and machine learning techniques to assist orthopedic professionals in identifying bone fractures, joint degeneration, or other orthopedic abnormalities.

The initial step of the CAD system involves preprocessing the medical image to eliminate any noise or artifacts that may affect the analysis's accuracy. The image is segmented to identify the regions of interest, such as bones, cartilage, or soft tissue. From the segmented regions, relevant features, such as shape, size, and density of bones or the extent of cartilage wear and tear, are extracted. These features are then utilized to classify the region as normal or abnormal. Machine learning algorithms, such as deep learning, decision tree, k-nearest neighbor, or random forests, are used to analyze the extracted features and make a diagnosis.

The CAD system may match the extracted features to a database of known orthopedic cases, leading to a more accurate diagnosis. The output is a visual representation of the analyzed orthopedic regions, highlighting the abnormal areas for further examination by the orthopedic professional. The system may also provide a numerical score or probability of disease to aid in making the final diagnosis.

CAD systems have improved the accuracy and efficiency of orthopedic diagnosis, particularly for identifying early-stage disease or subtle changes in bone or joint structure. Despite their advantages, CAD systems have limitations, and false positives and false negatives can occur. As a result, trained orthopedic professionals should always make the final diagnosis because CAD systems are not a substitute for human expertise. Overall, CAD systems have many practical applications in medical diagnostics, improving the accuracy and efficiency of diagnosis for a wide range of diseases and conditions. With ongoing advancements in machine learning and imaging technology, CAD systems are likely to become even more effective and widespread in the future.

4.6 Performance evaluation of computer-aided diagnosis system

Fig. 4.3 shows the ways to evaluate the performance of a CAD system.

Speed: The speed of a CAD system is an important metric to consider. The faster a CAD system can process data and generate drawings, the more efficient it will be. This can be measured in terms of how quickly the system can open files, display models, perform calculations, and render images.

User interface: A CAD system with a user-friendly interface can improve productivity and reduce user errors. An intuitive interface with easy-to-use tools and commands can help users create drawings quickly and efficiently.

Compatibility: A CAD system that is compatible with other software applications and file formats can help streamline workflows and increase productivity. Compatibility with industry standards and file formats is important for sharing and collaborating on designs.

Reliability: A CAD system that is reliable and stable is important to prevent data loss, crashes, and downtime. The system should be able to handle large and complex models without performance degradation or system crashes.

Scalability: The ability of a CAD system to handle larger and more complex models is important as the size and complexity of designs increase over time. Scalability can be evaluated by measuring the performance of the system when working with larger models.

Support: A CAD system should come with adequate support to help users troubleshoot problems and get the most out of the software. Good support includes technical documentation, online forums, and customer support.

Classifier performance: The heart of CAD systems is its classification model, and the overall performance of CAD system depends on the performance of classification models. Here are some common methods for evaluating the performance of a machine learning model:

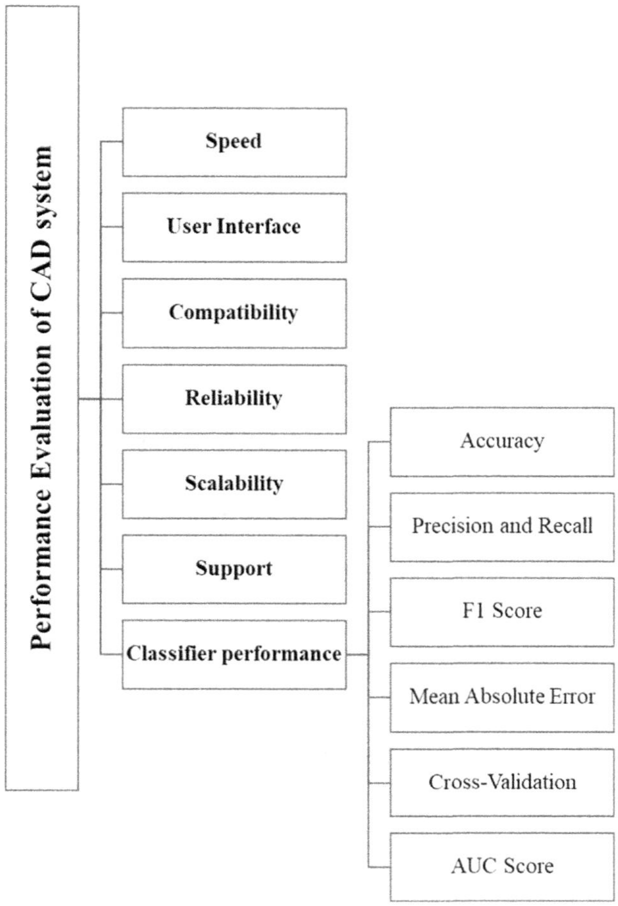

FIGURE 4.3 Performance evaluation of CAD system.

Accuracy: This is the most basic metric for evaluating classification models. It measures the percentage of correct predictions made by the model on a test dataset. But, it can be misleading if dataset is imbalanced or the cost of making incorrect predictions is high.

Precision and recall: When dealing with an imbalanced dataset, precision and recall are valuable metrics for assessing classification models. Precision quantifies the proportion of correct positive predictions relative to all positive predictions, whereas recall quantifies the proportion of correct positive predictions relative to all true positives. A well-performing model should exhibit high precision and high recall.

F1 score: The F1 score is the harmonic mean of precision and recall. It provides a balanced evaluation of a model's performance for binary classification problems.

Mean absolute error and root mean squared error: These parameters are commonly used to assess regression models. Mean absolute error measures the average absolute difference between the predicted and actual values. Root mean squared error measures the square root of the average squared difference between the predicted and actual values. A good model should have low values for both metrics.

Cross-validation: It is a technique for assessing the performance of a model on multiple test datasets. It involves dividing the dataset into k-folds and training the model on k-1 folds while evaluating it on the remaining fold. Iteration of this process are performed k times, and the results are averaged to get an overall evaluation of the model's performance.

ROC curve and AUC (area under the curve) score: These parameters are useful for evaluating classification models when the cost of false positives and false negatives is not equal. The ROC curve plots the rate of true positives against the rate of false positives at various thresholds. The AUC score represents the area under the ROC curve, and a good model should have a high AUC score.

By considering these metrics, you can evaluate the performance of a machine learning model and select the one that performs the best on the problem at hand. In addition, the performance of a CAD system can be evaluated and selected as per the need.

4.7 Conclusion

In conclusion, CAD systems have revolutionized the field of medical diagnosis by providing clinicians with an automated, efficient, and accurate tool for analyzing medical images and making diagnoses. CAD systems are widely used in various medical domains, including cancer diagnosis, cardiovascular disease diagnosis, neurological disorder diagnosis, dental diagnosis, and orthopedic diagnosis.

The performance evaluation of CAD systems is crucial for assessing their accuracy and reliability in assisting clinical decision-making. A rigorous evaluation of CAD systems involves the use of appropriate parameters like positive predictive value, negative predictive value, accuracy, sensitivity, and specificity. The evaluation process should also consider the impact of factors such as data quality, feature selection, and algorithm selection on the performance of the CAD system.

The application of CAD systems in medical diagnosis is constantly evolving, with new algorithms and technologies being developed to improve the accuracy, efficiency, and reliability of these systems. However, it is important to note that CAD systems should be used as a tool to assist in clinical decision-making and not as a replacement for human expertise. A trained medical professional, who can take into account the patient's medical history, physical examination, and other clinical factors, should always make the final diagnosis.

This book aimed to provide an overview of the concepts, techniques, and applications of CAD systems in medical diagnosis and performance evaluation. It is our hope that this book will serve as a useful reference for medical professionals, researchers, and students who are interested in the field of CAD and performance evaluation.

4.8 Strengths and limitations of computer-aided diagnosis

The strengths of CAD include its ability to analyze large amounts of data quickly and accurately, reducing the likelihood of human error and variability. CAD algorithms can also learn from past data and improve their performance over time, making them useful in identifying subtle changes in medical images that may not be easily detected by human observers. Additionally, CAD has the potential to enhance the efficiency of medical imaging services by reducing the workload of radiologists and other healthcare professionals.

However, CAD also has limitations, including the potential for false positives and false negatives. CAD algorithms may sometimes misinterpret normal structures as abnormal or fail to detect subtle abnormalities that require further investigation. In addition, the accuracy of CAD algorithms may be affected by factors such as the quality of the medical image, patient variability, and the expertise of the healthcare professional using the system.

4.9 Future directions

Future research in CAD will likely focus on addressing the limitations of current algorithms and improving their accuracy and performance. One potential approach is to incorporate multi-modal imaging techniques, such as combining MRI and PET scans, to provide a more comprehensive view of the patient's condition. Another approach is to integrate clinical data, such as patient demographics and medical history, into CAD algorithms to improve their diagnostic accuracy and reduce false positives.

References

[1] G.C. Sutton, Computer-aided diagnosis: a review, Br. J. Surg. 76 (1) (1989) 82–85. Available from: https://doi.org/10.1002/bjs.1800760126.

[2] D.H. Wilson, D.J. Hewett, F.T. de Dombal, Computer aided diagnosis of acute abdominal pain: a multicentre study, Br. Med. J. (Clin. Res. Ed.) 293 (6550) (1986) 800–804. Available from: https://doi.org/10.1136/bmj.293.6550.800.

[3] H.P. Chan, K. Doi, S. Galhotra, C.J. Vyborny, H. Macmahon, P.M. Jokich, Image feature analysis and computer aided diagnosis in digital radiography. I. Automated detection of microcalcifications in mammography, Med. Phys. 14 (4) (1987) 538–548. Available from: https://doi.org/10.1118/1.596065.

[4] K. Doi, H. MacMahon, S. Katsuragawa, R.M. Nishikawa, Y. Jiang, Computer-aided diagnosis in radiology: potential and pitfalls, Eur. J. Radiol. 31 (2) (1999) 97–109. Available from: https://doi.org/10.1016/S0720-048X(99)00016-9.

[5] M.L. Giger, Computer-aided diagnosis of breast lesions in medical images, Comput. Sci. Eng. (2000) 39–45. Available from: https://doi.org/10.1109/5992.877391.

[6] K. Kanazawa, Y. Kawata, N. Niki, H. Satoh, H. Ohmatsu, R. Kakinuma, et al., Computer-aided diagnostic system for pulmonary nodules using helical CT images, Lecture Notes in Computer Science 1496 (1998) 449–456. Available from: https://doi.org/10.1007/bfb0056230.

[7] G.D. Tourassi, Journey toward computer-aided diagnosis: role of image texture analysis, Radiology. 213 (2) (1999) 317–320. Available from: https://doi.org/10.1148/radiology.213.2.r99nv49317.
[8] B. Van Ginneken, B.M. Ter Haar Romeny, M.A. Viergever, Computer-aided diagnosis in chest radiography: a survey, IEEE Trans. Med. Imaging. 20 (12) (2001) 1228–1241. Available from: https://doi.org/10.1109/42.974918.
[9] M. L. Giger, K. Suzuki. Computer-aided diagnosis. Biomedical information Technology. Academic Press, 2008. 359-XXII. https://doi.org/10.1016/B978-012373583-6.50020-7
[10] K. Doi, Current status and future potential of computer-aided diagnosis in medical imaging, Br. J. Radiol. 78 (SPEC. ISS.) (2005) 21–30. Available from: https://doi.org/10.1259/bjr/82933343.
[11] B. Van Ginneken, C.M. Schaefer-Prokop, M. Prokop, Computer-aided diagnosis: how to move from the laboratory to the clinic, Radiology. 261 (3) (2011) 719–732. Available from: https://doi.org/10.1148/radiol.11091710.
[12] H.P. Chan, L.M. Hadjiiski, R.K. Samala, Computer-aided diagnosis in the era of deep learning, Med. Phys. 47 (5) (2020) e218–e227. Available from: https://doi.org/10.1002/mp.13764.
[13] J. Shiraishi, Q. Li, D. Appelbaum, K. Doi, Computer-aided diagnosis and artificial intelligence in clinical imaging, Semin. Nucl. Med. 41 (6) (2011) 449–462. Available from: https://doi.org/10.1053/j.semnuclmed.2011.06.004.
[14] J. Yanase, E. Triantaphyllou, A systematic survey of computer-aided diagnosis in medicine: past and present developments, Expert. Syst. Appl. 138 (2019) 112821. Available from: https://doi.org/10.1016/j.eswa.2019.112821.
[15] M.R.K. Mookiah, U.R. Acharya, C.K. Chua, C.M. Lim, E.Y.K. Ng, A. Laude, Computer-aided diagnosis of diabetic retinopathy: a review, Comput. Biol. Med. 43 (12) (2013) 2136–2155. Available from: https://doi.org/10.1016/j.compbiomed.2013.10.007.
[16] K. Atrey, Y. Sharma, N.K. Bodhey, B.K. Singh, Breast cancer prediction using dominance-based feature filtering approach: a comparative investigation in machine learning archetype, Braz. Arch. Biol. Technol. 62 (2019) 1–15. Available from: https://doi.org/10.1590/1678-4324-.
[17] H. Fujita, AI-based computer-aided diagnosis (AI-CAD): the latest review to read first, Radiol Phys. Technol. 13 (1) (2020) 6–19. Available from: https://doi.org/10.1007/s12194-019-00552-4.
[18] Z. Xu, F.R. Sheykhahmad, N. Ghadimi, N. Razmjooy, Computer-aided diagnosis of skin cancer based on soft computing techniques, Open Med. 15 (1) (2020) 860–871.
[19] C. Tang, J. Ji, Y. Tang, S. Gao, Z. Tang, Y. Todo, Engineering applications of artificial intelligence: a novel machine learning technique for computer-aided diagnosis, Eng. Appl. Artif. Intell. 92 (2020) 103627. Available from: https://doi.org/10.1016/j.engappai.2020.103627.

Section III

Medical image description and representation

Chapter 5

Feature extraction and feature selection in medical images

Padma Selvaraj[1] and S. Sivaprakash[2]

[1]*Department of Computer Science & Technology, Madanapalle Institute of Technology & Science, Madanapalle, Andhra Pradesh, India,* [2]*Department of Software Systems, School of Computer Science and Engineering, Vellore Institute of Technology, Vellore, Tamil Nadu, India*

5.1 Introduction

Healthy people make a healthy society. The medical field faces exponential challenges in saving human lives. Currently, it is difficult for the medical field to handle the enormous data generated. A variety of medical devices is involved in disease diagnosis. These devices produce data comprising numbers, strings, signals, and images. Medical devices generate images like X-rays, positron emission tomography (PET), magnetic resonance image (MRI), fundus images, computed tomography (CT) images, and ultrasound images, as mentioned in Fig. 5.1. Adding to this, radiation ionization and nuclear medicine also produce images. It is possible to retrieve the necessary information from medical images utilizing image processing. In the current medical upheaval, computer-aided diseases diagnoses (CADD) play a vital role in diagnosing. CADD aims to identify diseases based on the medical image as input at reduced cost, improved precision, and people satisfaction.

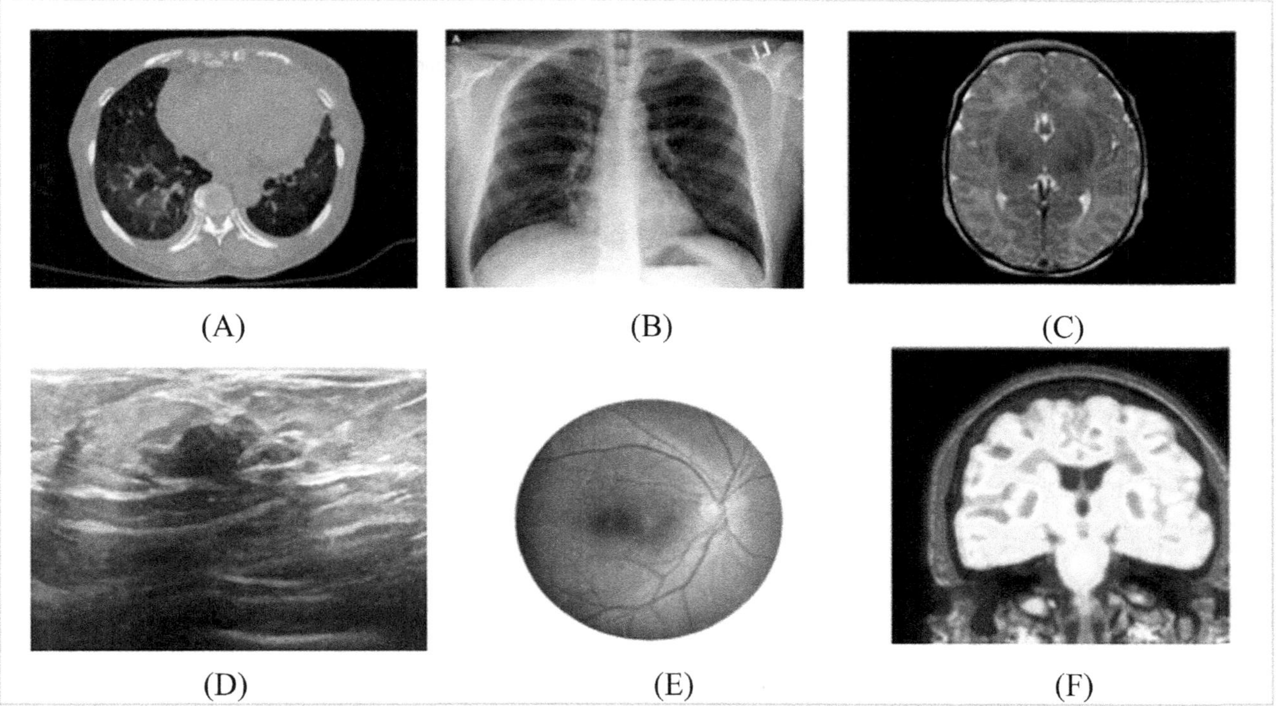

FIGURE 5.1 Types of medical images. (A) CT scan image, (B) X-ray image, (C) MRI image, (D) ultrasound image, (E) fundus image, and (F) PET image.

Intelligent Computing Techniques in Biomedical Imaging. DOI: https://doi.org/10.1016/B978-0-443-15999-2.00008-6

Image processing deals with image inputs, whereas output may be images or picture features. The feature can be an attribute, a characteristic of the entire data, and a column. The so-called features decide the output of the algorithm. Therefore it becomes essential to improve the feature significance through preprocessing, segmentation, feature extraction, and feature selection, as given in Fig. 5.2.

Images usually contain extraneous and undesirable chunks. Preprocessing is required to remove those chunks, which helps increase the image quality. Many authors have proposed different preprocessing techniques suitable for their research. According to our knowledge, we have categorized it as in Fig. 5.3, namely (1) normalization, (2) skewness, (3) scaling, (4) removal of noise, (5) grayscale image conversion, and (6) binarization. Bringing the image within a specified range is termed normalization. Skewness to correct or check the image distribution. Scaling refers to a change in the size of the image. Eliminating noise from the original image strengthens the quality of the image. Converting red, green, blue (RGB) images to shades of gray is the next step. Making color images into black and white (binary) is binarization.

The next step in image processing could be dimensionality reduction, termed feature extraction. Feature extraction extracts the required information from the original image. When the original image is of high dimension, it increases the execution time [1]. Feature extraction reduces the so-called sizes and generates feature vectors to enhance the productivity of the concerned algorithm or model. The researchers improve image processing by following many styles of feature extraction—upcoming sections of the chapter elaborate on feature extraction methods.

Feature selection in image processing plays a vital role. Only some of the features of an image may contain meaningful information suitable for classification. Processing meaningless components affect the efficiency of the algorithm. A subset of important elements from the original feature set describes feature selection. They start from image acquisition to classification using machine learning (ML) and deep learning (DL), as mentioned in Fig. 5.4. This chapter explains the various feature selection techniques that improve the algorithm's efficiency.

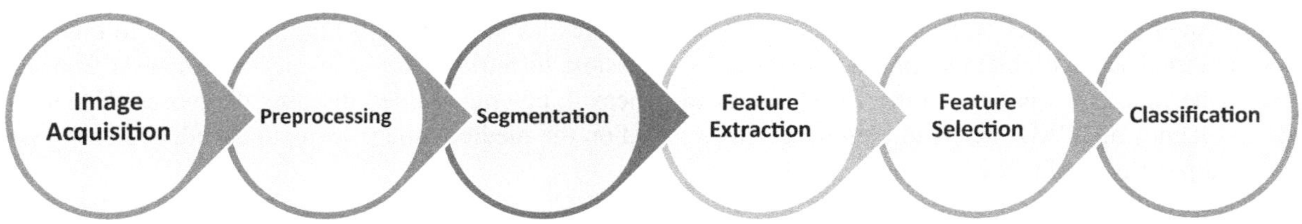

FIGURE 5.2 Steps in image processing.

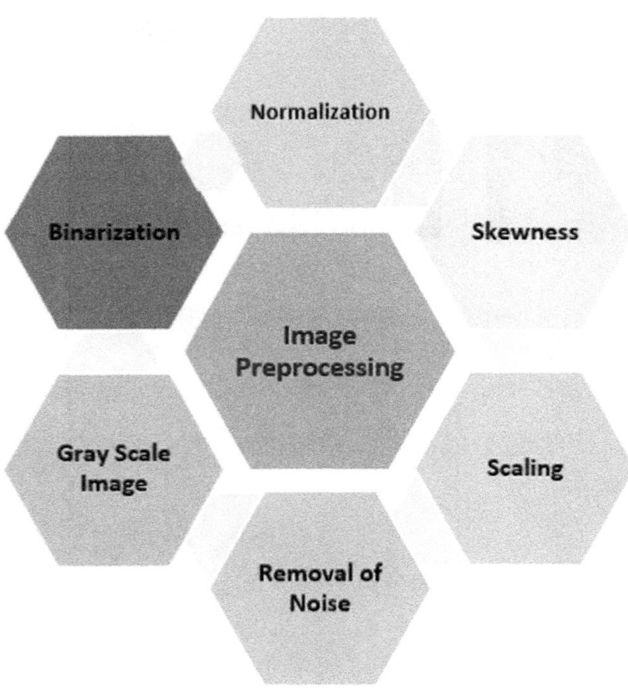

FIGURE 5.3 Categorization of image preprocessing.

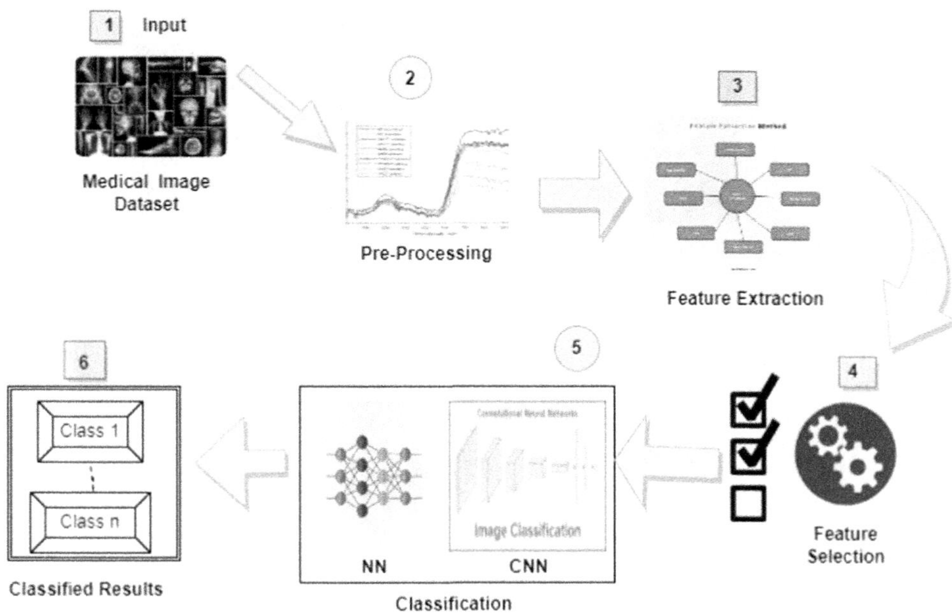

FIGURE 5.4 Steps of image classification.

The final phase of image processing is classification. Instead of using simple algorithms, developing a classifier or a model suitable for the images sound better. Medical image processing expects superior results. ML or DL models provide promising results with suitable preprocessing, feature extraction, and feature selection.

5.2 Feature extraction

Image processing applications aim to mine important structures from image records; the machine can provide an explanation, clarification, or understanding of the image. It is becoming normal to work with datasets of numerous features. Overfitting problem may arise if extracted features contain redundant values [1]. To overcome this problem, regularization or feature extraction can be the solution.

Common advantages of feature extraction are:

- Improved accuracy.
- Reduced overfitting.
- Reduced training time.
- Improved data visualization.
- Easy to understand the model.

The image processing medical field includes numerous types of images, including tomography X-ray technique, low dose mammography, radiography (X-Ray images), cardiogram of the heart muscle, and CT scans. Physicians use CT scan to detect deviations in the patient's body by comparing it with normal conditions. Using the computer-generated image processing analysis, a sequence of steps will be taken as mentioned in the introduction part. Among these phases, the feature extraction is significant to detect regions of interest from CT-scan images. The feature extraction phase applies various methods in identifying different image characteristics. Here we have presented a comprehensive appraisal of the feature extraction suitable for medical images.

Before entering into the feature extraction and selection topic, let us discuss a few methods involved in image segmentation.

1. Edge detection: It increases image contrast areas. Edge is the border of the objects in the image where the strength of the pixel changes sharply. Fig. 5.5 shows an example of an MRI image after Canny edge detection and Sobel edge detection.
2. Thresholding: This method selects pixels within an identified range with a particular value or arc. If an object's brightness level (or content) is determined, it can be used to locate it within an image. Fig. 5.6 shows the segmentation matching of a single threshold segmentation of thoracic images.

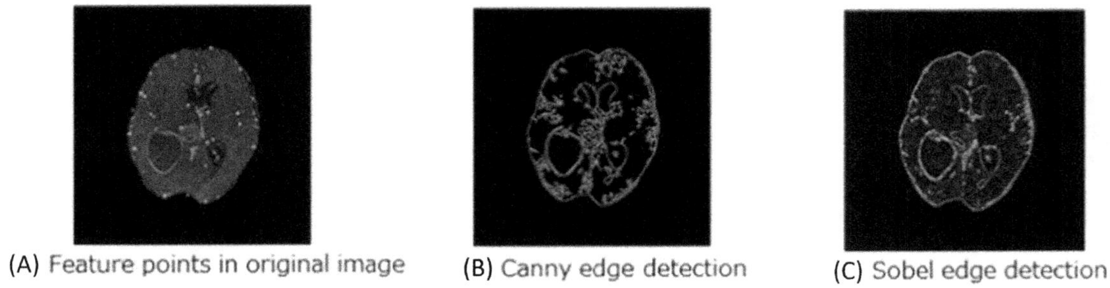

FIGURE 5.5 (A) Representing feature point in the original image, (B) image by Canny edge detection, and (C) Sobel edge detection image.

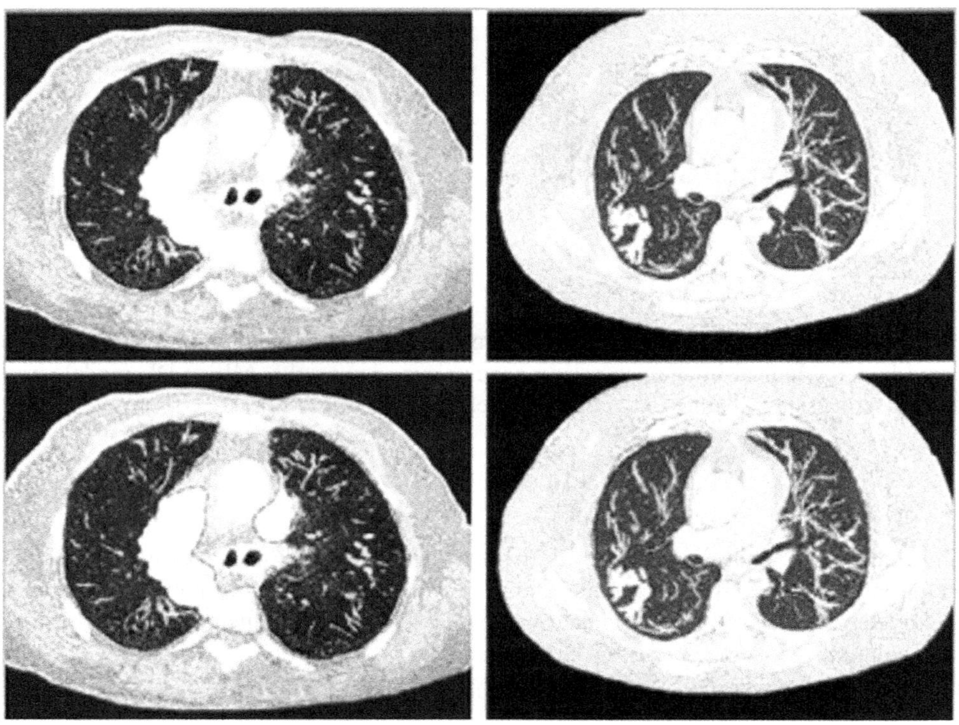

FIGURE 5.6 Threshold segmentation.

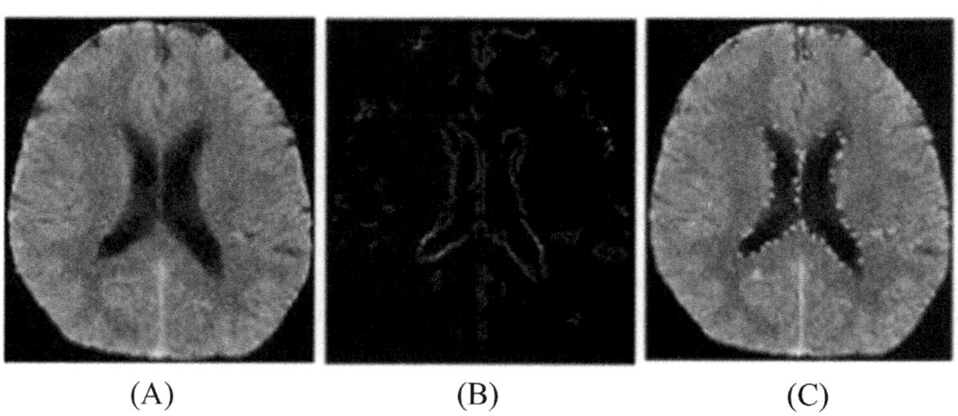

FIGURE 5.7 Image for corner detection. (A) Gray image (GI), (B) MTI of the GI, and (C) detected corners mapped to the GI.

3. Detecting image curvature (corner extraction): Parametric form of a planar curve defines the curvature of an image. Corner detection helps to segment the image. Fig. 5.7 shows the magnetization transfer imaging (MTI) applied for corner detection in an MRI image and the gastrointestinal tract of the same image.
4. Region/patch analysis: The region of the images is composed of pixel collections. The patch analysis method helps to identify specific areas based on algorithms.

5. Hough transform: An effective method that suits for binary template matching. This method is used to find imperfections in the image instances of objects within a certain class of images by a voting procedure: lines and quadratic forms and arbitrary.
6. Image motion detection: Identifying the difference among images by means of various movements of the respective images.
7. Histogram: The histogram intensity shows the brightness levels occupied in an image individually. A range of brightness levels contained in a histogram invokes further processing of images.
8. Haar wavelets: A binary-based function with a wide range of functions. Binary components are better mapped using discrete signals.
9. Texture extraction: A pattern arrangement with regular intervals in the image. Few methods are available to extract texture from the image, such as cooccurrence matrices, local binary patterns (LBP), and Fourier transform.

So far we have addressed about various digital image processing techniques with respect to feature extraction.

5.2.1 Feature extraction process

It is a process to extract new feature sets from the existing features using functional mapping.

Choosing the best technique for feature extraction in medical imaging is impossible. Yet only some common feature extraction techniques can manage all the attributes correctly. In order to select consistent feature mining, the simple things to remember are the types of criminal activities that happened, data availability and dimensions, data labeling and supervised learning, and unsupervised learning methods. Once the foundations for these steps are recognized, we can choose a suitable feature extraction technique among the various available algorithms. As discussed, Fig. 5.8 shows the categories of feature extraction.

5.2.1.1 Local binary pattern

In LBP, rays were originally described for local neighborhoods with special emphasis. The LBP operator first forms a binary code explaining a regional texture pattern in neighboring pixels [2]. Binary codes are found by thresholding the gray value of the neighborhood center. Also, binary codes revert to decimals representing LBP codes. Therefore the grayscale invariant texture statistics method is used in the LBP.

5.2.1.2 Grayscale cooccurrence matrix

Grayscale cooccurrence matrix (GLCM) is a feature extraction method that originally manages statistical techniques. It considers the spatial relationship of pixels as a grayscale coexistence matrix or a grayscale spatial dependence matrix. The spatial relationships in the grayscale coexistence matrix are defined for the pixels of interest. Horizontally adjacent pixels can signal extra space between two pixels. GLCM elements can be calculated through the sum of the number of times a pixel value appears in a specified spatial relationship with a pixel in the input image.

5.2.1.3 Grayscale run-length method

The encoding grayscale run-length method represents an image matrix containing symbol strings. In this case, a range of gray levels is nothing but collection of adjacent pixels with similar grayscale value. Feature extraction is carried out through run-length matrix.

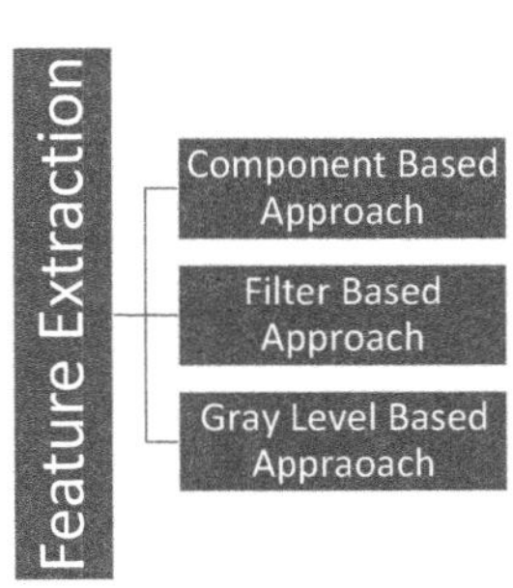

FIGURE 5.8 Categories of feature extraction.

5.2.1.4 Gabor texture function

Gabor function extraction depends on the Gabor filter. These filters are planned with human viewing characteristics in mind and are associated with Gabor wavelets. It has been proven to be most suitable for texture analysis. A linear filter is mainly used for edge detection. Gabor's texture features are a set of wavelets. Later, each wavelet that captures life force emerges with precise shape and precision. Once the energy distribution is collected, texture features can be extracted on behalf of the image.

5.2.1.5 Learning vector quantization

A supervised learning technique as per vector quantization that can be used when the author labels the input data. The reference vectors in the study were restructured to reduce the cost function using steep descent method. For improving the decision region classification quality shift vectors are made to store information about the classes.

5.2.1.6 Symbolic dynamic filtering

Low-dimensional features in time series data are extracted. It includes preprocessing through wavelet transform to suppress noise and to enable time-frequency localization. The existence of wavelet transforms implies modifying additional policy constraints, such as wavelet bases and scale sets. It is also a tool for extracting spatiotemporal features from fixed time series data. The primary work is that it can detect anomalies in complex dynamic systems early.

5.2.2 Principal components analysis

One of the most used methods among the various linear dimensionality reduction techniques. While using principal components analysis (PCA), the real data are considered as the input, and we are trying to form the group of input features which can be used in reducing the original dimensions of the actual data. PCA works on two aspects: (1) variance maximization and (2) reduction of reconstruction error [3]. In PCA, our actual data project into a group of orthogonal axes, displaying the axis rank per their importance. PCA is an unsupervised learning algorithm.

It is a predominant algorithm for feature extraction from the high-dimensional dataset. Using iterative algorithms or Eigenvalue problem helps to overcome the high-dimensional problems faced during feature extraction. An orthogonal transformation is transforming a group of correlated values into uncorrelated. The new values become primary values for execution. There will be increased or decreased number of values after using PCA. PCA can be of (1) nonlinear kernel PCA, (2) probabilistic PCA, (C) higher-order singular value decomposition (SVD), and (D) independent component analysis. Fig. 5.9 represents a snippet of code written in Python to execute this technique.

5.2.3 Independent component analysis

It is a linear dimensionality reduction technique that executes by accepting input from a group of independent components. It aims to recognize the removal unwanted noise exists in the data. Consider two independent input features if their linear and nonlinear dependence equals zero. Apply the independent component analysis (ICA) code as shown in Fig. 5.10.

```
# Applying PCA function on training
# and testing set of X component
from sklearn.decomposition import PCA
pca = PCA(n_components = 2)
X_train = pca.fit_transform(X_train)
X_test = pca.transform(X_test)
explained_variance = pca.explained_variance_ratio_
```

FIGURE 5.9 Code for principal components analysis.

```
# Applying ICA
#import necessary libraries
ICA = FastICA(n_components=2)
ICV = ICA.fit_transform(X)
```

FIGURE 5.10 Code for independent component analysis.

5.2.4 Linear discriminant analysis

It performs a dual role of both dimension reduction and classification. The primary work of it is to increase the distance between the mean of each class and decrease the movement within the class. Linear discriminant analysis (LDA) uses metrics within and between classes. It supports in getting better results by maximization of distance between the class mean and executing the same in a space with reduced dimensions.

5.3 Feature selection

Features play a key role in classification. A medical image, irrespective of its type, comprises multidimensional features, whereas it does not mean that all features are important. Feature selection (FS) is a technique that selects a subset of original features by removing redundant and unimportant features [4]. Feature subset selection (FSS) generates a subset of important features from the dataset. It makes clear that the performance of the classifier depends on feature selection. FSS helps improve the generalization of the ML model once it is included in the preprocessing steps. FSS understands the dataset, the importance of every feature, and the feature correlation and avoids redundant data that, in turn, improves the performance of ML algorithms like [3]:

1. Better performance—accuracy.
2. Feature reduction—resource-saving.
3. Algorithm generalization improved—storage space and execution time.
4. Understanding the data—data visualization.

5.3.1 Characteristics of feature selection algorithms

The algorithms tend to search subsets of features from feature space. Four steps [5] that affect the subset search are:

1. Initial point: Selecting a point to start the search is complicated. In the first case, it is a forward process where the search can begin with zero features and add features. The second case is a backward process of removing features consecutively from the feature space. The third case is to search from the middle of the feature space to create a subset.
2. Search method: The effectiveness of searching depends on the number of features in the feature space. Feature space holding fewer features is comfortable with the normal search method, whereas heuristic search is required with N features. Still, finding an optimal search method is quite difficult.
3. Evaluation: Methods like a filter, embedded, and wrapper are used for evaluating the feature search, which was explained in the forthcoming topics.
4. End point: While searching, the feature selector has to stop the search at a point. This is possible by the evaluation methods. In addition, make sure that the selector is not reaching the opposite end of the feature space.

FS covers two categories, as mentioned in Fig. 5.11: supervised and unsupervised feature selection. Supervised FS employs for the labeled, whereas unsupervised FS for unlabeled data. Supervised feature selection comprises (1) filter methods, (2) embedded methods, and (3) wrapper methods.

Filter methods contain feature ranking and feature subset evaluation. Feature ranking works by assigning rank/weight to each feature according to the significance of the end class. In the other case, feature subset evaluation executes by considering the importance and correlation among the features. This method retrieves the redundant feature and filters it. As per Ref. [3], performance of the classifier increases when redundant features are removed from a

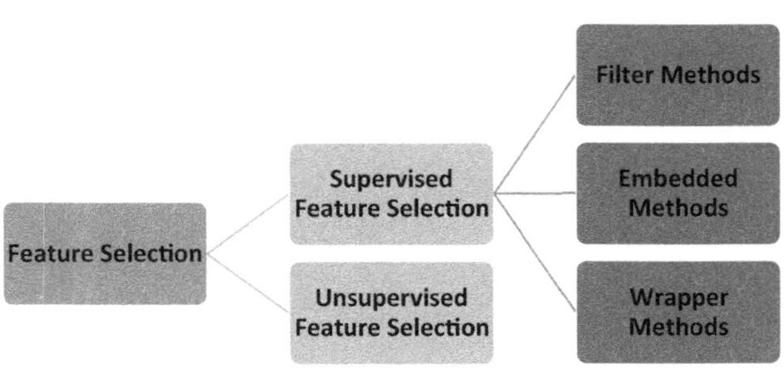

FIGURE 5.11 Feature selection categories.

dataset. Feature subset evaluation helps the researcher to yield a better result [6]. Fig. 5.12 gives the feature selection using the filter method. The first step covers the entire feature space, the second identifies a subset, the third uses a learning algorithm to learn the features, and the fourth filters qualified features.

The wrapper method searches the subset in the space set. Searching requires a set of features (space set), a beginning state, an ending state, and a classifier (search engine). The wrapper method of FS utilizes any of the learning algorithms in defining the quality of the features. Compared to the other methods, time consumption and utilization of computing resources are more for wrapper methods. However, researchers still show interest in using the wrapper method through particular classifiers for learning the features [6].

Fig. 5.13 explains selecting feature subsets using the wrapper method. The input and the output are the same as filter methods. The major part observed here is the iteration between subset selection and the learning algorithm. This increases the time taken to identify the qualified features.

Embedded methods consume the advantages of both filter and wrapper methods. It is an iterative method for retrieving qualified features from the space set. The selection of suitable features is embedded with the learning algorithm, providing a more accurate subset. Decision tree [7] and SVM [8,9] perform well for embedded methods.

As mentioned in Fig. 5.14, embedded methods work in creating qualified features using learning algorithms after receiving the input from the feature space.

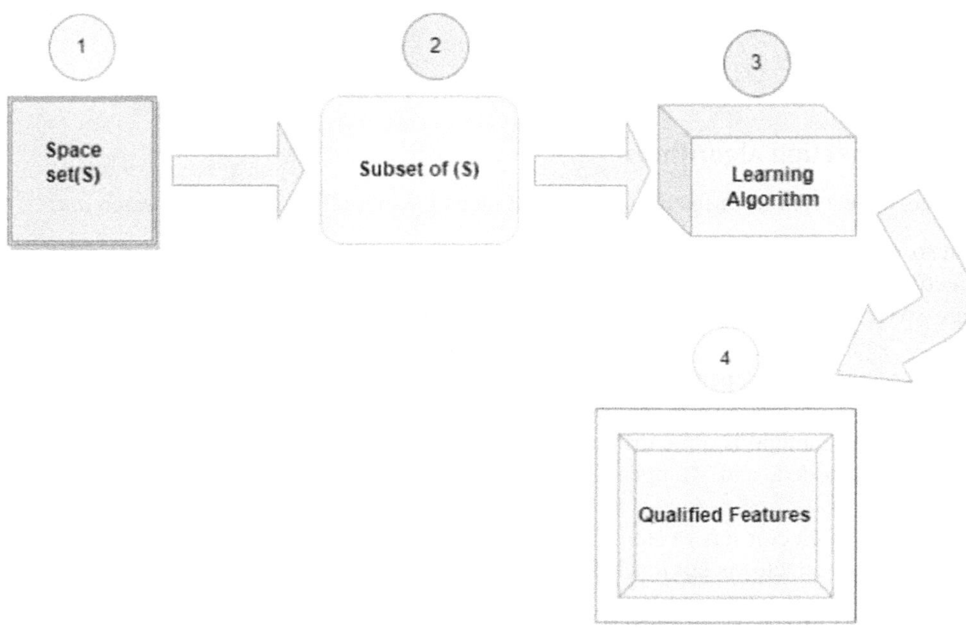

FIGURE 5.12 Feature selection using filter method.

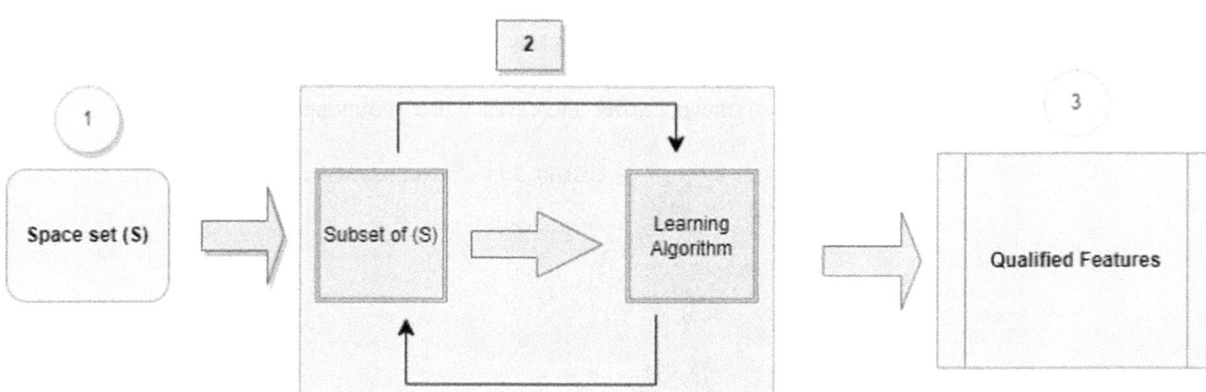

FIGURE 5.13 Feature selection using wrapper method.

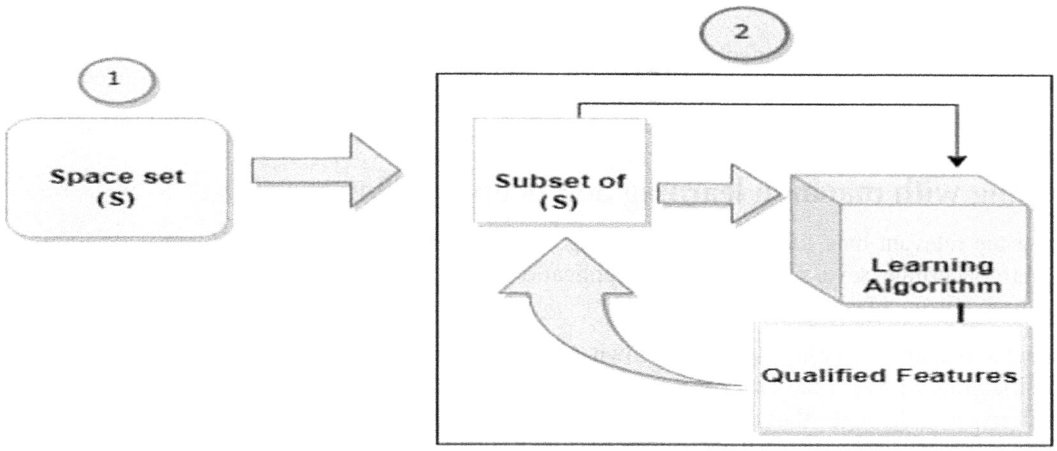

FIGURE 5.14 Feature selection using embedded method.

TABLE 5.1 Filter versus wrapper versus embedded methods.

Parameters	Filter	Wrapper	Embedded
Learning model	General methods	Optimized features from particular ML algorithms	Fixing features during model building
Time complexity	Fast execution	More time consumption for more features	Normal execution
Challenges	Sometimes it fails to give the best features	Suffers from overfitting	Finding a suitable classifier

5.3.2 Key differences between filter, wrapper, and embedded methods

All the feature selection methods mentioned above can generate qualified features. The relationship among these methods is observed in Figs. 5.12–5.14. Table 5.1 delivers the differences among these methods on important parameters to understand this topic better.

5.3.3 Popular feature selection algorithms

This topic will discuss a few popular feature selection algorithms that cover filter, embedded, and wrapper methods [6,10].

1. Information gain (IG): a method widely used to find entropy-based feature selection. It is a univariate filter method. It calculates the relationship/importance of each feature in the class. Finally gives a ranking of features suitable for execution.
2. ReliefF: a multivariate filter advanced algorithm for feature selection also suitable for handling missing data.
3. Recursive feature elimination (RFE): a wrapper-based model. This algorithm uses any of the ML algorithms and is wrapped with RFE. After ranking the features, features with minimum rank are eliminated recursively.
4. LASSO (least absolute shrinkage and selection operator) regression: an embedded model which uses a statistical formula. It gives good results for linear problems.

The above mentioned selection methods helps the researchers bring out quality features suitable for classification. Genetic algorithms (GA) improve feature selection models' effectiveness by optimizing the parameters' value at different levels. It is a mostly used optimization technique. As the chapter deals with processing medical images, producing accurate results weighs more. Forwarding these features as input for the ML and DL models serves the medical field in a better way.

5.3.4 Datasets

Before entering into the execution of medical images using classifiers, we want to share the most common repositories of medical image datasets, as in Table 5.2.

5.4 Working with machine learning and deep learning

After preparing the relevant data for processing, finding a suitable classifier for the problem is desirable. ML and DL ruins the world concerning several classification applications. We have considered a few important algorithms that work effectively for image classification [11].

1. Support vector machines (SVM) is a well-known algorithm for classification and regression problems. Speaking about classification separates the classes with a linear boundary. It completely uses hyperplanes for separation, and the real efficiency of the algorithm depends on the kernel. Linear, Gaussian, and polynomial are the most commonly used kernels of SVM—a snippet of Python code for SVM classification, as mentioned in Fig. 5.15.

   Normally, SVM uses hyperplanes to separate the data linearly, as shown in Fig. 5.16. Observe the binary class label separation and the margin width between the linear lines.
2. Decision tree (DT): A supervised learning classifier works according to the tree data structure. The DT follows the hierarchy-based execution, and the decision taken at each node results in the control flow. Remember these three points while constructing DT: (1) node partition, (2) terminal node finding, and (3) node with a class label. Sample snippet Python code for DT is shown in Fig. 5.17.

   Fig. 5.18 explains the working of a DT. It contains the root node, terminal nodes, and branches with respective values per the cases.
3. K nearest neighbor (KNN): a simple ML algorithm that classifies based on their neighbors. It identifies the distance between the feature vectors, and the nearest distance features are classified. Fig. 5.19 shows the sample snippet of python code for KNN.

   Fig. 5.20 delivers the working of KNN classifier, identification of neighbors as per k values, and class separation.
4. Artificial neural networks (ANN): executes as per the working style of a biological neuron. It contains nodes where weights and activation functions help to find the exact classes. ANN performs well for both binary and multiclassification. Fig. 5.21 shows a sample snippet code for ANN.

TABLE 5.2 List of a few medical image dataset repositories.

S. no.	URL	Remarks
1.	https://www.nibib.nih.gov/medical-imaging-and-data-resource-center	Medical imaging and data resource center
2.	https://www.image-net.org/	Images suitable for advanced computer vision and deep learning research
3.	https://archive.ics.uci.edu/mL/index.php	A huge repository where all types of datasets are available
4.	https://aimi.stanford.edu/medical-imagenet	An image repository maintained by Stanford University
5.	https://www.cancerimagingarchive.net/	Cancer image datasets
6.	https://www.bcsc-research.org/	Breast cancer datasets
7.	https://adni.loni.usc.edu/data-samples/access-data/	Alzheimer neuroimaging dataset

```
svm = SVM()
svm.fit(X_train, y_train)
y_pred = svm.predict(X_test)
confusion_matrix(y_test, y_pred)
```

FIGURE 5.15 Sample support vector machine code.

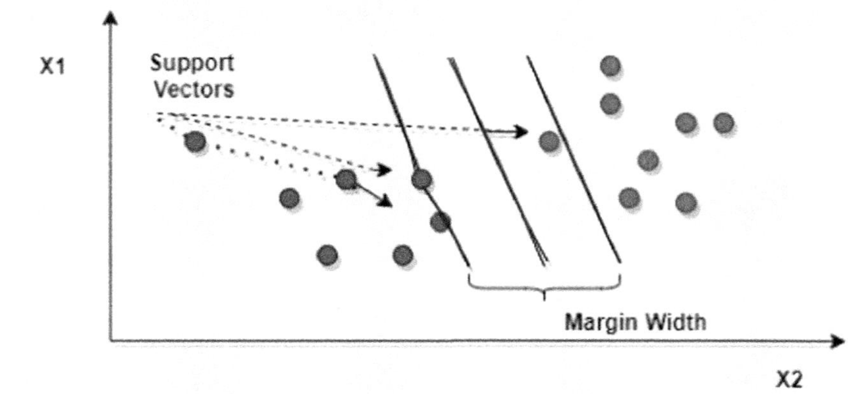

FIGURE 5.16 Support vector machine—sample.

```
dt = DecisionTreeClassifier(criterion = "entropy", random_state = 100,max_depth=3, min_samples_leaf=5)
dt.fit(feat_train,np.argmax(y_train,axis=1))
dt.score(feat_test,np.argmax(y_test,axis=1))
```

FIGURE 5.17 Sample decision tree code.

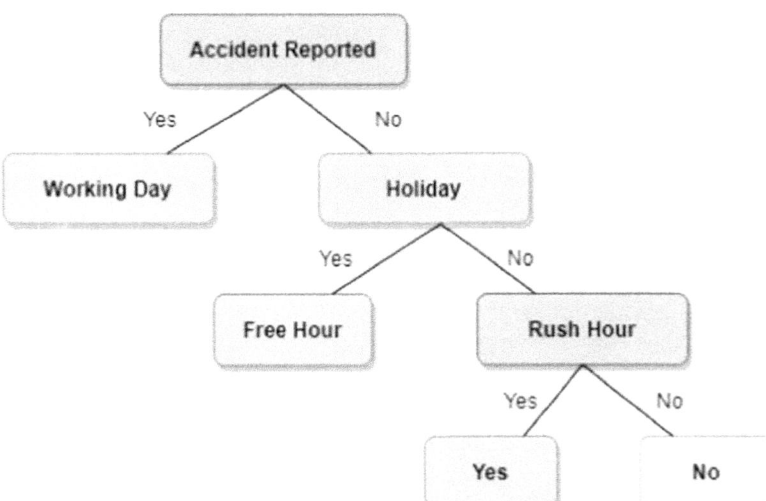

FIGURE 5.18 Sample—decision tree.

```
knn = KNeighborsClassifier(n_neighbors=12)
knn.fit(feat_train,np.argmax(y_train,axis=-1))
knn.score(feat_test,np.argmax(y_test,axis=1))
```

FIGURE 5.19 Sample K nearest neighbor code.

Fig. 5.22 describes the working of ANN. We can observe the input layer, hidden layers, and output layer. The input layer contains rows of input (neurons) with necessary features adding some necessary weights and bias. The hidden layer comprises an activation function to activate the neurons of the input layer as per their importance. It is possible to add some weights to the neurons if necessary. This activation helps the neuron properly classify in the output layer with the support of other suitable activation functions.

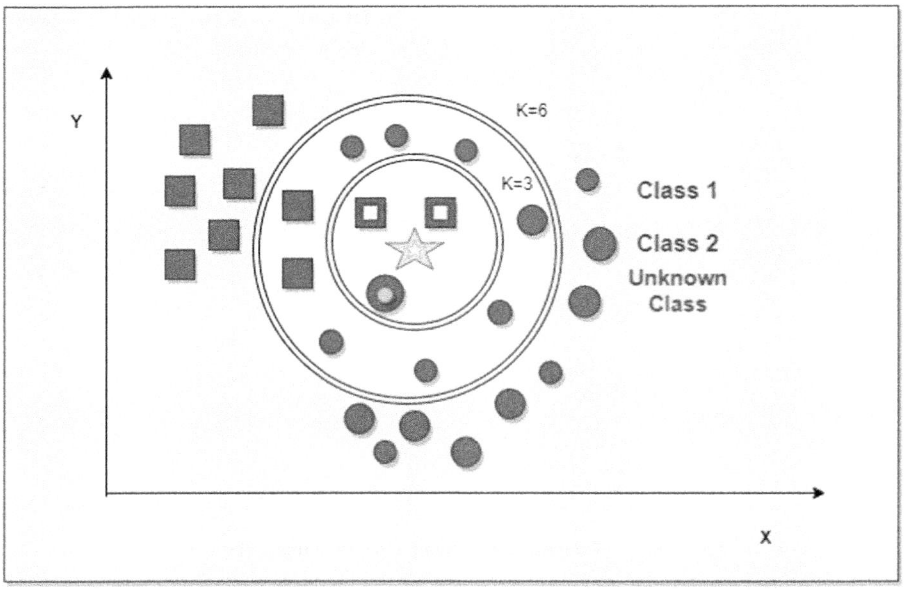

FIGURE 5.20 K nearest neighbor classifier—sample.

```
model_ann = Sequential()
model_ann.add(Dense(16, input_shape=X_train.shape[1:], activation='relu'))
model_ann.add(Dropout(0.4))
model_ann.add(Dense(32, activation='relu'))
model_ann.add(Dropout(0.6))
model_ann.add(Flatten())
model_ann.add(Dense(2, activation='softmax'))
model_ann.compile(loss='binary_crossentropy', optimizer='adam',
metrics=['accuracy'])
```

FIGURE 5.21 Sample artificial neural network code.

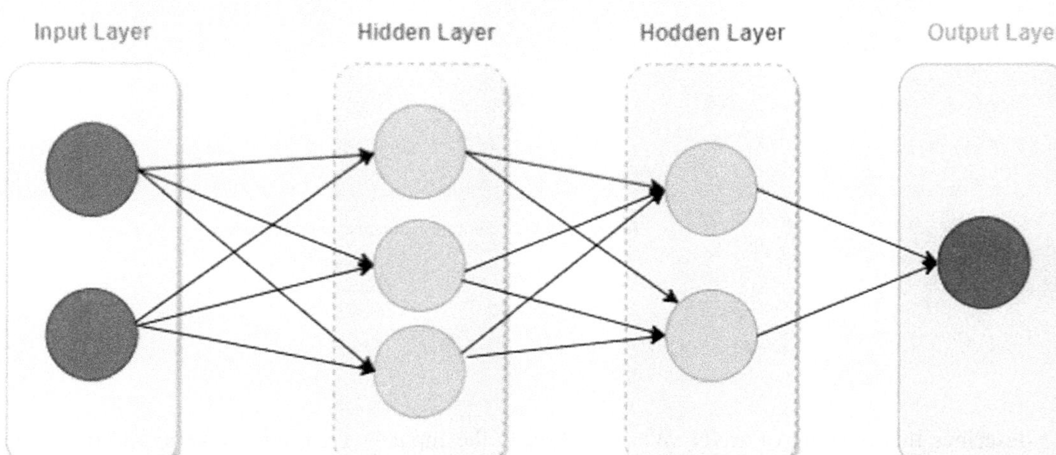

FIGURE 5.22 Sample—artificial neural network.

```
model = Sequential()
model.add(Conv2D(16,(5,5),padding='valid',input_shape = X_train.shape[1:]))
model.add(Activation('relu'))
```

FIGURE 5.23 Sample convolutional neural network code.

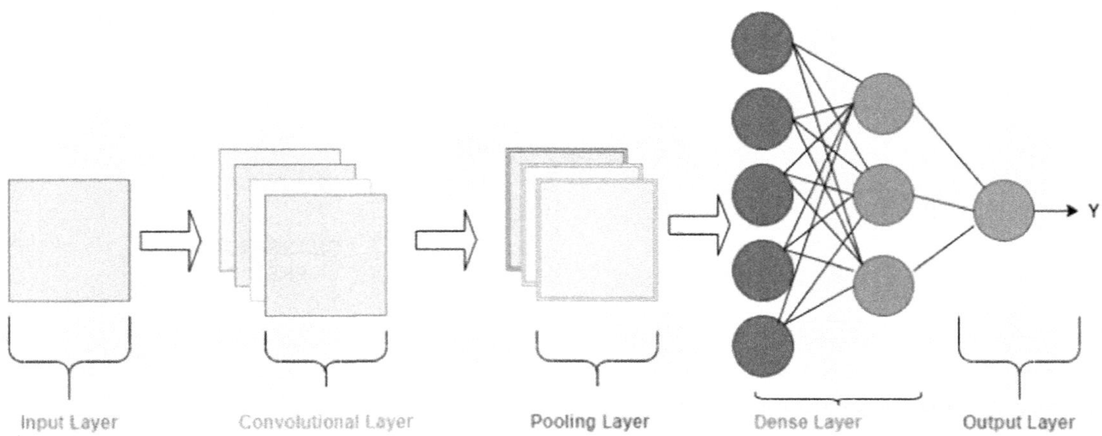

FIGURE 5.24 Sample—convolutional neural network.

5. Convolutional neural network (CNN): A special ANN structure suitable for computer vision activities. It includes convolution, pooling, and fully connected layers. Image filtering occurs in the convolutional layer; dimensionality reduction occurs in the pooling layer and classifies the image. Fig. 5.23 for sample snippet code for ANN.

   Fig. 5.24 explains the CNN architecture. It includes the input dataset (layer), followed by convolutional and pooling meant for feature extraction and dense layer as done in ANN, and finally, the output layer.

So far, in this chapter, we have covered the basics of images, why the medical image is considered, and various image-processing methods. We have also seen the importance of feature extraction and feature selection with their types. We have highlighted the classifiers and their working style with a few sample codes. This could not be the end of the solution. An important task called performance evaluation to measure the performance of the classifiers ends the work. We know the execution of ML/DL is an iterative process; once the evaluation does not meet the expected result, it goes for verification.

Commonly used evaluation metrics are confusion matrix, accuracy, precision, recall, ROC/AUC curves, etc. Only some metrics are required to evaluate the classifier, considering more than one metric improves the classifier's quality.

From Fig. 5.25, one can understand the process of an image. The figure represents the segmentation part, which undergoes feature extraction where dimension reduction takes place. In feature selection, we deal with feature space subset which is optimized through parameters at various levels. Once these images are taken for classification or prediction, good results are observed.

We have gone through various methods for dealing with feature extraction and feature selection of medical images. After a clear analysis we have identified the major advantages, disadvantages, and challenges, as listed in Table 5.3. Each method is having its own pros and cons, a user has to identify a suitable method convenient for their dataset.

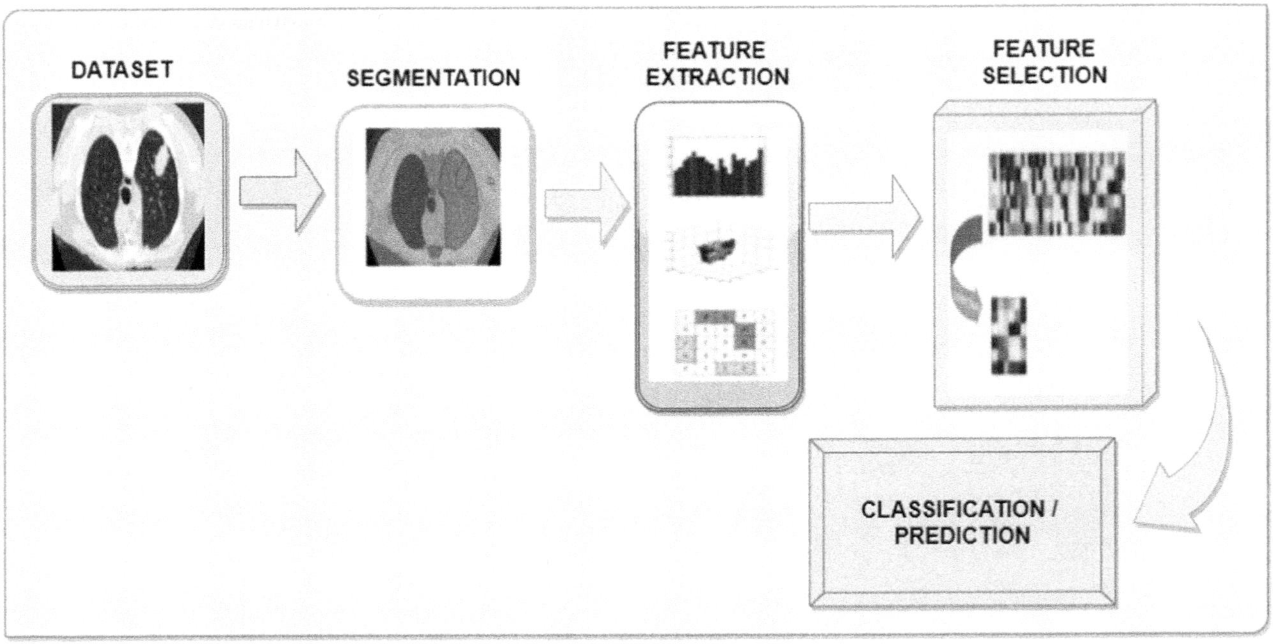

FIGURE 5.25 Demonstration of feature extraction/selection.

TABLE 5.3 Advantages and challenges.

Type	Advantages	Disadvantages	Challenges
Feature extraction	• Improved performance of the model in terms of accuracy • Reduction in model complexity • Reduces overfitting issues	• May suffer from data interpretation • Data transformation is expensive • Chances for noisy and redundant data if noninformative features are extracted	• Extracted features are not human readable • Large datasets are difficult to handle
Feature selection	• Efficient computation • Precision is high • Low cost	• Increases overfitting • Domain knowledge required • Chances for information loss due to selection of wrong features	• Complexity increases as dataset size increases • Identification of a suitable hybrid model

5.5 Conclusion

This chapter provides knowledge on how to work with medical images. As described earlier, how important is the task of accessing medical images and the expectation of the medical field from the technology people? We have given a systematic in dealing with images. Feature extraction division describes the categories and their nature of it. It also reveals with few examples of its working nature. The next part discusses the nature of feature selection and its benefits in image processing. We have seen the different types of feature selection methods. Completing image processing ends in classification; we explored a few important ML and DL algorithms in this regard. Metrics for performance evaluation assess the classifier.

References

[1] G. Kumar, P.K. Bhatia, A detailed review of feature extraction in image processing systems, in: 2014 Fourth International Conference on Advanced Computing & Communication Technologies. Presented at the 2014 Fourth International Conference on Advanced Computing & Communication Technologies (ACCT), IEEE, Rohtak, India, 2014, pp. 5–12. https://doi.org/10.1109/ACCT.2014.74.

[2] N. Noreen, S. Palaniappan, A. Qayyum, I. Ahmad, M. Imran, M. Shoaib, A deep learning model based on concatenation approach for the diagnosis of brain tumor, IEEE Access. 8 (2020) 55135–55144. Available from: https://doi.org/10.1109/ACCESS.2020.2978629.

[3] I. Guyon, A. Elisseeff, An introduction to feature extraction, in: I. Guyon, M. Nikravesh, S. Gunn, L.A. Zadeh (Eds.), Feature Extraction: Foundations and Applications, Studies in Fuzziness and Soft Computing, Springer, Berlin, Heidelberg, 2006, pp. 1–25. Available from: https://doi.org/10.1007/978-3-540-35488-8_1.

[4] S. Jain, A.O. Salau, An image feature selection approach for dimensionality reduction based on kNN and SVM for AkT proteins, Cogent Eng. 6 (2019) 1599537. Available from: https://doi.org/10.1080/23311916.2019.1599537.

[5] S. Lei, A feature selection method based on information gain and genetic algorithm, in: 2012 International Conference on Computer Science and Electronics Engineering. Presented at the 2012 International Conference on Computer Science and Electronics Engineering (ICCSEE), IEEE, Hangzhou, Zhejiang, China, 2012, pp. 355–358. https://doi.org/10.1109/ICCSEE.2012.97.

[6] R. Kohavi, G.H. John, Wrappers for feature subset selection, Artif. Intelligence, Relevance 97 (1997) 273–324. Available from: https://doi.org/10.1016/S0004-3702(97)00043-X.

[7] R.O.D. Duda, Pattern Classification, second ed., Wiley [WWW Document], 2001. URL https://www.wiley.com/en-us/Pattern + Classification, + 2nd + Edition-p-9780471056690 (accessed 3.21.23).

[8] J. Weston, S. Mukherjee, O. Chapelle, M. Pontil, T. Poggio, V. Vapnik, Feature Selection for SVMs. NIPS, 2000.

[9] I. Guyon, J. Weston, S. Barnhill, V. Vapnik, Gene selection for cancer classification using support vector machines, Mach. Learn. 46 (2002) 389–422. Available from: https://doi.org/10.1023/A:1012487302797.

[10] R.J.S. Raj, S.J. Shobana, I.V. Pustokhina, D.A. Pustokhin, D. Gupta, K. Shankar, Optimal feature selection-based medical image classification using deep learning model in internet of medical things, IEEE Access. 8 (2020) 58006–58017. Available from: https://doi.org/10.1109/ACCESS.2020.2981337.

[11] E.M. Senan, M.E. Jadhav, T.H. Rassem, A.S. Aljaloud, B.A. Mohammed, Z.G. Al-Mekhlafi, Early diagnosis of brain tumour MRI images using hybrid techniques between deep and machine learning, Comput. Math. Methods Med. 2022 (2022) 1–17. Available from: https://doi.org/10.1155/2022/8330833.

Chapter 6

Deployment, feature extraction, and selection in computer vision and medical imaging

Pankaj Kumar Jain[1,2], Kalyan Tadepalli[1], Shivansh Kaushik[3], Sonam Jain[4] and Neeraj Sharma[2]

[1]*Artificial Intelligence and Data Science, Jio Institute, Navi Mumbai, Maharashtra, India,* [2]*School of Biomedical Engineering, Indian Institute of Technology (BHU), Varanasi, Uttar Pradesh, India,* [3]*Department of Chemistry, University of California Riverside, Riverside, CA, United States,* [4]*Department of Product Management, Medicare Imaging, Navi Mumbai, Maharashtra, India*

6.1 Introduction

Feature extraction is a critical step in computer vision and medical imaging, as it facilitates the recognition and retrieval of relevant image information. This information can be used to solve particular medical issues, such as object recognition, segmentation, and classification. Subsequently, these solutions lead to the diagnosis, monitoring, and planning of treatment for disease. This chapter discusses the significance of feature extraction methods, feature selection, and feature deployment in computer vision and medical imaging.

There is a large corpus of established strategies that may be used to extract meaningful features from imaging data. Each of these approaches is suited to certain applications and kinds of data. For instance, in the case of brain imaging like fMRI scans, MRI sequences and CT scans, features like the size and shape of a brain tumor, the intensity of gray matter, and the distribution of white matter fiber tracts would typically be extracted [1,2]. Similarly, in pulmonary imaging, the shape and size of the lungs, as well as the intensity and texture of the lung parenchyma and the distribution of air gaps, are extracted [3,4]. In dermatologic imaging, features such as lesion borders, gradient between the lesion color and the surrounding skin, and visual texture of the lesion and color may be extracted [5]. Additionally, in carotid ultrasound imaging, characteristics such as calcification, lipid content, necrotic core, and intraplaque hemorrhage are essential for recognizing atherosclerotic plaque tissue [6,7].

The nature of the image data, the problem that needs to be solved, and the outcome that is desired are all important considerations that play a role in the decision of which feature extraction method to use. Color histograms, gradient orientation histograms (HOG), scale-invariant feature transform (SIFT), speeded-up robust features (SURF), and texture features are some of the most common feature extraction methods used in computer vision. In medical imaging, the extracted features are typically associated with the underlying anatomy or pathology, such as the structure's shape, texture, and intensity.

Several steps precede the deployment of features in computer vision and medical imaging, including feature selection, feature dimensionality reduction, and feature classification. Feature selection entails selecting the most pertinent extracted features to aid in the resolution of a specific problem. The technique of feature dimensionality reduction reduces the number of features while maintaining vital information. This is done to decrease the computational cost and enhance the efficiency of the algorithms. Feature classification is the categorization of objects using machine learning algorithms such as support vector machines (SVMs) and deep neural networks (DNNs). Recently, numerous innovative techniques have been developed to extract the most information possible from extracted features using the fewest resources. A review of deep learning applications in healthcare can be seen here [8].

In the medical imaging decision-making process, the explainability of the model comes last. Explainable models are gaining popularity and significance in crucial applications such as healthcare, where decisions can have significant repercussions.

Intelligent Computing Techniques in Biomedical Imaging. DOI: https://doi.org/10.1016/B978-0-443-15999-2.00009-8

Consequently, an explicable model offers insight into the decision-making procedure. Grad-CAM, Grad-CAM++, and local interpretable model-agnostic explanations (LIME) are well-known techniques for displaying the insights of DL models.

The second section describes feature extraction techniques and their applications in various medical images. Different feature selection strategies, dimensionality reduction, and their deployment in machine learning and deep learning applications are discussed in Section 6.3. The fourth section describes how to use features for an explainable model.

6.2 Feature extraction methods

In the field of medical imaging, the utilization of features is essential for the resolution of a wide variety of clinical issues, including diagnosis, the course of disease, and the planning of treatment. In the field of medical image analysis, for instance, properties such as shape, texture, and intensity can be retrieved from images and utilized for the purpose of categorizing various tissues and structures. In addition, characteristics can be used to track the changes in disease development over time, which can contribute to a better understanding of the underlying pathology and help make treatment decisions that are more informed.

6.2.1 Statistical, texture, and histogram features

6.2.1.1 Color histograms

One of the most popular feature extraction techniques in computer vision uses color histograms. The distribution of colors (RGB color space) in an image is represented by color histograms, which are straightforward but effective features. The color histograms are plotted using by calculating the number of pixels in each color space. They are especially helpful for issues like object segmentation and recognition that call for color-based image retrieval.

Consider an image I(M × N), with RGB color space and pixel coordinates (x,y) is represented by

$$I(x, y) = [R(x, y), G(x, y), B(x, y)]$$

Now the R, G, and B pixels are counted by, $r = R(x,y)/256 \times K$; $g = G(x,y)/256 \times K$; $b = B(x,y)/256 \times K$. Thus the color histogram is represented by H(r,g,b). The normalized histogram is represented by Eq. (6.1).

$$H_{normalized} = H(r, g, b)/(M \times N) \tag{6.1}$$

6.2.1.2 Histogram of oriented gradients features

In computer vision, HOG are another popular technique for feature extraction. They are frequently used for object detection and tracking and capture the gradient information of an image. The feature extraction follows major steps such as gradient computation, cell division, orientation calculation, and block normalization. HOG features are particularly useful for images with complex textures and shapes, as they are robust to changes in lighting conditions and rotation.

6.2.1.3 Scale-invariant feature transform and speeded-up robust features

The feature extraction technique known as SIFT is one of the most reliable feature extraction techniques in computer vision and is frequently used for object recognition and image matching. This method is invariant to changes in scale, rotation, and illumination in an image. It detects distinctive local features or the keypoints from images, which are uniform to geometric and photometric transformations. The method deploys scale-space extrema detection using difference of Gaussian pyramid method for local features extraction. Further, locations of these local features are determined. Now local features are assigned an orientation based on the dominant gradient direction and finally, a descriptor is calculated for each local features which carries local gradients such that it is invariant to scale, rotation, and affine transformations.

Another feature extraction technique, similar to SIFT but computationally faster, is called SURF. SURF is frequently used in real-time applications like video surveillance and autonomous navigation and is commonly used for object recognition and image matching. Unlike SIFT, SURF uses a fast Hessian matrix for local features extraction at multiple scales and locations of an image. These local features are further assigned a dominant orientation based on the Haar wavelet response in their neighborhood. At last, box filters and integral images are applied to for fast computation of feature vectors. An application of these feature descriptors can be seen here [9] for bone age assessment on radiograph images, when used along with SVM classification. These local features or the keypoints are given below.

6.2.1.4 Statistical method of texture feature extraction

An image's texture features can be extracted statistically using the gray level cooccurrence matrix (GLCM) matrix. The joint occurrence of the gray level values at a specific pixel distance and direction is described. Numerous texture features, including energy, entropy, contrast, correlation, mean, variance, and homogeneity, can be calculated using the GLCM matrix.

The GLCM matrix $C(x, y)$ can be calculated from an image I by using the following Eq. (6.2):

$$C(x,y) = \sum_{i=1}^{n}\sum_{j=1}^{y}\begin{cases} 1 if \quad I \quad (i,j) = x and I(i+\Delta i, j+\Delta j) = y \\ 0 Otherwise \end{cases} \tag{6.2}$$

where, x and y are the pixel values, i and j are spatial positions in the image plane, Δx and Δy are offset for matrix.

Similar to the GLCM, GLRLM method analyzes the length and orientations of consecutive runs if the pixels are of the same intensity. In other words, a run is a sequence of consecutive pixels of same gray-level along a specific direction in an image. Thus GLRLM matrix is constructed by counting the runs of each length and orientation in an image. Various texture features calculated from GLRLM are short run emphasis (SRE), long run emphasis (LRE), run length nonuniformity (RLNU), and run entropy (RE). These features contain the texture properties related to the lengths and run distributions in an image.

These features are helpful in the medical imaging applications where texture play a vital role in segmentation and classification. A successful demonstration of atherosclerotic plaque tissue characterization for development of sRAS (stroke risk assessment system) can be seen here [7].

The restricted availability of labeled data is one of the obstacles that must be overcome during the process of feature extraction and deployment. The challenging and time-consuming process of generating annotated data is required in various medical imaging applications, which can hinder the effectiveness of machine learning algorithms. Unsupervised learning strategies like K-means, hierarchical clustering, and dimensionality reduction are examples of some approaches that have been developed to address this difficulty.

Feature extraction is paramount in computer vision, especially when applied to medical imaging. Features can solve an impressive array of problems by identifying and extracting relevant information from data. But the two stumbling blocks continue to be the lack of labeled data and the high dimensionality of the feature space. Addressing these two is an ongoing area of research and is necessary to make feature extraction and deployment more effective. Computer vision is making tremendous strides, and it is highly likely that new approaches that are more successful for feature extraction will continue to develop.

6.2.2 Deep learning-based feature extraction

6.2.2.1 Convolutional-based features

Convolutional neural network (CNNs) are becoming increasingly popular as an image classification and feature extraction architecture for deep learning. They are put to use in the process of extracting hierarchical characteristics from pictures, which are then put to use in a variety of computer vision applications. Regarding DNNs, features normally are not hand-engineered by a human but are learned from the data during training. The input layer, the hidden layer, the activation function, the convolutional layers, the pooling layers, and the fully connected layers are all involved in feature extraction in DNNs. After being processed by the input layer, the data are next handled by one or more hidden layers, turning it into higher-level representations or features. This occurs after the data have passed through the input layer. These hidden layers use activation functions like ReLU (rectified linear unit) or sigmoid to inject nonlinearity into the model and help it learn complicated correlations between the input data and output labels. ReLU stands for "rectified linear unit." In CNNs, also known as ConvNets, the process of extracting local characteristics from picture data is handled by convolutional layers. These layers process the incoming data through a series of filters in order to identify the existence of particular patterns or features. The size of a feature map after the application of convolutional layer is determined by Eq. (6.3):

$$S_{FM} = (W - k - 2P)/S - 1 \tag{6.3}$$

where W = input feature k = filter size, P = zero padding, and S = stride. Pooling layers are used to reduce the spatial size of the data and control overfitting by aggregating the activations of nearby neurons. Based on selecting the values from the pooling filter, the mechanism is characterized as Maxpooling (Eq. 6.4), average pooling (Eq. 6.5), and hybrid pooling (Eq. 6.6).

$$x_{out} = \arg\ \max\left(x_{ij}\right) for\ all(i,j) \in R \tag{6.4}$$

$$x_{out} = \frac{1}{N} \sum_{(i,j)\in R} x_{ij} \tag{6.5}$$

$$x_{out} = \alpha(x_{out})_{maxpool} + (1 - \alpha)(x_{out})_{average} \tag{6.6}$$

The final layer of the network is typically a fully connected layer, where the output of the hidden layers is combined and used to make a prediction or classification. By modifying and refining the raw inputs through numerous hidden layers and activation functions, DNNs automatically extract features from the data. A loss function, which gages the discrepancy between the model's predictions and the actual data labels, is optimized to direct this process. The network changes its weights and biases throughout the training time to reduce the loss and improve its predictions. The next section explains the major architectures and methods which deploy these convolution layers to extract the features.

6.2.2.2 *Major deep learning network architecture*

Experimenting with various network topologies can assist in determining which architecture is most suited to a specific task, which can lead to enhanced feature extraction. For instance, employing deeper networks that contain more layers can assist in extracting more complicated characteristics, while wider networks that hold more neurons in each layer can assist in capturing more detail in the features. Researchers have built a number of different network designs for a variety of different medical imaging applications. The architectures that are discussed here are limited because of the applications. Recurrent neural networks, often known as RNNs, are specific kinds of neural networks intended to handle sequential input, like sequences of images. By processing image sequences over the course of a disease's progression, it is possible to extract features from medical images using these tools and use them for feature extraction.

6.2.2.3 *Unsupervised feature extraction*

Generative adversarial networks, often known as GANs, are a type of architecture for deep learning that can be utilized for unsupervised feature learning. A generator and a discriminator network make up their structure, respectively. The generator creates synthetic images, while the discriminator is in charge of determining whether or not the images are genuine. A detailed review of various GAN architectures and their applications in medical imaging is done by Ref. [10]. Autoencoders are a special kind of neural network that may be utilized for both the extraction of features and the reduction of dimensionality. They are made up of an encoder and a decoder, both of which learn a condensed data representation by collaborating. Modern UNet architecture is the enhanced version of the autoencoder.

6.2.2.4 *Supervised feature extraction*

The UNet architectures can be broken down into their parts, which are autoencoders and skip connections. In recent years, hybrid designs have become increasingly relevant in the segmentation and classification of medical images [11,12]. These hybrid models include changes in their parent architecture through novel feature extraction techniques such as attention module [13], multiscale feature extraction [14], fire module, fractal module, and inception module [11], or sometimes by mixing two or more models. These hybrid approaches have the potential to improve specific characteristics. These modules are not exhaustive; many are available but not discussed here due to space constraints. Few special feature extraction modules are shown in Fig. 6.1. These modules are special arrangements of convolutional and activation layers to extract vital features from images.

6.3 Feature selection, normalization, and dimensionality reduction

The high dimensionality of the feature space presents a significant impediment for the process of extracting and deploying features. Because the feature space is high-dimensional in so many applications for medical imaging, it can be challenging to view and comprehend the data in these applications. In order to circumvent this obstacle, feature dimensionalities reduction methods such as principal component analysis (PCA) and linear discriminant analysis (LDA)

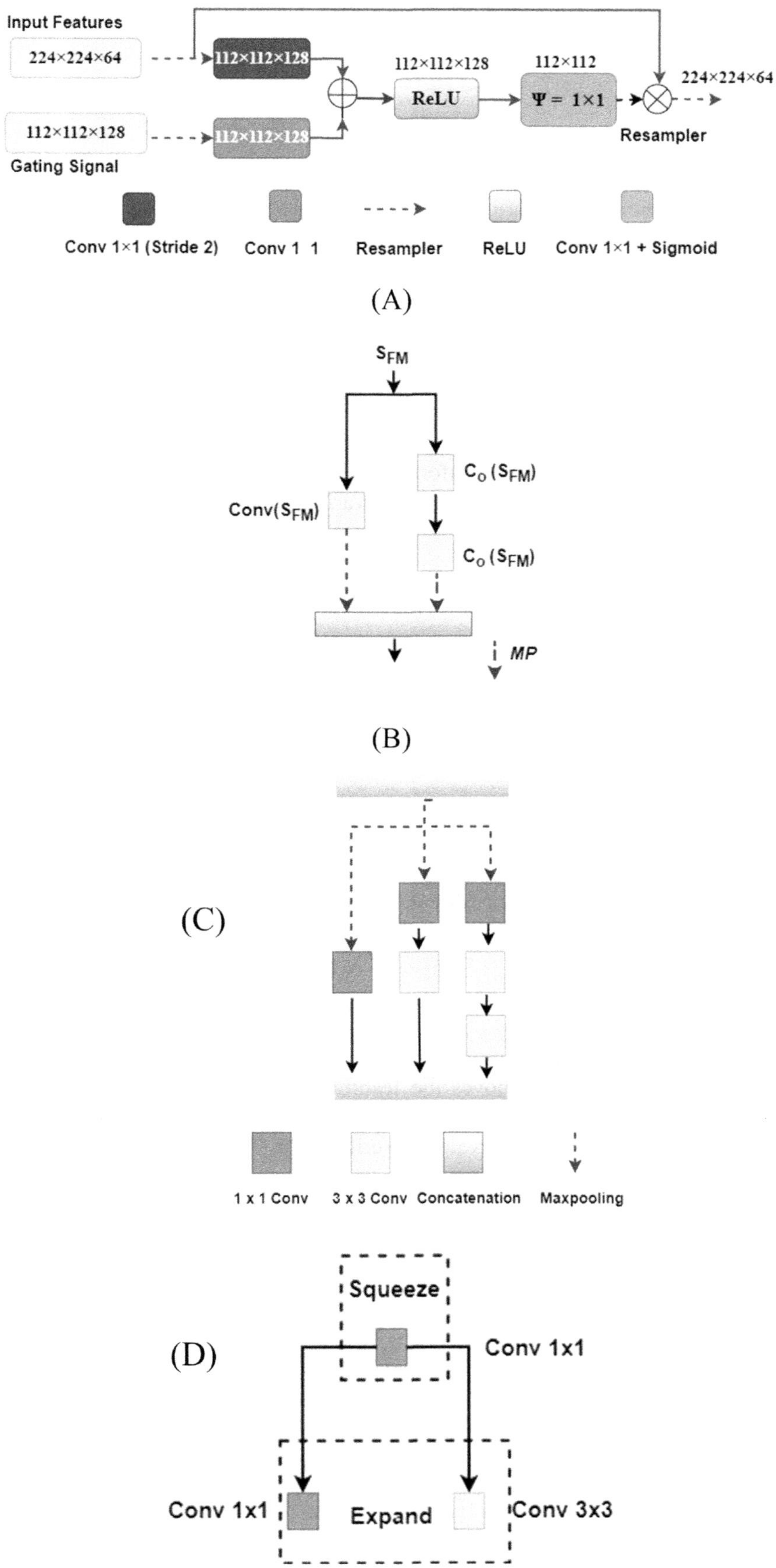

FIGURE 6.1 Spatial feature extraction modules: (A) attention module [13], (B) fractal module [11], (C) inception module, (D) fire module [11].

have been developed. These methods are designed to decrease the number of dimensions comprising the feature space while keeping the critical information intact.

6.3.1 Features selection

In machine learning, the process of selecting a subset of the most relevant characteristics from the complete collection of available features to utilize in creating a model is known as feature selection. The prime objective of using this procedure is to reduce instances of overfitting while improving the model's performance. By selecting a subset of features, the danger of overfitting can be decreased. Overfitting happens when a model is very sophisticated and fits the training data too well, which results in poor generalization performance when applied to new data. Fig. 6.2 describes the training graphs for overfit, underfit, and optimal fit models.

When there is a need to illustrate how the model arrived at its predictions, decreasing the number of features can be beneficial because it makes the model easier to read and understand. Furthermore, training a model that has a relatively large array of features can be a computationally complex process. The training time can be cut down significantly and the model can become more scalable if only a subset of the features is used. In addition, feature selection can improve the performance of a machine learning model by selecting only the most informative characteristics and removing those features that either generate noise or do not contribute to the prediction. There are various approaches to taking care of feature selection, including, but not limited to the following.

6.3.1.1 Filter methods

Filter methods organize features into a ranking using specific statistical criteria, such as correlation or information gain, and then choose the features with the highest ranking. This technique assesses each characteristic independent of the model itself; the features with the highest scores are retained, while those with the lowest scores are removed.

6.3.1.2 Wrapper methods

In contrast to the filter-based strategy, wrapper methods consider the feature subset by training the model and then evaluating how well it does its job. During this procedure, various potential permutations of the features are tried out, and the feature subset that delivers the best overall performance is chosen. Although this approach requires more time and effort to compute, it may produce superior results.

6.3.1.3 Recursive feature elimination

Recursive feature elimination (RFE) is a feature selection technique used to select most important features in machine learning. Similar to wrapper method, this technique recursively removes the features from the dataset and train the model on remaining features. The performance of model is evaluated after each iteration, and features with lowest importance is removed. Wrapper method uses model's performance as evaluation criteria for feature selection; however, RFE selects the features based on their score or ranking.

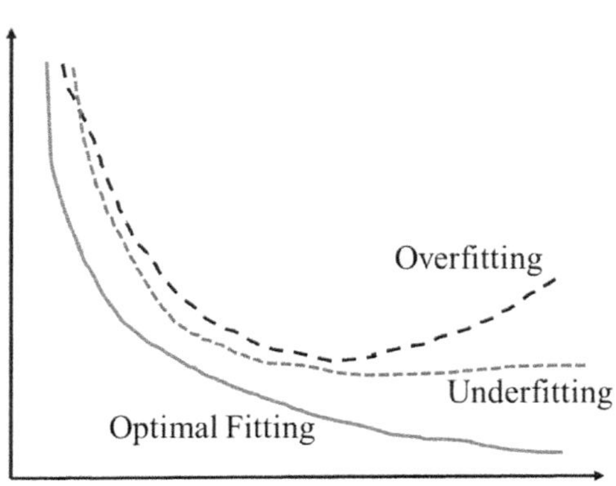

FIGURE 6.2 Overfitting, underfitting, and optimal fitting image.

6.3.1.4 Embedded methods

Embedded methods are a combination of filter and wrapper methods. Embedded methods, such as Lasso regression, perform feature selection as part of the model training process. A review of various feature selection algorithms for breast cancer ultrasound images is done in Ref. [15].

It is important to observe here that feature selection is not always mandatory. Its necessity is contingent upon the specific contours of the problem being attempted, as well as the profile of the data being utilized. In some scenarios, integrating all the features can yield better outcomes, whereas feature selection might be a better choice in others.

6.3.2 Dimensionality reduction

It could be argued that dimensionality reduction and feature selection are similar in places in their intent. Choosing a subset of original characteristics to model is feature selection. Feature selection selects helpful and informative features while discarding noisy or irrelevant ones. Dimensionality reduction reduces the features by translating data into a lower-dimensional representation. Projecting the data into a new space that captures the most relevant information while lowering dimensions is customary. Dimensionality reduction techniques include PCA, singular value decomposition (SVD), and LDA. Recently, Refs. [16,17] showed the applications of SVD and LDA for determining the few shot learning inference in medical data. Previously, dimensionality reduction is performed using PCA method on the 16 extracted feature (GLCM, GLRLM, chaotic features) from common carotid artery (CCA) ultrasound image database [18].

Feature selection selects a subset of features, whereas dimensionality reduction reduces data dimensions. Both methods may increase machine learning model performance, although they are utilized for distinct goals and situations. Feature selection is used to choose the most informative characteristics among several. Dimensionality reduction is used to simplify high-dimensional data and save the computational cost of building a model.

6.3.3 Feature normalization

Feature normalization or scaling refers to transforming a feature's values to a specific range, usually between 0 and 1. The purpose of normalization is to eliminate any scale inconsistencies that may emerge across features and to ensure that all characteristics are measured on a comparable scale. As certain algorithms for machine learning are sensitive to the size of the features being learned, normalization may be used as a tool to help enhance the performance of these algorithms.

6.3.3.1 Min–max normalization

This method transforms the values of a feature to a specific range, usually between 0 and 1. The transformed feature value is calculated by subtracting the minimum value from the original value and then dividing it by the difference between the maximum and minimum values of the feature as shown in Eq. (6.7).

$$x_{norm} = \frac{x - Min(x)}{Max(x) - Min(x)} \tag{6.7}$$

6.3.3.2 Mean normalization

Mean normalization is another method of normalization shown by Eq. (6.8) where $\overline{x} =$ mean(x) value is subtracted instead of Min(x), from the original value x in Eq. (6.7).

$$x_{norm} = \frac{x - \overline{x}}{Max(x) - Min(x)} \tag{6.8}$$

6.3.3.3 Z-score normalization

This method standardizes the values of a feature by subtracting the mean of the feature from each value and then dividing by the standard deviation of the feature. This standardization method results in a new feature with a mean of 0 and a standard deviation of 1. It is represented mathematically as Eq. (6.9):

$$x_{norm} = \frac{x - \overline{x}}{\sigma} \tag{6.9}$$

6.3.3.4 *Decimal scaling normalization*

This method scales the values of a feature by dividing each value by a power of 10, such as 10, 100, or 1000, depending on the magnitude of the values.

6.3.3.5 *Log normalization*

This method applies the natural logarithm function to the values of a feature. Log normalization can help to normalize skewed data and improve the performance of machine learning algorithms.

Above-mentioned techniques are used in Ref. [19] to evaluate the classifier performance when used for breast tumor classification in ultrasound images. It is essential to remember that normalizing the data is not always required since this is data and problem-statement specific. In some circumstances, normalizing the data may assist in increasing the performance of machine learning algorithms; however, in other circumstances, normalizing the data may have little effect or may even have a negative impact on performance.

6.3.4 Deep learning-based feature selection and enhancement

6.3.4.1 *Data augmentation*

Data augmentation is a technique that involves transforming the input information into various forms to augment the quantum of the training data and facilitate the network in learning more robust features. Among others, this may include flipping, scaling, rotation, and so forth.

6.3.4.2 *Transfer learning*

Transfer learning is a method in which a pretrained DNN is used as an initial point for a new task. This is especially useful for feature extraction since the pretrained network has already learned to extract features from a previous dataset and can now be fine-tuned for the new task.

6.3.4.3 *Regularization*

Regularization is a technique used to prevent overfitting, which happens when a model becomes too closely fitted to the training data and fails to generalize well to new data. Regularizing the model and enhancing feature extraction can be done by techniques like dropout and weight decay.

6.3.4.4 *Layer-wise pretraining*

Layer-wise pretraining is a technique in which each network layer is trained one at a time, starting with the lowest-level features and advancing to the highest-level features. This can aid the network in learning more robust features that are transferable to new tasks.

6.4 Features deployment

In machine learning, feature deployment is the process of utilizing the features of a model for specific purposes to harness it for real-world applications. Broadly, the goals of feature deployment cluster around generating model parameters, maximizing performance while minimizing resource utilization, and discovering hitherto novel objects. We discuss some strategies for optimizing the performance of a deep learning model.

6.4.1 Transfer learning

Transfer learning is a deep learning technique that employs previously trained models to carry out new tasks with limited labeled data. It allows for the transfer of learned features from one task to another by training models on large datasets, thereby reducing the training data needed for new tasks. Fine-tuning is a popular transfer learning technique that modifies the weights of a previously trained model to adapt to new tasks. Another transfer learning strategy involves freezing the lower layers of the pretrained model and training only the higher layers with new data. Authors here [20] used 10 transfer learning models for atherosclerotic plaque tissue classification.

6.4.2 Reinforcement learning

Reinforcement learning is a deep learning technique that teaches agents to maximize reward signals in a given environment. Reinforcement learning problems can be modeled as Markov decision processes, in which an agent interacts with its environment by performing actions and receiving rewards. Learning a strategy that maximizes the anticipated cumulative reward over time is the objective of reinforcement learning. Q-learning, policy gradient methods, and actor-critic methods are common reinforcement learning algorithms. This method has found its application here [21] in analyzing the social anxiety by facial expression.

6.4.2.1 Graph convolutional networks

Graph convolutional network (GCN) is a type of deep learning model that can operate on graph-structured data. GCNs extend CNNs to non-Euclidean domains such as social networks, citation networks, and protein interaction networks. GCNs aggregate information from a node's neighbor using a graph convolution operation, which is a learnable function that computes a weighted sum of neighbor node features. GCNs have been successfully applied to various tasks, such as node classification, link prediction, and graph classification. A combination of CNN and GCN is recently used by Ref. [22] for breast histopathological image classification.

6.4.3 Capsule networks

These are a type of neural network architecture modeling spatial relationships between objects. Capsule networks represent object properties such as pose, deformation, and viewpoint using "capsules." Lower-level features activate capsules, which then learn to anticipate higher-level features. Compared to traditional CNNs, capsule networks can better model viewpoint invariance and disentangle object attributes.

6.4.4 Attention mechanisms

Attention mechanisms are deep learning mechanism that focuses on particular portions of an input sequence, such as a sentence or an image. Different parts of the input sequence are given weights by attention mechanisms, allowing the model to focus on pertinent data. Attention mechanisms have been successfully applied to a variety of tasks, such as machine translation, image captioning, and question-answering. Attention mechanism was deployed in conjunction with UNet model by Ref. [13] for atherosclerotic plaque segmentation in ultrasound images and by Ref. [23] for various medical images.

6.4.5 Adversarial training

The models are trained to be resistant to adversarial attacks using this technique. Adversarial attacks are malicious inputs intended to deceive models into producing inaccurate predictions. It involves the addition of small perturbations to training data in order to produce adversarial examples. Models can learn to be more resilient against these attacks by being trained on adversarial examples.

6.4.6 Federated learning

This technique enables the training of models across distributed devices without the exchange of raw data. Federated learning involves training models on local data on individual devices, aggregating model updates on a central server and distributing updated models back to individual devices. Federated learning enables privacy-preserving machine learning, in which users can train models on their local data without disclosing it to third parties.

6.4.7 Zero-shot learning

It allows models to generalize to new classes without labeled data. In order to learn to recognize new classes, zero-shot learning employs additional semantic information, such as class attributes or text descriptions. Image classification, natural language processing, and speech recognition are just a few of the tasks that zero-shot learning has successfully applied to.

6.5 Model explainability

Explainable deep learning, also abbreviated as XDL, is the method for constructing deep learning models that are able to provide explanations that are intelligible to humans for the predictions and choices they choose to make. To phrase it another way, the objective of XDL is to provide a simpler way to understand and comprehend the findings of deep learning models. There are various ways to accomplish XDL.

6.5.1 Model-agnostic explanations

These methods offer global explanations for all machine learning models, including deep learning models, which they can apply to. For instance, LIME generates an explanation for a prediction by first training a simple linear model on a perturbed version of the input data.

6.5.2 Network dissection

This approach sheds light on the activations and representations that are picked up by individual neurons while participating in the deep learning model. It determines the extent to which individual neurons respond to particular semantic elements, such as the components of an object or its surface texture.

6.5.3 Prototype explanation

These methods search through the training data to find prototype samples that are analogous to the input, then use those samples to generate explanations. For instance, the P-DEX (prototype deep explanations) algorithm takes advantage of KNN to locate the prototypes that are the most similar to the target. Then it incorporates that information regarding their classes to explain the prediction.

6.5.4 Gradient-weighted class activation mapping

Grad-CAM and Grad-CAM++ are methods for generating heatmaps that highlight the regions of an input image most important for a prediction made by a DNN. Heatmaps are used to highlight the regions of an image that are most important for a prediction made by a DNN. While the basic philosophy is similar, Grad-CAM suffers from the severe limitation of using activations from only a single layer to generate the heatmap. In contrast, Grad-CAM++ utilizes a weighted combination of activations from multiple layers to generate its heatmap. While this ensures that Grad-CAM++ can capture more fine-grained information about the features used by the network to make its predictions, it also means that the incorporation of GradCAM++ is computationally more intensive. However, in areas like medical imaging, where explainability occupies a higher niche in order of priorities, Grad-CAM++ would be a first choice. Authors here [24] used Grad-CAM method to demonstrate the segmented infected lesions on Covid chest radiographs. Fig. 6.3 shows few medical imaging modalities and disease marked by Grad-CAM. Another recent article showed atherosclerotic plaque classification and the hyperactive plaque regions using various Grad-CAM based DL models [25].

6.6 Challenges and limitations

So far, this chapter discussed feature extraction and selection techniques in medical images. Feature extraction techniques involve statistical, texture, and histogram features for machine learning applications and convolutional features for deep learning applications. One of the significant challenges in medical image feature extraction and selection is the complex and heterogeneous nature of medical data. Medical images are acquired from different sources, equipment, ethnic groups, and specifically diagnosed patients. These variations make the feature extraction process a complex task. Further, noise and artifacts in images are a major setback to the meaningful feature extraction process. Additionally, manual feature extraction is tedious and time consuming process which, however, overcome by convolutional feature extraction. Convolutional feature extraction is highly dependent on high-performance computing resources, which is a major limitation of the deep learning-based systems. Another major shortcoming of convolutional features is lack of transparency in deep learning applications due to its black box nature. Explainable methods resolve this interpretability issue. Overall, feature extraction process is subjective to the medical data and its conditions which must be generalized for clinical practice.

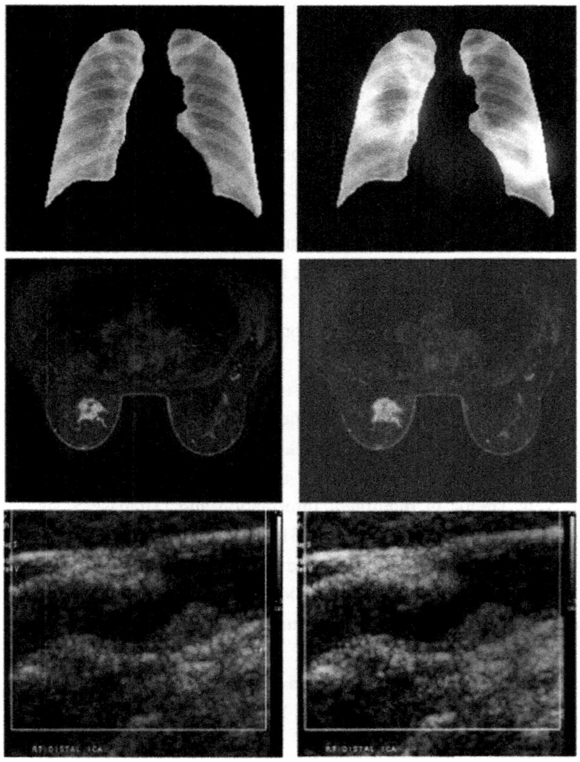

FIGURE 6.3 Heatmap images of segmented medical images (chest X-ray, breast MRI, and carotid ultrasound).

6.7 Conclusion

Feature extraction plays a critical role in computer vision and medical imaging by identifying and extracting relevant information from images. The selection and deployment of feature extraction algorithms depend on factors such as the nature of the image data, the problem being addressed, and the desired outcome. Once the features have been extracted, they are utilized to develop solutions to specific challenges encountered in computer vision and medical imaging. The deployment process typically involves feature selection, feature dimensionality reduction, and feature categorization, all of which are essential for producing accurate results. In addition, explainable models are crucial for improving the transparency and accountability of machine learning models. This is particularly important in areas such as medical diagnosis, financial services, and autonomous systems, where the consequences of inaccurate prediction can be dire, and the penalties thereof, severe. Therefore it is essential to develop reliable and transparent models, allowing for increased trust in the predictions they produce.

References

[1] S. Roy, D. Bhattacharyya, S.K. Bandyopadhyay, T.-H. Kim, An iterative implementation of level set for precise segmentation of brain tissues and abnormality detection from MR images, IETE J. Res. 63 (6) (2017) 769–783. Available from: https://doi.org/10.1080/03772063.2017.1331757.

[2] S. Roy, K.I. Shoghi, Computer-aided tumor segmentation from T2-weighted MR images of patient-derived tumor xenografts. in Lecture Notes in Computer Sciencein: F. Karray, A. Campilho, A. Yu (Eds.), Image Analysis and Recognition, 11663, Springer International Publishing, Cham, 2019, pp. 159–171. Available from: http://doi.org/10.1007/978-3-030-27272-2_14.

[3] N.M. Noor, et al., Automatic lung segmentation using control feedback system: morphology and texture paradigm, J. Med. Syst. 39 (3) (2015) 22. Available from: https://doi.org/10.1007/s10916-015-0214-6.

[4] J.C.M. Than, et al., Lung disease stratification using amalgamation of Riesz and Gabor transforms in machine learning framework, Comput. Biol. Med. 89 (2017) 197–211. Available from: https://doi.org/10.1016/j.compbiomed.2017.08.014.

[5] V.K. Shrivastava, N.D. Londhe, R.S. Sonawane, J.S. Suri, Computer-aided diagnosis of psoriasis skin images with HOS, texture and color features: a first comparative study of its kind, Comput. Methods Prog. Biomed. 126 (2016) 98–109. Available from: https://doi.org/10.1016/j.cmpb.2015.11.013.

[6] K. Lekadir, et al., A convolutional neural network for automatic characterization of plaque composition in carotid ultrasound, IEEE J. Biomed. Health Inform. 21 (1) (2017) 48–55. Available from: https://doi.org/10.1109/JBHI.2016.2631401.

[7] T. Araki, et al., Stroke risk stratification and its validation using ultrasonic echolucent carotid wall plaque morphology: a machine learning paradigm, Comput. Biol. Med. 80 (2017) 77–96.

[8] S. Roy, T. Meena, S.-J. Lim, Demystifying supervised learning in healthcare 4.0: a new reality of transforming diagnostic medicine, Diagnostics 12 (10) (2022) 2549. Available from: https://doi.org/10.3390/diagnostics12102549.

[9] M. Kashif, T.M. Deserno, D. Haak, S. Jonas, Feature description with SIFT, SURF, BRIEF, BRISK, or FREAK? A general question answered for bone age assessment, Comput. Biol. Med. 68 (2016) 67–75. Available from: https://doi.org/10.1016/j.compbiomed.2015.11.006.

[10] X. Yi, E. Walia, P. Babyn, Generative adversarial network in medical imaging: a review, Med. Image Anal. 58 (2019) 101552. Available from: https://doi.org/10.1016/j.media.2019.101552.

[11] P.K. Jain, N. Sharma, M.K. Kalra, A. Johri, L. Saba, J.S. Suri, Far wall plaque segmentation and area measurement in common and internal carotid artery ultrasound using U-series architectures: an unseen artificial intelligence paradigm for stroke risk assessment, Comput. Biol. Med. 149 (2022) 106017. Available from: https://doi.org/10.1016/j.compbiomed.2022.106017.

[12] P.K. Jain, N. Sharma, A.A. Giannopoulos, L. Saba, A. Nicolaides, J.S. Suri, Hybrid deep learning segmentation models for atherosclerotic plaque in internal carotid artery B-mode ultrasound, Comput. Biol. Med. 136 (2021) 104721.

[13] P.K. Jain, et al., Attention-based UNet deep learning model for plaque segmentation in carotid ultrasound for stroke risk stratification: an artificial intelligence paradigm, J. Cardiovasc. Dev. Dis. 9 (10) (2022) 326. Available from: https://doi.org/10.3390/jcdd9100326.

[14] R. Su, D. Zhang, J. Liu, C. Cheng, Msu-net: multi-scale u-net for 2d medical image segmentation, Front. Genet. 12 (2021) 639930.

[15] K. Verma, B.K. Singh, P. Tripathi, A.S. Thoke, Review of feature selection algorithms for breast cancer ultrasound image. in Studies in Computational Intelligencein: D. Barbucha, N.T. Nguyen, J. Batubara (Eds.), New Trends in Intelligent Information and Database Systems, 598, Springer International Publishing, Cham, 2015, pp. 23–32. Available from: http://doi.org/10.1007/978-3-319-16211-9_3.

[16] F.-C. Hsu, et al., A few-shot learning approach assists in the prognosis prediction of magnetic resonance-guided focused ultrasound for the local control of bone metastatic lesions, Cancers 14 (2) (2022). Available from: https://doi.org/10.3390/cancers14020445. Art. no. 2.

[17] J. Liu, K. Fan, X. Cai, M. Niranjan, Few-Shot Learning for Inference in Medical Imaging With Subspace Feature Representations. arXiv, 2023 [Online]. Available: http://arxiv.org/abs/2306.11152. (accessed 15.07.23).

[18] T. Araki, et al., PCA-based polling strategy in machine learning framework for coronary artery disease risk assessment in intravascular ultrasound: a link between carotid and coronary grayscale plaque morphology, Comput. Methods Prog. Biomed. 128 (2016) 137–158. Available from: https://doi.org/10.1016/j.cmpb.2016.02.004.

[19] B. KumarSingh, K. Verma, A.S. Thoke, Investigations on impact of feature normalization techniques on classifiers: performance in breast tumor classification, Int. J. Comput. Appl. 116 (19) (2015) 11–15. Available from: https://doi.org/10.5120/20443-2793.

[20] S.S. Sanagala, et al., Ten fast transfer learning models for carotid ultrasound plaque tissue characterization in augmentation framework embedded with heatmaps for stroke risk stratification, Diagnostics 11 (11) (2021) 2109. Available from: https://doi.org/10.3390/diagnostics11112109.

[21] M.L. Beltzer, K.E. Daniel, A.R. Daros, B.A. Teachman, Examining social reinforcement learning in social anxiety, J. Behav. Ther. Exp. Psychiatry (2023) 101810. Available from: https://doi.org/10.1016/j.jbtep.2022.101810.

[22] Z. Gao, Z. Lu, J. Wang, S. Ying, J. Shi, A convolutional neural network and graph convolutional network based framework for classification of breast histopathological images, IEEE J. Biomed. Health Inform. 26 (7) (2022) 3163–3173. Available from: https://doi.org/10.1109/JBHI.2022.3153671.

[23] D. Pal, P.B. Reddy, S. Roy, Attention UW-Net: a fully connected model for automatic segmentation and annotation of chest X-ray, Comput. Biol. Med. 150 (2022) 106083. Available from: https://doi.org/10.1016/j.compbiomed.2022.106083.

[24] Nillmani, et al., Segmentation-based classification deep learning model embedded with explainable AI for COVID-19 detection in chest X-ray scans, Diagnostics 12 (9) (2022) 2132. Available from: https://doi.org/10.3390/diagnostics12092132.

[25] S. Singh, P.K. Jain, N. Sharma, M. Pohit, S. Roy, Atherosclerotic plaque classification in carotid ultrasound images using machine learning and explainable deep learning, Intell. Med. (2023). Available from: https://doi.org/10.1016/j.imed.2023.05.003.

Section IV

Intelligent techniques for medical image analysis

Chapter 7

Image segmentation using template matching method for automated classification of knee osteoarthritis grades ☆

Deepak Saini[1], Ashima Khosla[2], Trilok Chand[2], Devendra K. Chouhan[3] and Mahesh Prakash[4]

[1]*Department of Information System, LM Thapar School of Management, Derabassi Campus, TIET Patiala, Punjab, India,* [2]*Department of Computer Science Engineering, Punjab Engineering College (Deemed to be University), Chandigarh, Punjab, India,* [3]*Department of Orthopedics, Post Graduate Institute of Medical Education & Research, Chandigarh, Punjab, India,* [4]*Department of Radiology, Post Graduate Institute of Medical Education & Research, Chandigarh, Punjab, India*

7.1 Introduction

Osteoarthritis (OA) of the knee joint known as knee osteoarthritis (KOA), is the most common and prevalent type of OA. KOA also known as "Degenerative Joint Disease," is one of the most common musculoskeletal disorders. It majorly affects the weight-wearing joints like the spine, hip, knee, feet, and fingers [1]. The knee joint is composed of two bones – the femur and the tibia. Thick cartilage is sandwiched between the bones which provides cushioning to the bones and helps in comfortable movement of the knee joint. This thick cartilage is known as the meniscus. The meniscus is critical to the knee joint as it helps in the smooth movement of the two bones. Cartilage volume decreases with age due to other contributing factors and this decrease in cartilage volume/thickness leads to narrowing of the gap between bones. Eventually, the gap reduces to the extent that the two bones start rubbing with each other and the patient feels a grating sensation in the joint. The grating sensation is the primary symptom of KOA. Other symptoms may include pain, formation of bone spurs, tenderness, stiffness, and loss of flexibility. Joint replacement surgery remains the only alternative [2,3] at the advanced stage of KOA. Therefore an early diagnosis of the disease increases the possibility for an individual to get treated at its onset, which may reduce the financial burden and delay the disability issues [4].

A large proportion of people aged 65 and up have OA in at least one joint, and with more than a quarter of Americans anticipated to reach this age by 2030, the risk of having OA is enormous [5]. The global prevalence of OA reached roughly 528 million individuals in 2019, representing a 113% rise from 1990. Notably, 73% of people affected by OA are above the age of 55, with 60% being female [6]. With 365 million instances, the knee is the most usually affected joint, followed by the hip and hand [7]. Approximately 344 million people in the United States have moderate or severe OA, which may benefit from rehabilitation [8]. The global prevalence of OA is predicted to rise due to aging populations, rising obesity rates, and an increase in the number of injuries. It is critical to underline that OA is not an inescapable result of aging. KOA, the most common type of arthritis, is predicted to rise further as life expectancy and obesity rates rise. Symptomatic KOA affects approximately 13% of women and 10% of men aged 60 and older, with the prevalence increasing to 40% in those over 70. Notably, not everyone with radiographic evidence of KOA suffers symptoms, with about 15% experiencing symptoms. The annual incidence of symptomatic KOA, regardless of age, is estimated to be roughly 240 cases per 100,000 people [9,10].

☆ Research article.

Intelligent Computing Techniques in Biomedical Imaging. DOI: https://doi.org/10.1016/B978-0-443-15999-2.00016-5

Physicians usually inspect the scanned radiographs and grade the severity of KOA ranging from 0 to 4 based on Kellgren and Lawrence (KL) grades [11,12]. Grade 0 signifies a normal or healthy knee and grade 4 being the worst case of KOA. When a person is detected with grade 4 OA, then total knee replacement is the only option left, but at early stages, progression of the disease can be controlled or even the disease can be treated. So, this implies that early detection and diagnosis of KOA is extremely required for reducing the progression of the disease.

In the field of medical imaging [3], various technologies such as magnetic resonance imaging [1,13,14], computed tomography scan [15,16], ultrasound [17,18], and plain radiographs (X-rays) are available for the diagnosis of KOA. The advantages like cost-effectiveness, less exposure to radiations, and easy availability make X-ray the versatile and most widely used imaging modality for the diagnosis of KOA [19–21].

Manual inspection of KOA from radiographic images is a time-consuming process and suffers from a problem of near-grade misclassification. Moreover, the result analysis may get affected with the experience of the experts. In the recent years, computer-aided diagnosis (CAD) for KOA severity grading from radiograph images has gained a widespread attention. Most of the studies in the existing literature [22–29] have worked toward the classification of KOA severity grades (KL grades) using a variety of preprocessing methods combined with deep learning architectures. These studies have achieved good classification performance, but there is a wider scope for improvement in this research area. For the scope of improvement, we have explored different segmentation methods that can be used for extraction of region of interest (RoI) from the knee joint images.

Segmentation plays a vital role in extracting the RoI from the given images, thus it helps in improving the classification performance of the deep learning architectures. Template matching [21,25,30] is one of the most efficient methods that can be employed to segment the RoIs from the given image. The template matching method is a foundational approach in knee image analysis for segmentation and classification. Starting with the creation of a representative template capturing distinctive features of interest within the knee joint, this systematic process involves preprocessing the images for consistency, selecting a matching metric, and employing a sliding window technique. The matching score map generated helps identify regions closely resembling the template, and a subsequent thresholding step distinguishes relevant structures from background noise, producing a binary segmented region. Postprocessing, such as morphological operations, may refine the segmentation. Finally, the segmented area can be subjected to classification using extracted features, contributing to the automated diagnosis of knee conditions. This method, when validated and optimized, proves valuable in enhancing the accuracy of knee image analysis, especially in the context of musculoskeletal disorders like OA. In the light of this, the main contributions of our proposed work have been defined as follows.

1. An automated segmentation of KOA radiograph images has been performed using template matching approach.
2. A transfer learning-based visual geometry group (VGG) 16 architecture has been used for automated KOA severity classification.
3. Apart from accuracy, the performance of the proposed work has been evaluated using other three metrics too, namely, precision, F1 score, and recall.
4. Our model is able to achieve an excellent classification accuracy of 91.44%, showing an improvement in the classification performance over the existing studies using different methods of segmentation.

7.2 Existing methods and motivation

Knee joint area can be extracted by using one of the three segmentation methods: manual, semiautomatic, and automatic. In the first two methods, human intervention is required at all or at some place, while in later case human intervention is not required and the computer-based system automatically performs the RoI extraction.

In manual image segmentation methods, trained physicians and expert radiologists visually analyze the joint and carefully map the structure of the joint employing CAD-based methods, simple vision-based tools, and image processing techniques. A number of studies [31,32] have used manual segmentation for knee area localization from X-ray images. Manual segmentation is widely accepted as ground truth above all the other segmentation techniques but it has certain limitations too. It is a very time-consuming and laborious task especially when the data set is huge. Since this type of segmentation is completely dependent on the expertise of the radiologists and clinicians, there are greater chances of inter and intraobserver variability and subjectivity [31–33].

Second category of segmentation is known as semiautomatic methods (or interactive methods) in which certain steps of contour extraction are automated followed by manual checking/inspection of the segments and sometimes even editing the segment contours or boundaries. A number of studies [12,34–38] have used semiautomatic methods of segmentation for knee area localization from X-ray images. Semiautomatic methods are also known as interactive methods in which certain steps of contour extraction are automated followed by manual checking/inspection of the segments and sometimes even editing the segment contours or boundaries. This type of segmentation has also few disadvantages, such as, whenever required or necessary, a human expert's intervention is needed for plotting or refining the boundary lines of the segmented image, which in turn may result in variability due to human error. Interobserver and intraobserver variations in measurements of various features happen which makes the results of segmentation "nonreproducible" [12,30,35,39].

Third category is automatic image segmentation that includes the trained algorithms used to segment an image and is becoming an essential part of clinical decision support systems and CAD [40]. Automatic methods of segmentation are relatively fast, accurate, and precise as compared to the other two methods and at the same time beneficial in clinical trials and pathology. Multiple attempts have been made till date to localize the knee joint automatically without any human interventions. A huge number of studies [21,22,24,25,40–50] are available which have worked toward different methods for automatic segmentation of knee X-ray images. Out of the available studies based upon automatic segmentation, only few [22,24,25,34,44,45,48,51] have worked toward the problem of automated multiclass classification of KOA severity grades. There is still a wider scope for improvement in terms of classification performance in this area of research. Therefore the main focus of our work is to develop an automated framework for knee joint segmentation and multiclass classification of OA severity grades from knee radiograph images.

In our work, we have used template matching approach for automated segmentation and VGG 16 architecture for the purpose of classification. One of the studies [46] has used a template matching algorithm to segment the knee joint from the given knee X-ray image, but the scope of the study is limited to binary classification (healthy knee joint vs OA affected knee joint). Antony et al. [24] have used template matching to localize the knee joint area from the knee radiograph, but a large template size of 300 × 300 pixels has been chosen in the respective study, thereby leading to the degradation in the classification performance of the model. Therefore we have chosen the smaller size of the template so that the final size of the output image gets reduced to 224 × 124 pixels. Lower size of the template-matched window leads to better localization of the knee joint, thereby contributing to higher accuracy of the diagnostic process. Our proposed work shows the improvement in the classification performance over the existing studies using different methods of segmentation.

7.3 Proposed methodology

The proposed methodology has been graphically represented in Fig. 7.1 and has been explained in detail in the following subsections.

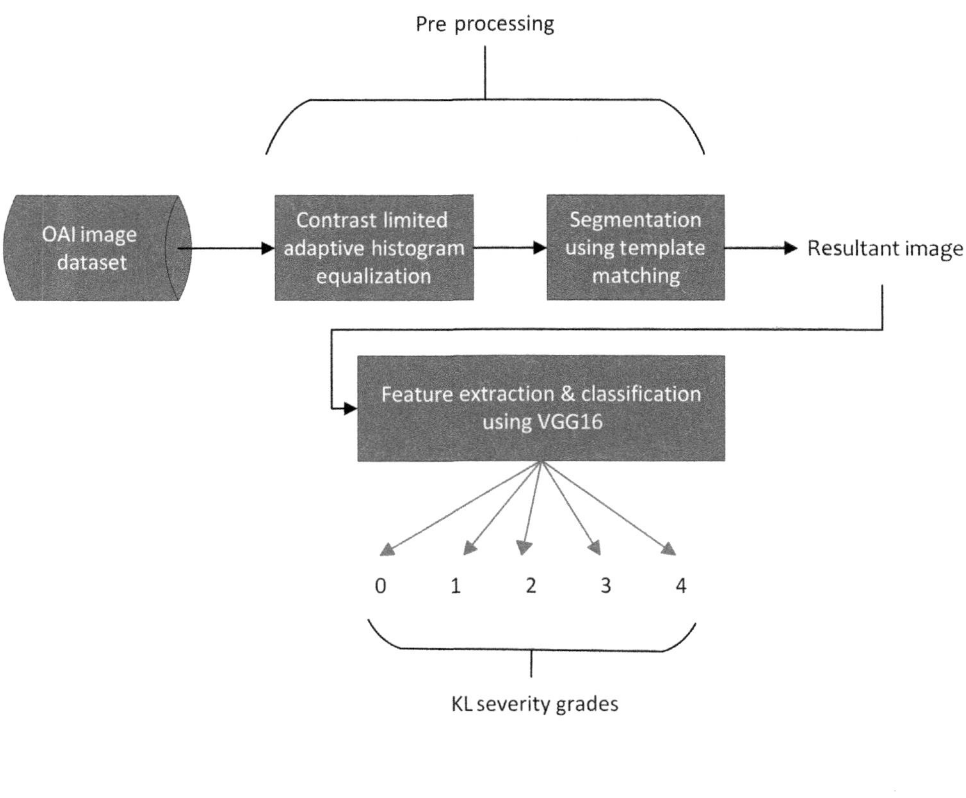

FIGURE 7.1 Flow diagram for the proposed work.

7.3.1 Dataset from osteoarthritis initiative (OAI)

In our work, the data has been collected from Mendeley datasets under the name of "Knee Osteoarthritis Severity Grading Dataset" being published in 2018 [52,53]. It contains the X-ray images for both of the knees with anteroposterior (AP) view. The given dataset included knee joint detection as well as knee KL grading. Knee joint area has already been localized in the given dataset. It is being organized by OAI data releases. In total, 3696 radiographs have been used in our study. The data distribution under different KL grades has been shown in Fig. 7.2.

A total of 3000 number of radiographs are used for the testing and training purposes and rest of the data is used for the validation. The distribution of the data is shown under test, train, and valid sets as represented through Fig. 7.3.

7.3.2 Data preprocessing

In the given dataset, the joint area is localized with a region size of 224 × 224 pixels. All the models used in the study are trained for the right knee and the entire left knee X-rays are horizontally flipped to generate the more training dataset. All the knee images in the dataset are taken under fixed flexion with an AP view of the knees. The aim of the preprocessing is an improvement of the image data that suppresses undesired distortion or enhances some image features relevant for further processing and analysis tasks. Initially X-ray image is read and converted to gray-scale image.

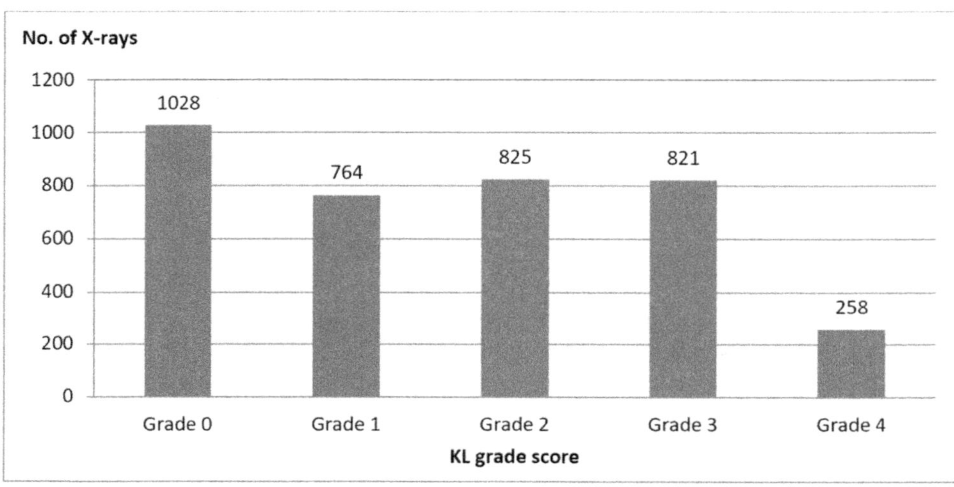

FIGURE 7.2 Distribution of dataset under Kellgren and Lawrence grades.

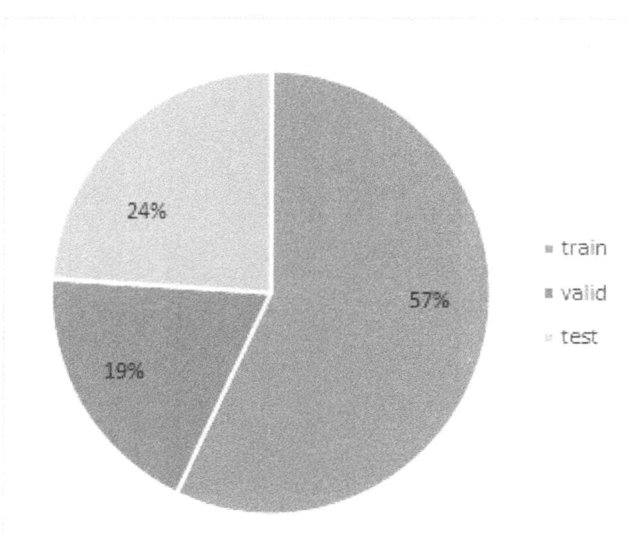

FIGURE 7.3 Distribution of data under train, validation, and test sets.

Then the image is converted to uint8 image and then further enhanced using the contrast stretch function with the help of contrast-limited adaptive histogram equalization (CLAHE) [54].

7.3.2.1 Image enhancement using contrast-limited adaptive histogram equalization

CLAHE is used for intensity level normalization that limits the noise amplification and improves the visibility level of the light and blur images. In this method, image is processed using the technique of median filtering. It is a nonlinear method of digital filtering method that involves the removal of noise while preserving the edges. It operates on small areas called tiles and contrast is enhanced for each tile. It is the baseline method that is adopted in most of the studies as a preprocessing technique [55,56].

Fig. 7.4 shows the original image (A) and the enhanced image after applying contrast-limited adaptive histogram equalization (B).

7.3.2.2 Knee joint area localization using template matching approach

The preprocessed dataset obtained so far still requires some improvements which have been explained as follows.

- The image size is quite large which leads to storage overhead and increases the computational time of the model.
- The knee radiograph image consists of three regions: femur bone, tibia bone, and knee joint. In numerous studies, it has been found out that not all the three regions of the knee radiographic image contribute equally to train the deep convolutional neural networks. So, the presence of unwanted image data not only increase the computational time but also affects the performance and accuracy. The unwanted image data can also be termed as *noisy data* or noise embedded in the image.

The RoI is the joint space only between the two knee bones. So, the images are segmented for the extraction of RoI. Knee joint area localization has been performed using template matching approach.

7.3.2.2.1 Template matching

For better quantification of the knee severity, the localized knee area is cropped out with the help of template matching and stored in different directories. In this method, a template image is created and is used as ground truth, as shown in Fig. 7.5. This method is widely used for object detection [57].

A template image is made for each grade of OA as shown in the sample template (Fig. 7.5), which consists of the gap between the tibia and femur bones. The template matching has been implemented on the entire dataset and the knee joint area has been extracted for all five grades. The resultant image dataset is then used to train the VGG 16 network and perform the classification task. Fig. 7.6 shows a view of dataset after applying template matching.

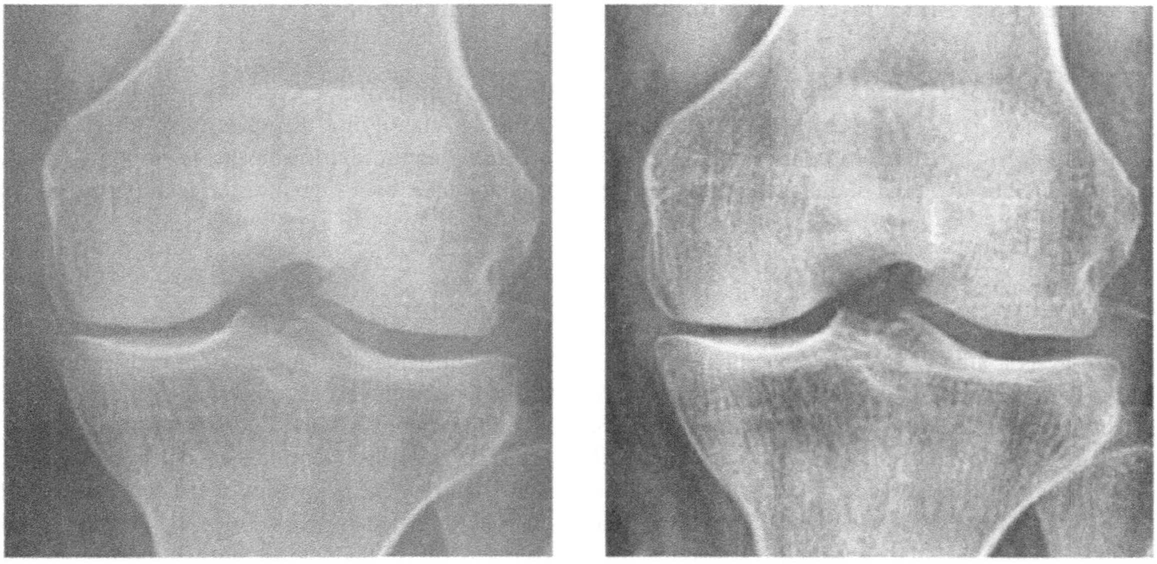

(A) (B)

FIGURE 7.4 Preprocessing using contrast-limited adaptive histogram equalization. (A) Original image. (B) Enhanced image after preprocessing.

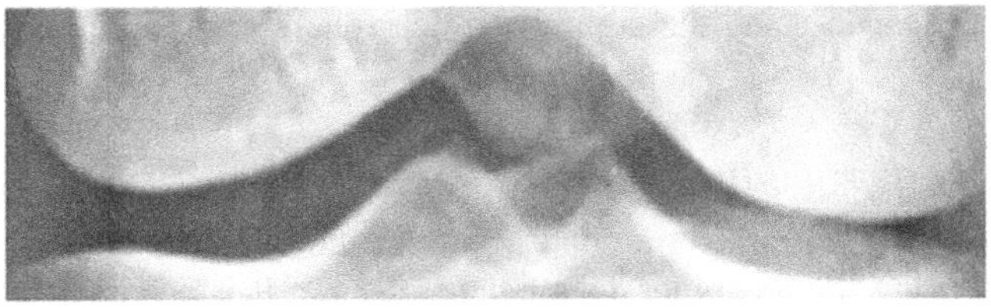

FIGURE 7.5 Sample template image used for knee joint localization.

FIGURE 7.6 Dataset images for grade 0 after applying template matching.

7.3.3 Feature extraction and classification using transfer learning-based visual geometry group 16 architecture

After the extraction of the RoI, the next step is the feature extraction and further classification of knee X-ray images under various grades. The assessment of the knee damage is done by classifying the severity of knee under various grades using KL scoring system. Most important part of the system is feature extraction. In our proposed work along with the classification, the task of feature extraction has been automated by using transfer learning-based VGG 16 architecture.

7.3.3.1 Transfer learning

Transfer learning is a kind of optimization approach in deep learning in which learning of an already trained model is used for learning the new task. This approach of transfer learning is popular in various computer vision and natural language processing tasks [58]. In this, source model is selected, then the model is reused according to the requirements and fine-tuned with our dataset. During the fine-tuning, the output layer of the pretrained model is replaced with the layer which recognizes the number of categories according to our requirement. Convolutional neural network (CNN) trained from scratch requires a lot of labeled data and is computationally expensive. In medical applications, due to the limited availability of labeled data, training a neural network from scratch is challenging. So, the promising alternative is to use transfer learning on pretrained models fine-tuned with new dataset. Transfer learning-based VGG 16 architecture has been used for the purpose of automated KOA severity grading from X-ray images.

7.3.3.2 Visual geometry group 16

VGG 16 has been pretrained for general classification of images on a very huge dataset of ImageNet LSVR dataset [59]. This dataset contains more than 1.2 million images having 1000 classes. Fig. 7.7 shows the design layout of the transfer learning-based VGG 16 architecture tuned on our dataset.

The model is pretrained on ImageNet dataset and is further fine-tuned for KOA image classification. This approach is motivated by transfer learning. In this, network is being trained on some other data and the weights of the initial layers are further used by target network. In the VGG 16 architecture, the network consists of 13 convolution layers and 3 fully connected layers. It consists of reoccurring 2D convolutional layers (they apply a filter to an input to create a feature map) and Max pooling layers (its function is to progressively reduce the spatial size of the representation with the aim to reduce the number of parameters and computation in the network) with number of units increasing as we move towards the bottom of the model. Fully connected layers are added and backpropagation is employed for fine-tuning the weights. Last layer, that is, dense layer or fully connected layer with 1000 classes is replaced with new fully connected layer with five categories as we need to classify the severity under five grades. The last layer, that is, prediction layer (dense layer) is replaced with dense_1 with five classes. VGG 16 is used for multiclass classification of different severity grades. Fig. 7.8 shows the schematic diagram of VGG 16 architecture. The network has 13 convolution layers and three fully connected layers along with five max-pool layers.

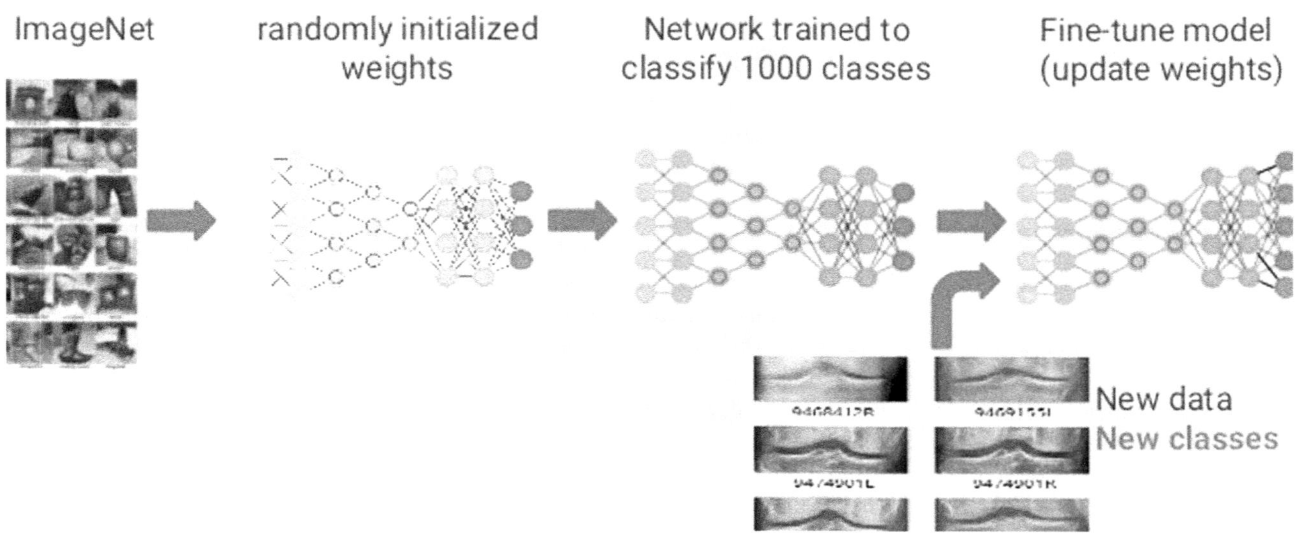

FIGURE 7.7 Design layout of visual geometry group 16 model tuned for new dataset.

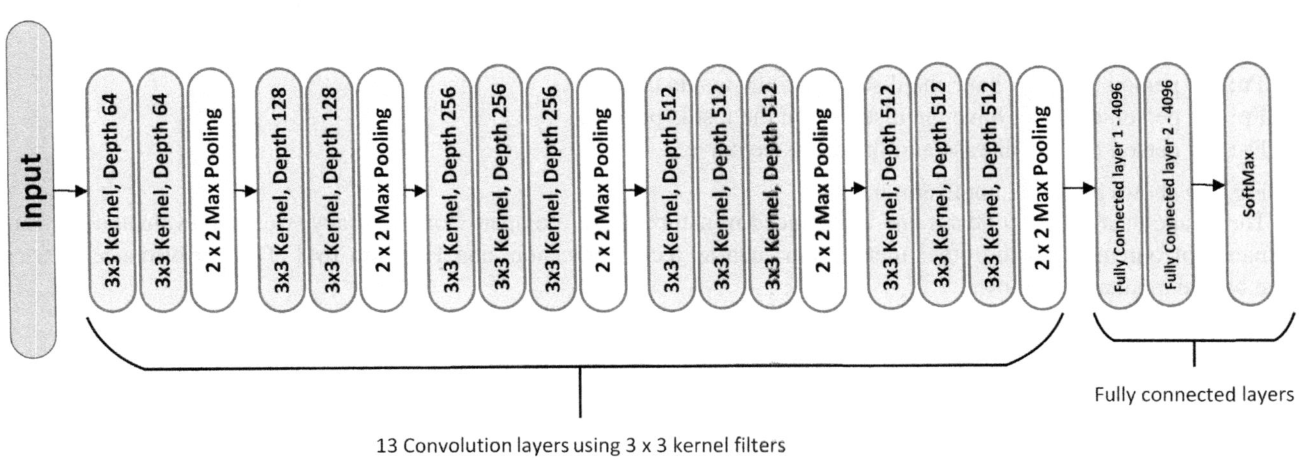

FIGURE 7.8 Schematic diagram of visual geometry group 16 model.

TABLE 7.1 Hyper-parameter values for visual geometry group 16.

Parameters	Value
Learning rate	0.001
Optimizer	Adam optimizer
Regularization type	L2
Batch size	32
Number of epochs	60
Loss function	Categorical

For input layer, the size of the input shape is $224 \times 224 \times 3$, number of filters is 64, and kernel size is 3×3. To preserve the spatial dimensions of the volume, value of "padding" is kept same. The activation function used is "ReLU." ReLU function is represented through Eq. (7.1).

$$f(x) = \begin{cases} 0 & x < 0 \\ x & x \geq 0 \end{cases} \tag{7.1}$$

During the training of the model, the hyperparameters have been optimally chosen as summarized in Table 7.1.

7.4 Performance evaluation

This section briefly describes the implementation environment and various performance metrics used to measure the efficiency of the proposed work. A VGG 16 pretrained model has been used for severity grade classification. Also, the section includes the code snippets from the working environment.

7.4.1 Performance metrics

To evaluate the performance of any deep learning model, choice of appropriate metric is very crucial. Varoius performance metrics are used to evaluate the performance of machine learning models in different applications [60]. It is worth remembering that metric is very different from loss function. A loss function shows the measure of the model performance and are used to train a machine learning model (using some kind of optimization). Whereas, metrics are used to monitor/measure the performance of model during training and testing.

7.4.1.1 Confusion matrix

It is the easiest way to measure the performance of any classification algorithm. There may be two or more than two classes in which the given input has to be classified. A confusion matrix is basically a table with two dimensions, viz., "Actual" and "Predicted." Before we dig further into the concept of other performance evaluators, we need to understand some terminologies such as true positive (tp), false negatives (fn), false positive (fp), and true negative (tn).

- **Tp:** the person has the disease and the test is positive.
- **Tn:** the person does not have the disease and the test is negative.
- **Fp:** the person does not have the disease and the test is positive.
- **Fn:** the person has the disease and the test is negative.

Fig. 7.9 shows a sample confusion matrix.

The image dataset is multiclass and imbalanced in nature, therefore apart from accuracy, we have evaluated the performance of our model using other metrics too such as recall, precision, and F1 score [61]. The performance metrics have been explained as follows.

1. **Accuracy**

   It can be defined as the ratio of correct predictions for the test data to the total number of predictions. Mathematically, accuracy can be represented through Eq. (7.2) as follows.

$$\text{Accuracy} = \frac{\text{tp} + \text{tn}}{\text{tp} + \text{tn} + \text{fp} + \text{fn}} \tag{7.2}$$

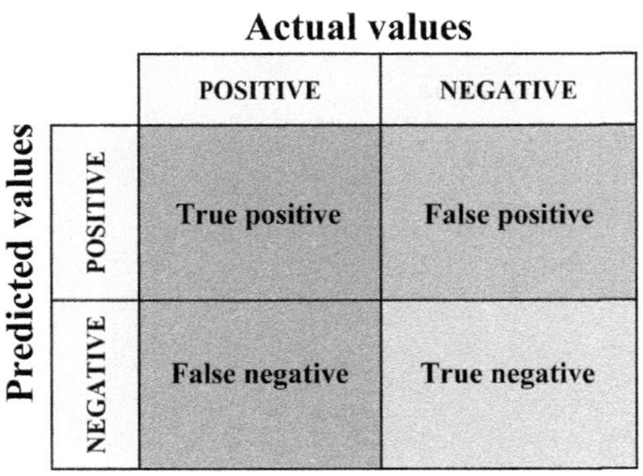

FIGURE 7.9 Sample confusion matrix.

2. Precision

It is the fraction of appropriate occurrences amongst the retrieved occurrences. It is equated as follows in Eq. (7.3).

$$\text{Precision} = \frac{\text{tp}}{\text{tp} + \text{fp}} \tag{7.3}$$

3. Recall

It is the fraction of appropriate occurrences that have been retrieved over the overall sum of appropriate occurrences and it is equated as follows in Eq. (7.4).

$$\text{Recall} = \frac{\text{tp}}{\text{tp} + \text{fn}} \tag{7.4}$$

4. F1 score

It is the harmonic mean of precision and recall. F_β is a factor which indicates how much more important recall is than precision. For example, if recall is considered twice as important as precision, then value of "β" would be 2. The standard value of "β" is 1. It is equated as follows in Eq. (7.5).

$$F_\beta = (1 + \beta^2)\frac{\text{precision} \times \text{recall}}{\beta^2 \cdot \text{precision} + \text{recall}} \tag{7.5}$$

Therefore F1 score can be defined through Eq. (7.6) as follows.

$$\text{F1 score} = 2 \times \frac{\text{precision} \times \text{recall}}{\text{precision} + \text{recall}} \tag{7.6}$$

7.4.2 Implementation environment and code snippets

Fine-tuned VGG 16 has been used for the multiclass classification (grades 0–4) of KOA severity grades. The model has been trained for 60 epochs with learning rate of 0.001. Adam optimizer, categorial cross entropy, and accuracy are parameters involved in the training of the model. The performance of the model has been examined by studying loss and accuracy curves. In general, increase in accuracy and decrease in loss are signs of effective fine-tuning.

The significant code snippets explaining the fine-tuning of the model have been captured from implementation environment (Jupyter Notebook) and have been shown in Fig. 7.10. To fine-tune the model, we replaced the last dense layer (in the original VGG 16 model for 1000 classes classification) with our custom dense layer which uses "Softmax" as an activation function. The implementation details in Fig. 7.10 have been explained as follows:

- Fig. 7.10A shows the model summary of VGG 16 model before removing the last dense layer.
- Fig. 7.10B shows the removal of last dense layer from the presented model.
- Fig. 7.10C shows an addition of new dense layer with two parameters- "5" denotes that the newly fine tuned model will categorize the data into five grades using "Softmax" as the activation function.
- Fig. 7.10D shows the model summary after adding the custom dense layer.

```
flatten (Flatten)            (None, 25088)          0
_________________________________________________________________
fc1 (Dense)                  (None, 4096)           102764544
_________________________________________________________________
fc2 (Dense)                  (None, 4096)           16781312
_________________________________________________________________
predictions (Dense)          (None, 1000)           4097000
=================================================================
```

(A)

```
In [14]: model = Sequential()
         #for layer in vgg16_model.layers:
         #    model.add(layer)
         for layer in vgg16_model.layers[:-1]: # just exclude last layer from copying
             model.add(layer)
```

(B)

```
In [18]: model.add(Dense(5, activation='softmax'))
```

(C)

```
flatten (Flatten)            (None, 25088)          0
_________________________________________________________________
fc1 (Dense)                  (None, 4096)           102764544
_________________________________________________________________
fc2 (Dense)                  (None, 4096)           16781312
_________________________________________________________________
dense_1 (Dense)              (None, 5)              20485
=================================================================
```

(D)

FIGURE 7.10 Jupyter Notebook snippets to fine-tune VGG 16 model for multiclass classification. (A) Original model summary. (B) Code to remove last dense layer. (C) Code to add dense layer for multiclass classification. (D) Model summary after adding the custom dense layer.

7.4.3 Classification results and performance analysis

Fine-tuned VGG 16 has been used for multiclass classification of KOA severity from grade 0 to 4. The model has been trained for 60 epochs using a learning rate of 0.001 with Adam optimizer, categorical cross-entropy loss, and accuracy as metrics. As already discussed, dataset is divided into train, valid, and test sets. Firstly the performance of VGG 16 model has been evaluated for the images that have been preprocessed using only baseline method (CLAHE). Then the VGG 16 has been trained for our proposed method using baseline method as well as method of knee joint area localization (based on template matching approach). The performance of the models has been evaluated using various performance metrics such as accuracy, recall, precision, and F1 score.

7.4.3.1 Classification performance on preprocessed dataset (without template matching)

Learning curves for the multiclass classification for grades 0–4 are shown in Fig. 7.11. Fig. 7.11A shows the accuracy curve and Fig. 7.11B shows loss curve of training and validating data for 60 epochs. Curve in blue is for training set while curve in orange is for validation set.

Fig. 7.12 shows the confusion matrix for five class data for VGG 16 model (without template matching). The classification performance of VGG 16 for multiclass classification of KOA severity grades (without template matching) in terms of various performance metrics has been summarized in Table 7.2. From the classification results, it can be analyzed that although VGG 16 model shows good performance in terms of accuracy metrics but shows poor performance in terms of rest of the metrics. Accuracy does not reflect the true capability of the model especially for imbalanced and

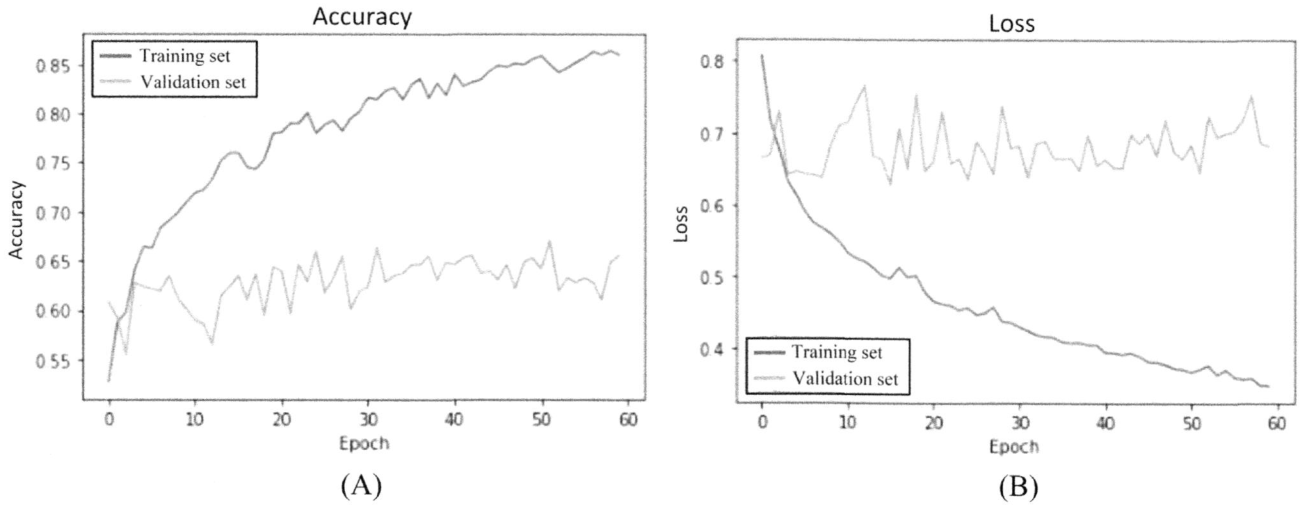

FIGURE 7.11 Learning curves for visual geometry group 16 for multiclass classification of knee osteoarthritis severity grades (without template matching). (A) Training and validation accuracy curve. (B) Training and validation loss curve.

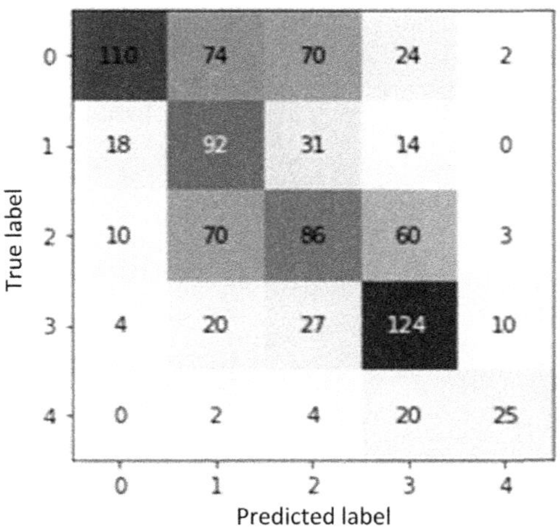

FIGURE 7.12 Confusion matrix for visual geometry group 16 model (without template matching).

TABLE 7.2 Classification performance results of visual geometry group 16 for multiclass classification of knee osteoarthritis severity grades (without template matching).

Classes	Precision	Recall	F1 score	Accuracy
Class 0	0.774647887	0.392857143	0.521327014	0.775555556
Class 1	0.356589147	0.593548387	0.445520581	0.745555556
Class 2	0.394495413	0.375545852	0.384787472	0.694444444
Class 3	0.512396694	0.67027027	0.580796253	0.801111111
Class 4	0.625	0.490196078	0.549450549	0.954444444

multiclass scenarios. The model has poor performance in terms of precision, recall, and F1 score, reflecting the ineffectiveness of the model in performing multiclass classification of KOA severity grading. The classification performance of VGG 16 for multiclass classification of KOA severity grades (without template matching) in terms of various performance metrics has been graphically represented in Fig. 7.13.

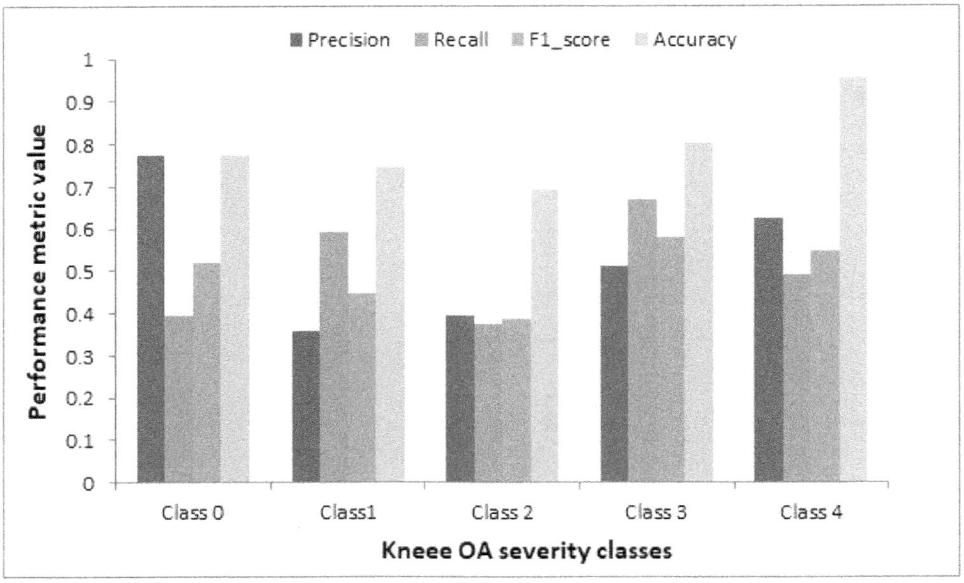

FIGURE 7.13 Classification performance results of visual geometry group 16 for multiclass classification of knee osteoarthritis severity grades (without template matching).

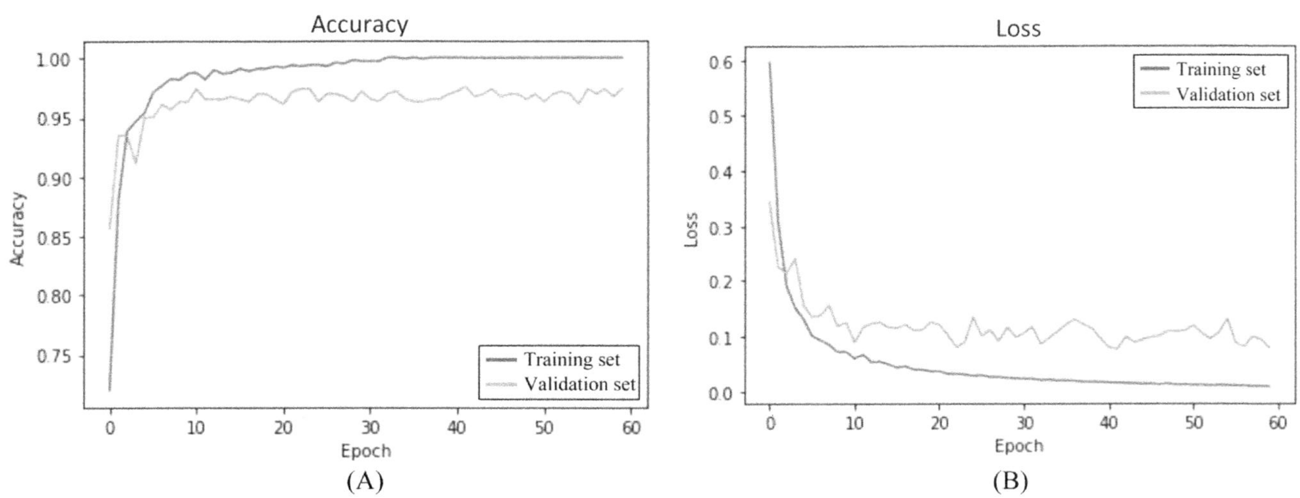

FIGURE 7.14 Learning curves for visual geometry group 16 for multiclass classification of knee osteoarthritis severity grades (with template matching). (A) Accuracy curve. (B) Training and validation loss curve.

7.4.3.2 Classification performance on preprocessed dataset (with template matching)

Learning curves for the multiclass classification for grades 0–4 are shown in Fig. 7.14. Fig. 7.14A shows the accuracy curve and Fig. 7.14B shows loss curve of training and validating data for 60 epochs. Curve in blue is for training set while curve in orange is for validation set.

Fig. 7.15 shows the confusion matrix for five class data for VGG 16 model (with template matching). The classification performance of VGG 16 for multiclass classification of KOA severity grades (with template matching) in terms of various performance metrics has been summarized in Table 7.3. From the classification results, it can be analyzed that VGG 16 model shows excellent performance in terms of accuracy metric. The model has good performance in terms of precision, recall, and F1 score as well, reflecting the improved performance of the model in performing multiclass classification of KOA severity grading as compared to the model using only CLAHE method. The classification performance of VGG 16 for multiclass classification of KOA severity grades (with template matching) in terms of various performance metrics has been graphically represented in Fig. 7.16. The good performance results show the effectiveness of the model to be deployed in clinical use for KOA severity classification.

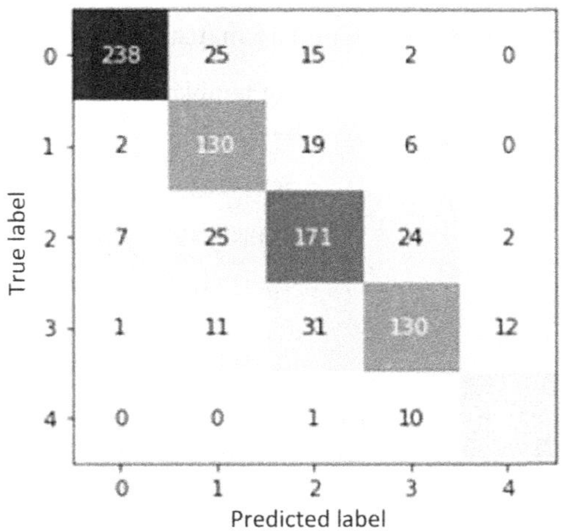

FIGURE 7.15 Confusion matrix for visual geometry group 16 model (with template matching).

TABLE 7.3 Classification performance results of visual geometry group 16 for multiclass classification of knee osteoarthritis severity grades (with template matching).

Classes	Accuracy	Recall	Precision	F1 score
Class 0	0.942350333	0.85	0.959677419	0.901515152
Class 1	0.902439024	0.828025478	0.680628272	0.747126437
Class 2	0.862527716	0.746724891	0.721518987	0.733905579
Class 3	0.892461197	0.702702703	0.755813953	0.728291317
Class 4	0.972283814	0.784313725	0.740740741	0.761904762

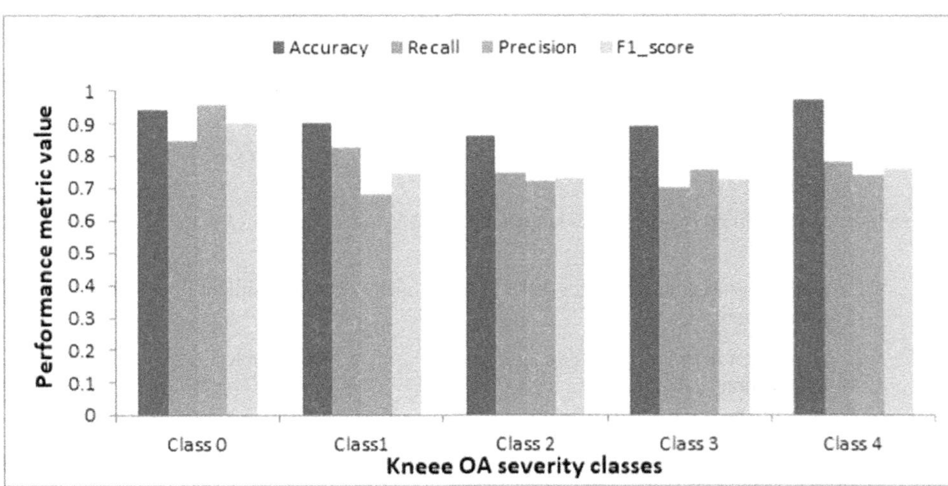

FIGURE 7.16 Classification performance results of visual geometry group 16 for multiclass classification of knee osteoarthritis severity grades (with template matching).

7.4.4 Performance comparsion of visual geometry group 16 model (with and without template matching)

The performance comparison of VGG 16 model for two preprocessing methods (with and without template matching) has been made in Table 7.4. The performance has been compared in terms of various evaluation metrics such as precision, F1 score, accuracy, and recall. From the results, it can be analyzed that our proposed work using template matching approach shows a lot of improvement for multiclass classification of KOA severity grades as compared to that of using only CLAHE method.

TABLE 7.4 Performance comparsion of visual geometry group 16 model (with and without template matching).

Performance metrics	Without template matching	With template matching
Precision	0.532625828	0.771675875
Recall	0.504483546	0.782353359
F1 score	0.496376374	0.774548649
Accuracy	0.794222222	0.914412417

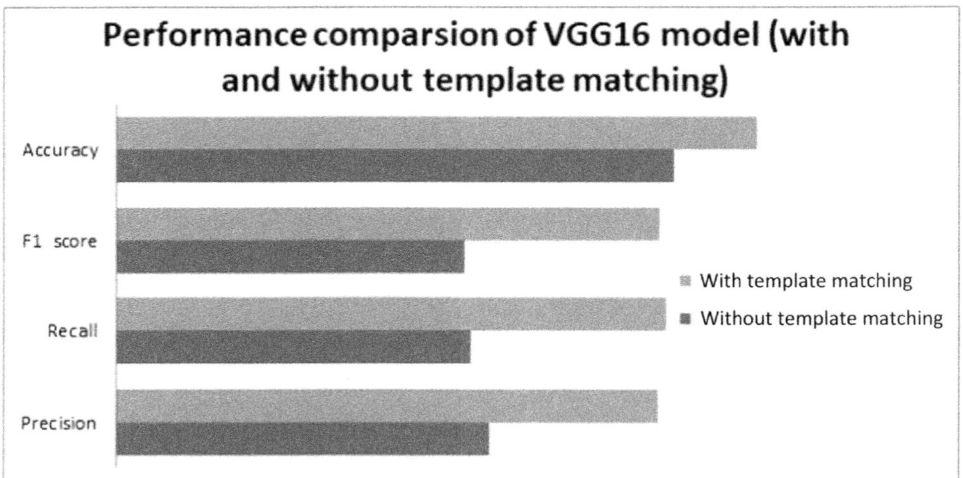

FIGURE 7.17 Performance comparsion of visual geometry group 16 model (with and without template matching).

TABLE 7.5 Comparison of proposed work with the existing studies for multiclass classification of knee osteoarthritis severity grades.

Research study	Dataset	Segmentation method	Classifier	Accuracy
Antony et al. [24]	OAI	Linear SVM on horizontal image gradients	BVLC Net	57.6%
Antony et al. [25]	OAI and MOST	A fully convolutional neural network for automatic localization of knee joint.	6 layer CNN	63.6%
Liu et al. [45]	Custom collected	Modified faster R-CNN with large anchors (FLA)	Faster R-CNN (FLA)	82.5%
Bany Muhammad and Yeasin [62]	OAI	Single shot multibox detector (SSD) fo knee joint localization.	Ensemble of CNN	87.0%
Mikhaylichenko and Demyanenko [63]	OAI	Single shot detector (SSD) for knee joint localization	Ensemble	71%
Proposed work	OAI	Template matching	CNN (VGG 16)	91.44%

OAI, Osteoarthritis initiative; *VGG 16*, visual geometry group; *CNN*, convolutional neural network.

The graphical representation for performance comparison of VGG 16 model for two methods has been shown in Fig. 7.17.

7.4.5 Comparison of proposed work with the existing studies

The comparison of our proposed work with the existing studies for multiclass classification of KOA severity grades has been presented in Table 7.5. From the table, it can be analyzed that our proposed work outperforms the existing studies in terms of the classification performance. Our model is able to achieve an accuracy of 91.44%, showing the improvement of our work over the existing methods.

Our work has an advantage of automated segmentation, therefore avoiding the need for user assistance for manual segmentation of the knee X-ray images. Instead of KOA detection, we have combined the automated method of segmentation with VGG 16 architecture for the multiclass classification problem of KOA severity grades.

7.5 Conclusion and future scope

In this work, we have developed an automated framework for knee joint segmentation and multiclass classification of OA severity grades from knee radiograph images. In our work, we have used template matching approach for automated segmentation and VGG 16 architecture for the purpose of classification. The template size has been selected in such a way that the knee radiograph image size reduces to 224×124 pixels as compared to 224×224 pixels (original size). Our results show a lot of improvement in terms of various performance metrics for the VGG 16 model using template matching approach as compared to that of using only the CLAHE method of preprocessing (without template matching approach). Our model is able to achieve an accuracy of 91.44%, showing the improvement of our work over the existing methods. It can be concluded that our work is a major contribution towards developing the automated framework for multiclass classification of KOA severity grades in clinical decision-making.

In future, the classification of the model can be evaluated for the segmented images using templates of different sizes. Our work is not able to achieve 100% efficiency, therefore efforts can be made toward the development of automated classification frameworks with higher efficiency. This can be done by exploring a variety of efficient methods for the automatic segmentation of knee radiograph images.

References

[1] S.M. Eijgenraam, A.S. Chaudhari, M. Reijman, S.M. Bierma-Zeinstra, B.A. Hargreaves, J. Runhaar, et al., Time-saving opportunities in knee osteoarthritis: T_2 mapping and structural imaging of the knee using a single 5-min mri scan, Eur. Radiol. 30 (4) (2020) 2231–2240.

[2] A. Tiulpin, S. Saarakkala, Auomatic grading of individual knee osteoarthritis features in plain radiographs using deep convolutional neural networks, Osteoarthr. Cartil. 28 (2020) S308.

[3] P.P. Kawathekar, K.J. Karande, Severity analysis of osteoarthritis of knee joint from X-ray images: a literature review, 2014 International Conference on Signal propagation and Computer Technology (ICSPCT 2014), IEEE, 2014, pp. 648–652.

[4] M. Cross, E. Smith, D. Hoy, S. Nolte, I. Ackerman, M. Fransen, et al., The global burden of hip and knee osteoarthritis: estimates from the global burden of disease 2010 study, Ann. Rheum. Dis. 73 (7) (2014) 1323–1330. Available from: https://doi.org/10.1136/annrheumdis-2013-204763, https://ard.bmj.com/content/73/7/1323.full.pdf, https://ard.bmj.com/content/73/7/1323.

[5] An aging nation: the older population in the united states – health & environmental research online (hero) – US EPA. https://hero.epa.gov/hero/index.cfm/reference/details/reference_id/2990744, (Accessed 25 November 2023).

[6] T. Vos, S.S. Lim, C. Abbafati, K.M. Abbas, M. Abbasi, M. Abbasifard, et al., Global burden of 369 diseases and injuries in 204 countries and territories, 1990–2019: a systematic analysis for the global burden of disease study 2019, Lancet 396 (10258) (2020) 1204–1222.

[7] H. Long, Q. Liu, H. Yin, K. Wang, N. Diao, Y. Zhang, et al., Prevalence trends of site-specific osteoarthritis from 1990 to 2019: findings from the global burden of disease study 2019, Arthritis Rheumatol. 74 (7) (2022) 1172–1183.

[8] A. Cieza, K. Causey, K. Kamenov, S.W. Hanson, S. Chatterji, T. Vos, Global estimates of the need for rehabilitation based on the global burden of disease study 2019: a systematic analysis for the global burden of disease study 2019, Lancet 396 (10267) (2020) 2006–2017.

[9] K. Magnusson, A. Turkiewicz, M. Englund, Nature vs nurture in knee osteoarthritis-the importance of age, sex and body mass index, Osteoarthr. Cartil. 27 (4) (2019) 586–592.

[10] J.-S. Li, T.-Y. Tsai, M.M. Clancy, G. Li, C.L. Lewis, D.T. Felson, Weight loss changed gait kinematics in individuals with obesity and knee pain, Gait Posture 68 (2019) 461–465.

[11] J.H. Kellgren, J.S. Lawrence, Radiological assessment of osteo-arthrosis, Ann. Rheum. Dis. 16 (4) (1957) 494–502. Available from: https://doi.org/10.1136/ard.16.4.494arXiv, https://ard.bmj.com/content/16/4/494.full.pdf, https://ard.bmj.com/content/16/4/494.

[12] H. Oka, S. Muraki, T. Akune, A. Mabuchi, T. Suzuki, H. Yoshida, et al., Fully automatic quantification of knee osteoarthritis severity on plain radiographs, Osteoarthr. Cartil. 16 (11) (2008) 1300–1306.

[13] A.J. Barr, B. Dube, E.M. Hensor, S.R. Kingsbury, G. Peat, M.A. Bowes, et al., The relationship between three-dimensional knee mri bone shape and total knee replacement- a case control study: data from the osteoarthritis initiative, Rheumatology 55 (9) (2016) 1585–1593.

[14] D. Hayashi, F.W. Roemer, A. Guermazi, Magnetic resonance imaging assessment of knee osteoarthritis: current and developing new concepts and techniques, Clin. Exp. Rheumatol. 37 (120) (2019) S88–S95.

[15] S. Hamai, N.J. Dunbar, T.-a Moro-oka, H. Miura, Y. Iwamoto, S.A. Banks, Physiological sagittal plane patellar kinematics during dynamic deep knee flexion, Int. Orthop. 37 (8) (2013) 1477–1482.

[16] L. Holzer, M. Kraiger, E. Talakic, G. Fritz, A. Avian, A. Hofmeister, et al., Microstructural analysis of subchondral bone in knee osteoarthritis, Osteoporos. Int. 31 (2020) 1–9.

[17] H. Majidi, F. Niksolat, K. Anbari, Comparing the accuracy of radiography and sonography in detection of knee osteoarthritis: a diagnostic study, Open. Access. Maced. J. Med. Sci. 7 (23) (2019) 4015.

[18] P. Desai, I. Hacihaliloglu, Knee-cartilage segmentation and thickness measurement from 2d ultrasound, J. Imaging 5 (4) (2019) 43.

[19] Z. Lovrenovic, M. Doumit, Development and testing of a passive walking assist exoskeleton, Biocybern. Biomed. Eng. 39 (4) (2019) 992–1004.

[20] A. Aprovitola, L. Gallo, Knee bone segmentation from mri: a classification and literature review, Biocybern. Biomed. Eng. 36 (2) (2016) 437–449.

[21] M. Saleem, M.S. Farid, S. Saleem, M.H. Khan, X-ray image analysis for automated knee osteoarthritis detection, Signal Image Video Process. 14 (6) (2020) 1–9.

[22] A. Tiulpin, J. Thevenot, E. Rahtu, P. Lehenkari, S. Saarakkala, Automatic knee osteoarthritis diagnosis from plain radiographs: a deep learning-based approach, Sci. Rep. 8 (1) (2018) 1727.

[23] K.A. Thomas, Ł. Kidziński, E. Halilaj, S.L. Fleming, G.R. Venkataraman, E.H. Oei, et al., Automated classification of radiographic knee osteoarthritis severity using deep neural networks, Radiol. Artif. Intell. 2 (2) (2020) e190065.

[24] J. Antony, K. McGuinness, N.E. O'Connor, K. Moran, Quantifying radiographic knee osteoarthritis severity using deep convolutional neural networks, 2016 23rd International Conference on Pattern Recognition (ICPR), IEEE, 2016, pp. 1195–1200.

[25] J. Antony, K. McGuinness, K. Moran, N.E. O'Connor, Automatic detection of knee joints and quantification of knee osteoarthritis severity using convolutional neural networks, International Conference on Machine Learning and Data Mining in Pattern Recognition, Springer, 2017, pp. 376–390.

[26] Y. Feng, J. Liu, H. Zhang, D. Qiu, Automated grading of knee osteoarthritis x-ray images based on attention mechanism, 2021 IEEE International Conference on Bioinformatics and Biomedicine (BIBM), IEEE, 2021, pp. 1927–1932.

[27] S. Kondal, V. Kulkarni, A. Gaikwad, A. Kharat, A. Pant, Automatic grading of knee osteoarthritis on the kellgren-lawrence scale from radiographs using convolutional neural networks, Advances in Deep Learning, Artificial Intelligence and Robotics, Springer, 2022, pp. 163–173.

[28] M. Górriz, J. Antony, K. McGuinness, X. Giró-i Nieto, N.E. O'Connor, Assessing knee oa severity with cnn attention-based end-to-end architectures, International Conference on Medical Imaging with Deep Learning, Proceedings of Machine Learning Research, 2019, pp. 197–214.

[29] S. Moustakidis, E. Christodoulou, E. Papageorgiou, C. Kokkotis, N. Papandrianos, D. Tsaopoulos, Application of machine intelligence for osteoarthritis classification: a classical implementation and a quantum perspective, Quantum Mach. Intell. 1 (3) (2019) 73–86.

[30] S.S. Gornale, P.U. Patravali, R.R. Manza, Detection of osteoarthritis using knee x-ray image analyses: a machine vision based approach, Int. J. Comput. Appl. 145 (1) (2016).

[31] T. Woloszynski, P. Podsiadlo, G. Stachowiak, M. Kurzynski, A signature dissimilarity measure for trabecular bone texture in knee radiographs, Med. Phys. 37 (5) (2010) 2030–2042.

[32] J. Hirvasniemi, J. Thevenot, V. Immonen, T. Liikavainio, P. Pulkkinen, T. Jämsä, et al., Quantification of differences in bone texture from plain radiographs in knees with and without osteoarthritis, Osteoarthr. Cartil. 22 (10) (2014) 1724–1731.

[33] M.P. Starmans, S.R. van der Voort, J.M.C. Tovar, J.F. Veenland, S. Klein, W.J. Niessen, Radiomics: data mining using quantitative medical image features, Handbook of Medical Image Computing and Computer Assisted Intervention, Elsevier, 2020, pp. 429–456.

[34] A. Brahim, R. Jennane, R. Riad, T. Janvier, L. Khedher, H. Toumi, et al., A decision support tool for early detection of knee osteoarthritis using x-ray imaging and machine learning: data from the osteoarthritis initiative, Comput. Med. Imaging Graph. 73 (2019) 11–18.

[35] A.C. Marijnissen, K.L. Vincken, P.A. Vos, D. Saris, M. Viergever, J. Bijlsma, et al., Knee images digital analysis (kida): a novel method to quantify individual radiographic features of knee osteoarthritis in detail, Osteoarthr. Cartil. 16 (2) (2008) 234–243.

[36] M. Subramoniam, V. Rajini, Statistical feature based classification of arthritis in knee x-ray images using local binary pattern, 2013 International Conference on Circuits, Power and Computing Technologies (ICCPCT), IEEE, 2013, pp. 873–875.

[37] M. Subramoniam, S. Barani, M. Rajini, A non-invasive computer aided diagnosis of osteoarthritis from digital x-ray images, Biomed. Res. 26 (4) (2015).

[38] A. Brahim, R. Riad, R. Jennane, Knee osteoarthritis detection using power spectral density: Data from the osteoarthritis initiative, Proceedings of International Conference on Computer Analysis of Images and Patterns, Springer, Salerno, Italy, 2019, pp. 480–487.

[39] J. Duryea, J. Li, C. Peterfy, C. Gordon, H. Genant, Trainable rule-based algorithm for the measurement of joint space width in digital radiographic images of the knee, Med. Phys. 27 (3) (2000) 580–591.

[40] A. Tiulpin, J. Thevenot, E. Rahtu, S. Saarakkala, A novel method for automatic localization of joint area on knee plain radiographs, Scandinavian Conference on Image Analysis, Springer, 2017, pp. 290–301.

[41] L. Shamir, S.M. Ling, W. Scott, M. Hochberg, L. Ferrucci, I.G. Goldberg, Early detection of radiographic knee osteoarthritis using computer-aided analysis, Osteoarthr. Cartil. 17 (10) (2009) 1307–1312.

[42] P. Podsiadlo, M. Wolski, G. Stachowiak, Automated selection of trabecular bone regions in knee radiographs, Med. Phys. 35 (5) (2008) 1870–1883.

[43] L. Shamir, S.M. Ling, W.W. Scott Jr, A. Bos, N. Orlov, T.J. Macura, et al., Knee X-ray image analysis method for automated detection of osteoarthritis, IEEE Trans. Biomed. Eng. 56 (2) (2008) 407–415.

[44] T.K. Yoo, D.W. Kim, S.B. Choi, J.S. Park, Simple scoring system and artificial neural network for knee osteoarthritis risk prediction: a cross-sectional study, PLoS One 11 (2) (2016) e0148724.

[45] B. Liu, J. Luo, H. Huang, Toward automatic quantification of knee osteoarthritis severity using improved faster r-cnn, Int. J. Comput. Assist. Radiol. Surg. 15 (3) (2020) 457–466.

[46] L. Anifah, I.K.E. Purnama, M. Hariadi, M.H. Purnomo, Automatic segmentation of impaired joint space area for osteoarthritis knee on x-ray image using gabor filter based morphology process, IPTEK J. Technol. Sci. 22 (3) (2011) 159–165.

[47] J. Thomson, T. O'Neill, D. Felson, T. Cootes, Automated shape and texture analysis for detection of osteoarthritis from radiographs of the knee, International Conference on Medical Image Computing and Computer-Assisted Intervention, Springer, 2015, pp. 127–134.

[48] S. Suresha, L. Kidziński, E. Halilaj, G. Gold, S. Delp, Automated staging of knee osteoarthritis severity using deep neural networks, Osteoarthr. Cartil. 26 (2018) S441.

[49] B. Norman, V. Pedoia, A. Noworolski, T.M. Link, S. Majumdar, Applying densely connected convolutional neural networks for staging osteoarthritis severity from plain radiographs, J. Digit. Imaging 32 (3) (2019) 471–477.

[50] R. Bindushree, S. Kubakaddi, N. Urs, Detection of knee osteoarthritis by measuring the joint space width in knee x ray images, Int. J. Electron. Commun. 3 (4) (2015) 2321–5984.

[51] B. Guan, F. Liu, A. Mizaian, S. Demhri, T. Neogi, A. Guermazi, et al., Deep learning approach to predict radiographic knee osteoarthritis progression, Osteoarthr. Cartil. 27 (2019) S395–S396.

[52] P. Chen, L. Gao, X. Shi, K. Allen, L. Yang, Fully automatic knee osteoarthritis severity grading using deep neural networks with a novel ordinal loss, Comput. Med. Imaging Graph. 75 (2019). Available from: https://doi.org/10.1016/j.compmedimag.2019.06.002.

[53] P. Chen, Knee osteoarthritis severity grading dataset, Mendeley Data V1 (2018). Available from: https://doi.org/10.17632/56rmx5bjcr.1.

[54] S.M. Pizer, E.P. Amburn, J.D. Austin, R. Cromartie, A. Geselowitz, T. Greer, et al., Adaptive histogram equalization and its variations, Comput. Vis. Graphics Image Process. 39 (3) (1987) 355–368.

[55] R.S. Hegadi, U.P. Chavan, D.I. Navale, Identification of knee osteoarthritis using texture analysis, Data Analytics and Learning, Springer, 2019, pp. 121–129.

[56] L. Anifah, I.K.E. Purnama, M. Hariadi, M.H. Purnomo, Osteoarthritis classification using self organizing map based on gabor kernel and contrast-limited adaptive histogram equalization, Open Biomed. Eng. J. 7 (2013) 18.

[57] R. Brunelli, Template Matching Techniques in Computer Vision: Theory and Practice, John Wiley & Sons, 2009.

[58] K. Leung, B. Zhang, J. Tan, Y. Shen, K.J. Geras, J.S. Babb, et al., Prediction of total knee replacement and diagnosis of osteoarthritis by using deep learning on knee radiographs: data from the osteoarthritis initiative, Radiology 296 (3) (2020) 192091. Available from: https://doi.org/10.1148/radiol.2020192091.

[59] O. Russakovsky, J. Deng, H. Su, J. Krause, S. Satheesh, S. Ma, et al., Imagenet large scale visual recognition challenge, Int. J. Comput. Vis. 115 (3) (2015) 211–252.

[60] M. Kayest, S.K. Jain, An incremental learning approach for the text categorization using hybrid optimization, Int. J. Intell. Comput. Cybern. 12 (3) (2019) 333–351. Available from: https://doi.org/10.1108/IJICC-12-2018-0170.

[61] A. Khosla, P. Khandnor, T. Chand, Eeg based automatic multi-class classification of epileptic seizure types using recurrence plots, Expert Syst. 39 (5) (2021) e12923. Available from: https://doi.org/10.1111/exsy.12923.

[62] M. Bany Muhammad, M. Yeasin, Interpretable and parameter optimized ensemble model for knee osteoarthritis assessment using radiographs, Sci. Rep. 11 (1) (2021) 1–11.

[63] A. Mikhaylichenko, Y. Demyanenko, Automatic grading of knee osteoarthritis from plain radiographs using densely connected convolutional networks, International Conference on Analysis of Images, Social Networks and Texts, Springer, 2020, pp. 149–161.

Chapter 8

Applications of intelligent techniques in pulmonary imaging

Vaibhav Koshta[1], Bikesh Kumar Singh[1], Ajoy K. Behera[2] and Ranganath T. Ganga[2]

[1]Department of Biomedical Engineering, National Institute of Technology Raipur, Raipur, Chhattisgarh, India, [2]Department of Pulmonary Medicine and TB, AIIMS, Raipur, Chhattisgarh, India

8.1 Introduction

Intelligent computing is a field of computer engineering designed to imitate human intelligence in the computers or machines to perform a specific task. Artificial intelligence (AI) is often called as "Intelligent Computing" because of its ability to process big data in order to extract complex and important information to be processed for the prediction of diseases. AI, machine learning (ML), and deep learning (DL) are generally used in similar context. However, it is important to understand the relationship between these terminologies to appropriately apply them to the database as per the requirement. AI is a broad class involving imitating the human intelligence using computer systems. On the other hand, ML is a sub category in AI which provides information using its capabilities to learn and analyze the big data whereas DL is a subset of ML, allowing the algorithm to train itself from the data and apply it to perform a desired task on the new data (Fig. 8.1). As the medical data is increasing nonlinearly, it has become very difficult for the doctors and medical practitioners to analyze them manually. Artificial intelligence possesses the capability to have significant impact on medical image as well as data analysis [1].

The contributions of AI in the diagnosis of various diseases by identifying the lung images have been crucial in the treatment as well as predicting the nature of disease. Few of the software's and devices used in medical imaging to classify the lung diseases are listed (Table 8.1) as approved by United States and European committees.

The effect of disease causes the downfall in the health of the patients due to alterations in the environmental and climate change, lifestyle, and other associated lifestyle factors. As per the records of World Health Organization (WHO), worldwide about 262 million population was affected by asthma in 2019 which causes 455,000 deaths. Similarly, chronic obstructive pulmonary disease (COPD) caused 3.23 million deaths in the year 2019 thereby becoming the third largest factor of deaths. The WHO report on mental health of the person worldwide cited pandemic as a contributing factor for worsening of the mental health which accounted for one death in 100 persons of an age below 50 years. About 5% of the population worldwide is affected by the sickle cell anemia thereby resulting about 300,000 infants born with sickle cell anemia each year. On the basis of above reports by WHO, it become very essential to develop technologies which can predict in advance the probability of a person or a child being affected with such disease.

To aid the doctor's in order to detect such chronic diseases, there is a high demand of intelligent techniques to automatically predict the data produced from biological sources or images. In recent years, the availability of huge amount of data led to the development of intelligent techniques utilizing DL and ML algorithms to analyze the patients. Automated biological signal analysis is the necessity of present time for quick and accurate diagnose as well as to reduce the human intervention and time. In the field of medical, analysis of biological data is very difficult. However, human beings are capable of analyzing the data accurately although they are subjective to make errors when the same experiment was performed two or more times on same data. The inaccuracy in the diagnostic process of these data can cause high risk to the life of a patient which is a motivation to the development of computer-aided technology. The biological sources generate huge amount of data which can be explored to acquire the useful information. The processing of such a huge data cannot be accomplished by the humans within specified time. Most of the computer-aided diagnose is done through AI- and DL-based ML models.

Intelligent Computing Techniques in Biomedical Imaging. DOI: https://doi.org/10.1016/B978-0-443-15999-2.00007-4

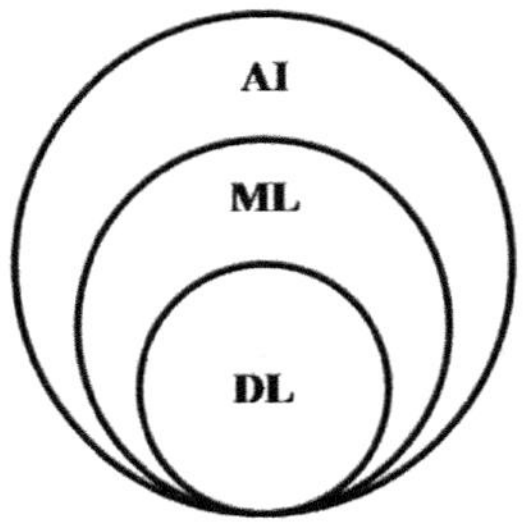

FIGURE 8.1 Artificial intelligence, machine learning, and deep learning class subset.

TABLE 8.1 Description of software and artificial intelligence device.

S. no.	Disease, disease software and methodology	Description
1	Chronic obstructive pulmonary disease (COPD) analysis using the chest CT imaging techniques involved in lung density analysis and lungQ software	It performs 3D segmentation of the lungs and evaluates lung volumetric and density parameters. On the other hand, lungQ is utilized to perform the lung volume and airway morphological parameter analysis.
2	Interstitial lung disease (ILD) identification from chest CT imaging utilizing lung print and lung texture analysis	Lung print software is used for evaluating lung tissue and airway. The lung texture analysis converts the chest CT images into a map object to represent the area or the extent of ILD infection in the lungs.
3	Lung infection detection from zero contrast chest CT images using Icolung and InferRead CT pneumonia detection device	A zero contrast images of the chest CT have been utilized to detect the symptoms of corona virus. Whereas, InferRead CT device is used for real-time classification of pneumonia.
4	Lung swelling or tumor can be detected from chest X-ray utilizing various devices such as clear-read CT, syngo CT, AI Rad, etc.	A computer aid detection method is employed in syngo device which integrates CNN for the detection of solid pulmonary nodules. While, the segmentation and recognition of lung injury can be done with the help of AI Rad. An asymptomatic population and superposition of new image on old image are some other methods which are deployed in temporal comparison and clear-read CT imaging techniques.

Intelligence computing is a new and emerging era of the computer science which is reshaping and promoting revolution in the field of big data, AI, internet of things, etc. introducing new concepts, architectural modifications, redesigning of systems with enhanced capabilities, improved methodology, and new areas of applications. It has widened the reach of computing to perceptual, cognitive, autonomous and human–computer integration intelligence. Intelligent computer is not only capable of making intelligent decisions but it can also learn from its past data or actions. The concepts involved in the designing of DL models are discussed in Section 8.3.

This chapter can help the medical practitioners and researchers to think in a direction to utilize the DL technologies in the detection lung diseases such COPD, asthma, lung fibrosis, etc. and assist in detecting lung infection with high efficiency and effectively to protect many susceptible subjects hence reducing the rate of lung damage caused by asthma, COPD, etc.

8.2 Literature review

Despite development of initial computer-aided development software in 1980 for detecting lung tumors or swelling or infections in the lungs are having issues with it due to nonavailability of proper computing materials to be used in the development of image processing methods at that time. The development of graphical processing unit (GPU) and convolutional neural network (CNN) has increased the performance of CAD (for lung disease diagnosing) and the demand for automated decision-making algorithms. In the past, a lot of work has been done in designing of ML and DL models for predicting lung diseases.

The work presented in [2] focuses on the detection of thorax diseases. The dataset used under study consists of 108948 front X-ray images out of which 24,636 images are affected with diseases and remaining 84,312 images are from healthy patients. The author calculated eight different pathologies viz. atelectasis, cardiomegaly, effusion, infiltration, mass, nodule, pneumonia, and pneumothorax and tested the performance of different CNN models AlexNet, GoogleNet, VGGNet-16, ResNet-50. A three-dimensional (3D) deep CNN was proposed in [3] to detect the lung nodules with multiscale prediction strategies. Lung nodules analysis 2016 dataset was used for the study having excluded a slice thickness greater than 2.5 mm. Hence, the database consists of total 888 CT scan images. A fully connected CNN was proposed in [4]. It utilizes Luna 16 dataset containing CT scan images which is used to train the CNN model to reduce the chances of wrong diagnosis and false positives in classification of lung tumors or nodules.

Automatic lung cancer diagnosis system using CT scans was designed on LIDC-IDRI dataset [5]. The author implemented two different methods for nodule detection and classification. The author in [5] proposed 3D faster regions with convolutional neural net (R-CNN) with 3D dual path block for nodule detection and U-net like encoder—decoder structure to learn the lung tumor features. For the purpose of classification, gradient boosting machine with 3D dual path network was utilized. The importance of AI in the classification of chest X-ray and analysis is discussed in [6]. In [7], a structure for analyzing lung pneumonia and cancer was proposed and implemented using two DL algorithms to solve the problem. The first technique proposes a modified AlexNet using support vector machine (SVM) classifier to predict the chest X-ray images into normal and pneumonia classes. The output was compared with the softmax layer, further the performance was validated with VGG16, VGG19, AlexNet, and ResNet models. The second technique implements the fusion of handcrafted and learned feature in order to improve the accuracy of the system.

A hybrid technique involving Gabor filter for image enhancement, marker-controlled watershed for segmentation purpose, and random forest for classification of lung cancer was proposed in [8]. Comprehensive studies are described in [9] on the detection of consolidation of chest X-ray images according to DenseNet121 and VGG 16. The work done in [10] observed 100 patients including 47 patients with significant pulmonary nodules or masses and 53 patients without pulmonary nodules. The DL algorithm was utilized to observe the performance in four nodules (heat map, abnormal probability, nodule probability, and mass probability). The DL algorithms demonstrated in [11] utilizing various pretrained networks viz. DenseNet121, AlexNet, Inception V3, etc. A CNN model was designed and deployed in [12] which works on a dataset from CheXpert consisting of 224,316 chest images obtained from 65,240 patients [13] for classifying the chest images into healthy and unhealthy.

Moreover, the currently available big datasets are assumed to have images with objects that are identified lightly. Hence, several methods have been designed and developed which can perform object identification and segmentation effectively. Some of the powerful methods are fully convolutional network (FCN) and faster region-based convolutional neural network (F-RCNN) [14,15]. The further improvement in fast region-based CNN network is known as the mask region CNN (mask RCNN) which has greater accuracy and efficiency as compared to fast RCNN. The authors in [16] uses mask R-CNN method for dividing the object into different groups of diseases detection on COCO 2016 dataset as well as compares the proposed algorithm with other algorithms designed by various authors and proved that described algorithm is superior from other algorithms mentioned in the literature [17,18]. MixNet which is a combination of two or more networks is developed [19] for recognizing lung tumors or swelling uses gradient boosting method for solving classification problem in an image dataset available in LUNA16 and LIDC-IDRI. The study showed that there is a demand of research in the designing of new techniques and algorithms for solving the lung disease classification and detection.

8.3 Applications of artificial intelligence in pulmonary imaging

8.3.1 Application of artificial intelligence for identification of covid-19 using lung X-ray image

The outbreak of corona virus named covid-19 was declared as the pandemic worldwide in March 11, 2020 by WHO which causes a huge number of deaths around the world. This makes the government of all the countries to think on taking precautions remedial steps to stop the spread of covid-19 virus and research was made to develop the technology which will detect the severity in lung infection of the affected people so that there life can be saved. On the other hand, it is also necessary to keep a track and record the details of a person come in contact with the covid-19 affected person to begin early treatment of infected persons. The major symptoms involved in covid-19 infection are respiratory disorder, kidney disorder, fluid accumulation in the lungs, etc. The most common and effective method to detect the whether a person is affected with covid-19 is reverse transcriptase PCR (RT-PCR). It involves collecting the sample and test it in a laboratory. This process requires a 2—3 days of time to generate the report as sample has to be tested in the laboratory. To overcome this disadvantage of RT-PCR test, lung images have been utilized along with AI and DL algorithm to predict the corona virus infection hence, lung imaging is very crucial in screening the covid-19 cases. In the literature it has been cited that computed tomography (CT) or chest X-ray can effectively and efficiently detect the covid-19 symptoms.

AI- or DL- based ML models can precisely diagnose the covid-19 cases through the lung damage indications seen in CT or X-ray images. The methodology for the detection of covid-19 infected patients using the lung X-ray images was proposed and block diagram is shown in Fig. 8.2 [20].

The author of the proposed work has used 40 contrast-enhanced lung X-ray images having 512 × 512 resolution which comprises of 15 normal lung images and 25 infected lung images. In the research work, multilevel thresholding is utilized to split the images into several distinct areas with contrast intensities corresponding to one background and several objects. SVM is implemented to separate the images into normal and covid-19 affected classes.

8.4 Application in the area of predicting lung cancer

Lung cancer is the leading cause of death from cancer around the world. Disease progression and treatment of lung cancer vary widely from patients to patients. Hence, an accurate diagnosis for the selection of treatment for lung cancer patient is the need of present scenario. Hematoxylin and eosin (H&E) is the most common method of tissue highlighting which includes whole slide imaging (WSI) has become now-a-days a regular clinical procedure generating a large number of medical images consisting of diseased and healthy persons historical and histological details. The bounded capacity of diseased image processing algorithms is a constrain in the field of digitalization to have a significant clinical impact.

The ease of utilizing physically visible features in diseased images of lung cancer provides an opportunity to modify and update the existing technology to enhance the efficacy in predicting the lung cancer using intelligent computing. Intelligent computing integrates DL which is showing a remarkable improvement in field of medical image analysis which leads to an increase in the demand for AI-based computer-aided technologies. With the advancement in DL algorithms-based image analysis and classification, it has solved the challenges in diagnosing lung diseases like identifying neoplasia, detecting tumor metastasis, determining mitoses, and inflammation, etc. A list of lung image analysis methods that have been developed is shown in Table 8.2.

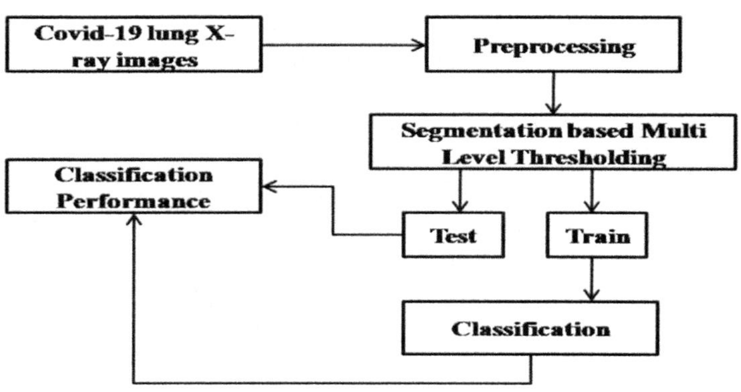

FIGURE 8.2 Block diagram of detecting covid-19 from X-ray images.

TABLE 8.2 List of DL-based lung cancer analysis methods.

Cancer	Method	Deep learning model	Objective
Adenocarcinoma (ADC)	Lung cancer detection and classification	Convolutional Neural Network (CNN), Fast Convolutional Neural Network (FCN) and Mask fast region CNN	To classify malignant and nonmalignant lung images and to detect the lung nodules
Adenocarcinoma (ADC) and squamous cell carcinoma (SCC)	Lung cancer detection and classification	Convolutional Neural Network (CNN)	To classify ADC and SCC images
squamous cell carcinoma (SCC), nonsmall cell lung cancer (NSCLC), and small cell lung cancer (SCLC)	Lung cancer detection	Convolutional Neural Network (CNN)	To classify malignant and nonmalignant lung images

8.5 Deep learning application for detecting lung diseases using X-ray images

Here, DL algorithms used for the detection of lung infection using X-ray images is discussed in detail. There are various image processing techniques developed based on DL and ML like—CNN, vanilla neural network, visual geometry group neural network (VGG), capsule network, etc. for the lung diseases classification. In this paper, the author proposed hybrid DL framework combining VGG, data augmentation and spatial transformer network (STN) with CNN. The model is implemented on NIH chest X-ray image dataset is utilized which is available in Kaggle. The author is solving a binary classification problem parameters viz. age, X-ray images, gender, view position, which predict the output class as "1" or "0" for affected and healthy patients, respectively [21]. The block diagram of vanilla CNN and proposed VGG Data spatial transformer network (VDSNet) architecture are shown in Figs. 8.3 and 8.4, respectively.

The basic structure of proposed VDSNet consists of three spatial transformer layers with first part as lambda indicating default features in lung X-ray. Second layer is a batch normalization layer and third layer is a spatial transformer layer for eliminating the most significant features.

A pretrained VGG16 model with 13 convolutional layers, 5 max pooling layers, and 3 dense layers has been utilized for feature extraction. An output layer with two dense layers, a dropout layer and a flatten layer is used for the continued reduction in the depth of VGG16 layers [22].

The capsule net model as proposed by the author is shown in Fig. 8.5. It consists of convolutional layer with 256 filters having stride = 2, kernel size = 9 and ReLU activation [22]. The dimensions of primary capsule are stride = 2, kernel size = 9, and number of channels = 32. Diagnosis capsule is having number of classes set equal to the number of the capsules and dimension of capsule = 16.

The performance of the proposed method and existing model was compared from the obtained results which showed that the authors proposed model, that is, VDSNet outperforms all the existing models discussed in the literature. For the purpose of comparison, the author has implemented following models and listed the accuracies obtained under each model [22] (Table 8.3).

8.6 Application to quantitative analysis of lung diseases

The quantifiable analysis of a very minute part in CT of the chest showed continuously rising interest in the field of clinical evaluation, treatment, and management of lung diseases like asthma, COPD, emphysema etc. The presently

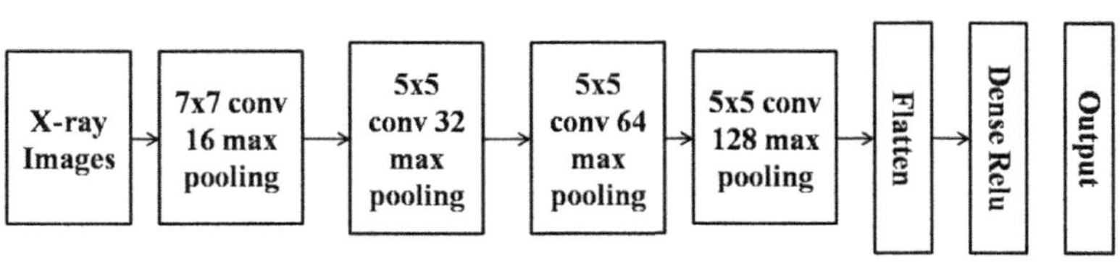

FIGURE 8.3 Vanilla convolutional neural network block diagram.

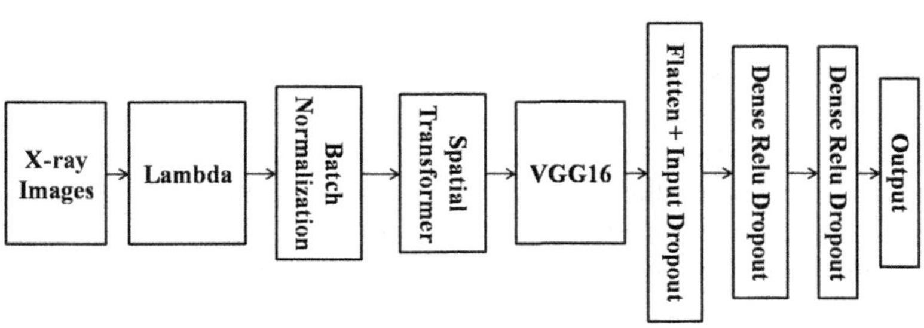

FIGURE 8.4 VGG data spatial transformer network architecture block diagram.

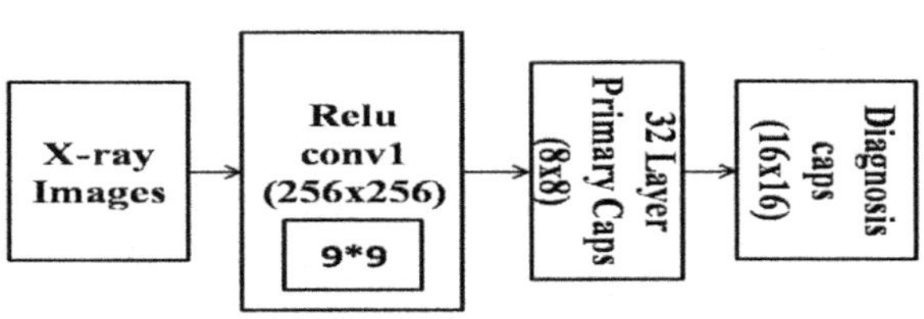

FIGURE 8.5 Capsule net block diagram.

TABLE 8.3 Deep learning models and their accuracies.

Dataset	Model	Accuracy (%)
Full dataset	VGG data spatial transformer network (VDSNet)	73%
	vanilla gray	67.8%
	vanilla RGB	69%
	hybrid CNN and VGG	69.5%
	modified capsule network	63.8%
Sample dataset	VGG Data spatial transformer network (VDSNet)	70.8%
	Vanilla gray	50.7%
	Vanilla RGB	51.8%
	Hybrid CNN and VGG	68%
	Modified capsule network	59%

available tools for quantitative analysis of lung diseases can accurately and efficiently performs the diagnosis, disease severity assessment through quantification of characteristics properties of the diseases thereby helping the practitioner in efficient treatment. The major challenges faced in the development and application of present analysis in the real world includes unavailability of standard format for training data and criteria to prove the authenticity of the results obtained. Apart from the unavailability of properly formatted training data and standard criteria there are few other artifacts which affects and limits the application of quantitative tools in image analysis viz. patient movement, differences in inspiration rate, variation in image acquisition and reconstruction techniques, inefficient preprocessing etc. Despite these challenges, quantitative analysis is developing into a feasible and viable tools for the assessment, evaluation and treatment of lung diseases [23]. After reading this article, the readers will be able to learn:

- The basic concepts involved in statistical methods and ML principles.
- The applications of quantitative analysis for the detection of lung diseases and associated effect on the lung structure or functioning.
- To find potential drawbacks in the automated interpretation from the results obtained after preprocessing and treatment of real world data.

8.7 Clinical applications

8.7.1 Methods used for the detection chronic obstructive pulmonary disease

A study of infinitesimal small section of CT scans of emphysema showed a rise in the low-fading region thereby representing destruction caused by the lung infection. One method to determine the extent of emphysema is to find value of total lobar lung parenchyma below cut off value of −950 HU which is often called as emphysema index. It is a powerful tool to represent the extent of disease effect on lung. The lower value of emphysema index represents less extensive disease and vice versa.

Another method used to detect emphysema is the 15th percentile which is based on complete lung histogram method. A lower value of 15th percentile represents high severity of disease and vice versa. A high value of basic emphysema has been proven to be significant compared to low value basic lung function, increased reduction in lung functioning, and an increased rate of developing airway obstruction. Perc15 is utilized to evaluate the percentage of lung tissue damage. However, for the subjects accompanying fibrosis, an emphysema index is not able to accurately diagnose emphysema compared with other measures such as mean lung density.

Threshold measure can also be utilized to evaluate the degree of air trapping which is calculated as a percentage of total lung pixels having fading of −856 HU or below. The cutoff value of −856 HU is selected such that to have attenuated healthy lung measured at inspiration CT images. The area of healthy lung tends to grow in expiratory images; on the other hand, the region of air trapping shows no changes. These air-trapping measures have been proven to accurately correlate with spirometer findings than the emphysema index for patients having infected from COPD. Likewise, for the asthma subjects, it is capable of identifying a subset with high lung damage [23].

In order to differentiate between low-attenuation areas corresponding to air trapping and emphysema, respectively, a parametric response (PR) mapping method can be utilized which is a mixture of threshold measures at inspiration–expiration CT scan images. In the expiration images, the region of emphysema and air confine-affected lung can be separated from healthy lung using the air-trapping threshold measure. On the other hand, abnormal areas from inspiration images can be detected using the emphysema threshold measure called as emphysema index. It has been reported in the literature that the regions of air confinement possess normal fading on inspiration images, while the regions of emphysema have shown abnormal glow. The incorporation of PR mapping method as discussed above has derived measures that can be fed to predictive models of COPD in order to increase their discriminative value [23].

The parameters which impacted the threshold measure based as well as first-order statistical methods are creating problem in the application of CT accession and rebuild technologies. However, the available methods can be utilized to show the similarity between physiology and disease severity, among those threshold measure-based methods shows an identifiable variation in technical measures like average volume, small CT division thickness, various noisy images with respect to varying amount of radiation dose and CT image reconstruction methods like corners improvement or smoothing [23]. In order to overcome the drawbacks of threshold measure method, higher-order texture analysis method is used to distinguish the local separation and morphology of low-fading areas. Hence, it is used to propose particular types of emphysema or to distinguish between visually and pathological low-attenuation areas.

8.7.2 Detection algorithms for idiopathic pulmonary fibrosis lung disease

In the detection of idiopathic pulmonary fibrosis (IPF) an infinitesimal small section of CT plays a vital role in examining and predicting the intensity as well as the rate of increase in the disease. As compared to emphysema where the extent of disease diffusion in the lung is directly related with decrease in lung density as a result it can be evaluated using attenuation-based threshold measures hence calculating the fibrotic variations is comparatively challenging task. A complete lung histogram-based method finds the kurtosis or the acuteness of the maximum value of attenuation histogram. Generally, the maximum value has to be sharp thereby reciprocating parenchymal similarity. IPF is characterized by high fading in fibrosis and low-fading destructive change in honeycombing which extends the tails of histogram hence decreasing the maximum value which represents abnormally low kurtosis. It is a measure of predicting mortality in the subjects however; kurtosis is insensitive to temporal changes and is not a good means of detecting the disease damage.

In the literature various texture-based analysis methods have been proposed to evaluate the fibrotic reticular change. A supervised learning-based classification algorithm using an SVM has been developed in Chen et al. [23] to investigate various lung parenchymal behavioral patterns consisting of changes difficult to be addressed by normal vision. A quantitative ILD (QILD) score is utilized to find the reticular, honeycombing, and ground-glass opacity changes. A texture-based study for automatically classifying and representing reduction in lung functioning have been proposed in Chen et al. [23] which uses supervised ML models to illustrate the effectiveness in choosing a cutoff value 22.05% or less. This classification system serves as a tool to search for an increase in disease progression.

8.7.3 Application in chronic hypersensitivity pneumonitis detection (CHP)

Like emphysema, which mainly displays increased low-fading areas, chronic hypersensitivity pneumonitis is comparatively difficult to evaluate using complete lung summary as it has heterogeneous presentation to honeycombing, and possess complicated as well as composite parenchymal behavioral patterns. It is seen that the variations in lung density shows the amount of destruction caused by the emphysema while lung infection severity is difficult to capture in chronic hypersensitivity pneumonitis as the large fading fibrotic and low-faded cystic regions may cancel out each other. An infinitesimal small part of CT is capable of effectively reflecting diseases includes the percentage of highly faded region having an attenuation greater than −200 HU, low-faded value having an attenuation less than −960 HU and kurtosis. A higher-order texture analysis statistical tool can accurately captures the mixed nature of CHP. A quantitative analysis model is discussed in Jacob et al. [24] showed that the results obtained are comparable to the pulmonary function test based models for CHP prediction of mortality.

8.8 Future work

Most of the currently available quantitative infinitesimal-section CT tools look for capturing known parenchymal behaviors. As we deep dive into the concepts involved in ML and AI, various different features have been discovered that may have practical clinical applications in the field of imaging. On the other hand, in some diseases the vascular structure feature extracted may have uncorrelated predictive value with respect to fatality and possess the potential to predict future pulmonary disorder more accurately as compared to parenchymal features.

The enormous development in field of ML algorithms and processing power with GPU or intelligent computing put in medical imaging. As ML and DL expands to serve a wide area of chest X-Ray imaging, it is likely to anticipate the possibility that many more novel features can be extracted and tested for clinical observations that humans are unable to see with respect to clinical relevance. There is multiple quantitative analysis tools available for studying and classifying diseases viz. COPD, IPF, etc., however, research for other lung diseases like Langherhan cell histiocytosis or sarcoidosis are limited, and future work can be done in these areas to design new algorithms and new features need to be invented and tested to predict the severity of lung diseases.

References

[1] S. Choudhury, A. Chohan, R. Dadhwal, A.P. Vakil, R. Franco, P.T. Taweesedt, Applications of artificial intelligence in common pulmonary diseases, Artif. Intell. Med. Imaging 3 (2022) 1–7.

[2] X. Wang, Y. Peng, L. Lu, Z. Lu, M. Bagheri, R.M. Summers. Chestx-ray8: hospital-scale chest x-ray database and benchmarks on weakly-supervised classification and localization of common thorax diseases. In: Proceedings of the IEEE conference on computer vision pattern recognition; 2017. p. 2097–2106.

[3] Y. Gu, X. Lu, L. Yang, B. Zhang, D. Yu, Y. Zhao, et al., Automatic lung nodule detection using a 3D deep convolutional neural network combined with a multi-scale prediction strategy in chest CTs, Comput. Biol. Med. 103 (2018) 220–231.

[4] A.A.A. Setio, A. Traverso, T. De Bel, M.S.N. Berens, C. Van Den Bogaard, P. Cerello, et al., Validation, comparison, and combination of algorithms for automatic detection of pulmonary nodules in computed tomography images: the LUNA16 challenge, Med. Image Anal. 42 (2017) 1–13.

[5] W. Zhu, C. Liu, W. Fan, X. Xie. Deeplung: deep 3d dual path nets for automated pulmonary nodule detection and classification. In: 2018 IEEE winter conference on applications of computer vision; 2018. p. 673–81.

[6] K. Kallianos, J. Mongan, S. Antani, T. Henry, A. Taylor, J. Abuya, et al., How far have we come? Artificial intelligence for chest radiograph interpretation, Clin. Radiol. 74 (2019) 338–345.

[7] A. Bhandary, G.A. Prabhu, V. Rajinikanth, K.P. Thanaraj, S.C. Satapathy, D.E. Robbins, et al., Deep-learning framework to detect lung abnormality—a study with chest X-ray and lung CT scan images, Pattern Recognit. Lett. 129 (2020) 271–278.

[8] S. Bharati, P. Podder, P.K. Paul, Lung cancer recognition and prediction according to random forest ensemble and RUSBoost algorithm using LIDC data, Int. J. Hybrid Intell. Syst. 15 (2019) 91–100.

[9] H. Behzadi-Khormouji, H. Rostami, S. Salehi, T. Derakhshande-Rishehri, M. Masoumi, S. Salemi, et al., Deep learning, reusable and problem-based architectures for detection of consolidation on chest X-ray images, Comput. Methods Programs Biomed. 185 (2020) 105162.

[10] C.-H. Liang, Y.-C. Liu, M.-T. Wu, F. Garcia-Castro, A. Alberich-Bayarri, F.-Z. Wu, Identifying pulmonary nodules or masses on chest radiography using deep learning: external validation and strategies to improve clinical practice, Clin. Radiol. 75 (2020) 38–45.

[11] V. Chouhan, S.K. Singh, A. Khamparia, D. Gupta, P. Tiwari, C. Moreira, et al., A novel transfer learning based approach for pneumonia detection in chest X-ray images, Appl. Sci. 10 (2020) 559.

[12] S. Rajaraman, S.K. Antani, Modality-specific deep learning model ensembles toward improving TB detection in chest radiographs, IEEE Access 8 (2020) 27318–27326.

[13] T. Takaki, S. Murakami, R. Watanabe, T. Aoki, T. Fujibuchi, Calculating the target exposure index using a deep convolutional neural network and a rule base, Phys. Medica. 71 (2020) 108–114.

[14] M. Kholiavchenko, I. Sirazitdinov, K. Kubrak, R. Badrutdinova, R. Kuleev, Y. Yuan, et al., Contour-aware multi-label chest X-ray organ segmentation, Int. J. Comput. Assist. Radiol. Surg. 15 (2020) 425–436.

[15] S. Ren, K. He, R. Girshick, J. Sun, Faster r-cnn: Towards real-time object detection with region proposal networks, Adv. Neural Inf. Process. Syst. 28 (2015).

[16] J. Long, E. Shelhamer, T. Darrell. Fully convolutional networks for semantic segmentation. In: Proceedings of the IEEE conference on computer vision and pattern recognition; 2015. p. 3431–40.

[17] H. He, H. Xu, Y. Zhang, K. Gao, H. Li, L. Ma, et al., Mask R-CNN based automated identification and extraction of oil well sites, Int. J. Appl. Earth Obs. Geoinf. 112 (2022) 102875.

[18] J. Huang, V. Rathod, C. Sun, M. Zhu, A. Korattikara, A. Fathi, et al. Speed/accuracy trade-offs for modern convolutional object detectors. In: Proceedings of the IEEE conference on computer vision and pattern recognition; 2017. p. 7310–11.

[19] J. Sang, M.S. Alam, H. Xiang, Automated detection and classification for early stage lung cancer on CT images using deep learning, Pattern Recognit. Track. XXX, SPIE, 2019, pp. 200–207.

[20] L.N. Mahdy, K.A. Ezzat, H.H. Elmousalami, H.A. Ella, A.E. Hassanien, Automatic x-ray covid-19 lung image classification system based on multi-level thresholding and support vector machine, MedRxiv (2020) 2003–2020.

[21] S. Wang, D.M. Yang, R. Rong, X. Zhan, J. Fujimoto, H. Liu, et al., Artificial intelligence in lung cancer pathology image analysis, Cancers (Basel), 11, 2019, p. 1673.

[22] S. Bharati, P. Podder, M.R.H. Mondal, Hybrid deep learning for detecting lung diseases from X-ray images, Informatics Med. Unlocked 20 (2020) 100391.

[23] A. Chen, R.A. Karwoski, D.S. Gierada, B.J. Bartholmai, C.W. Koo, Quantitative CT analysis of diffuse lung disease, Radiographics 40 (2020) 28–43.

[24] Joseph Jacob, et al., Mortality prediction in idiopathic pulmonary fibrosis: evaluation of computer-based CT analysis with conventional severity measures, Eur. Respir. J. 49 (2017). Available from: https://doi.org/10.1183/13993003.01011-2016.

Chapter 9

Digital breast tomosynthesis versus two-dimensional mammography: a comparison on breast imaging for cancer screening

M. Jagannath[1], K. Adalarasu[2], N. Nathiya[3], P. Vijayakumar[1] and R. Keerthik Dhivya[1]

[1]*School of Electronics Engineering, Vellore Institute of Technology (VIT), Chennai, Tamil Nadu, India,* [2]*School of Electrical and Electronics Engineering, SASTRA Deemed to be University, Thanjavur, Tamil Nadu, India,* [3]*School of Advanced Sciences, Vellore Institute of Technology (VIT), Chennai, Tamil Nadu, India*

9.1 Introduction

Breast cancer has been steadily increasing in India, and it is now the most common cancer among Indian women. Approximately 2–2.5 million cancer cases had been reported in India last decade [1]. Breast cancer is increasing at a significant rate in both developed and developing countries. The rate of occurrence of breast cancer is high at the age of 50 in developed countries and 40 in developing countries [2]. According to the World Health Organization (WHO), breast cancer primarily arises from the epithelial cells lining the ducts or lobules in the glandular tissue of the breast [3]. The symptoms such as lump in the breast, change in the morphology of the breast or areola, skin reddening, and unusual discharge from breast indicate the breast cancer. Early detection seems to be the better solution to reduce the mortality rate due to cancer. Breast cancer screening is a process of examining breast tissue for the detection of cancer before there are signs or symptoms of disease. Two-dimensional (2D) mammography and three-dimensional (3D) mammography are discussed in the upcoming subsections.

Breast cancer screening plays a vital role in early detection and treatment of breast cancer, significantly improving the chances of successful outcomes and reducing mortality rates. Regular screening can identify breast cancer at its earliest stages, often before any noticeable symptoms or physical changes occur. This allows for early intervention and increases the likelihood of less aggressive treatments being required.

Imaging modalities are essential tools in breast cancer screening and early detection. The primary imaging techniques used in breast cancer screening include mammography, ultrasound, and magnetic resonance imaging (MRI). Each modality has its specific strengths and is used based on various factors, including the woman's age, risk factors, and breast density.

Mammography: Mammograms are the most commonly used imaging technique for breast cancer screening. They involve low-dose x-rays that create detailed images of the breast tissue. Mammograms can detect abnormal areas or lesions, including microcalcifications or masses that may indicate the presence of cancerous cells.

Ultrasound: Ultrasound uses high-frequency sound waves to produce images of the breast. It is often used as a complementary tool to mammography, particularly for women with dense breast tissue or younger women. Ultrasound can help further evaluate suspicious findings detected on mammograms and provide additional information about the nature of a breast abnormality.

MRI: MRI is a powerful imaging technique that uses a magnetic field and radio waves to generate detailed images of the breast. It is primarily recommended for women at high risk of breast cancer or for further evaluation of suspicious findings detected on mammograms or ultrasounds. MRI can provide valuable information about the size, location, and characteristics of breast tumors.

These imaging modalities aid in the early detection of breast cancer by identifying abnormalities that may be indicative of malignancy. When breast cancer is detected at an early stage, treatment options are typically more effective, less

Intelligent Computing Techniques in Biomedical Imaging. DOI: https://doi.org/10.1016/B978-0-443-15999-2.00014-1

invasive, and associated with better prognoses. Early detection through screening can also help identify precancerous lesions, allowing for preventive measures or close monitoring to prevent the development of invasive cancer. It is important for women to follow the recommended breast cancer screening guidelines based on their age, risk factors, and family history. Regular screenings and discussions with healthcare providers about the most appropriate imaging modalities can greatly contribute to early detection, reducing the impact of breast cancer and saving lives.

9.1.1 Two-dimensional mammography

Mammography plays a crucial role in breast cancer screening that uses x-rays for creating a 2D image which is based on the ionizing radiation principle and differential absorption of x-rays by different tissues in the breast. As the x-rays pass through the breast, it travels across various tissues and hence the x-rays get attenuated which results in differences in the intensity of the x-ray beam. Images of breast can be captured either by analog film or digital detectors. Digital detectors provide faster image acquisition, enhanced images, and effortless readability of images. It can be even used along with computer-aided detection (CAD) system, to analyze the breast cancer and hence it assists radiologists. 2D mammography is unable to detect tiny tumors in denser tissue of breast leading to misdiagnosis [4].

While 2D mammography has been a standard screening tool for breast cancer, it does have some challenges and limitations:

False positives and false negatives: 2D mammograms can produce false-positive results, meaning they detect abnormalities that are not cancerous. This can lead to unnecessary follow-up tests, such as biopsies, causing anxiety and inconvenience for patients. On the other hand, 2D mammography can also miss certain cancers, resulting in false-negative results. This can lead to delayed diagnoses and potentially more advanced stages of cancer when it is eventually detected.

Breast density: Breast density can affect the accuracy of 2D mammography. Dense breast tissue, which is common in younger women, can appear white on mammograms, making it difficult to detect tumors that may also appear white. This can result in missed or delayed diagnoses in women with dense breasts.

Overlapping tissue: In 2D mammograms, the breast tissue can overlap, making it challenging to differentiate between normal tissue and potential abnormalities. This overlapping effect can reduce the clarity of the images and make it harder to detect small tumors.

Limited views: 2D mammography provides 2D images of the breast from two angles only. This limited perspective can make it challenging to accurately characterize the size, shape, and location of a tumor, potentially affecting treatment planning.

Radiation exposure: Although the radiation dose in mammography is generally considered low, repeated exposure over time can still carry a small risk. Minimizing unnecessary radiation exposure is a concern, particularly for women who undergo regular screening.

It's worth noting that advancements have been made to address some of these limitations. One significant advancement is the introduction of 3D mammography, also known as digital breast tomosynthesis (DBT). DBT creates a 3D image of the breast by taking multiple low-dose x-ray images from different angles. This technology helps reduce the impact of overlapping tissue, improves cancer detection rates, and reduces false positives. The adoption of newer imaging technologies like DBT has helped to overcome some of these limitations and improve the detection and diagnosis of breast cancer.

9.1.2 Digital breast tomosynthesis

DBT, also known as 3D mammography, is an advanced imaging technique that builds upon the principles of 2D mammography but captures images of the breast from multiple angles to create a 3D representation of the breast tissue (Fig. 9.1). It helps to reduce the misdiagnosis and it is helpful for breast with dense tissues to discriminate the normal tissue from the cancerous tissue.

Digital mammography led to drastic changes in radiologist's performance levels in reviewing breast tissues especially in case of thick (more dense) tissues [5,6]. But mammography may not perform well in case of overlapping tissues which reduces specificity and sensitivity. Then the actual cancer will be hidden inside these tissues and more chances to interpret such a case as false negative [7]. Breast tomosynthesis is a technique that solves this overlap issue by taking images from multiple angles. This technique uses a 3D digital technology which is approved by US Food and Drug Administration (FDA) [8,9].

DBT can be employed as a valuable tool in this process. Its 3D imaging capabilities aid in assessing the size, shape, and characteristics of lesions, facilitating accurate diagnosis and guiding treatment planning [8,10]. Numerous studies have conducted on recall rates and false prediction values for various screening protocols. Among them, DBT shows high accuracy, sensitivity, and specificity compared to conventional mammography. The combination of DBT and digital mammography has been found to enhance the accuracy and sensitivity of breast cancer detection [11]. But these

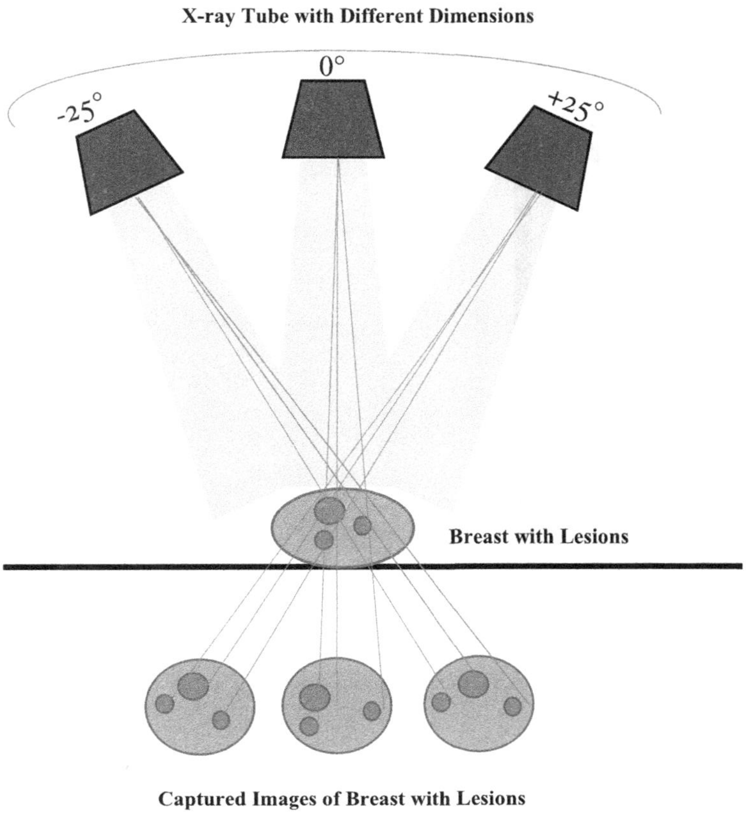

FIGURE 9.1 Schematic representation of principle of digital breast tomosynthesis.

studies did not demonstrate a factually huge distinction in the recognition rate for noninvasive diseases [12]. It is helpful to analyze the demonstrative performances of both DBT and digital mammography on behalf of lesion/tumor characterization and size of different breast cancer tissues [13].

Tivnan et al. [14] proposed a technique for position retrieval of lesion using the DBT and the dielectric properties of microwave radar. Then the subtraction of dielectric distributions of healthy lesion obtained from DBT from the abnormal lesion that aims to detect changes in tissue electrical properties associated with breast abnormalities. A multimodal fusion technique comprising 2D ultrasound and 3D tomosynthesis for breast cancer detection was proposed by Sinha et al. [15]. The image volumes from both tomosynthesis and ultrasound at same geometry are fused using in-house software. Later radiologists can use a process called image correlation that involves drawing a rectangular 3D box around the abnormal lesion in one image set and then correlate with the other set. On the prima facie, it is evident that tomosynthesis has advantages over mammography and can provide additional information than later one. The study gives a comparison between DBT and mammography of breast imaging on cancer detection.

9.2 Materials and methods

9.2.1 Participants and study protocol

All DBT examinations used as a part of this study met all the ethical guidelines established by the Institute. In the study, a total of 130 participants between the ages of 29 and 67 years were included. The DBT examinations were obtained for the consented participants on an exploratory DBT setup. Screening protocol details are explained to the participants before commencement of the procedure. Tomosynthesis is a unique form of mammogram that delivers a 3D breast image by utilizing a few low measurement x-ray beams [15,16]. During a mammogram, the breast is positioned and compressed between two plates or paddles to immobilize it during image acquisition. Tomosynthesis has also same positioning, but the x-ray beam will move around the breast in a circular manner. The information contained in the x ray will be given to computer and converts it into a 3D image. Tomosynthesis acquisitions for this study executed at low radiation dose. Approximately a dose of 2 mGy per view is usual for tomosynthesis. The tomosynthesis data set was created by remodeling the acquired data into 50 and 90 parallel slices with a 1-mm thickness [17].

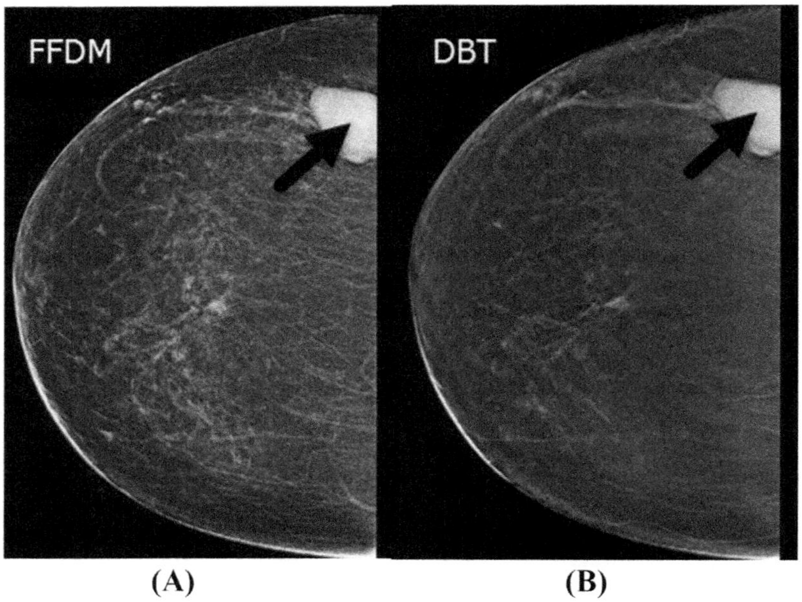

FIGURE 9.2 Mammogram images obtained from a 66-year-old female. (A) Full-field digital mammogram and (B) digital breast tomosynthesis, which show the cancerous lymph node indicated with arrows.

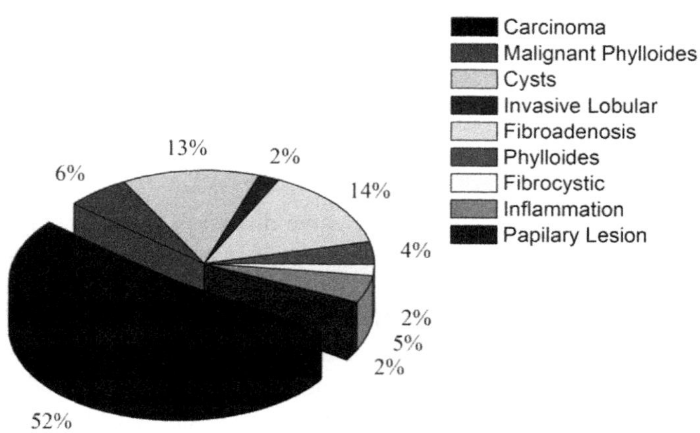

FIGURE 9.3 Common benign and malignant breast diseases.

The study was consented by participants having unusual mass, architectural defects or asymmetric lesions and later underwent biopsy (n = 22) or identified as an abnormal cyst (n = 3). This study incorporated normal mammography; DBT; and incisional biopsy for a lump, architectural distortion, or diffuse/focal asymmetry. Two statistical tests, analysis of variance (ANOVA) and paired t-test were employed to analyze various mammographic parameters using SPSS 15.0 for Windows (SPSS Inc. Chicago IL). Participants encountered screening with 3D tomosynthesis, 2D mammography, and ultrasonography (USG) and corresponding results were interpreted with the help of breast imaging reporting and data system (BI-RADS) (Fig. 9.2). Comparison of these three screening techniques was based on size and margin of lesion and breast configurations.

Of these 130 cases, 37 cases were described as architectural distortion, 18 cases were described as diffuse/focal asymmetry, 40 cases as multiple/fiber cysts, and 35 cases were spotted as masses/lumps. Microcalcifications cases were not included. Ninety-four cases were recommended to do biopsy for the confirmation of cancer. The common benign and malignant breast diseases were observed on the participants under study (Fig. 9.3).

9.2.2 Data interpretation

Clinical radiologists would decide the number of recalls after screening mammography. These radiologists would interpret 2D mammography and 3D tomosynthesis in a diagnostic environment. The radiologists analyzed the tomosynthesis examinations by viewing both 1-mm parallel slices and images on a digital prototype. First step in analysis procedure was the comparison between DBT and 2D mammography through direct examination. Then tomosynthesis was reviewed in screening platform and made decision on recall. Finally, BI-RADS ratings for benign, high-risk and cancer cases were collected from all readers.

9.3 Results

9.3.1 Digital breast tomosynthesis and mammography findings on different breast compositions and lesion types

Utilization of tomosynthesis made it conceivable to assess its performance in various breast configurations and furthermore, different kinds of lesions, for example, calcifications, asymmetries, distortions, and margins of lesions. Some results would exhibit how the utilization of tomosynthesis may influence the administration of indicative patients. A parametric test, paired t-test on parameters such as margin of lesion, calcification, diffuse asymmetry, focal asymmetry, and additional lesion, was used to determine the significant difference between mammography and DBT. Significant level is reported at $P < .05$. Clinical trial based on tomosynthesis for different configurations shows significant differences in each category except in diffuse asymmetry. After mammography, most of the recalls were due to asymmetries. Asymmetries correspond to breast tissue overlapping while compression. The statistical analysis conducted in the study revealed a significant difference ($P < .05$) between mammography and DBT in asymmetry evaluation. This finding suggests that the two imaging techniques yielded distinct results when assessing asymmetry in breast images. For evaluating these asymmetries in mammography, several projections were needed. But tomosynthesis would help to eliminate the usage of projections [18,19]. Architectural distortions were the main causes for false negatives in 2D mammography. Tomosynthesis showed high detection rate of these anomalies rather than 2D mammography [20]. Tomosynthesis had better calcification detection sensitivity ($P < .05$) than mammography.

9.3.2 Digital breast tomosynthesis and mammography findings on size and margin of different lesions

Most of the digital systems had size of a lesion as one of the features for classification between benign (normal) and malignant (abnormal) tissues. In breast cancer detection, size of tumor would help in cancer prophecy and prediction of stage. In this study trial, size of lesions was determined by using USG, mammography, and tomosynthesis/DBT with the help of post-hoc Tukey's HSD (Honestly Significant Difference) analysis. Studies have shown that USG significantly overestimated ($P < .05$) the size of lesion in contrast with DBT and mammography. The statistical analysis conducted in the study did not find a significant difference ($P < .05$) between mammography and DBT in size estimation. This means that the two imaging modalities yielded similar results when it came to estimating the size of lesions or abnormalities in breast images. During biopsies, margin of lesions was marked for clarification about the edge of the tissue that must remove [21]. DBT was more sensitive than mammography in outlining margin especially in dense breast. The study revealed a significant difference ($P < .05$) was found between mammography and DBT in margin calculation. Figs. 9.4 and 9.5 show the margin of lesions calculated using 2D mammography and 3D DBT, respectively.

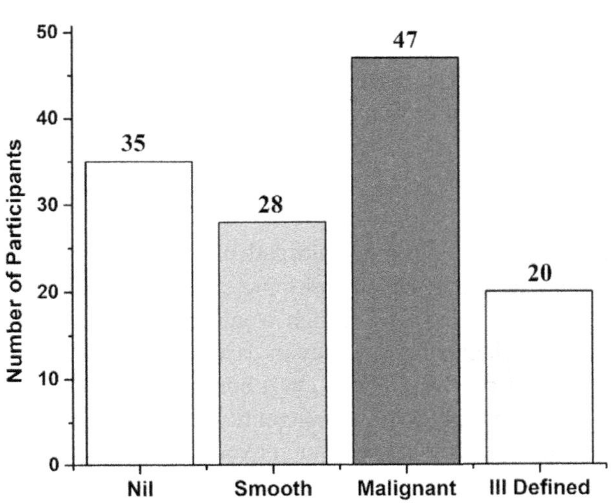

FIGURE 9.4 Characteristics of margin of lesions in 2D mammography.

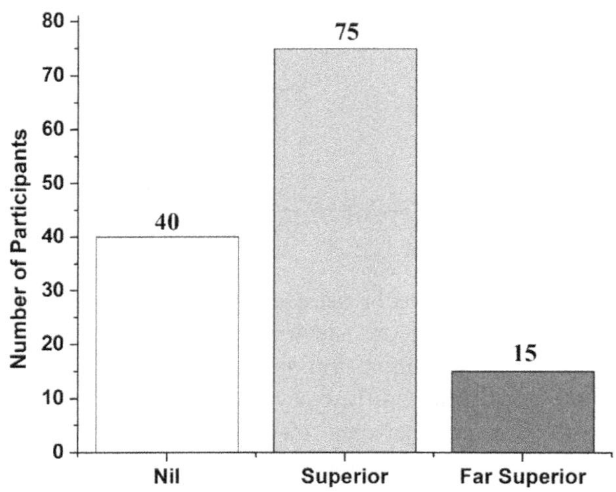

FIGURE 9.5 Characteristics of margin of lesions in 3D digital breast tomosynthesis.

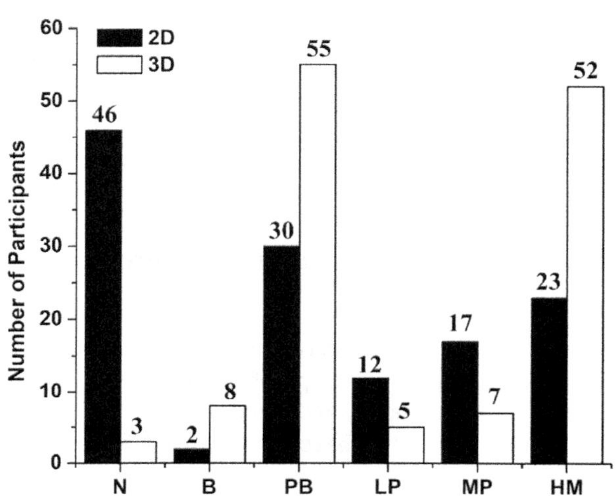

FIGURE 9.6 Distribution of breast imaging reporting and data system rating from all participants. *N*, Nil; *B*, benign; *PB*, probably benign; *LP*, low probability; *MP*, moderate probability; *HM*, high malignancy.

9.3.3 Analysis on breast imaging reporting and data system ratings

The American College of Radiology (ACR) established the BI-RADS. It is a rating system widely using for breast cancer diagnosis in mammogram screening. BI-RADS will classify cancer cases into well-defined categories numbered from 0 to 6 [22]. Repeated measure of ANOVA test was employed for identifying significant difference between mammography and DBT for mammographic parameters. Post-hoc Tukey's HSD analysis was utilized and set the significance at $P < .05$. BI-RADS was really beneficial for radiologists who were analyzing mammogram (both 2D and DBT). Out of 130 participants, above 65 participants were detected with cancer cases and almost 40 benign cases have been observed. Fig. 9.6 shows BI-RADS classification of participants underwent this study trial.

9.4 Discussion

The performance in describing lesion size and margin was identical between DBT and normal mammography [23]. DBT plus 2D mammography combined eliminated the problem of overlapping tissues thereby increased the specificity and sensitivity. Also, the margin of lesions could be clearly figured out by using DBT which would reduce redundant views [24,25]. Zuley et al. [26] proven that DBT had comparatively high diagnostic accuracy in case of microcalcification lesions. Thus unnecessary mammographic analysis could be avoided by using DBT. On a contrary, Hayes et al. [27] conducted a literature review on breast cancer screening by considering DBT versus conventional mammography. The review came up with a conclusion that DBT did not show much better performance than conventional mammography unless combining it with conventional one.

There were not much data published in DBT screening as a primary image screening technique. This study trial was carried out with an objective of using DBT widely in clinical practices. DBT is one of the whole breast images screening methods instead of checking a specific part of breast. Mammography screening practice is different from one organization to the other. Also, the analysis by readers or diagnosticians may vary widely from one another. Further examination of abnormality areas found from mammography was through breast USG. Proper usage of USG adjunct to normal mammography enhanced specificity and sensitivity of cancer analyses [28,29], especially women having thick breast and younger age group [30].

A quantitative analysis for cancer evaluation was not possible through this study because it was not designed with this objective. According to BI-RADS classification, tomosynthesis performed well than mammography in detecting mucinous carcinoma. During the initial mammogram screening it was diagnosed as a benign case. Later carcinoma was revealed in tomosynthesis. But in two invasive cases—ductal and lobular, both tomosynthesis and mammogram assessments were same. For ductal carcinoma, BI-RADS rating was 5 and BI-RADS rating was 4 for lobular carcinoma. Margin speculation and shape of lesion was featured by tomosynthesis better than mammography that leads to an objective diagnosis.

This contemplative study reveals some of the restrictions that are reviewed from the existing literature. First one was the size of population. The size and participant characteristics might vary performance analysis of DBT and mammography. In this trial, most of the participants were malignant and number of benign cases was relatively small. Second, for USG, B-mode ultrasound images were used. So in clinical point of view, this study was not perfect. Instead of B-mode images, Doppler images were more reliable and would contain detailed information about characteristics of lesion. Third, interobserver and intraobserver variability would be there because the radiologists who participated in this study were from diverse fields. This retrospective review shown that DBT is a perfect tool in screening and diagnostic platforms in which an abnormal lump or architectural asymmetry is identified. Hence, the DBT is also considered as a useful tool for women having dense breast.

DBT and 2D mammography are two different imaging techniques used in breast cancer screening. Here are the key differences between them and how these impact image quality and diagnostic capabilities:

Image acquisition: In 2D mammography, two x-ray images of each breast are taken from different angles (top-to-bottom and side-to-side). These images are then combined to create a 2D representation of the breast tissue. On the other hand, DBT captures multiple low-dose x-ray images as the x-ray tube moves in an arc over the breast. These images are reconstructed into thin slices, creating a 3D representation of the breast tissue.

Image clarity and detail: Due to overlapping breast tissue in 2D mammography, it can be challenging to differentiate between normal structures and potential abnormalities. In DBT, the images are obtained from various angles, reducing the impact of overlapping tissue. This improves image clarity and detail, allowing radiologists to better visualize and evaluate breast abnormalities.

Lesion visibility: DBT has been shown to improve the visibility of lesions, particularly small masses or architectural distortions, compared to 2D mammography. With DBT's 3D images, radiologists can scroll through the slices and examine the breast tissue layer by layer, which can enhance the detection and characterization of lesions.

Breast density: Breast density can impact the accuracy of mammography. Dense breast tissue appears white on mammograms, making it more challenging to detect tumors that may also appear white. DBT has shown advantages in women with dense breasts, as it reduces the masking effect of dense tissue, leading to improved cancer detection rates.

Radiation exposure: Both 2D mammography and DBT involve radiation exposure, but the radiation dose in DBT is slightly higher due to the multiple images acquired. However, the dose is still within safe limits, and the benefits of improved detection and reduced false positives outweigh the minimal increase in radiation.

Overall, DBT offers several advantages over 2D mammography, including improved image clarity, enhanced lesion visibility, reduced false positives, and better detection in women with dense breasts. These differences contribute to improved diagnostic capabilities and help radiologists in detecting breast cancer at earlier stages, leading to more accurate diagnoses and appropriate treatment planning.

9.5 Conclusion

In the diagnostic environment, ultrasound and 2D mammography were traditionally used in imaging and screening workup. Tomosynthesis is an advanced version of conventional full-field mammography. One of the main advantages is its high spatial resolution at the same time having low radiation doses within the permissible limits. The DBT has high-grade image quality compared to conventional 2D mammography for finding abnormalities and characterization of asymmetries and distortions. The DBT has put forward a mass change in screening and diagnosing environment with better performance in comparison with traditional 2D mammography (improved sensitivity and specificity). Recall reduction rate can be achieved by DBT and thereby a reduction in cost. This preliminary study concludes that DBT has high diagnostic accuracy in the assessment of abnormalities in breast.

The attempt of combining 3D and 2D mammography is a welcoming approach for cancer detection with enhanced accuracy and lessen recalls.

References

[1] B.B. Rewari, S. Gupta, Supportive Care in Oncology: Historical Perspectives and Future Options, in: S.P. Agarwal, S. Gupta (Eds.), Fifty Years of Cancer Control in India, National Cancer Control Programme, Government of India, 2002, pp. 155–183.

[2] A Nandakumar, Consolidated Report of the Population Based Cancer Registries Incidence and Distribution of Cancer: 1990-96, Coordinating Unit, National Cancer Registry Programme, Indian Council of Medical Research, Bangalore, 2001, pp. 1–238.

[3] Breast cancer. https://www.who.int/news-room/fact-sheets/detail/breast-cancer; n.d. [accessed 10.03.23].

[4] F. U, Mammography I Radiology Key. https://radiologykey.com/mammography-5/; n.d. [accessed 9.03.23].

[5] E.D. Pisano, C. Gatsonis, E. Hendrick, M. Yaffe, J.K. Baum, S. Acharyya, et al., Diagnostic performance of digital versus film mammography for breast-cancer screening, New Engl. J. Med. 353 (2005) 1773–1783. Available from: https://doi.org/10.1056/NEJMOA052911.

[6] S. Destounis, A. Arieno, A. Santacroce, Comparison of cancers detected by screening breast ultrasound and digital breast tomosynthesis, Acad. Radiol. 29 (2022) 339–347. Available from: https://doi.org/10.1016/J.ACRA.2021.01.008.

[7] Screening Performance Benchmarks: BCSC. https://www.bcsc-research.org/statistics/screening-performance-benchmarks; n.d. [accessed 10.03.23].

[8] M.A. Helvie, Digital mammography imaging: breast tomosynthesis and advanced applications, Radiol. Clin. North Am. 48 (2010) 917. Available from: https://doi.org/10.1016/J.RCL.2010.06.009.

[9] R. Murakami, N. Uchiyama, H. Tani, T. Yoshida, S. Kumita, Comparative analysis between synthetic mammography reconstructed from digital breast tomosynthesis and full-field digital mammography for breast cancer detection and visibility, Eur. J. Radiol. Open 7 (2020) 100207. Available from: https://doi.org/10.1016/J.EJRO.2019.12.001.

[10] M.P. Jeong, E.A. Franken, M. Garg, L.L. Fajardo, L.T. Niklason, Breast tomosynthesis: present considerations and future applications, Radiographics 27 (2007) Suppl 1. Available from: https://doi.org/10.1148/RG.27SI075511.

[11] A. Smith, E. Rafferty, L. Niklason. Breast tomosynthesis reduces radiologist performance variability compared to digital mammography. In: Proceedings of the European congress of radiology annual meeting. Vienna, Austria: European Congress of Radiology. 2009.

[12] I. Andersson, D.M. Ikeda, S. Zackrisson, M. Ruschin, T. Svahn, P. Timberg, et al., Breast tomosynthesis and digital mammography: a comparison of breast cancer visibility and BI-RADS classification in a population of cancers with subtle mammographic findings, Eur. Radiol. 18 (2008) 2817–2825. Available from: https://doi.org/10.1007/S00330-008-1076-9.

[13] M.L. Marinovich, D. Bernardi, P. Macaskill, A. Ventriglia, V. Sabatino, N. Houssami, Agreement between digital breast tomosynthesis and pathologic tumor size for staging breast cancer, and comparison with standard mammography, Breast 43 (2019) 59–66. Available from: https://doi.org/10.1016/J.BREAST.2018.11.001.

[14] M. Tivnan, C. Rappaport, M. Lambert, D. Lesselier. A modified gradient descent reconstruction algorithm for breast cancer detection using microwave radar and digital breast tomosynthesis. In: 2016 10th European conference on antennas and propagation (EuCAP).2016. https://doi.org/10.1109/EUCAP.2016.7481582.

[15] S.P. Sinha, M.A. Roubidoux, M.A. Helvie, A.V. Nees, M.M. Goodsitt, G.L. LeCarpentier, et al., Multi-modality 3D breast imaging with x-Ray tomosynthesis and automated ultrasound, Ann. Int. Conf. IEEE Eng. Med. Biol. Soc. 2007 (2007) 1335–1338. Available from: https://doi.org/10.1109/IEMBS.2007.4352544.

[16] D. Gur, G.S. Abrams, D.M. Chough, M.A. Ganott, C.M. Hakim, R.L. Perrin, et al. Digital breast tomosynthesis: observer performance study. 2012;193:586–91. https://doi.org/10.2214/AJR.08.2031.

[17] H.R. Peppard, B.E. Nicholson, C.M. Rochman, J.K. Merchant, R.C. Mayo, J.A. Harvey, Digital breast tomosynthesis in the diagnostic setting: indications and clinical applications, Radiographics 35 (2015) 975–990. Available from: https://doi.org/10.1148/RG.2015140204.

[18] R. Gartner Roth, A.D.A. Maidment, S.P. Weinstein, S. Orel Roth, E.F. Conant. Digital breast tomosynthesis: lessons learned from early clinical implementation. 2014: 34. https://doi.org/10.1148/RG.344130087.

[19] E.F. Conant, Clinical implementation of digital breast tomosynthesis, Radiol. Clin. North Am. 52 (2014) 499. Available from: https://doi.org/10.1016/J.RCL.2013.11.013.

[20] N. Houssami, Evidence on synthesized two-dimensional mammography versus digital mammography when using tomosynthesis (three-dimensional mammography) for population breast cancer screening, Clin. Breast Cancer 18 (2018) 255–260. Available from: https://doi.org/10.1016/J.CLBC.2017.09.012. e1.

[21] J.A. Baker, J.Y. Lo, Breast tomosynthesis: state-of-the-art and review of the literature, Acad. Radiol. 18 (2011) 1298–1310. Available from: https://doi.org/10.1016/J.ACRA.2011.06.011.

[22] R. Ranjan, L. Singh, S.K. Arava, M.K. Singh, Margins in skin excision biopsies: principles and guidelines, Indian J. Dermatol. 59 (2014) 567. Available from: https://doi.org/10.4103/0019-5154.143514.

[23] T.M. Svahn, D.P. Chakraborty, D. Ikeda, S. Zackrisson, Y. Do, S. Mattsson, et al., Breast tomosynthesis and digital mammography: a comparison of diagnostic accuracy, Br. J. Radiol. 85 (2012) e1074. Available from: https://doi.org/10.1259/BJR/53282892.

[24] M. Noroozian, L. Hadjiiski, S. Rahnama-Moghadam, K.A. Klein, D.O. Jeffries, R.W. Pinsky, et al., Digital breast tomosynthesis is comparable to mammographic spot views for mass characterization, Radiology 262 (2012) 61–68. Available from: https://doi.org/10.1148/RADIOL.11101763.

[25] E.A. Rafferty, J.M. Park, L.E. Philpotts, S.P. Poplack, J.H. Sumkin, E.F. Halpern, et al., Assessing radiologist performance using combined digital mammography and breast tomosynthesis compared with digital mammography alone: Results of a multicenter, multireader trial, Radiology 266 (2013) 104–113. Available from: https://doi.org/10.1148/RADIOL.12120674/-/DC1.

[26] M.L. Zuley, A.I. Bandos, M.A. Ganott, J.H. Sumkin, A.E. Kelly, V.J. Catullo, et al., Digital breast tomosynthesis versus supplemental diagnostic mammographic views for evaluation of noncalcified breast lesions, Radiology 266 (2013) 89–95. Available from: https://doi.org/10.1148/RADIOL.12120552.

[27] C. Hayes, D. Jackson, P.M. Davidson, T. Power, Medication errors in hospitals: a literature review of disruptions to nursing practice during medication administration, J. Clin. Nurs. 24 (2015) 3063–3076. Available from: https://doi.org/10.1111/JOCN.12944.

[28] K. Flobbe, A.M. Bosch, A.G.H. Kessels, G.L. Beets, P.J. Nelemans, M.F. Von Meyenfeldt, et al., The additional diagnostic value of ultrasonography in the diagnosis of breast cancer, Arch. Intern. Med. 163 (2003) 1194–1199. Available from: https://doi.org/10.1001/ARCHINTE.163.10.1194.

[29] S.H. Lee, J.M. Chang, N. Cho, H.R. Koo, A. Yi, S.J. Kim, et al., Practice guideline for the performance of breast ultrasound elastography, Ultrasonography 33 (2014) 3. Available from: https://doi.org/10.14366/USG.13012.

[30] N. Houssami, L. Irwig, J.M. Simpson, M. McKessar, S. Blome, J. Noakes, Sydney breast imaging accuracy study: comparative sensitivity and specificity of mammography and sonography in young women with symptoms, AJR Am. J. Roentgenol. 180 (2003) 935–940. Available from: https://doi.org/10.2214/AJR.180.4.1800935.

Chapter 10

Study on Alzheimer's detection using machine learning and deep learning techniques

Arun Kumar Dhuraisamy and G. Subashini
Department of Robotics and Automation, PSG College of Technology, Coimbatore, Tamil Nadu, India

10.1 Introduction

In this era of rapid enhancement in technological aspects, created a vacuum in deploying these advancements in various nuances in our day-to-day life. These nuances could be considered from manufacturing to educational fields. In the long list of fields lies a prominent field involved in revolutionizing the human life, it's medical field. The emerging technologies like machine learning (ML), deep learning (DL), computer vision, natural language processing, big data analytics, etc., could be imparted into the medicinal routine to aid in effective healthcare of the patients. This chapter aims in aiding the diseases that are caused in the primal area, the brain. The brain diseases are often unnoticed. It is usually very difficult for the doctors to detect and treat patients with brain diseases. The conventional method of medicine is often prone to failure when it comes to treating of issues of the brain. Usually, the brain diseases are majorly categorized as the following:

1. Alzheimer's disease
2. Parkinson disease
3. Multiple sclerosis
4. Epilepsy
5. Huntington's disease
6. Brain tumors
7. Stroke

In this work, Alzheimer's is considered as a test case and the various aspects of applying ML, DL, and biomedical imaging solutions to find a better care for the healthcare system.

Alzheimer's, a progressive neurodegenerative disease, that reduces the human being's capability of cognitive skills and behavioral habits. The people of age after 60 and up to 80 are highly prone to Alzheimer's. The probability of the elderly population getting affected to Alzheimer's increasing exponentially twice in every 5 years and is progressive. From Fig. 10.1, one can outlay how the Alzheimer's progresses along the time in the brain. Basically, the neural regions of the brain shrink losing its functional capability. Cerebral cortex of the brain in addition to the hippocampus shrinks with the massive enlargement of the ventricles. The disease is capable of rendering the human to lose his mental state.

So, now understanding the sensitivity of the disease calls upon these modern technologies to help us in getting better care. The growing high-end technologies in ML and DL are capable of capturing even tiniest pattern in the growth progression of the disease. These ML and DL techniques accelerated by the modern computer vision and biomedical imaging techniques that help in better identification pattern or feature in the brain. These pattern and feature in the Alzheimer's growth can only be noticed by an expert doctor in the field of neurology. At least, the probability of detecting won't be much effective as it solely depends on the physician's capability. These techniques aid in helping out the physicians in pointing out the disease progression and aiding a far more accurate and precise care of the patients. These techniques are also being implemented for brain tumor detection, breast cancer detection, heart disease detection, etc.,

Intelligent Computing Techniques in Biomedical Imaging. DOI: https://doi.org/10.1016/B978-0-443-15999-2.00019-0

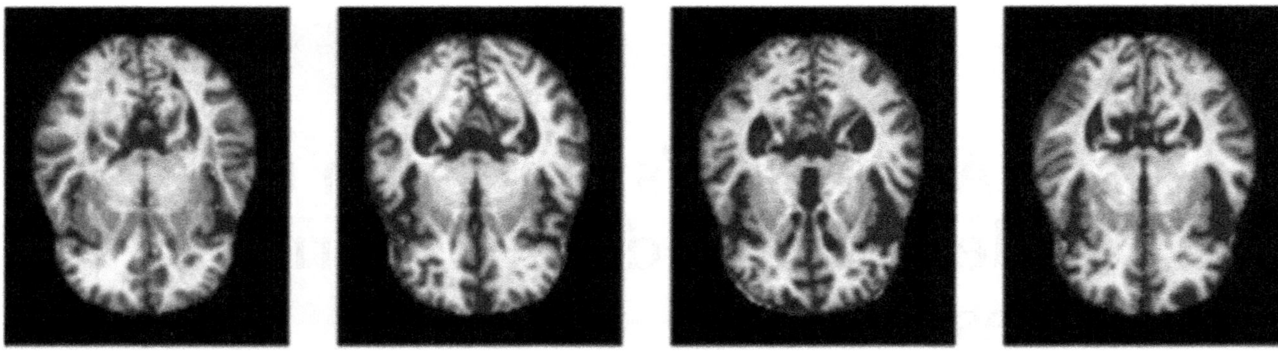

(A) (B) (C) (D)

FIGURE 10.1 Progression of Alzheimer's disease: (A) none demented, (B) moderately demented, (C) mildly demented, and (D) very mildly demented.

extending the scope to this case. To diagnose the Alzheimer's in the affected patients, the magnetic resonance imaging (MRI) scan is being considered. The MRI dataset is labeled with the four conditions of the Alzheimer's is provided to the learning model. The four conditions of the people suffering from Alzheimer's are mildly demented, moderately demented, nondemented, and very mildly demented. The proper preprocessing techniques and feature extraction are being carried out in order to diagnose the patients in a much more effective way. The main cause behind this chapter is to identify an effective learning, efficient preprocessing procedures, and proper feature extraction techniques to detect Alzheimer's disease detection.

10.1.1 Current challenges

Alzheimer's is a very common brain disease that causes the cognitive decline, memory loss, behavioral changes, motor disabilities (in final progression of the disease) and lack of spatial awareness. These symptoms signify the need for earlier diagnosis of the diseases. The conventional method of diagnosis has various cons that deems it undesirable keeping patient's early recuperation in mind. The conventional method of diagnosis is more subjective and variable. The diagnosis is the result of interpreting the test results of the diagnostic tests. These results are based on the interpretation of the medical professional. The results may vary from professional to professional, the underlying issue for this is that Alzheimer's is a progressive disease so the chance of diagnosing the disease at an earlier stage is uncertain for the healthcare professional. Uncertainty could be other probable diseases so to be certain for a medical professional is difficult at the earlier stage. It could be dementia or age-related decline in cognitive ability which could be very tough for a human being to distinguish when the symptoms are very minimum. The need for more factors weighs the accuracy of the diagnosis. But the probability of diagnosing increases exponentially as the disease progression increases which undercuts the notion of putting the patients care on top. So, a need for a tool that could diagnose Alzheimer's at the earlier stage making it objective based on the pattern in the diagnostic test results. Early detection of the Alzheimer's is very difficult due to the very less symptoms. So, the chance of misdiagnosing is very high making the patient undergo wrong medication which could be very harmful. Chances of the misdiagnosing are very high and happens a lot when it comes to localities with less advanced medical technology exposure. Often dementia and Alzheimer's are misdiagnosed such uncertainties could cause a great deal of mental stress to the patient and his medical care. Another vital reason is that the lack of expert specialists in the field. Even, a fair number of specialists are present but they are not in immediate access to most of the people. So due to lack of experienced specialists the patients settle for intermediate care they can get at their immediate convenience. This majorly deviates from the notion of patient care and earlier recuperation.

Incorporation of the ML and DL approaches help objective diagnosing of Alzheimer's disease. Learning models process large datasets of the relevant disease to find underlying patterns and combination of biological markers in the MRI scans and provide a diagnosis. The models have much more ability to note small patterns in the scans that could be normal for a human being's perception. This increases the accuracy and the effectives of the diagnosis. The models are able to distinguish in the disease in the earlier stage make the early diagnosing better for the patient's treatment. The models are also helping the doctors understand the comprehensive view of the patient's medical chart that gives a logical interpretation to approach the treatment of the patient. The uncertainties are reduced in a very large scale and having a human supervision on top of this tool helps the diagnosis more accurate. This could not only revolutionize the patient's care but also increase the understanding on Alzheimer's and the human body itself to the healthcare population which leads to an era of objective medical treatment.

10.1.2 Objective

To approach the needs to satisfy the patient's care and to get a much more understanding of Alzheimer's and the effective method of diagnosing, it's at most to define the scope of this study. The main objective is to devise a tool which has early and accurate diagnosing of the Alzheimer's disease. The earlier detection proves very prominent in the healthier rehabilitation of the patient. The other aspect is to have an objective model with biomarker identification from MRIs to find patterns and relationship with the distinct feature points and patterns of Alzheimer's. As mentioned in the challenges, the subjective approach of diagnosing patients by healthcare professional has very low accuracy of correct diagnosis. The tool with the high accuracy will automatically influence in helping people to believe that AI could be a trustworthy tool in medicinal field. This brings to the approach for the future that a multimodel diagnostic approach where the multimodel diagnostic AI tools could detect multiple diseases. The futuristic scope of the paper is contributing the understanding of diagnosing Alzheimer's so that the development of such multimodel disease diagnosing tools could be developed with better results. To conclude the study aims in leveraging the ML and DL techniques to find an objective ground to have a better diagnosis tool which should be more reasonable in providing better and effective results.

10.2 Literature review

Before listing the effective method to detect Alzheimer's, it's high time that the existing technologies are being understood. The below passages depict a detailed description of the existing systems available being used in the medical industry.

In this reference [1], the author considers a 2D-weighted T1-MRI. In the T1 type of MRIs the fatty tissues are being enhanced. The system utilizes shallow convolutional neural network (CNN) architecture. The paper entails the normalization of the images since the MRI data are highly prone to noise. The resizing is also done to make the images in the correct size ensuring optimal processing. The optimal resized and normalized image is provided across the shallow CNN architecture. The proposed system by the author got an accuracy of 100% and testing accuracy of 99.68%. The author compares the results of the training and testing across the various predefined architectures like DenseNet121, ResNet50, VGG 16, EfficientNetB7, and InceptionV3. The shallow CNN proved to more effective and accurate than the other transfer learning models.

The reference [2] entails how to increase the accuracy of the model with the help of data augmentation. The author notably points out that with the help of clinical and laboratory tests the probability of detecting the Alzheimer's is at only 77%. This signifies the necessity ML and DL techniques powered by advanced computer vision and biomedical imaging techniques. The author considers a CNN-based ResNet50 model. This learning model was able to provide F1 score of 89%. To increase the accuracy, data augmentation is implemented. GANs—generative adversarial networks (GANs) are used as they generate augmented data in a large scale with entailing the features in the real source data. The reason for GANs widely used for data augmentation with two neural networks imparted which acts as generator and discriminator. The author uses cycle GAN for data augmentation. After the usage of cycle GAN, the F1 score of the model increased to 95% making it an effective method for data augmentation.

The reference [3] brings about the ML techniques that are used in the medical imaging for Alzheimer's. Usually, the neural network-based models have high computational cost. The implementation of ML techniques could simplify the issue. The DL models identify the feature without the need of feature extraction whereas the ML models ought to be extracted with the features. The author uses an adaptive mean filter to reduce the overall noise from the MRI which is followed by histogram equalization to improve the image quality. The Haar wavelet transform is used for capturing the features in the MRI data. The author trains and tests MRI data against the least square-support vector machine-radial basis function (LS-SVM-RBF), support vector machine (SVM), K-nearest neighbour (KNN), and random forest classifier learning models. From the experiment performed the author was able to identify that the LS-SVM-RBF offers high accuracy comparative to the other ML models.

This paper [4] revolves around improving the accuracy of the detection as the conventional CNN-based architecture. The author uses the DenseNet169- and ResNet50-based CNN architecture to improve the accuracy. The author uses the regular data augmentation to improve the quantity of the data thereby increasing the model training and testing rate. The accuracy of DenseNet169 and ResNet50 is 97.7% and 88.7% which is far more improved from the conventional CNN models.

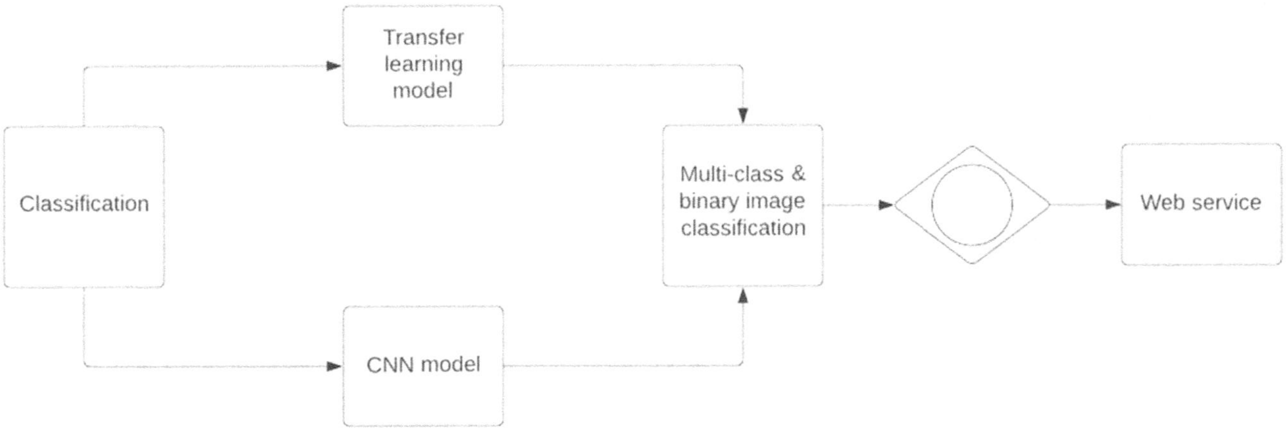

FIGURE 10.2 Classification and web layer of the reference system.

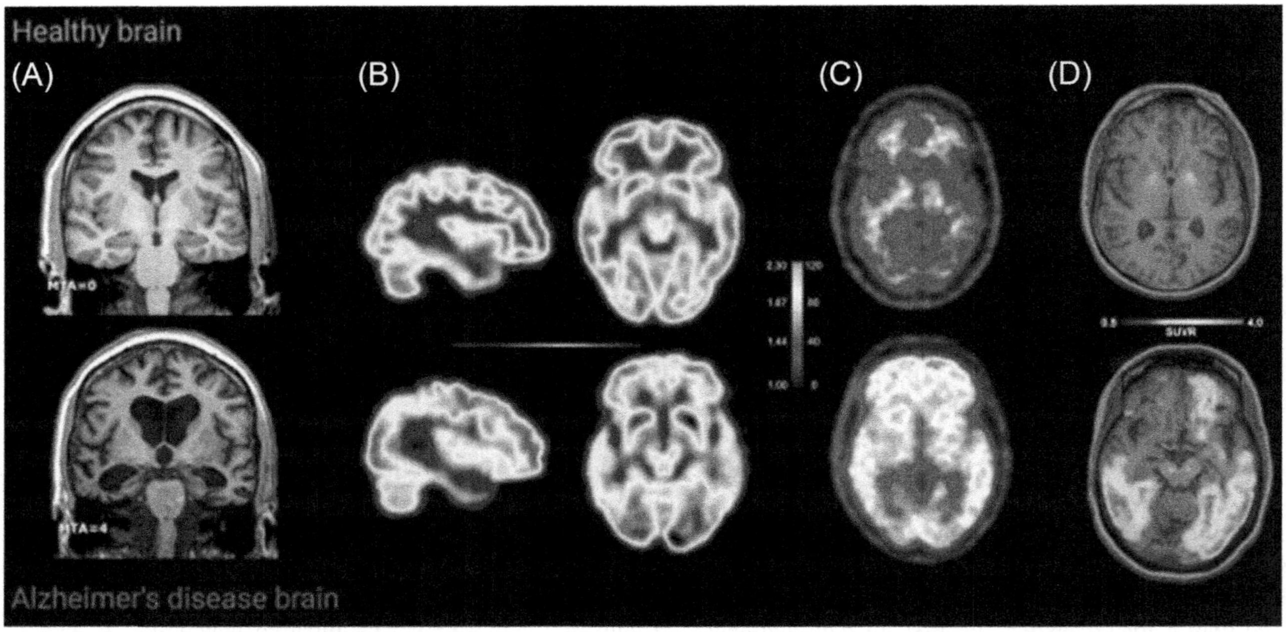

FIGURE 10.3 Neuroimages of the healthy versus the Alzheimer's disease (AD) brain. Neuroimaging with (A) structural magnetic resonance imaging, (B) Fluorodeoxyglucose positron emission tomography (FDG-PET), (C) amyloid-PET with Pittsburgh compound-B (PiB), and (D) tau PET with 18F-AV1451 in both healthy and AD brains.

This paper [5] provides a brief description on detection of Alzheimer's using VGG19 model and also extensively exploring the 2D- and 3D-based image scan detection in addition to the web service-based real-time detection. The author considers dataset which is being gathered from the MRI machines and these obtained data are being converted into .jpg formant. The conventional preprocessing method, that is, data augmentation is used. The dataset has four classes of data and the classification step involves two stages. The first stage involves the CNN-based architecture and the second stage is all about classification using VGG19 model. The system uses a web-based deployment for the real-time analysis. The system uses python framework for the back-end processing and the front-end is being implemented using HTML, CSS, and JavaScript (Fig. 10.2).

This paper [6] briefly discusses the dataset that is required for the diagnosing of Alzheimer's disease. The dataset can either be obtained from MRI or positron emission tomography (PET). The reference paper debates on the effectiveness of both these scans which provides a better accuracy. The schematic diagram (Fig. 10.3) provides the procedure for this experiment

The PET scan image is usually the one in RGB format and MRI is in black and white. From the reference, the MRI scan suits to be better one for detection of Alzheimer's detection as PET's are very expensive and are computational costly for processing. So, the use of MRIs in biomedical imaging is very efficient and cost effective.

10.3 Proposed system

The significance of the Alzheimer's has been established and the existing systems are also been studied. This significance and the literature study would help in designing a more effective and efficient system. The proposed system is initially separated into various modules, these modules are studied and the optimal technique has been selected to have a system with high accuracy and reliability. The various modules of the proposed system are listed below:

1. Data acquisition
2. Preprocessing
3. Feature extraction
4. Model selection
5. Model training
6. Model testing

The proposed system follows the below-mentioned workflows:

- The data from the MRIs need to be properly acquired from the MRI machine. The noise, database, and the patient's posture are all to be monitored and set in the proper way to get high defined data.
- The dataset is considered from Kaggle which has four classes namely
  - Mildly demented
  - Moderately demented
  - Nondemented
  - Very mildly demented
- The data is preprocessed by using adaptive mean filter to remove the noise and histogram equalization for enhancing the quality of the data.
- The features captured using the Haar wavelet transform to capture the features in the MRI to identify the pattern in Alzheimer's affected brain.
- The model is then selected, in this case the CNN with 5-layer architecture is selected as it was optimal and did not over fit.
- The model is trained using CNN architecture for a standard epoch of 100 with Adam optimizers. The training accuracy obtained by the CNN architecture was 100% which means the model fits very accurately and precisely.
- The model is then tested against the testing dataset for the performance and found to be efficient.

10.4 Data acquisition

The role of data is very essential when it comes to the implementation of any ML or DL algorithm. Though the quality of the input data plays an eminent role in predicting the target variables, the data acquisition technique carried out to get the data is more important. The role of the acquisition system is very pivotal as the data obtained is corrupted in the acquisition process then the whole system fails. Ultimately, this stage of the learning model deployment should be done with care and precision. The data is gathered in the form of MRI scans. These MRI scans work with the coupling effect of high magnetic field and radio waves. Using this coupled phenomena, one can generate the MRI scans of the desired body part, in this case the brain. During the data acquisition the possibility of many data acquisition failure is prone to occur. The quality of the data is smudged by this mishap. For instance, let's consider the patient is moving continuously while taking scan, this will affect the quality of the scan as the data gathered won't be much accurate. To avoid this Head immobilization devices and padding is provided for undesired motion of the patient during MRI scans. The chance of the machine malfunction is also to be considered so it is highly noted that the scanner and imaging protocol should be carefully calibrated and tested. The data management center is also a prime area of concern while considering the MRI scans. The scanned data should be stored in a proper database which cannot be corrupted in any case. The corruption of data in the database level will lead to the entire failure of the system (Fig. 10.4).

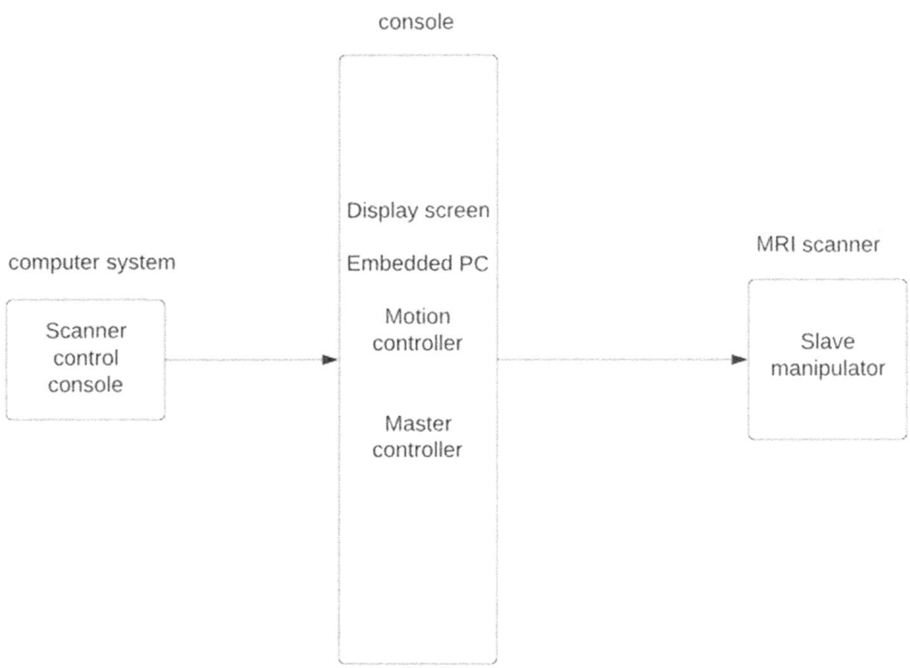

FIGURE 10.4 Working of an MRI machine.

The MRI machine uses magnet, gradient coil, RF coil, and operational amplifiers which are connected with a digital control logic and analog to digital converter. This basic skeleton circuit design of the MRI should be present in order to obtain an effective MRI scan.

10.5 Prepocessing

The preprocessing stage is applying any processing or computation on the data before providing it to the learning model. The preprocessing is the most essential for cleaning the source raw dataset to get an accurate predicting model. The Alzheimer's disease uses MRI scans to detect the disease. These MRI scans are monochrome in nature. The data might be noisy or image can be unclear. Usually, the most effective filter to apply to MRI scans is the adaptive mean filter which removes the noise in the data (Fig. 10.5).

The image obtained by the filters is not a qualitative one, so histogram equalization is applied on MRI scan. The histogram equalization is subjected on the scan to adjust the image contrast. This method also enhances the quality of the image. Another advantage of histogram equalization is that the method normalizes the pixel values of the image which ultimately the high intensity images as it is might lead to high computation. The equalization enhances the quality of the MRI thereby making the scan even clearer where various features could be extracted to find the Alzheimer's (Fig. 10.6).

The other main data preprocessing step is data augmentation. The technique aims in increasing the dimensionality of the dataset so as to increase the accuracy of the final deployment. Data augmentation is conventionally done by carrying out the basic translation, rotational, zooming in and zooming out of the image. Though these technologies are being used in the various instances of the medical imaging, the widely proven effective technique for data augmentation is GAN. The GAN's learning model has two neural networks, namely,

1. Generator—generates images, videos which look exactly like the real source data.
2. Discriminator—scrutinizes the images produced by the generator to make it look real.

Based on this mechanism, a very effective data augmentation is obtained unlike the conventional methods (Fig. 10.7).

10.6 Feature extraction

Now that image is all preprocessed it could be directly provided to the learning model but it's very essential to always implement feature extraction to the preprocessed dataset before giving it to the training model. The feature extraction is used to highlight the unique pattern in the data, so it will be the useful for the model to learn the data based on this

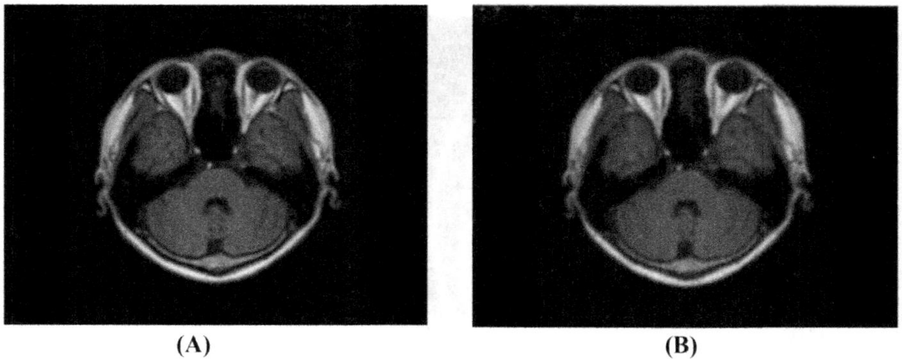

FIGURE 10.5 (A) Original image, (B) image with adaptive mean filter.

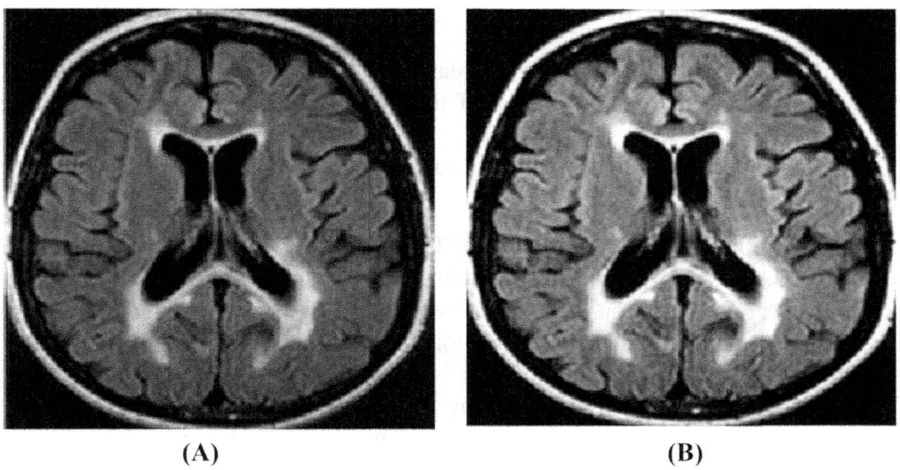

FIGURE 10.6 (A) Original image, (B) image with histogram equalization.

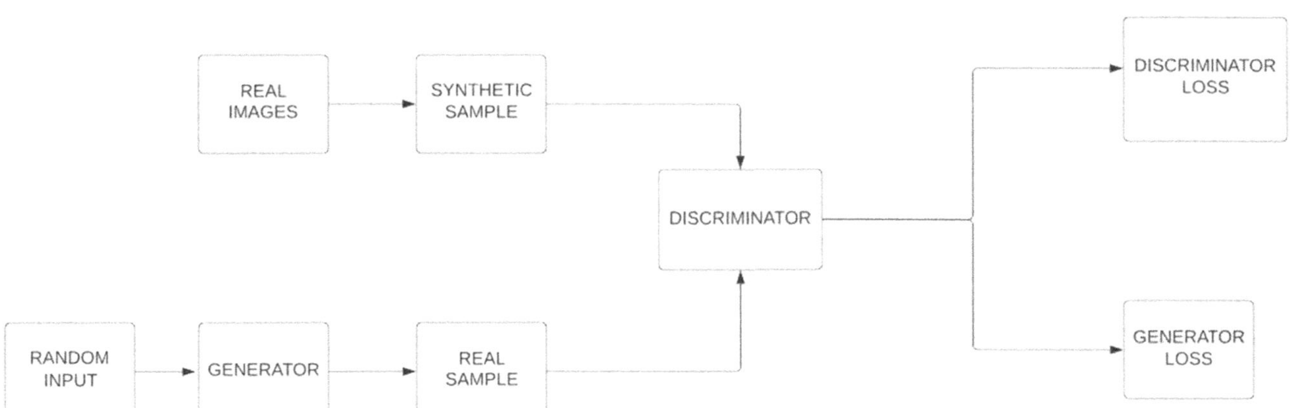

FIGURE 10.7 Data augmentation using generative adversarial network.

feature. In medical imaging the feature extraction for the brain disorder will highlight the portion of the infected area so that the model could use it as learning pattern to classify while being deployed. The various types of feature extraction techniques in MRI scans are provided below:

1. Intensity-based features rely on pixel intensity of the image, such as mean intensity, standard deviation, and entropy.
2. Texture-based features entails the spatial arrangement in pixel intensities.

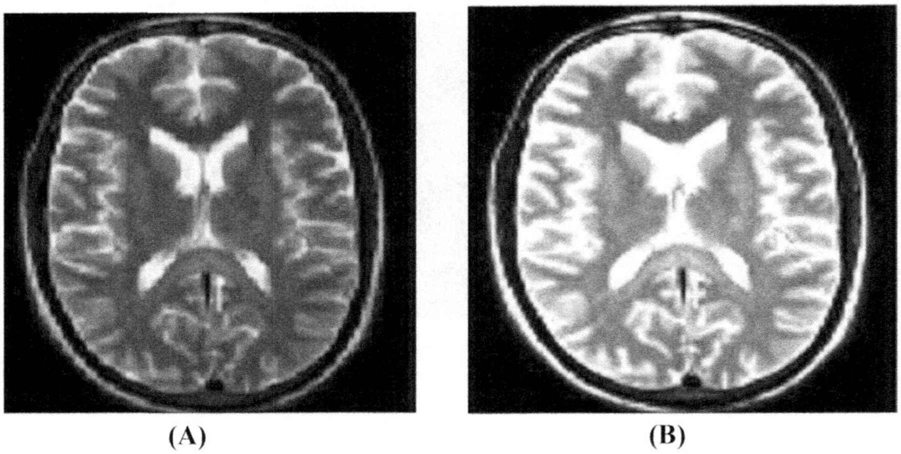

FIGURE 10.8 (A) Original image, (B) image with Haar wavelet transform.

3. Shape-based features signifies the shape and geometry of objects within the image, such as area, perimeter, and circularity.
4. Wavelet-based features captures with reference to the decomposition of the image into multiple frequency bands using wavelet transforms.
5. DL-based features are extracted using deep neural networks that learn automatically to identify relevant features from the raw image data.

Based on the need, suitable method could be opted to extract the feature. The Haar wavelet transform is one of the predominantly used extraction technique for Alzheimer's. Texture and structure of an image can be studied using the Haar wavelet transform in the context of MRI feature extraction. It operates by segmenting the main image into a number of smaller images, each of which represents a distinct scale and orientation. The image is divided into two parts at each scale: an approximation image and a detail image.

The detail image depicts the high-frequency information, whereas the approximation image depicts the low-frequency information. By examining wavelet coefficients at various scales and orientations, the Haar wavelet transform can be used to extract information from MRI images. Following that, a variety of applications, including picture segmentation, registration, and classification, can make use of these attributes (Fig. 10.8).

10.7 Model selection

Now that the raw source data are acquired, preprocessed and features are being extracted, the next step is selecting a model architecture to diagnose Alzheimer's disease. Normally, the model being selected is done after a careful study of the all the available models that are appropriate for the given dataset. From the literature study conducted, it can be concurred that if the approach is an ML model, then LS-SVM-RBF is more effective. If the approach is a DL model, then CNN is more effective. So, now the decision is to be made whether to opt for ML approach or DL approach. So, usually the approach is decided by the dataset available to us. If the dataset is small and less complex, then the ML approach is being opted. If the dataset is large and complex, then the DL approach is being opted. In this case, the dataset is vast as the dataset consists of four classes so opting CNN is logical and efficient. The best part of CNN is that the model is used in classification, segmentation, and reconstruction. The model can learn features in the data without any feature extraction (Fig. 10.9).

10.8 Model training

The model training is one of primitive step when it comes to implementing learning models. The training step basically trains the model with the given preprocessed and feature extracted data to which it trains itself. The training architecture is very important. The architecture determines the effectiveness of the algorithm. Here, the model chosen is CNN – Convolutional Neural Network. The CNN architecture considered is a 5-layer model with maxpooling layer and Relu as the activation function. Batch Size of the training model is 40. The Learning Rate of the model is given as 0.001. The Optimizer used is Adam. The Model is trained for a total Epochs of 100. The dataset for Alzheimer's disease classification has four classes (Figs. 10.10 and 10.11).

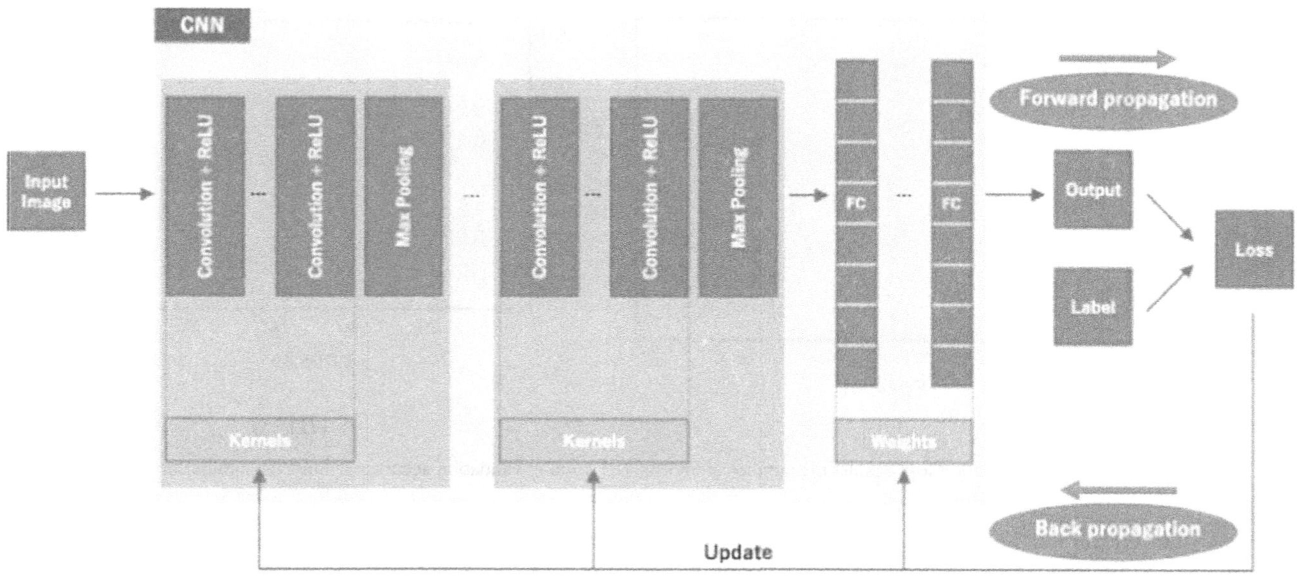

FIGURE 10.9 Convolutional neural network architecture.

Layer	Convolution	Maxpooling	Dropout
First layer	filters = 64, kernel_size = (5,5), padding = 'Same', activation = 'relu'	pool_size = (2,2)	(0.25)
Second Layer	filters = 128, kernel_size = (3,3), padding = 'Same', activation = 'relu'	pool_size = (2,2), strides = (2,2)	(0.25)
Third Layer	filters = 128, kernel_size = (3,3), padding = 'Same', activation = 'relu'	pool_size = (2,2), strides = (2,2)	(0.3)
Fourth Layer	filters = 128, kernel_size = (2,2), padding = 'Same', activation = 'relu'	pool_size = (2,2), strides = (2,2)	(0.3)
Fifth Layer	filters = 256, kernel_size = (2,2), padding = 'Same', activation = 'relu'	pool_size = (2,2), strides = (2,2)	(0.3)
Batch Size		40	
Learning Rate		0.001	
Optimizer		Adam	
No. of Epochs		100	
Total Parameters		4,619,524	
Trainable Parameters		4,619,524	
Non-Trainable Parameters		0	

FIGURE 10.10 Training summary.

From the training and the validation loss/accuracy, one can concur that the CNN architecture works more effective on the considered dataset. The model is not to overfitted, it's optimal. In the training dataset, 80% is considered for training and 20% is taken for network testing. From the 80% considered for training, 16% is taken as validation set.

10.9 Model testing

Now that the model is trained completely for the given CNN architecture and the dataset. The testing is to be done to check the accuracy and precision of the model in detecting Alzheimer's. The testing data is considered and given to the model that was trained. Later, the predicted results and the ground truth are compared to form a conclusion on how the model is effective. The model provided a 100% accuracy it in training. The testing accuracy is 99.68%. From this testing accuracy result, it could be concluded that the detection of Alzheimer's it was very accurate with help of the proposed system flow and techniques. From this observation, it could be inferred that the model is fine-tuned, not over

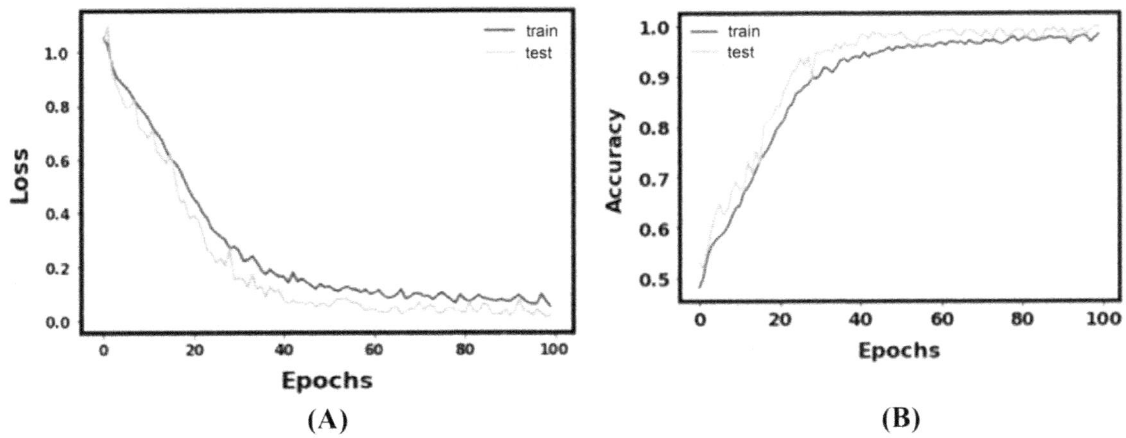

FIGURE 10.11 (A) Training validation loss against the number of epochs, (B) training validation accuracy against the number of epochs.

fitted and very accurate. The accuracy of model is very high due to the feature extraction and preprocessing techniques majorly. The CNN model also uses its feature extraction and high robust model to diagnose the Alzheimer's.

10.10 Limitations

The methodology adopted in this paper provides an approach that is prominent and effective in diagnosing Alzheimer's yet there are other factors that might hinder the ability of the approach to provide better results. The main notion of this approach depends on the dataset that is considered. But to diagnose diverse and different stages of the Alzheimer's, the need for diverse and various stage depicting dataset is much needed. The dataset of such dimensionality and feature is not yet aggregated so without that the it would be much difficult to solve this approach. The other factor being that ML and DL techniques diagnose based on the feature markers present in the data but there is no logical reasoning behind such diagnosis. So, there is always a stance of uncertainty present in such approaches. The model is trained and tested on the dataset that is available but it isn't always true that such dataset will be generalized for all real-world scenario. The model needs to be transferable and reproducible to the real-world data. There is a class imbalance in the data collection for the dataset. To be clear, the people suffering from Alzheimer's is much lesser than the people without suffering from Alzheimer's. So, due to this need for a class balanced data for the model is much high to be deployed in the real-world aspects. The certainty of people to trust the tool very doubted. So irrespective of the study methodologies to have a better diagnosing tool, people should be in a stance to trust the results so that this could be deployed in the day-to-day situations.

10.11 Future scope

Detection system of Alzheimer's disease has been developed and deployed but the scalability of the system is very essential. Presently, a clear diagnosis of Alzheimer's can only be made by an investigation on the brain tissue after death. But, improvements in imaging technology might make it possible to identify the condition early. For instance, beta-amyloid plaques, telltale signs of Alzheimer's, can be found in living brains using PET scans. The identification of Alzheimer's disease in its early stages may become more precise and dependable with further advancements in PET imaging. Researchers can better understand how Alzheimer's disease develops and impacts the brain by using biomedical imaging. MRI, for instance, can reveal alterations in brain structure and function over time. Researchers may be able to create more efficient medications by examining these alterations. Alzheimer's sufferers may benefit from individualized care thanks to biomedical imaging. For instance, doctors determine which patients are more likely to respond to specific treatments by examining PET scans. This might result in treatment that are more precise and successful. Another application for biomedical imaging is to evaluate how well Alzheimer's disease treatments are working. For instance, using PET scans to monitor changes in beta-amyloid levels over time could reveal if a certain medication is slowing or halting the disease's progression. These are the various researches that are being currently researched in the field of biomedical imaging for Alzheimer's disease. Overall, biomedical imaging has great promise for the treatment and preventative measures of Alzheimer's disease. Further developments in this field may have a big impact on these issues.

The prominent futuristic scope of this study can be leveraged to develop a multimodel diagnostic tool which can diagnose multiple disease from the patient data. This tool will revolutionize the medicinal field in diagnosing the patients.

The medical care would be very easy to access and the patient well-being and treatment could be improved much better. The other aspect is to improvise this methodology on top of the class balanced and diverse dataset so that it could be incorporated for a much more generalized real-world situation. The dataset should be evaluated very distinctly to support the agreement of providing an effective diagnosing tool. The ML and DL techniques used guarantees the diagnosis based on the patterns and featured biomarkers but there is no underlying explanation that being a effective grounds for diagnosis. So, the future scope could move towards having a tool that gives you a fair explanation on the diagnosis so that the uncertainties in the results could be avoided. These futuristic paths could be prominent in solving the issue of having an effective diagnostic system.

10.12 Conclusion

people all around the globe are affected by this neurological condition known as Alzheimer's. It currently has no known cure, but early detection and treatment can enhance patients' quality of life and decrease the course of illness. A possible method for diagnosing Alzheimer's disease is biomedical imaging. The defining characteristics of Alzheimer's disease, such as structural abnormalities in the brain, beta-amyloid plaques, and neurofibrillary tangles, have been identified using a variety of biomedical imaging techniques. In Alzheimer's research, imaging techniques like computed tomography, PET, and MRI are frequently used. The accuracy and reliability of Alzheimer's disease identification by biomedical imaging have been improved by the development of advanced image analysis techniques, including ML and DL. These methods may evaluate enormous picture datasets and spot minor alterations in the structure and function of the brain that might be a sign of Alzheimer's disease.

CNNs have been utilized to evaluate MRI and PET scans with good accuracy rates for the diagnosing of Alzheimer's. One instance of study showed that employing a CNN to identify Alzheimer's disease from brain MRI scans had a detection rate of above 95%, according to the journal NeuroImage. CNN was utilized in a different study that was published in the journal Alzheimer's and Dementia to accurately identify Alzheimer's from PET images with over 90% accuracy. With earlier treatment and better patient outcomes, biomedical imaging offers the potential to enhance early identification and diagnosis of Alzheimer's disease. Additionally, it can aid in the development of more efficient treatments by allowing researchers to better understand the course and effects of Alzheimer's disease on the brain. However, there are still several issues that need to be resolved when using biomedical imaging to determine Alzheimer's. Lack of defined imaging methods is one issue, which can cause variances in image consistency and quality. Overall, the application of biomedical imaging for Alzheimer's disease detection is a fast-developing field with enormous promise to significantly advance the understanding, management, and prevention of this tragic condition. Biomedical imaging is predicted to play a bigger part treatment of Alzheimer's when more sophisticated imaging methods and processing algorithms are developed.

References

[1] M. EL-Geneedy, H. El-Din Moustafa, F. Khalifa, H. Khater, E. AbdElhalim, An MRI-based deep learning approach for accurate detection of Alzheimer's disease, Alex. Eng. J. 63 (2023) 211–221. ISSN 1110-0168.

[2] A.B. Shaker Hammam Badr, MRI-based diagnosis of Alzheimer's disease using deep learning with cycleGAN for data augmentation, Int. J. Health Sci. Res. 4 (2022) 1–7.

[3] K. Mustafa, R. Pratap, Adimulam, N. Mohd, Z. Abu, P. Nancy, et al., Machine learning and image processing enabled evolutionary framework for brain MRI analysis for Alzheimer's disease detection, Comput. Intell. Neurosci. 2022 (2022) 1–8. Available from: https://doi.org/10.1155/2022/5261942.

[4] W. Al Shehri, Alzheimer's disease diagnosis and classification using deep learning techniques, PeerJ Comput. Sci. 8 (2022) e1177. Available from: https://doi.org/10.7717/peerj-cs.1177.

[5] H.A. Helaly, M. Badawy, A.Y. Haikal, Deep learning approach for early detection of Alzheimer's disease, Cogn. Comput. 14 (2022) 1711–1727. Available from: https://doi.org/10.1007/s12559-021-09946-2.

[6] A. Akhilesh, V. Sourabh, C. Prasun, C. Tulika, E. Ahmed, N. Mohammad, et al., A Systematic Review on Machine Learning and Deep Learning Techniques in the Effective Diagnosis of Alzheimer's Disease, 09 September 2022, PREPRINT (Version 1) (2022). Available from: https://doi.org/10.21203/rs.3.rs-2028945/v1.

Section V

Modern applications (Case studies)

Chapter 11

Early detection of pediatrics renal disease using ultrasound imaging

Moushmi Kar and Susmita Ghosh Mazumadar
Department of Electronics, PTSNS University, Shahdol, Madhya Pradesh, India

11.1 Introduction

Compact, transportable, and high-quality ultrasound equipment have been made possible in recent years by technology advancements [1].

Renal diseases are a common cause of morbidity and mortality in children. Renal ultrasound is a noninvasive imaging modality that can be used to identify a wide range of renal diseases in children.

In pediatric nephrology and urology patients, renal ultrasonography is frequently used as the initial screening exam to detect vascular anomalies, urolithiasis, nephrocalcinosis, renal fusion, bladder outlet obstruction, cystic kidney diseases, and congenital anatomical variations of the kidney and urinary tract [2,3].

Several ultrasound techniques are available for the identification of renal diseases in children, including gray ultrasound, 2D ultrasound, Doppler ultrasound, and contrast-enhanced ultrasound (CEUS). In this literature review, we will explore the use of different ultrasound techniques for the identification of renal diseases in children. Gray-scale ultrasound is the most commonly used technique in renal imaging. It can identify abnormalities in the size, shape, and echogenicity of the kidneys. In a study by Lin et al. [4], the kidney and the dilated pelvicalyceal system, as shown in Fig. 11.1, gray-scale ultrasound was found to have a sensitivity of 90% and specificity of 100% for the detection of hydronephrosis in children. In the same way, Doppler ultrasound is used to assess blood flow in the renal arteries and veins. In a study by Makhija et al. [5], Doppler ultrasound was found to be useful in identifying renal artery stenosis in children with hypertension. CEUS involves the injection of a contrast agent to improve the visualization of renal blood vessels. In a study by Dong et al. [6], CEUS was found to be a reliable method for the diagnosis of renal cortical necrosis in children.

Three-dimensional ultrasound enables the visualization of the kidneys in three dimensions, which enhances the precision of measuring renal volume and identifying any abnormalities in the kidneys. In a study by Vester et al. [7,8], three-dimensional ultrasound was found to be useful in the diagnosis of renal cystic diseases in children. Therefore

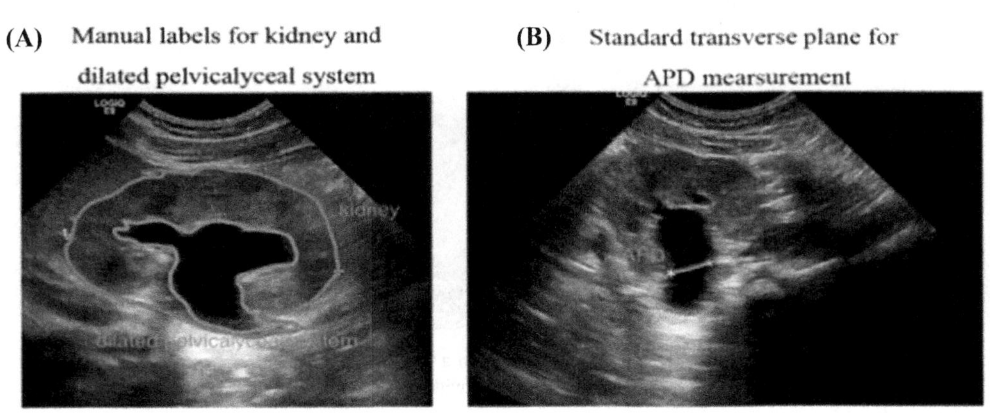

FIGURE 11.1 Images of the kidneys taken with a convex probe and B-mode ultrasonography. (A) The labels for the kidney and dilated pelvicalyceal system were drawn by hand. The area fraction was measured in this view as the projected fluid-to-kidney-area ratio. (B) The usual anterior-posterior diameter measurement is shown in this image from the middle transverse plane [4].

Intelligent Computing Techniques in Biomedical Imaging. DOI: https://doi.org/10.1016/B978-0-443-15999-2.00013-X

different ultrasound techniques can be used for the identification of renal diseases in children. Gray-scale ultrasound is the most commonly used technique, but other techniques such as Doppler ultrasound, CEUS, and three-dimensional ultrasound can also be useful in specific clinical scenarios.

To fully understand the function of ultrasound imaging in the early diagnosis of renal disease, this review paper will give a thorough overview of the subject. In particular, the paper will go through the many ultrasound detection methods, the different kidney disorders that can be found using ultrasound imaging, and the advantages and disadvantages of using ultrasound for early detection. For patients at risk of developing renal disease, the chapter will also include guidelines for the frequency and timing of ultrasonography scans. This review study intends to add to the expanding body of knowledge on the successful management of renal disease by emphasizing the significance of early identification and the possible advantages of ultrasound imaging.

11.2 Ultrasound

11.2.1 2D ultrasound

2D ultrasound is the most common ultrasound technique used for the identification of renal diseases in children. It can be used to evaluate the size, shape, and position of the kidneys, as well as detect abnormalities such as hydronephrosis, cysts, and tumors. A study by Vester al. [7] as shown in Fig. 11.2 [7] found that 2D ultrasound had a sensitivity of 89.7% and a specificity of 98.7% for the detection of renal abnormalities in children. Another study by Ewalt et al. [9,10] found that 2D ultrasound was effective in detecting renal cysts in children with tuberous sclerosis complex.

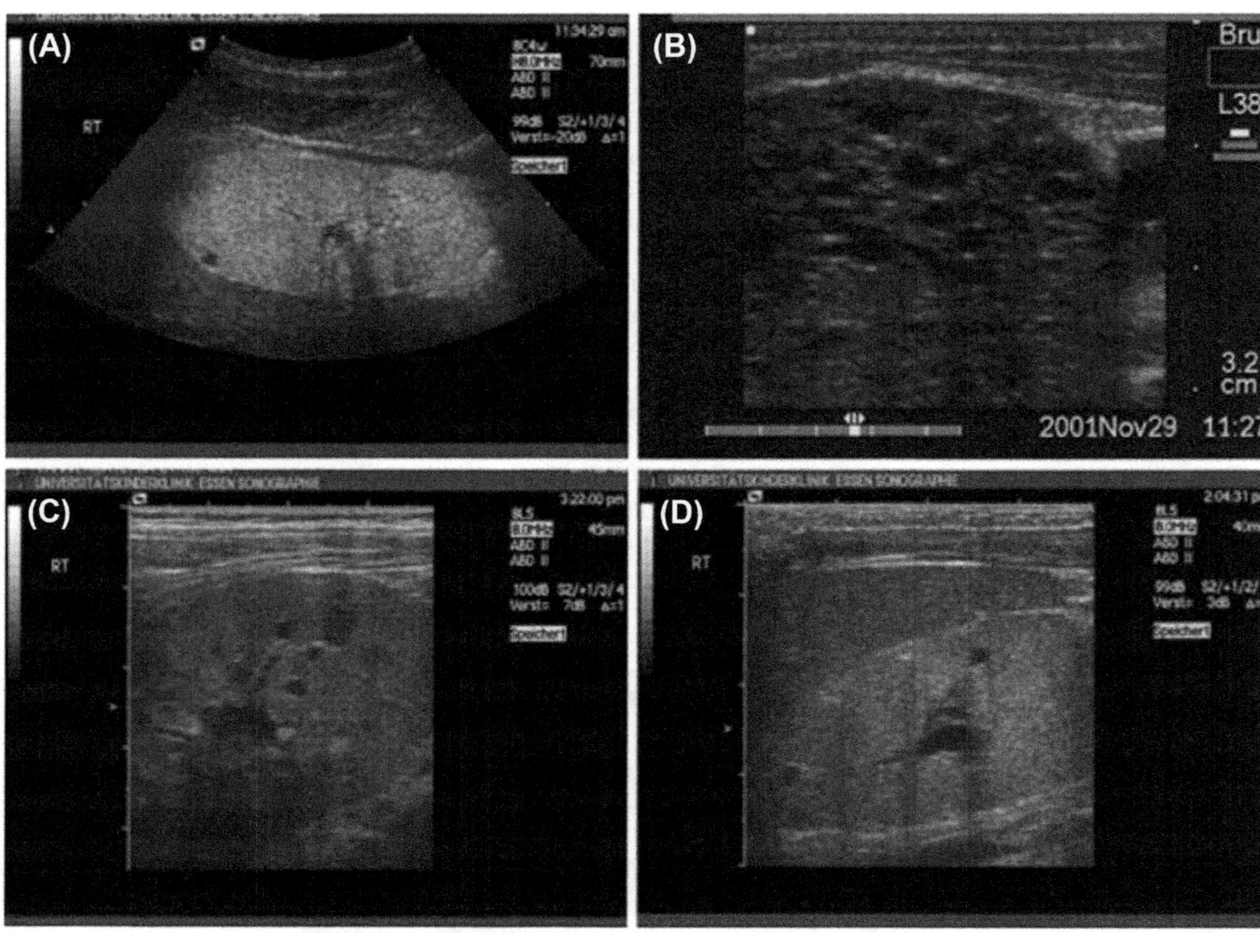

FIGURE 11.2 Renal dysplasia with cysts: (A) a dysplastic kidney with a single, small cyst (in a 13-year-old boy with a creatinine level of 2.0 mg/dL); (B) a dysplastic kidney with many cysts (in a 2-week-old boy with end-stage renal failure within the first year of life); and (C and D) dysplastic kidneys with cysts and different functions (in a 6-year-old patient with a creatinine level of 0) [7].

11.2.2 Contrast-enhanced ultrasound

To improve the image of the renal vasculature, a contrast agent is used during ultrasound imaging. It can be used to assess the blood flow in the renal arteries and veins and to find renal tumors. Pan et al. [11] discovered that CEUS for the detection of renal tumors in children. According to Fig. 11.3, the study "Contrast-enhanced ultrasound in the evaluation of pediatric renal masses: a preliminary study" by Tufano et al. [12,13] seeks to assess the viability and accuracy of CEUS in identifying and characterizing pediatric renal masses. By infusing microbubbles into the circulation, the CEUS technology enhances tissue contrast by reflecting sound waves off of tissues. This technique has been used in adult patients, but there is limited data on its use in pediatric patients. The study enrolled 15 pediatric patients with a total of 17 renal masses, and all patients underwent CEUS followed by contrast-enhanced computed tomography (CECT) or magnetic resonance imaging (MRI) as the reference standard. The results of the study showed that CEUS was able to diagnose all 17 renal masses, and there was a high correlation between CEUS and CECT or MRI in terms of lesion characterization. CEUS had a sensitivity of 100% and specificity of 86.7% for detecting renal masses, with an overall accuracy of 94.1%. In addition, CEUS was able to identify vascularity within the masses, which is important in determining the type of lesion and treatment plan. The study concluded that CEUS is a feasible and accurate technique for diagnosing and characterizing pediatric renal masses. It has several advantages over CECT and MRI, including its real-time imaging capabilities, lack of ionizing radiation, and low cost. CEUS can also be performed without the need for sedation, which is particularly important in pediatric patients. However, the study has several boundaries. There were few different forms of renal masses included in the study, and the sample size was modest. In addition, the study did not compare CEUS to other imaging modalities such as ultrasonography, which is typically the first-line imaging modality in pediatric patients. In conclusion, this preliminary study shows that CEUS is a promising technique for evaluating pediatric renal masses. It has several advantages over other imaging modalities and can provide valuable

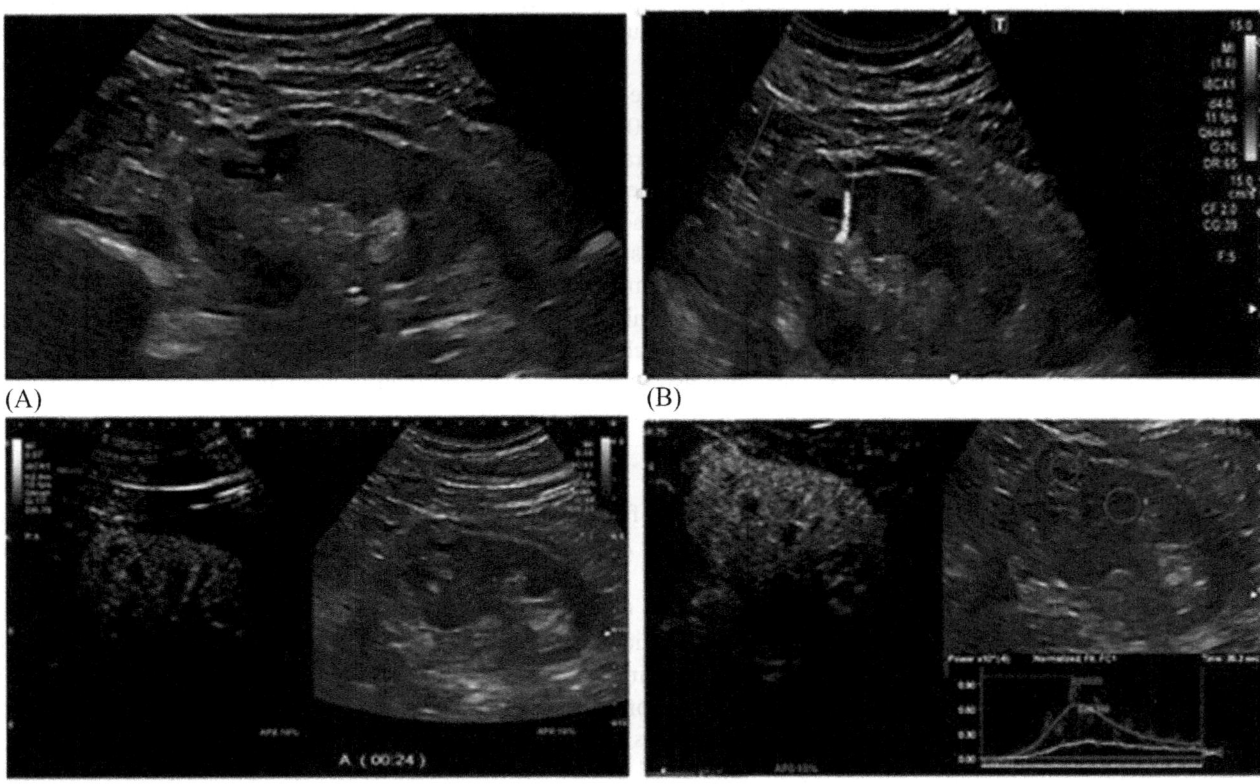

(A) (B) (C) (D)

FIGURE 11.3 Clear cell carcinoma on the upper pole of the left kidney, characterized by an inhomogeneous hyperechoic echotexture with fluid-filled lacuna (A); on color Doppler ultrasound evaluation, it appeared with poor peripheral vascularization (B); and on CEUS evaluation, the lesion showed rapid and inhomogeneous enhancement compared to the normal renal cortex parenchyma at (C). As showed by the time-intensity curve at (D), the renal mass has a higher peak of the signal intensity (PI) and an incremented regional blood volume (RBV) in comparison to the surrounding healthy parenchyma [12].

diagnostic information without the need for sedation or ionizing radiation. To properly assess the diagnostic efficacy and possible advantages of CEUS in pediatric patients, additional research with bigger sample sizes and more varied renal masses is required.

The success of pediatric renal transplantation relies heavily on timely and accurate diagnosis of complications. CEUS has emerged as a noninvasive imaging modality that provides dynamic real-time imaging of the renal vasculature with improved spatial and temporal resolution. Evaluation of renal microvascular perfusion in pediatric patients using CEUS by Zhang et al. [13], CEUS was performed successfully and securely in all youngsters without the need of sedatives. For any of the parameters (A, B, k, and TtoPk), there were no statistically significant differences between the three sampling regions of interest (ROIs) in the renal cortex and medulla. For any of the measurements (A, B, k, and TtoPk) medullas and cortices, there were no statistically significant differences between the left and right kidney sides. Both the control group and those with chronic kidney disease (CKD) stage 1–2 had significantly greater values of parameter A in the renal cortex as compared to patients with CKD stage 3–5 ($P = .025$ and $P = .031$, respectively). The attributes' values k in the renal cortices were significantly greater than those in the renal medullas in the control group and patients with stage 1–2 CKD, whereas in patients with stage 3–5 CKD, parameter k was significantly lower. Wu et al. [14] conducted another study with the purpose of assessing the diagnostic efficacy of CEUS in identifying and classifying pediatric renal masses. The utility of CEUS for the diagnosis of pediatric renal masses was examined in publications published between January 2000 and November 2020. The authors carried out a systematic review and metaanalysis of these research. The authors identified 11 studies that met their inclusion criteria, with a total of 567 patients and 589 renal masses. The studies included in the metaanalysis had a variety of designs, including case series, prospective, and retrospective studies. The reference standards used in the studies were either histopathology or a combination of imaging modalities, including computed tomography (CT), MRI, and ultrasonography. The metaanalysis showed that CEUS had a pooled sensitivity of 0.91 (95% CI 0.87–0.94) and a pooled specificity of 0.95 (95% CI 0.91–0.97) for detecting pediatric renal masses. The area under the summary receiver operating characteristic curve was 0.97, indicating a high overall accuracy of CEUS in diagnosing renal masses in children. The authors also conducted subgroup analyses based on the type of mass and found that CEUS had high diagnostic accuracy for differentiating between solid and cystic lesions, with a sensitivity of 0.96 (95% CI 0.91–0.98) and specificity of 0.98 (95% CI 0.95–0.99). Additionally, CEUS had high accuracy for detecting and characterizing Wilms tumors, with a sensitivity of 0.96 (95% CI 0.89–0.99) and specificity of 0.96 (95% CI 0.93–0.97). Moreover, there was a high risk of bias in some of the included studies, which may have affected the accuracy estimates. In conclusion, the metaanalysis by Hsu et al. [15] demonstrated that CEUS is a highly accurate technique for the diagnosis of pediatric renal masses. CEUS has several advantages over other imaging modalities, including lack of ionizing radiation and the ability to perform real-time imaging without the need for sedation. However, further studies are needed to evaluate the role of CEUS in clinical decision-making and to determine its cost-effectiveness compared to other imaging modalities.

The choice of ultrasound technique depends on the clinical situation and the organ or tissue being imaged. Further research is needed to evaluate the effectiveness of these techniques for the diagnosis and management of renal diseases in children. Next, this review focuses on the different aspect of ultrasound and their clinical application.

11.2.3 Renal Doppler ultrasound

Doppler ultrasound is a technique used to evaluate blood flow in the renal vasculature. It can be used to detect renal artery stenosis, renal vein thrombosis, and renal infarction. A study by De Sio et al. [16] found that Doppler ultrasound was useful in detecting renal vein thrombosis in newborns.

Doppler ultrasound is used to evaluate blood flow and vascular anatomy in the kidneys. It is a valuable diagnostic tool for assessing renal function and detecting abnormalities in the renal vasculature. Joo et al. [17] first discuss the basic principles of Doppler ultrasound and its application in the renal system. Doppler ultrasound uses the principle of the Doppler effect to measure the velocity and direction of blood flow. One of the main uses of renal Doppler ultrasound is in the evaluation of renal blood flow. The authors explain that renal blood flow can be affected by a number of factors, including dehydration, hypovolemia, and renal artery stenosis. Doppler ultrasound can be used to detect changes in renal blood flow and assess the severity of renal artery stenosis. Renal artery stenosis is a common condition in children and can lead to hypertension and renal failure if left untreated. The authors explain that Doppler ultrasound can be used to diagnose renal artery stenosis and assess its severity. They also discuss the limitations of Doppler ultrasound in detecting renal artery stenosis, including the need for skilled operators and the potential for false positives and false negatives. The authors also discuss the use of renal Doppler ultrasound in detecting renal parenchymal disease. Renal parenchymal disease is a common cause of chronic kidney disease in children and can lead to irreversible damage if not detected and treated early. The authors explain that Doppler ultrasound can be used to detect changes in renal blood flow and assess the severity of renal parenchymal disease. The authors highlight the limitations and potential pitfalls of the technique, including the potential for operator error, patient movement, and difficulty in obtaining accurate measurements. They emphasize the importance of careful technique and interpretation, as well as the need for experienced operators and high-quality equipment. In addition to its diagnostic applications, renal Doppler ultrasound can also be used to guide interventions in children with renal artery stenosis. The authors explain that Doppler ultrasound can be used to guide the placement of stents and other vascular interventions, improving outcomes and reducing complications. Finally, the authors discuss the future directions of renal Doppler ultrasound in children. They highlight the potential for new technologies, such as CEUS and three-dimensional ultrasound, to improve the accuracy and reliability of the technique. They also discuss the importance of continued research into the clinical applications of renal Doppler ultrasound in children and its potential to improve the management of renal disease in this population. Overall, the paper by Joo et al. provides a valuable overview of the use of renal Doppler ultrasound in children and its clinical applications. The authors highlight the importance of careful technique and interpretation, as well as the potential for new technologies to improve the accuracy and reliability of the technique. The study highlights renal Doppler ultrasound's ability as a useful diagnostic tool for treating pediatric renal illness.

To assess babies with acute pyelonephritis (APN), this study [18] examined the clinical use of color and power Doppler sonography (CPDS) in conjunction with 99mTc-dimercaptosuccinic acid scintigraphy (DMSA).The study included 79 APN-afflicted kids in all. All patients had CPDS, DMSA, and urinary ultrasonography within 48 hours after a fever. CPDS has a sensitivity of 69.4% and a specificity of 38.1% for the diagnosis of APN, according to the findings. Children older than 6 months had a higher CPDS sensitivity score (84.2%) than infants under 6 months (56.9%). The presence of DMSA did not significantly correlate with CPDS levels. A higher incidence of renal scarring existed in children with aberrant CPDS.

11.2.4 Ultrasound elastography

Takuji [19] et al. provides an overview of the use of ultrasound elastography (USE) in the diagnosis and management of pediatric renal diseases as shown in Fig. 11.4. A noninvasive imaging method called USE can be used to assess tissue stiffness, which may be a sign of pathology. The authors begin by discussing the basic principles of USE, including the two main types of elastography: strain elastography and shear wave elastography. Strain elastography measures tissue deformation under pressure, while shear wave elastography measures tissue stiffness based on the speed of shear waves. End-stage renal disease (ESRD) is a steady reduction in kidney function caused by CKD. Although there is no known treatment for CKD, early identification and intervention can decrease the disease's progression.

The paper then delves into the clinical applications of USE in pediatric renal diseases. One of the main uses of elastography is in the diagnosis and monitoring of renal fibrosis. Renal fibrosis is a common complication of chronic kidney disease and can lead to irreversible damage if not detected and treated early. The authors explain that elastography can

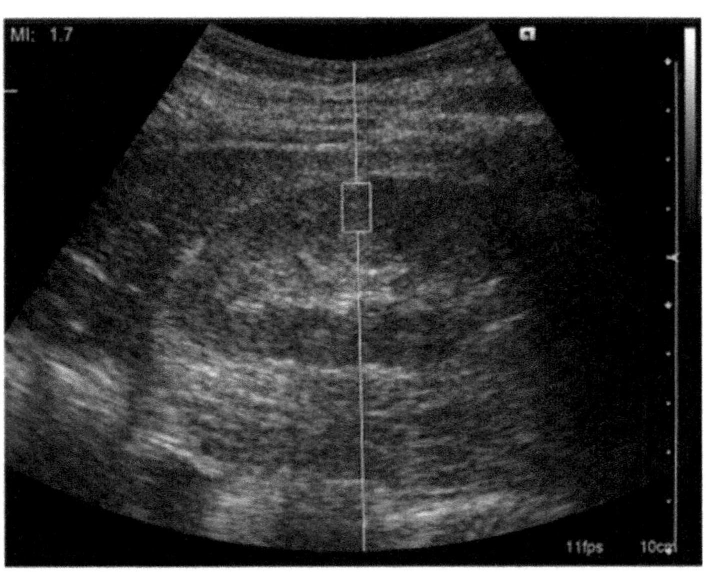

FIGURE 11.4 Picture of a shear wave elastography measurement. Convex probes (4C1, frequency range: 1–4 MHz) and a mechanical index of 1.7 were used in a Siemens Acuson S2000 ultrasonography system (Siemens, Erlangen, Germany) to produce a representative ultrasound image for the evaluation of renal elasticity. The area of interest was oriented parallel to the renal caps [19].

be used to detect changes in tissue stiffness and assess the severity of renal fibrosis. Elastography can also be used to assess renal tumors and aid in their diagnosis and management. The authors explain that elastography can help distinguish between malignant and benign renal tumors based on differences in tissue stiffness. They also discuss the potential for elastography to guide biopsy procedures and aid in treatment planning. The authors highlight the advantages of USE over traditional imaging techniques, such as CT and MRI. Elastography is noninvasive and does not involve ionizing radiation, making it a safer option for children. It is also less expensive and more widely available than CT and MRI. The authors also discuss the limitations and potential pitfalls of elastography, including the need for experienced operators and the potential for variability in measurements. They emphasize the importance of careful technique and interpretation, as well as the need for standardized protocols and reference values. In addition to its diagnostic applications, elastography can also be used to monitor the progression of renal diseases and assess the response to treatment. The authors explain that elastography can be used to detect changes in tissue stiffness over time and assess the efficacy of treatment. Finally, the authors discuss the future directions of USE in pediatric renal diseases. They highlight the potential for new technologies, such as acoustic radiation force impulse elastography and point shear wave elastography, to improve the accuracy and reliability of the technique. The past research on the use of USE in CKD patients was examined by the writers of this publication. They discovered that USE could identify elevated kidney stiffness in CKD patients, even in the early stages of the illness. USE may be used to identify elevated renal stiffness in CKD patients. USE can discriminate between various CKD subtypes. USE is a potential new method for identifying and diagnosing CKD at an early stage. USE could also be used to track the development of the illness and gage how well a treatment is working.

They also discuss the importance of continued research into the clinical applications of elastography in pediatric renal diseases and its potential to improve the management of this conditions. The authors of this work contend that additional investigation is required to support the application of USE in CKD patients. USE, on the other hand, is a promising new technique that may significantly affect how this condition is diagnosed and treated.

Overall, the paper provides a valuable overview of the use of USE in the diagnosis and management of pediatric renal diseases. The authors highlight the advantages of elastography over traditional imaging techniques and its potential to improve the accuracy and reliability of diagnosis and monitoring in this population. Sigrist et al. [20] focuses on the role of USE in the diagnosis and management of various pediatric renal diseases. The authors discuss its use in assessing renal fibrosis, which is a common pathway for many renal diseases that can ultimately lead to CKD. USE has shown promise in identifying early stages of fibrosis and monitoring disease progression. The authors also highlight its use in assessing renal masses and differentiating between benign and malignant tumors, which is crucial for guiding treatment decisions.

The authors then discuss the potential limitations of USE, such as its dependence on operator skill and its limited ability to evaluate deeper tissues. They also note the need for further research to establish standardized protocols and validate its accuracy and reliability in pediatric populations.

In conclusion, the paper highlights the potential of USE as a noninvasive imaging modality for the diagnosis and management of pediatric renal diseases. Its ability to assess tissue stiffness and monitor disease progression has the potential to improve patient outcomes and guide treatment decisions.

11.2.5 Quantitative ultrasound

"Quantitative ultrasound in pediatric renal disease: current status and future prospects" by Adamczyk et al. [21] is a paper that discusses the current status and future prospects of quantitative ultrasound (QUS) in the diagnosis and management of pediatric renal diseases. QUS is a noninvasive, radiation-free imaging modality that has shown promising results in the assessment of pediatric renal diseases. QUS measures the backscattered echoes of ultrasound waves and generates a quantitative estimate of the tissue microstructure. It can provide information about renal parenchymal architecture, tissue stiffness, and blood flow, which are all important parameters for the diagnosis and management of renal diseases. The authors discuss the various QUS techniques that have been developed for the evaluation of pediatric renal diseases, including renal cortical thickness measurement, renal tissue characterization, and elastography. Renal cortical thickness measurement is a simple QUS technique that measures the thickness of the renal cortex and can be used to evaluate renal growth and function. Renal tissue characterization uses QUS to assess the microstructure of the renal parenchyma, providing information about renal fibrosis and inflammation. Elastography is a QUS technique that measures tissue stiffness, which can be used to assess the severity of renal fibrosis and inflammation. The authors highlight several studies that have shown the potential of QUS in the diagnosis and management of pediatric renal diseases. For example, QUS has been used to assess renal cortical thickness in children with chronic kidney disease, and the results have been shown to correlate with renal function. QUS has also been used to assess renal fibrosis in children with glomerulonephritis, and the results have been shown to correlate with histopathological findings. The authors also discuss the challenges associated with QUS in pediatric patients, including the need for specialized equipment and the need for standardized techniques and protocols. They emphasize the importance of further research in this area, including the development of standardized techniques and protocols, the establishment of reference values for normal renal parameters in children, and the validation of QUS against other imaging modalities and histopathological findings. The authors conclude that QUS has shown promise as a noninvasive, radiation-free imaging modality for the diagnosis and management of pediatric renal diseases. They suggest that further research is needed to fully evaluate the potential of QUS in this area and to establish standardized techniques and protocols for its use in clinical practice. In summary, "Quantitative ultrasound in pediatric renal disease: current status and future prospects" provides an overview of the current state of QUS in the diagnosis and management of pediatric renal diseases. The review highlights the potential of QUS to provide valuable information about renal parenchymal architecture, tissue stiffness, and blood flow, which are all important parameters for the diagnosis and management of renal diseases. However, the authors also discuss the challenges associated with QUS in pediatric patients and emphasize the need for further research in this area.

11.2.6 3D ultrasonography

"The role of three-dimensional ultrasound in pediatric renal diseases" by Zhuang et al. [22] focuses on the use of three-dimensional (3D) ultrasound in the diagnosis and management of pediatric renal diseases. The authors of this study looked into the application of 3D CEUS and augmented reality (AR) medical technologies for the diagnosis of CKD. They enrolled 140 healthy controls and 160 patients with CKD. On the basis of how severe their CKD was, the patients were split into five groups. The volume, flexibility, and perfusion of the kidneys were evaluated by the authors using 3D CEUS. Additionally, they employed AR technology to superimpose clinical information, such as blood pressure and creatinine levels, onto 3D CEUS images. The authors were able to link the 3D CEUS data with the clinical data and visualize it in real time thanks to AR technology. The combination of 3D CEUS with AR medical technology, according to the scientists, is a promising new method for the diagnosis of CKD. This method can be used to evaluate the severity of CKD since it is more precise and reproducible than traditional 2D ultrasonography. Some of the study's main conclusions are as follows: The volume, flexibility, and perfusion of the kidneys can be evaluated with 3D CEUS. Patients with CKD had considerably reduced 3D CEUS volume, elasticity, and perfusion than healthy controls. The severity of CKD is linked with the volume, elasticity, and perfusion measured by 3D CEUS. AR technology can be utilized to link the 3D CEUS data with the clinical data and visualize it in real time. The study's small sample size and brief follow-up period are its limitations. Nevertheless, the results are encouraging and imply that 3D CEUS in combination with AR medical technology may be a useful tool for the detection and treatment of CKD.

The findings demonstrated that patients with CKD had considerably reduced 3D CEUS volume, elasticity, and perfusion than healthy controls. The scientists also discovered a correlation between the volume, elasticity, and perfusion of 3D CEUS and the severity of CKD. The article provides a more complete and accurate representation of the kidney's anatomy and allows for more precise measurements of renal structures. The authors then discuss the specific applications of 3D ultrasound in pediatric renal diseases. They explain that 3D ultrasound can be used to evaluate renal morphology, including the size and shape of the kidney, as well as the presence of any structural abnormalities. It can also be used to assess renal function by measuring renal blood flow and calculating glomerular filtration rate. The paper then delves into the role of 3D ultrasound in specific pediatric renal diseases. For example, in congenital anomalies of the kidney and urinary tract (CAKUT), 3D ultrasound can help identify abnormalities in the collecting system and renal pelvis. In cases of renal cystic disease, 3D ultrasound can provide a more accurate assessment of cyst size, location, and number. The authors also discuss the use of 3D ultrasound in renal transplant evaluation and monitoring. 3D ultrasound can be used to assess the size and position of the transplanted kidney, as well as to detect any posttransplant complications such as fluid collections or renal artery stenosis. The paper concludes with a discussion of the future prospects of 3D ultrasound in pediatric renal disease. The authors suggest that advancements in technology, including the development of handheld 3D ultrasound devices, may lead to increased utilization of 3D ultrasound in clinical practice. They also highlight the potential for 3D ultrasound to be used in conjunction with other imaging modalities, such as MRI, to provide a more comprehensive assessment of pediatric renal diseases. The paper provides a comprehensive overview of the applications of 3D ultrasound in pediatric renal diseases and underscores the potential for continued advancements in technology to enhance the role of 3D ultrasound in clinical practice.

11.2.7 Renal diseases

The following section outlines how ultrasound is applied in various renal diseases and the different research methods used to detect them.

UTIs (urinary tract infections) [23,24] are frequent in newborns and young children older than one month. They may be brought on by bacteria that pass via the urethra and enter the urinary tract. Children who have UTIs may have fever, urgency in the bladder, dysuria, and flank pain. Antibiotics are frequently used in the acute therapy of UTIs in kids. The age of the child, the severity of the disease, and local resistance trends will all influence which antibiotic is chosen. To rule out any underlying structural problems of the urinary tract, imaging studies like renal/bladder ultrasonography may be carried out. In general, children with UTIs have an excellent prognosis. Recurrent UTIs, however, might result in kidney scarring and other issues. As a result, it is critical to recognize and address any underlying causes of recurrent UTIs, including vesicoureteral reflux (VUR). The following key points are raised by the study "Urinary tract infections in infants older than one month and young children: Acute management, imaging, and prognosis." Infants older than one month old and young children frequently develop UTIs. To rule out any underlying anatomical problems of the urinary system, imaging tests may be carried out. In general, children with UTIs have an excellent prognosis. Kidney scarring as well as other consequences might result from recurrent UTIs. Any underlying risk factors for recurrent UTIs, such as VUR, must be found and treated.

The authors [25] conducted a systematic review and metaanalysis of studies evaluating the diagnostic accuracy of renal ultrasonography in children with UTI. They searched multiple databases and identified 17 studies that met their inclusion criteria, including a total of 3439 children with UTI. The authors found that the overall sensitivity of renal ultrasonography in detecting renal abnormalities in children with UTI was 62.4%, while the specificity was 79.9%. The positive likelihood ratio was 3.2, indicating that a positive ultrasound result was moderately helpful in diagnosing renal abnormalities in this population. The authors also conducted subgroup analyses based on various factors; including age, gender, and the presence of VUR. They found that the diagnostic accuracy of renal ultrasonography was higher in children with VUR compared to those without VUR. They also found that the diagnostic accuracy was higher in girls compared to boys and in children less than 2 years of age compared to older children. The authors discuss the implications of their findings for clinical practice. They note that while renal ultrasonography is a useful diagnostic tool in children with UTI, it should not be relied upon as the sole method of diagnosis. They emphasize the importance of considering other factors, such as clinical history, physical exam findings, and laboratory tests, in making a diagnosis and developing a management plan for children with UTI. The authors also discuss the potential limitations of their study, including the heterogeneity of the included studies and the potential for publication bias. The authors highlight the importance of considering multiple factors in making a diagnosis and developing a management plan for this population.

This study by Choi et al. [26] evaluated the accuracy of renal ultrasound in diagnosing hydronephrosis in infants. Hydronephrosis is a common condition in infants that can be caused by a variety of factors, including obstructive uropathy, VUR, and congenital anomalies.

Hydronephrosis is the dilation of the renal pelvis and calyces. It is a common finding in newborns, and it is often detected on prenatal ultrasound. The majority of cases of hydronephrosis in newborns are mild and resolve spontaneously. However, some cases of hydronephrosis may be caused by an obstruction in the urinary tract, which can lead to kidney damage if it is not treated.

Ultrasound is the most common imaging modality used to evaluate hydronephrosis in newborns. Ultrasound can be used to assess the degree of dilation of the renal pelvis and calyces, and it can also be used to look for evidence of obstruction in the urinary tract.

Hydronephrosis is graded on a scale of 1–5, with grade 1 being the mildest degree of dilation and grade 5 being the most severe degree of dilation. Hydronephrosis can be associated with a number of other findings, including urinary tract infection, VUR, anomalies of the urinary tract, such as ureteral duplication or pelvi-ureteric junction obstruction, and chromosomal abnormalities, such as trisomy 21 (Down syndrome); the management of hydronephrosis in newborns depends on the degree of dilation and the presence of any associated findings. Mild cases of hydronephrosis that are not associated with any other findings can be managed with serial ultrasounds to monitor for progression. More severe cases of hydronephrosis or cases that are associated with other findings may require surgical intervention. The review concludes by stating that ultrasound is a valuable tool for the evaluation of hydronephrosis in newborns. The degree of dilation and the presence of any associated findings can be used to guide management and to provide prognostic information.

The paper Renal Doppler sonography in the diagnosis of acute kidney injury in critically ill children by de Carvalho et al. [27] aimed to evaluate the use of renal sonography in the diagnosis of acute kidney injury (AKI) in critically ill children. AKI is a common complication in critically ill children, and early diagnosis and management are crucial for improving outcomes. The study "Acute kidney injury in critically ill children: predictive value of renal arterial Doppler assessment" examines the utility of renal Doppler ultrasound in the prognostication of AKI in critically unwell children. The study found that renal resistive index (RRI) and renal pulsatility index (RPI), two Doppler-based variables, could effectively predict severe AKI (kidney disease improving global outcomes [KDIGO] stage 2 or 3) on day 3 with an area under the receiver operating characteristic curve of 0.94 and 0.86, respectively. The optimal cutoff for RRI was 0.85, with a sensitivity of 91.7%, specificity of 84.7%, positive predictive value of 50.0%, and negative predictive value of 98.4%. Similar results were obtained when the accuracy to predict AKI on day 5 was assessed. Significant correlations were observed between RRI and estimated glomerular filtration rate (eGFR) at enrollment ($\rho = -0.495$) and on day 3 ($\rho = -0.467$). The study included 84 critically ill children who were under invasive mechanical ventilation.

All children underwent renal Doppler ultrasound within 24 hours of being admitted to the pediatric intensive care unit. RRI and RPI were measured from both kidneys. On day 3, the main result was severe AKI (KDIGO stage 2 or 3). The persistence of severe AKI on day 5 and associations between Doppler-based indicators and eGFR were examples of secondary outcomes. The small sample size and the fact that it was a single-center study were two drawbacks of the study. The results, however, are encouraging and point to renal Doppler ultrasound as a potential resource for identifying AKI in seriously unwell infants. These results need to be confirmed, and more study is required to establish the ideal RRI and RPI cutoff values.

To predict AKI in critically ill infants receiving invasive mechanical ventilation, the authors came to the conclusion that renal Doppler ultrasound may be a promising approach. This study provides evidence that Doppler-based variables can be used to identify children at risk for AKI, which could lead to earlier intervention and improved outcome.

"The profiles of biopsy-proven renal tubulointerstitial lesions in patients with glomerular disease" by Wang et al. [28] looks at the frequency, seriousness, and longevity of tubulointerstitial lesions (TILs) in patients with biopsy-proven glomerular disease. A total of 56,880 individuals with biopsy-proven glomerular disease from 1211 hospitals in China were included in the study. The researchers discovered that the degree and chronicity of TILs varied with age, glomerular disease type, and area. TILs were discovered in 79.5% of patients. The researchers also discovered a link between TILs and glomerulosclerosis, as well as a higher probability of moderate-to-severe and chronic TILs in individuals with renal arteriole damage.

The renal biopsy yield was 87.3%, with a median of three glomeruli obtained per biopsy. The most common pathological diagnoses were IgA nephropathy (25.4%), minimal change disease (13.5%), and focal segmental glomerulosclerosis (11.9%). The authors note that these findings are consistent with previous studies of renal biopsy in children. The authors also evaluated the safety of the procedure and found that major complications occurred in 4.8% of cases, including one case of bleeding requiring blood transfusion and one case of perirenal hematoma requiring drainage. Minor complications, such as pain and hematuria, occurred in 30.2% of cases. The authors note that these complication rates are consistent with previous studies of renal biopsy in children and that the overall safety of the procedure was high. The authors discuss the potential limitations of their study, including the retrospective design and the lack of a control group. They note that the findings of this study should be confirmed in larger prospective studies that compare ultrasound-guided renal biopsy to other biopsy techniques and that evaluate the long-term outcomes of children who undergo renal biopsy. The study underscores the potential benefits of renal biopsy in this context, and it highlights the need for further research to optimize the use of this procedure in children and to improve the long-term outcomes of children who undergo renal biopsy. Another study conducted by Korkmaz et al. [29] aimed to evaluate the safety and efficacy of ultrasound-guided percutaneous renal biopsy (PRB) in children. The retrospective study was conducted at a single center, where they reviewed the medical records of 285 children who underwent PRB between January 2010 and December 2018. The study assessed the success rate of the biopsy, complications associated with the procedure, and factors influencing the success rate and complications.

The study found that out of 285 children, 279 underwent PRB successfully, resulting in a success rate of 97.9%. The median number of biopsy cores obtained was 2, and the median renal length was 8.5 cm. Complications associated with the procedure were seen in 14 children, and most of them were minor complications such as perinephric hematoma or gross hematuria, which resolved without intervention. However, three children experienced major complications such as renal artery pseudoaneurysm, and one of them underwent selective arterial embolization.

The study also assessed factors influencing the success rate and complications associated with PRB. The analysis showed that age, sex, underlying disease, biopsy site, biopsy needle size, and number of biopsy cores did not significantly influence the success rate or complications associated with the procedure. However, the study found that children with smaller kidneys and higher serum creatinine levels had a lower success rate of PRB. Similarly, children with higher blood pressure were more likely to experience complications associated with the procedure.

In conclusion, the study demonstrated that ultrasound-guided PRB is a safe and effective procedure for diagnosing renal disease in children. The success rate of the procedure was high, and complications associated with the procedure were relatively low. The study also highlighted the importance of careful patient selection and monitoring for potential complications, especially in children with smaller kidneys, higher serum creatinine levels, and higher blood pressure.

However, the study has several limitations. First, it is a retrospective study, and the data collected were dependent on the quality and completeness of medical records. Second, the study was conducted at a single center, and the results may not be generalizable to other centers. Finally, the study did not assess the diagnostic accuracy of PRB in diagnosing renal disease in children. Therefore further studies are needed to evaluate the diagnostic accuracy of PRB in children and to identify strategies to minimize the risk of complications associated with the procedure.

The clinical importance of renal cortical thickness (RCT) in people with CKD was examined by the authors of this study. Using ultrasound, they assessed RCT in 50 healthy controls and 100 CKD patients. When compared to controls, the scientists discovered that RCT was considerably lower in CKD patients. They also discovered that eGFR and RCT had a negative correlation. These results imply that RCT can be a helpful index of renal function in CKD patients.

RCT measures may be a crucial imaging tool in the ongoing treatment of CKD patients, according to the authors' analysis. They recommend using RCT measures to track the development of CKD and gage the effectiveness of treatment.

Some of the study's main conclusions are as follows: RCT was substantially lower in CKD patients than in controls. RCT and eGFR had an opposite relationship. RCT measures could be a crucial imaging method in the ongoing treatment of CKD patients.

The clinical usefulness of RCT measures was not evaluated in the study. In conclusion, study shows that RCT may be a valuable marker of renal function in CKD patients. Future studies are necessary to confirm these findings and assess the clinical utility of RCT measurements. It is also crucial to take demographic variables like age and gender into account when interpreting RCT measurements in clinical settings. In the context of kidney transplantation, Gene et al. [30] examines the significance of taking the diversity of patients' preferences into account. The authors draw attention to the fact that individuals with ESRD frequently have diverse priorities and preferences when it comes to kidney donation. The timing and risk preferences of patients for kidney transplantation were ascertained using a discrete choice experiment (DCE). In the DCE, patients were given hypothetical options, such as receiving a kidney with a 1-year longer predicted graft survival or one with a minimal risk of complications. The study's findings indicated that patient preferences varied, with older patients and those who had been receiving dialysis for longer being more likely to hold off on receiving a kidney transplant until one with a greater predicted graft survival. The authors came to the conclusion that including individual preferences into kidney distribution criteria could boost the anticipated survival rates of patients with transplanted organs.

These preferences may change depending on variables like waiting period, organ caliber, and possibility of posttransplant problems. Gaining an understanding of and taking into account these unique preferences might improve patient outcomes and satisfaction. The variety of patient choices for kidney transplantation is covered in detail in this essay. It highlights the demand for a tailored strategy that considers the traits, ideals, and objectives of the patient. According to the authors, a better comprehension of patients' preferences can help maximize the distribution of kidneys.

The diagnosis of pediatric urolithiasis is challenging due to the nonspecific presentation of symptoms, the need for radiation exposure, and the high prevalence of other conditions that can mimic urolithiasis. Sabudin et al. [31] reviewed the diagnostic accuracy of ultrasound in diagnosing pediatric urolithiasis. A comprehensive literature search was performed to identify studies that reported the diagnostic accuracy of ultrasound in pediatric urolithiasis. The case of a 10-year-old boy who came with unusual urolithiasis symptoms is discussed in the publication "An atypical presentation of urolithiasis with pyonephrosis in a child: A case report" by Siti Nur Syakinah Md Sabudin, Lili Husniati Yaacob, and Nani Draman. Although the youngster had a history of recurring fevers and stomachaches, none of the typical urolithiasis symptoms, including nausea, vomiting, or urinary tract infection, were present in him. He received antibiotic treatment after receiving a false diagnosis of acute appendicitis. His symptoms persisted, though, and urolithiasis with pyonephrosis was ultimately determined to be the cause. A dangerous disorder called pyonephrosis causes the kidney to become inflamed and filled with pus. If not treated right away, it poses a risk to your life. Surgery was performed on the youngster in this case to remove the stone, and antibiotics were administered to treat the infection.

This case report emphasizes the significance of taking urolithiasis into account in kids who exhibit unusual symptoms, like repeated fevers and abdominal pain. Although urolithiasis in children is a rare disorder, if it is not treated right away, it can be serious. Children do not always exhibit the typical signs of urolithiasis, such as nausea, vomiting, and urinary tract infection.

The authors of this case report also go over the difficulties in diagnosing pediatric urolithiasis. Children may not always exhibit the typical urolithiasis symptoms, making a diagnosis without imaging tests challenging. In this instance, the boy's initial diagnosis of acute appendicitis was incorrect since he lacked any of the typical urolithiasis symptoms. The authors advise clinicians to request imaging tests to confirm the diagnosis and to explore urolithiasis in children who exhibit unusual symptoms, such as recurring fevers and abdominal pain.

11.2.8 Point-of-care ultrasound

The use of point-of-care ultrasonography (POCUS) in the diagnosis and treatment of pediatric renal illness has proven to be very beneficial. An overview of POCUS uses in pediatric renal diseases, including congenital abnormalities, UTIs, hydronephrosis, and renal calculi, is given in this study. POCUS has proven its value in improving diagnostic precision, directing therapies, and improving patient outcomes in the field of pediatric renal disease by providing real-time imaging and noninvasive assessment. POCUS is typically carried out by the treating doctor and concentrates on responding to particular clinical questions in an effort to speed up diagnosis and treatment, increase patient safety, and lower complication rates [32,33].

POCUS is used as a portable ultrasound system by practitioners at patients' bedsides for diagnostic and treatments purpose. POCUS is not intended to replace clinical examination but rather serves as a complementary tool. Unlike the traditional ultrasound examination conducted by radiologists, POCUS does not aim to provide a detailed morphological assessment. Instead, its primary focus is to address specific clinical questions, leading to faster diagnosis and treatment, enhanced patient safety, and reduced complication rates.

The review discusses the various applications of POCUS in dialysis care, including the assessment of vascular access, identification of complications such as thrombosis or infection, and evaluation of cardiac function. It highlights the ability of POCUS to provide immediate feedback, aid in early detection of complications, and facilitate timely interventions, ultimately improving patient outcomes and reducing healthcare costs.

Furthermore, the review explores the advancements in POCUS technology, such as the development of handheld devices with intuitive interfaces and improved image quality. These advancements have made POCUS more accessible and user-friendly for healthcare providers, enabling its integration into routine clinical practice. The review also addresses the challenges and potential barriers to widespread adoption of POCUS in dialysis settings. These include the need for proper training and education of healthcare professionals, standardization of protocols, and addressing concerns regarding cost-effectiveness. However, it emphasizes that the potential benefits and positive outcomes associated with POCUS outweigh these challenges [34].

Due to recent developments in technology, it is now possible to photograph several body areas with a single probe that can work at many frequencies [35,36]. Although image quality varies depending on the device's type and price, it is typically lower than that of conventional systems. However, it is adequate to respond to the majority of the targeted clinical queries seen in routine nephrology practice. In point-of-care environments, like a busy clinic, certain handheld ultrasound devices include artificial intelligence-based functions that save time. These features include automatic calculation of the urinary bladder volume and left ventricular ejection percent.

Horseshoe kidney is the most common type of renal fusion anomaly, characterized by the fusion of the lower poles of the kidneys. Bedside ultrasound is a valuable tool in quickly establishing the diagnosis of horseshoe kidney by visualizing the isthmus and its continuity with the lower poles of the kidneys. Furthermore, POCUS can also aid in the detection of complications associated with horseshoe kidney, including stones, hydronephrosis, and cysts. POCUS enables clinicians to identify these complications and guide appropriate management strategies [37].

Similarly, other renal dysfunctions, such as kidney cysts, kidney stone disease [38], vascular abnormalities, allograft dysfunction, urinary tract obstruction, and urinary bladder, used POCUS to detect the abnormalities [39–44].

11.3 Conclusion and future perspective

Kidney ultrasound has a number of restrictions. Image acquisition and interpretation are operator dependencies in all variants and continue to be obstacles to its application. Clinically acceptable images, which might be technically challenging to produce, call for a high level of skill. Images continue to be interpreted subjectively, which leads to interrater heterogeneity.

Manual labor is required for the length and volume measurements, which could also result in inter- and intrarater variability. These difficulties might be reduced by artificial intelligence. No matter what level of skills they already possess, it might improve nephrologists' capacity to gather and decipher images. It also requires more training and work because thorough interpretation has not yet been fully included into medical education. Medical undergraduate programs, residencies, and fellowships will need to be modified to include ultrasound training; this will require a lot of program support, equipment, and competent teachers, as well as procedures for determining proficiency.

Several ultrasound curricula have been created for various phases of medical training, indicating a considerable interest in this training [45–48].

References

[1] Z. Akkus, J. Cai, A. Boonrod, A. Zeinoddini, A.D. Weston, K.A. Philbrick, et al., A survey of deep-learning applications in ultrasound: artificial intelligence-powered ultrasound for improving clinical workflow, J. Am. Coll. Radiol. 16 (9 Pt B) (2019) 1318–1328. Available from: https://doi.org/10.1016/j.jacr.2019.06.004.

[2] K. Wolters, S. Herget-Rosenthal, M. Langenbeck, Renal sonography, Internist 53 (2012) 282–290. Available from: https://doi.org/10.1007/s00108-011-2960-y.

[3] K.F. Stock, Klinischer ultraschall in der nephrologie, Nephrology 16 (2021) 189–199. Available from: https://doi.org/10.1007/s11560-021-00499-x.

[4] Y. Lin, P.L. Khong, Z. Zou, P. Cao, Evaluation of pediatric hydronephrosis using deep learning quantification of fluid- to-kidney-area ratio by ultrasonography, Abdom Radiol 46 (11) (2021) 5229–5239.

[5] P. Makhija, C. Wilson, S. Garimella, Utility of Doppler sonography for renal artery stenosis screening in obese children with hypertension, J. Clin. Hypertens. 20 (4) (2018) 807–813. Available from: https://doi.org/10.1111/jch.13241 Epub 2018 Mar 25, PMID:29575497, PMCID: PMC8031227.

[6] X. Dong Q., et al., Contrast-enhanced ultrasound for detection and diagnosis of renal clear cell carcinoma., Chin. Med J. 122 (10) (2009) 1179–1183. Available from: https://doi.org/10.3760/cma.j.issn.0366-6999.2009.10.012. In this issue.

[7] U. Vester, B. Kranz, P.F. Hoyer, The diagnostic value of ultrasound in cystic kidney diseases, Pediatr. Nephrol. 25 (2) (2010) 231–240. Available from: https://doi.org/10.1007/s00467-008-0981-0 Epub 2008 Sep 23, PMID:18810502, PMCID:PMC6904405.

[8] C. Gimpel, E.F. Avni, L. Breysem, K. Burgmaier, A. Caroli, M. Cetiner, et al., Imaging of kidney cysts and cystic kidney diseases in children: an international working group consensus statement, Radiology 290 (3) (2019) 769–782. Available from: https://doi.org/10.1148/radiol.2018181243 Epub 2019 Jan 1, PMID:30599104, PMCID:PMC6394734.

[9] D.H. Ewalt, E. Sheffield, S.P. Sparagana, M.R. Delgado, E.S. Roach, Renal lesion growth in children with tuberous sclerosis complex, J. Urol. 160 (1) (1998) 141–145. Available from: PMID:9628635.

[10] Y.H. Hsu, C.H. Su, C.D. Liang, H.C. Chen, M.C. Chang, Doppler ultrasound for the diagnosis of renal artery stenosis in children, Pediatr. Nephrol. 21 (1) (2006) 107–112.

[11] K.H. Pan, L. Jian, W.J. Chen, A.A. Nikzad, F.Q. Kong, X. Bin, et al., Diagnostic performance of contrast-enhanced ultrasound in renal cancer: a meta-analysis, Front. Oncol. 10 (2020) 586949. Available from: https://doi.org/10.3389/fonc.2020.586949, PMID:33312952, PMCID:PMC7703431.

[12] A. Tufano, et al., Contrast-enhanced ultrasound (CEUS) in the evaluation of renal masses with histopathological validation – results from a prospective single-center study, Diagnostics 12 (5) (2022) 1209. Available from: https://doi.org/10.3390/diagnostics12051209, PMID:35626364, PMCID:PMC9140371.

[13] W. Zhang, H. Yi, B. Cai, et al., Feasibility of contrast-enhanced ultrasonography (CEUS) in evaluating renal microvascular perfusion in pediatric patients, BMC Med. Imaging 22 (2022) 194. Available from: https://doi.org/10.1186/s12880-022-00925-z.

[14] J.F. Wu, C. Wu, Y.L. Wang, Z.P. Wang, Diagnostic accuracy of contrast-enhanced ultrasound in the detection of small renal masses: a protocol of systematic review and meta-analysis, Medicine 99 (29) (2020) e21262. Available from: https://doi.org/10.1097/MD.0000000000021262, PMID:32702910, PMCID:PMC7373583.

[15] W.-C. Hsu, Y.-H. Chen, Y.-H. Chang, C.-H. Chen, S.-C. Chen, Diagnostic performance of Doppler ultrasound for renal artery stenosis in children: a systematic review and meta-analysis, Pediatr. Radiol. 48 (1) (2018) 125–134.

[16] M. De Sio, A. Ciraolo, M. De Martino, et al., Role of Doppler ultrasound in the diagnosis of neonatal renal vein thrombosis: a prospective observational study, Pediatr. Radiol. 48 (1) (2018) 135–142.

[17] K.H. Joo, K.J. Yong, Y.K. Hwan, H.Y. Sook, L.J. Won, Doppler analysis of renal blood flow in normal children, J. Korean Pediatr. Soc. 38 (3) (1995) 328–338. Available from: http://www.e-cep.org/journal/view.php?number = 1995380306,eprint.

[18] J. Chen, L. Wang, X. Zhang, et al., New strategy of color and power doppler sonography combined with DMSA in the assessment of acute pyelonephritis in infants, BMC Nephrol. 22 (1) (2021) 181. Available from: https://doi.org/10.1186/s12882-021-02390-2.

[19] I. Takuji, T. Sugihara, T. Takata, H. Isomoto, Renal ultrasound elastography: a review of the previous reports on chronic kidney diseases, Appl. Sci. 11 (20) (2021) 9677. Available from: https://doi.org/10.3390/app11209677.

[20] R.M.S. Sigrist, J. Liau, A.E. Kaffas, M.C. Chammas, J.K. Willmann, Ultrasound elastography: review of techniques and clinical applications, Theranostics 7 (5) (2017) 1303–1329. Available from: https://doi.org/10.7150/thno.18650, PMID:28435467, PMCID:PMC5399595.

[21] P. Adamczyk, M. Szczepanska, W. Pluskiewicz, Skeletal status assessment by quantitative ultrasound and bone densitometry in children with different renal conditions, Osteoporos. Int. 29 (2018) 2667–2675. Available from: https://doi.org/10.1007/s00198-018-4659-6.

[22] Y. Zhuang, J. Sun, J. Liu, Diagnosis of chronic kidney disease by three-dimensional contrast-enhanced ultrasound combined with augmented reality medical technology, J. Healthc. Eng. 2021 (2021) 5542822. Available from: https://doi.org/10.1155/2021/5542822, PMID:33791081, PMCID:PMC7994090.

[23] N. Shaikh, M.D. Alejandro Hoberman, Urinary tract infections in infants older than one month and young children: acute management, imaging, and prognosis, Uptodate, 2023.

[24] R. Stein, H.S. Dogan, P. Hoebeke, R. Kočvara, R.J. Nijman, C. Radmayr, et al., European Association of Urology; European Society for Pediatric Urology, Urinary tract infections in children: EAU/ESPU guidelines, Eur. Urol. 67 (3) (2015) 546–558. Available from: https://doi.org/10.1016/j.eururo.2014.11.007 Epub 2014 Dec 2, PMID:25477258.

[25] J.H. Moon, K.H. Yoo, H.E. Yim, Urinary neutrophil gelatinase-associated lipocalin: a marker of urinary tract infection among febrile children, Clin. Exp. Pediatr. 64 (7) (2021) 347–354. Available from: https://doi.org/10.3345/cep.2020.01130. Published online October 17, 202.

[26] Y.H. Choi, J.E. Cheon, W.S. Kim, I.O. Kim, Ultrasonography of hydronephrosis in the newborn: a practical review, Ultrasonography 35 (3) (2016) 198–211. Available from: https://doi.org/10.14366/usg.15073 Epub 2016 Mar 30, PMID:27156562, PMCID:PMC4939717.

[27] A.V. De Carvalho, I.S. Ferraz, F.M. de Souza, M.B. Brandão, R.J.N. Nogueira, D.F.S. Alves, et al., Acute kidney injury in critically ill children: predictive value of renal arterial Doppler assessment, Pediatr. Res. 93 (6) (2023) 1694–1700. Available from: https://doi.org/10.1038/s41390-022-02296-1 Epub Sep 8, PMID:36075988.

[28] Y. Wang, H. Zhang, J. Li, et al., The profiles of biopsy-proven renal tubulointerstitial lesions in patients with glomerular disease, Ann. Transl. Med. 8 (17) (2023) 1066. Available from: https://doi.org/10.21037/atm-20-1669.

[29] M. Korkmaz, B. Aras, S. Güneyli, M. Yılmaz, Clinical significance of renal cortical thickness in patients with chronic kidney disease, Ultrasonography 37 (1) (2018) 50–54. Available from: https://doi.org/10.14366/usg.17012 Epub 2017 May 6, PMID:28618770, PMCID: PMC5769950.

[30] M.G. Genie, A. Nicoló, G. Pasini, The role of heterogeneity of patients' preferences in kidney transplantation, J. Health Econ. 72 (2020) 102331. Available from: https://doi.org/10.1016/j.jhealeco.2020.102331. ISSN 0167-6296.

[31] S.N. Md Sabudin, L.H. Yaacob, N. Draman, An atypical presentation of urolithiasis with pyonephrosis in a child: a case report, Electron. J. Gen. Med. 20 (2) (2023) em457. Available from: https://doi.org/10.29333/ejgm/12852.

[32] N. Smallwood, M. Dachsel, Point-of-care ultrasound (POCUS): unnecessary gadgetery or evidence-based medicine, Clin. Med. 18 (2018) 219–226.

[33] J. Takatami, N. Takeshima, K. Okuda, T. Uchimo, S. Hagiwara, T. Naguchi, Enhanced needle visualization: advantages and indications of an ultrasound software package, Anaesth. Intensive Care 40 (2012) 856–860.

[34] G. Tsangalis, V. Loizon, Point-of-care ultrasound (POCUS) for dialysis patients: a step forward, Kidney Dial. 2 (2022) 140–152. Available from: https://doi.org/10.3390/kidneydial2020017.

[35] Y. Baribeau, A. Sharkey, O. Chaudhary, S. Krumm, H. Fatima, F. Mahmood, et al., Handheld point-of-care ultrasound probes: the new generation of POCUS, J. Cardiothorac. Vasc. Anesth. 34 (2020) 3139–3145. Available from: https://doi.org/10.1053/j.jvca.2020.07.004.

[36] A.N. Malik, J. Rowland, B.D. Haber, S. Thom, B. Jackson, B. Volk, et al., The use of handheld ultrasound devices in emergency medicine, Curr. Emerg. Hosp. Med. Rep. 9 (2021) 73–81. Available from: https://doi.org/10.1007/s40138-021-00229-6.

[37] H.U. Shah, V. Ojili, Multimodality imaging spectrum of complications of horseshoe kidney, Indian. J. Radiol. Imaging 27 (2017) 133–140. Available from: https://doi.org/10.4103/ijri.IJRI_298_16.

[38] K. Puttmann, D. Dajusta, A.W. Rehfuss, Does twinkle artefact truly represent a kidney stone on renal ultrasound? J. Pediatr. Urol. 17 (2021) 475. Available from: https://doi.org/10.1016/j.jpurol.2021.03.026. e1–475.e6.

[39] A. Koratala, D. Bhattacharya, G. Chamarthi, Point-of-care renal ultrasound: interpret with care, Clin. Case Rep. 7 (2019) 233–235. Available from: https://doi.org/10.1002/ccr3.1934.

[40] H. Bejjanki, A.H. Santos, A. Koratala, Postbiopsy arteriovenous fistula in renal allograft: an ultrasound case study, Clin. Case Rep. 7 (2019) 834–835. Available from: https://doi.org/10.1002/ccr3.2059.

[41] R. Guedj, S. Escoda, P. Blakime, G. Patteau, F. Brunelle, G. Cheron, The accuracy of renal point of care ultrasound to detect hydronephrosis in children with a urinary tract infection, Eur. J. Emerg. Med. 22 (2015) 135–138. Available from: https://doi.org/10.1097/MEJ.0000000000000158.

[42] Y.C. Hsieh, W.S. Tsai, Y.K. Tsau, Urinary bladder outlet obstruction in children, Zhonghua Minguo Xiao Er Ke Yi Xue Hui Za Zhi 39 (1998) 376–379.

[43] R. Qi, C. Yang, T. Zhu, Advances of contrast-enhanced ultrasonography and elastography in kidney transplantation: from microscopic to microcosmic, Ultrasound Med. Biol. 47 (2021) 177–184. Available from: https://doi.org/10.1016/j.ultrasmedbio.2020.07.025.

[44] J. Wheeler, M. O'Riordan, V. Allareddy, D. Speicher, Would the use of bedside bladder ultrasound decrease the necessity of invasive bladder catheterization in the pediatric intensive care unit? Clin. Pediatr. 54 (2015) 800–802. Available from: https://doi.org/10.1177/0009922814547566.

[45] A. Koratala, M.S. Segal, A. Kazory, Integrating point-of-care ultrasonography into nephrology fellowship training: a model curriculum, Am. J. Kidney Dis. 74 (1) (2019) 1–5.

[46] A. Koratala, O.A. Olaoye, B. Bhasin-Chhabra, A. Kazory, A blueprint for an integrated point-of-care ultrasound curriculum for nephrology trainees, Kidney360 2 (10) (2021) 1669–1676.

[47] I.W. Ma, P. Steinmetz, K. Weerdenburg, et al., The Canadian medical student ultrasound curriculum, J. Ultrasound Med. 39 (7) (2020) 1279–1287.

[48] N. Reisinger, A. Koratala, S. Goral, Use of point-of-care ultrasound to assess CKD, Am. J. Kidney Dis. 77 (1) (2021) A16–A19.

Chapter 12

Imaging in neurodevelopmental disorders

R.H. Talwekar and Priyanka Bibay Thakkar
Department of Electronics and Telecommunication Engineering, Government Engineering College, Raipur, Chhattisgarh, India

12.1 Introduction

Neurodevelopmental disorders (NDDs) refer to a group of conditions that affect the development and functioning of the nervous system, including the brain, spinal cord, and other nerves, resulting in difficulties with communication, social interaction, learning, and behavior. These disorders typically begin in infancy or childhood and can persist into adulthood.

Some examples of NDDs include attention-deficit hyperactivity disorder (ADHD), intellectual disability (ID), autism spectrum disorder (ASD), communication disorder and specific learning disorder (SLD). These disorders can impact a range of abilities, such as communication, socialization, attention, and cognitive functioning.

Neurodevelopmental diseases have a complicated etiology that may combine environmental, genetic, and other causes. The pathophysiology of these disorders involves alterations in brain development, structure, and function, leading to disrupted neural circuitry and communication. These alterations may be the result of genetic mutations, prenatal exposure to toxins or infection, or early-life trauma or stress.

The concept of NDDs has evolved over time, reflecting advances in our understanding of the brain and its development. Here is a brief overview of the history of NDDs:

- Early observations: Some of the earliest recorded observations of NDDs date back to ancient civilizations, where descriptions of individuals with IDs and other developmental abnormalities can be found in texts from Egypt, Greece, and Rome.
- Emergence of special education: In the 18th and 19th centuries, the first schools for individuals with IDs were established, marking the emergence of special education as a distinct field.
- Diagnostic classification: In the mid-20th century, the American Psychiatric Association developed the first diagnostic classification system for mental disorders, which included a category for "childhood disorders" that included conditions such as ID, autism, and ADHD.
- Advances in brain imaging: In the latter part of the 20th century, advances in brain imaging technology have enabled researchers to study the developing brain in more detail, providing insights into the neural basis of NDDs.
- Current research: Today, research on NDDs continues to advance our understanding of the underlying causes, genetic and environmental risk factors, and potential treatments for these conditions.

In conclusion, the history of NDDs spans centuries and reflects a growing recognition of the unique challenges faced by individuals with developmental abnormalities. Advances in our understanding of the brain and its development have enabled us to better diagnose and treat these conditions, improving outcomes for affected individuals and their families.

12.2 Types of neurodevelopmental disorder

There are several types of NDDs, each with its own set of symptoms and characteristics. Some of the most common types of NDDs are shown in Fig. 12.1.

1. ASD: It is a category of developmental abnormalities where the person had trouble interacting with others.
2. ADHD: A condition that often interferes with a person's ability to concentrate and finish tasks and is characterized by inattention, hyperactivity, and impulsivity.

Intelligent Computing Techniques in Biomedical Imaging. DOI: https://doi.org/10.1016/B978-0-443-15999-2.00004-9

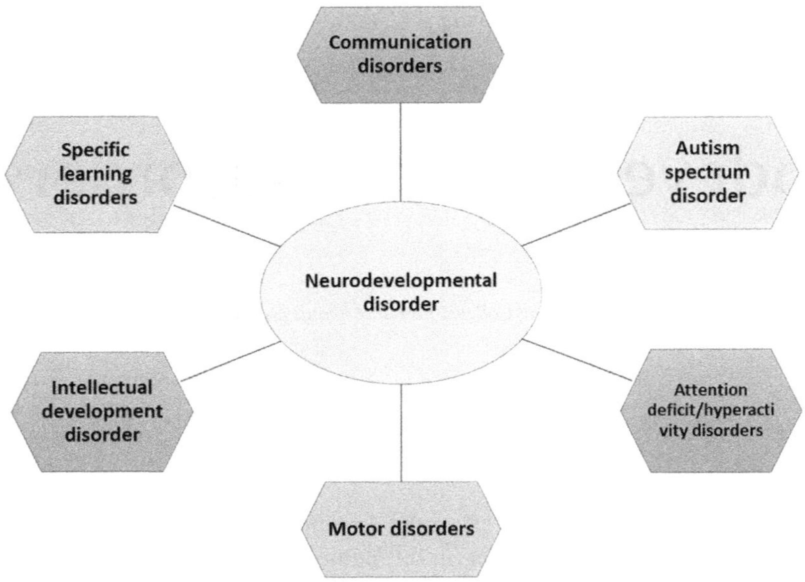

FIGURE 12.1 Types of neurodevelopmental disorder.

3. ID: A condition when adaptive behavior and intellectual functioning are significantly impaired.
4. SLD: A condition marked by challenges with writing, reading, or mathematics despite normal intelligence and access to proper education.
5. Developmental coordination disorder (DCD): A disorder marked by problems in motor planning and coordination.
6. Communication disorders: A class of disorders characterized by issues with language, speech, and interpersonal communication.

Other less common types of NDDs include Tourette syndrome, Angelman syndrome, Fragile X syndrome, Prader–Willi syndrome, and Down syndrome. Each of these disorders has its own set of symptoms, causes, and treatments.

12.3 Literature survey

In Ref. [1], the corpus callosum was examined in magnetic resonance imaging (MRI) scans of teenagers with Down and Williams syndromes as well as healthy controls. A number of metrics derived from the photos were evaluated for intrarater reliability in the study, including "global parameters" like area, r-length, r-height, and circularity as well as "regional parameters" like the sum of width in each region. These parameters were computed using stereology software, and two-tailed nonparametric statistical analyses were carried out. The investigation discovered that the highest reliability was also displayed by the regional parameters, whereas the lowest reliability for a single determination was $r = 0.977$. According to the scientists, studying the callosum in patient populations that are genetically determined can aid in understanding the connections between the brain and behavior.

This study's goal was to examine whether certain brain morphometric markers might tell whether a child is in the control group or is part of a mixed-type group with developmental dyslexia and ADHD [2]. A substantial discriminant function was created using six MRI measurements, including the Wechsler Intelligence Scale for Children-Revised (WISC-R) Full Scale Intelligent quotient (FSIQ), left planum length, left insula length, and the width of the right frontal area. The predictive discriminant analysis's overall classification accuracy increased from 60% to 87% when FSIQ and chronological age were taken into account. These findings suggest that neuroanatomical evaluations can differentiate between children with neurodevelopmental abnormalities and those who do not, and in this study, FSIQ was a significant predictor.

In paper [3], the impact of neurofibrillary tangles (NFT) on the neural mechanisms governing selective attention was examined using functional magnetic resonance imaging (fMRI) . The patients underwent fMRI scans while doing the Counting Stroop task both before starting NFT (Time 1) and 1 week after finishing the training (Time 2). Both groups displayed significant activation in the left superior parietal lobule during the Counting Stroop test during Time 1, while the anterior cingulate cortex (ACC) exhibited no evidence of activity. Both groups showed significant activation in the left superior parietal lobule during the Counting Stroop test at Time 2, but only the experimental group showed activation in the right ACC. According to these results, NFT might help the ACC, which is a crucial neuronal substrate for selective attention in kids with AD/HD, to operate normally.

In order to evaluate children with fetal alcohol spectrum disorders (FASDs) for any global or focal abnormalities and to distinguish between diagnostic subclassifications across the spectrum, this article used a thorough approach that included neuropsychological/psychiatric evaluations, MRI, fMRI, and MR spectroscopy [4]. Static encephalopathy/alcohol exposed (SE/AE), FAS/partial FAS [5], neurobehavioral disorder/alcohol exposed (ND/AE), and healthy peers were the four groups that participated in the study. In order to assess activity in seven different brain regions during N-back working memory tasks, fMRI was used in the study. The study discovered that working memory deficits and abnormal brain activity patterns were present in all diagnostic groups of children with FASD.

In order to establish a link between particular neurological symptoms and signs and abnormal scans, the MRI scan images of patients with NDDs were examined in this article [6]. As a result of the study's findings, it is recommended that all individuals with NDDs of unknown origin who exhibit neurological symptoms or signs, such as movement abnormalities, pyramidal disorders, epilepsy, or an abnormal head circumference, get an MRI scan. The authors further advise considering systematic re-assessment in a specialized facility for patients with NDDs of unclear origin following completion of genetic and metabolic screens in conjunction with MRI findings.

In terms of both basic and applied research, this article reviews studies that looked at the role of prefrontal cortex (PFC) networks in the behavior and cognitive abilities that are compromised in childhood NDDs [7]. The aim of this article is to explore the connection between these functions and circuit abnormalities found in various illnesses as shown by neuroimaging research. The results indicate that there is a significant relationship between the anatomical circuitry that is responsible for reduced functions in children with neuropsychiatric disorders and patterns of structural and functional abnormalities in the brain.

This article looks at how NDD models could benefit from the use of superresolution imaging techniques to learn more about how synapses are organized on a molecular level and how that organization impacts their function and plasticity [8]. The authors propose that genetic mutations can impair the dynamic organization of synapses and result in NDDs by causing errors in synaptic scaffolding and architecture. The study finds that synaptic function can be significantly impacted by mutations targeting synaptic proteins, frequently by producing modest changes in synapse nanoscale dynamics that may not be immediately evident.

This article provides a summary of the conceptual framework, research design, and metrics found in a dataset made up of 1445 teenagers who underwent multimodal neuroimaging as part of the study and ranged in age from 8 to 21 years old [9]. T1-weighted structural neuroimaging, resting state functional connectivity imaging, and functional imaging tasks involving working memory and emotion detection were only a few of the neuroimaging measurements that were obtained. In the article's conclusion, the data-sharing processes that make it possible for the PNC to act as a freely available resource for advancing our understanding of both normal and aberrant brain development are discussed.

The ontogeny, phylogeny, and function of the amygdala are briefly discussed in this article, which is followed by an analysis of the data from neurodevelopmental diseases that indicate dysregulation of this brain area [10]. The emphasis is on conditions where postmortem examinations, structural MRI analysis, or fMRI reveal signs of amygdala dysregulation. It is unclear if the contradictory outcomes of these researches are the consequence of methodological discrepancies or underlying heterogeneity. The amygdala, however, is a typical site for neuropathology in NDD, the paper claims, making it a possible target for therapies meant to relieve associated symptoms.

The significance of early diagnosis and intervention in developmental disorders and its connection to brain development are covered in this review article [11]. The difficulties of early ASD detection are discussed, and the vital importance of family involvement in early intervention is emphasized. The helpfulness of early intervention is summarized in the article for a number of groups, including infants at low-to-moderate risk, infants with or at high risk of cerebral palsy (CP), medically fragile newborns, and infants with or at high risk of ASD.

The neuroimaging research on schizophrenia and its causes is reviewed in this article [12]. In order to learn more about the onset and source of the brain abnormalities linked to schizophrenia, studies analyzed MRI and CT scan pictures. According to the findings, schizophrenia is not a degenerative condition but rather a developmental one. A role for genes that regulate neurodevelopment is suggested by the presence of ventriculomegaly and decreased hemisphere asymmetry in patients with familial schizophrenia and their relatives. Although early environmental insults like obstetric problems may interact with a genetic propensity to cause schizophrenia, genetic factors alone do not fully explain how the condition develops.

In order to identify prospective areas for future research, this study summarized the neuroimaging data concerning obsessive-compulsive disorder (OCD) in children and adolescents [13]. Up until October 1, 2008, pertinent publications were acquired from the Medline, PsycINFO, and reference lists with an emphasis on neuroimaging research involving kids and teens with OCD in order to accomplish this goal.

This study set out to look at the neurobiology of ASD in young children. The researchers used diffusion tensor imaging (DTI) and MRI to study the structural makeup of the brain [14]. While the latter was used to evaluate the consistency of white matter (WM) tracts, the former was utilized to measure total gray matter (GM) and WM volumes. They used voxel-based morphometry to assess regional changes. Based on their findings, the study hypothesized that neural architecture abnormalities in several brain regions, particularly in early childhood (2−3 years old), may have a role in the behavioral and cognitive deficits associated with ASD.

This review paper focuses on offering useful suggestions for carrying out clinical developmental investigations using MRI [15]. The article covers measures for measuring brain morphometry, typical adolescent developmental trajectories, and variations from those trajectories seen in a few neurodevelopmental diseases. The paper discusses methodological difficulties as well as recommended practices for study design, quality assurance, and picture acquisition and analysis. The authors argue for more thoroughness and transparency in the reporting of procedures while also discussing statistical analysis and result interpretation.

The use of brain imaging techniques offers valuable assistance in managing NDDs in children. By utilizing these imaging methods, it is possible to identify children who are at the highest developmental risk, diagnose and prognoses their conditions more accurately, and better communicate biologic information to their families. This article provides an overview of these neuroimaging techniques and their importance in caring for children with neurodevelopmental disabilities.

12.4 Famous personalities with neurodevelopmental disorder

There are many famous people throughout history who have been diagnosed with NDDs. Here are some examples:

1. Albert Einstein—It is believed that Einstein had Asperger's syndrome, which is a type of autism.
2. Temple Grandin—Grandin is a Professor and animal behaviorist who has autism. She is also an autism activist and advocate.
3. Charles Darwin—Darwin is believed to have had Asperger's syndrome.
4. Bill Gates—Gates has spoken publicly about having ADHD.
5. Dan Aykroyd—Aykroyd has been diagnosed with Asperger's syndrome.
6. Michelangelo—There is speculation that Michelangelo had ADHD.
7. Hans Christian Andersen—Andersen is believed to have had dyslexia.
8. Steven Spielberg—Spielberg has been diagnosed with ADHD.
9. Emily Dickinson—Dickinson is believed to have had Asperger's syndrome.
10. Vincent van Gogh—Van Gogh is believed to have had bipolar disorder.

It is important to note that diagnoses of NDDs were not available during the lifetimes of many of these individuals, so their conditions may not have been officially diagnosed or recognized. Additionally, having an NDD does not define a person's entire identity or limit their potential for success.

12.5 Diagnosis of neurodevelopmental disorder

Fig. 12.2 shows the general architecture for NDD diagnosis. Early diagnosis of NDDs is critical for improving outcomes and providing appropriate interventions for affected individuals. There are several approaches to early diagnosis of NDDs. One approach is to screen infants and young children for developmental delays using standardized assessment tools. For example, the Ages and Stages Questionnaires (ASQ) and the Denver Developmental Screening Test (DDST) are commonly used screening tools that assess various areas of development, such as gross motor skills, fine motor skills, problem solving, communication, and socialization.

Another approach is to monitor children's development closely and identify any early signs of NDDs. For example, parents and caregivers can look out for delays in language development, difficulty with social interaction, repetitive behaviors, and other signs that may indicate an NDD.

If a qualified healthcare professional, such as a pediatrician, neurologist, or developmental psychologist, doubts a child of having an NDD, further assessment and diagnosis may be necessary. This may involve a more comprehensive evaluation of the child's development and behavior, as well as medical and genetic testing to rule out other conditions that may have similar symptoms.

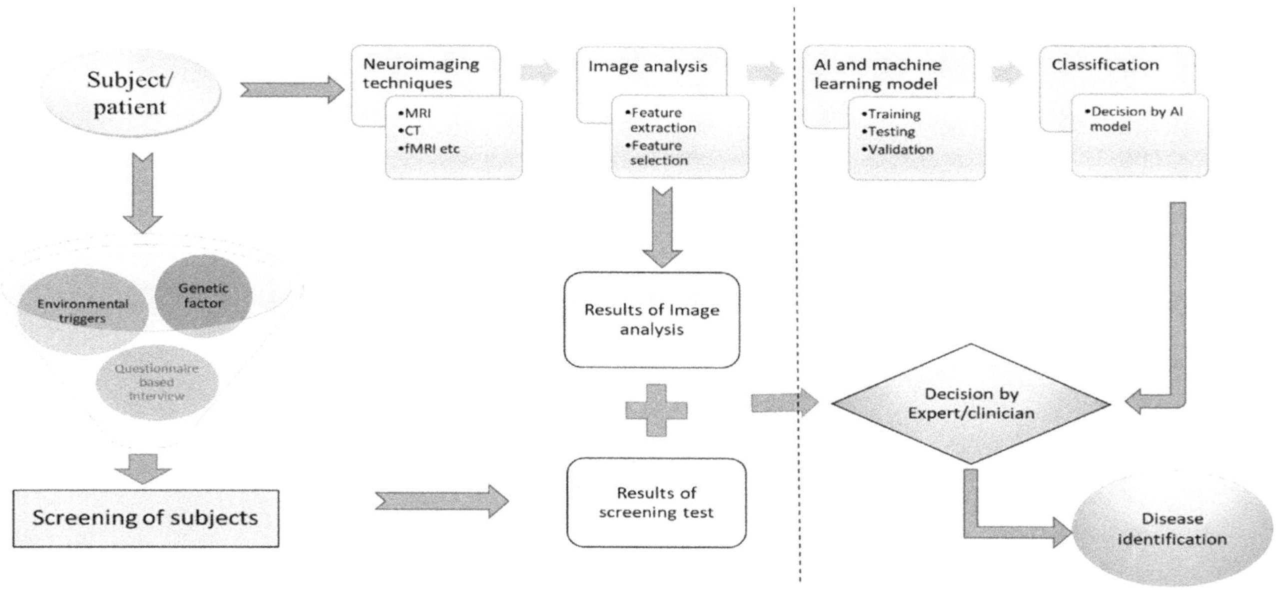

FIGURE 12.2 Architectural block diagram of diagnostic model for neurodevelopmental disorder.

Early diagnosis of NDDs can help ensure that affected individuals receive appropriate interventions and support as early as possible. This may include speech therapy, behavioral therapy, occupational therapy, and medication in some cases.

12.6 Tools used for the diagnosis of neurodevelopmental disorder

There are several tools and methods used for the diagnosis of NDDs. Here are some commonly used ones:

1. Developmental screening tools: These are brief, standardized questionnaires that are used to identify children who may be at risk for a developmental delay. Examples include the ASQ and the DDST.
2. Diagnostic interview: This is a structured interview that is conducted by a trained professional to assess a child's developmental and behavioral history. The Autism Diagnostic Interview—Revised (ADI-R) and the Diagnostic Interview for Social and Communication Disorders (DISCO) are examples of diagnostic interviews.
3. Observation: This involves observing the child's behavior and interactions with others. The Autism Diagnostic Observation Schedule (ADOS) is a commonly used observation tool for diagnosing ASD [16].
4. Neuropsychological assessment: This involves a battery of tests that assess numerous aspects of cognitive functioning including attention, memory, language, and problem solving. Neuropsychological assessment can help identify underlying neurological impairments that may contribute to a child's developmental difficulties.
5. Genetic testing: Some NDDs are caused by genetic mutations. Genetic testing can help identify these mutations and confirm a diagnosis. Examples of genetic tests include chromosomal microarray analysis and whole-exome sequencing.

It's essential to note that the diagnosis of NDDs should be made by a trained professional, such as a pediatrician, neurologist, or clinical psychologist, who has experience in diagnosing these disorders.

12.7 Advancement in the detection of neurodevelopmental disorder using imaging techniques

Fig. 12.3 shows different imaging techniques for NDD diagnosis. Advancements in imaging techniques have led to significant improvements in the diagnosis and management of NDDs. Here are some examples of how imaging techniques are used as a tool for the diagnosis of NDD:-

1. MRI and computed tomography (CT) scans have been used to detect structural abnormalities in the brain, such as malformations, tumors, and lesions, which can contribute to developmental delays, IDs, and other NDDs. These techniques provide high-resolution images of the brain and have become a standard tool in the diagnostic process.

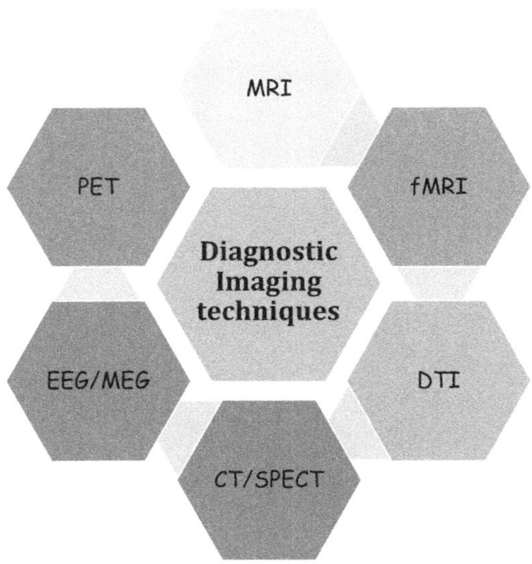

FIGURE 12.3 Different types of diagnostic imaging techniques.

2. DTI is a type of MRI technique that has been used to study the white matter tracts in the brain. DTI can identify abnormalities in the connectivity of brain regions.
3. fMRI has been used to study the brain's functional activity and connectivity. This technique allows researchers to study how different regions of the brain communicate with each other during specific tasks, providing insight into the underlying neural processes that contribute to NDDs.
4. Positron emission tomography (PET) scans have been used to measure brain metabolism and neurotransmitter activity. These scans can identify changes in brain chemistry that are associated with NDDs, such as abnormalities in dopamine and serotonin levels.
5. Electroencephalography (EEG) and magnetoencephalography (MEG) are techniques used to measure the electrical and magnetic activity of the brain, respectively. These techniques can identify abnormalities in brain activity that are associated with NDDs, such as epileptic seizures, which can cause developmental delays and cognitive impairments.
6. CT: CT scans use X-rays to create cross-sectional images of the brain. It can be used to detect abnormalities in the brain structure, such as tumors or bleeding.
7. Single-photon emission computed tomography (SPECT): SPECT scans use a small amount of a radioactive substance to create images of brain blood flow. It can be used to identify changes in blood flow and metabolism associated with NDDs.

In summary, the advancements in imaging techniques have led to significant improvements in the diagnosis and management of NDDs. These techniques provide valuable insights into the underlying neural processes that contribute to these disorders, allowing for earlier and more accurate diagnoses and improved treatment options.

12.8 Performance measure for the diagnosis of neurodevelopmental disorder using imaging techniques

The performance measure for the diagnosis of NDD using imaging techniques will depend on the specific imaging modality being used, the type of disorder being diagnosed, and the specific outcome being measured.

Diagnostic accuracy, which refers to the ability of an imaging technology to correctly identify individuals with and without the condition, is a commonly used performance statistic. Metrics like sensitivity (the percentage of people with the disorder who are correctly identified as having the disorder) and specificity (the percentage of people without the disorder who are correctly recognized as not having the disorder) can be used to quantify this.

Other performance metrics may include positive predictive value, which is the percentage of people who test positively for a disorder and actually have it, negative predictive value, which is the percentage of people who test negatively for a disorder but actually don't have it, and receiver operating characteristic (ROC) curves, which display the trade-off between sensitivity and specificity for various diagnostic cutoff points.

It is important to note that the performance of imaging techniques for diagnosing NDDs may vary depending on the population being studied and the specific imaging protocol used. Additionally, imaging techniques may be used in conjunction with other diagnostic tools, such as clinical assessments and behavioral measures, to improve diagnostic accuracy.

12.9 Performance evaluation of neuroimaging for neurodevelopmental disorder

Several neuroimaging modalities, such as MRI, fMRI, DTI, and PET, have been used to investigate the neurobiological underpinnings of these disorders. These techniques have provided valuable insights into the brain alterations associated with NDDs, which can potentially inform diagnosis, prognosis, and treatment planning.

Performance evaluation of neuroimaging for NDDs involves assessing the sensitivity, specificity, and reliability of these techniques in detecting brain alterations in affected individuals compared to typically developing individuals. It also involves evaluating the accuracy and validity of the derived biomarkers in predicting clinical outcomes and response to treatment.

Several challenges exist in the performance evaluation of neuroimaging for NDDs. These include the heterogeneity of the disorders, the variability in imaging protocols and analysis methods across studies, and the need for large sample sizes to ensure statistical power and generalizability.

Despite these challenges, neuroimaging holds promise as a valuable tool for the assessment and understanding of NDDs. Continued research and standardization of imaging protocols and analysis methods are needed to maximize the potential of these techniques in clinical practice.

References

[1] Paul P. Wang, et al., Callosal morphology concurs with neurobehavioral and neuropathological findings in two neurodevelopmental disorders, Arch. Neurol. 49 (4) (1992) 407–411.

[2] M. Semrud-Clikeman, S.R. Hooper, G.W. Hynd, K. Hern, R. Presley, T. Watson, Prediction of group membership in developmental dyslexia, attention deficit hyperactivity disorder, and normal controls using brain morphometric analysis of magnetic resonance imaging, Arch. Clin. Neuropsychol. 11 (6) (1996) 521–528.

[3] J. Lévesque, M. Beauregard, B. Mensour, Effect of neurofeedback training on the neural substrates of selective attention in children with attention-deficit/hyperactivity disorder: a functional magnetic resonance imaging study, Neurosci. Lett. 394 (3) (2006) 216–221. Available from: https://doi.org/10.1016/j.neulet.2005.10.100.

[4] S.J. Astley, et al., Functional magnetic resonance imaging outcomes from a comprehensive magnetic resonance study of children with fetal alcohol spectrum disorders, J. Neurodev. Disord. 1 (1) (2009) 61–80. Available from: https://doi.org/10.1007/s11689-009-9004-0.

[5] S.Y. Bookheimer, E.R. Sowell, Brain imaging in FAS: commentary on the article by Malisza et al., Pediatr. Res. 58 (6) (2005) 1148–1149. https://doi.org/10.1203/01.pdr.0000188720.59781.b3.

[6] H.M. Engbers *et al.*, The clinical utility of MRI in patients with neurodevelopmental disorders of unknown origin, Eur. J. Neurol. 17 (6) (2010) 815–822. https://doi.org/10.1111/j.1468-1331.2009.02927.x.

[7] A.F.T. Arnsten, K. Rubia, Neurobiological circuits regulating attention, cognitive control, motivation, and emotion: disruptions in neurodevelopmental psychiatric disorders, JAAC 51 (4) (2012) 356–367. Available from: https://doi.org/10.1016/j.jaac.2012.01.008.

[8] H.L. Zieger, D. Choquet, Nanoscale synapse organization and dysfunction in neurodevelopmental disorders, Neurobiol. Dis. 158 (July) (2021) 105453. Available from: https://doi.org/10.1016/j.nbd.2021.105453. Elsevier Inc.

[9] T.D. Satterthwaite, et al., Neuroimaging of the Philadelphia neurodevelopmental cohort, Neuroimage 86 (2014) 544–553. Available from: https://doi.org/10.1016/j.neuroimage.2013.07.064.

[10] C.M. Schumann, M.D. Bauman, D.G. Amaral, Neuropsychologia abnormal structure or function of the amygdala is a common component of neurodevelopmental disorders, Neuropsychologia 49 (4) (2011) 745–759. Available from: https://doi.org/10.1016/j.neuropsychologia.2010.09.028.

[11] M. Hadders-Algra, Early diagnostics and early intervention in neurodevelopmental disorders — age-dependent challenges and opportunities, J. Clin. Med. 10 (4) (2021) 861.

[12] S. Frangou, R.M. Murray. Imaging as a tool in exploring the neurodevelopment and genetics of schizophrenia, Br. Med. Bull. 52 (3) (1996) 587–596.

[13] C. Huyser, D.J. Veltman, E. De Haan, F. Boer. Paediatric obsessive – compulsive disorder, a neurodevelopmental disorder ? Evidence from neuroimaging, Neurosci. Biobehav. Rev. 33 (6) (2009) 818–830. https://doi.org/10.1016/j.neubiorev.2009.01.003.

[14] Z. Xiao, T. Qiu, X. Ke, X. Xiao, Autism spectrum disorder as early neurodevelopmental disorder: evidence from the brain imaging abnormalities in 2–3 years old toddlers, J. Autism Dev. Disord. 44 (7) (2014) 1633–1640. https://doi.org/10.1007/s10803-014-2033-x.

[15] L.L. Backhausen, M.M. Herting, C.K. Tamnes, N.C. Vetter, Best practices in structural neuroimaging of neurodevelopmental disorders, Neuropsychol. Rev. (2021). Available from: https://doi.org/10.1007/s11065-021-09496-2. 0123456789.

[16] S. Goh, Z. Dong, Y. Zhang, S. DiMauro, B. S. Peterson, Mitochondrial dysfunction as a neurobiological subtype of autism spectrum disorder: evidence from brain imaging, JAMA Psychiatry 71 (6) (2014) 665–671. https://doi.org/10.1001/jamapsychiatry.2014.179.

Chapter 13

A computational intelligence-based approach for detecting drowsiness in sleep apnea patients using multimodal bio-signals

Prateek Pratyasha and Saurabh Gupta
Department of Biomedical Engineering, National Institute of Technology Raipur, Raipur, Chhattisgarh, India

13.1 Introduction

Sleep apnea (SA) is a crucial sleep disorder that causes unwanted cease of breathing in repeated interval of time. This disorder is quite common in adults within the age group of 30–70 causing cardiac and neuro dysfunction. In a global survey, it is observed that 17% of women and 34% of men suffers from undiagnosed SA condition every year [1]. Some of the observed symptoms of SA are sleep fragmentation, oxidative stress, day time sleepiness, inflammation, and hypertension. However, drowsiness is most common consequences among all of them which ultimately lead for a poor quality of sleep. Biologically, sleep involves the interconnectivity of neurons of various brain regions to generate and integrate structural neural networks [2]. Any topographical alternation in the network structures can affect the quality of sleep.

Although sleep disruption is very common during an individual's sleep cycle, an intermittent sleep disruption for prolonged hours or days could be nuisance. Such chronic sleep deficit and be observed in younger as well as older patients. The consequences of day-time sleepiness often cause mood alteration, fatigue, depression, cognitive impairment, etc. SA associated with drowsiness is also responsible for nausea, obesity, hypertension, diabetes, frequent cardio-pulmonary and cerebro-vascular disorders [3]. Therefore it is essential to monitor the night-sleep cycle and day-sleep cycle of an individual to maintain a quality of life.

In general, SA is categorized into two types: central sleep apnea (CSA) in which the breathing stops in a cyclic or intermittent pattern due to neural dysfunctionality of central nervous system (CNS) [4]. However, obstructive sleep apnea (OSA) obstructs the upper airways during sleep that creates inadequate airflow for breathing [5]. This breathing obstruction can reduce the blood oxygen level that may lead to frequent arousal condition of the patient. Both of this sleep disorders can disturb rapid eye movement (REM) as well as non-rapid eye movement (NREM) sleep. CSA is very rarely observed in patients whereas OSA is quite common.

Recently, National Sleep Foundation (NSF), the United States has recommended a sleep chart for individuals ranging 1–70 years age group including newborns. According to the study, a newborn of 0–3 months requires 14–17 hours of sleep, infant of 4–11 months needs 12–15 hours of sleep. Similarly, toddler of 3–5 years requires 10–13 hours of sleep whereas kids of 6–14 years require 9–11 hours of sleep. However, the sleeping duration is reduced to 7–9 hours for youngsters of 18–25 years and adults of 26–60 years. For people more than 60 years are recommended to take 7–8 hours of minimum sleep [6]. All these studies have represented the night-time sleep duration including the day-time power nap of 30 minutes to 1 hour. However, the symptoms of drowsiness not only affect the day-time functions but also create cardio-metabolic disruptions and therapeutic responses. For the same subjects with same severity level of sleep apnea syndrome (SAS), the functionality of drowsiness is elusive and varies substantially.

Some study has represented drowsiness as a disorder in heterogeneous brain network at a structural or functional level by considering neuroimaging data. In [7], diffusion tensor imaging (DTI) is used to analyze the brain network at structural level and observed the neural connectivity driven by changes in white matter for the patients of SA. Another

Intelligent Computing Techniques in Biomedical Imaging. DOI: https://doi.org/10.1016/B978-0-443-15999-2.00010-4

study highlights the global topological characteristics of structural brain network involving the cortical volumes [8]. The authors have collected functional MRI data to study SA severity along with abnormal functional connectivity and cognitive dysfunction. In few articles, drowsiness associated with functional neural connectivity is framed in a default mode network (DMN) that plays a key role for cognition. The accumulation of β-amyloid in the DMN regions is also associated with the drowsiness symptoms [9–11]. In [12], the problem of residual sleepiness with SA patients are analyzed by observing the abnormality of white matter on DTI. However, it is still ambiguous to determine the mechanism of brain connectivity in detail while monitoring SA patients in association with drowsiness.

Most of the techniques for the analysis of structural and functional network have limitations for specifying SA severity and drowsiness as the relationship between them (i.e., neural connectivity) requires a continuous measurement on drowsiness scale to check the degree of apnea severity. Hence, there is a demand for advanced techniques that can extract linear characteristics from nonlinear complex model and to determine the exact location in neural connectivity that is responsible for drowsiness. This chapter approaches a deep learning algorithm as a computational technique to retrieve the corelation between feature input and target variables. Compared to conventional model-based methods, deep learning-based techniques are superior in learning features from complex data.

For finding the most susceptible neural connection between SA and drowsiness, the relationship between actual scores and predicted scores are maximized while minimizing their common errors. Generally, most of the deep learning model adjusts the weights the neural network according to the error obtained between actual and predicted scores. The weight parameters are continuously adjusted till the process is succeeded in reducing the error in order to result an appropriate output model. It has been observed that input features with potential gradient have better impact on the predicted score as compared to a nonweighted input feature. Moreover, the trained model obtained from individual features can perform satisfactorily on decision-making process by extracting the gradient information from error deviation.

The proposed study emphasizes on detecting drowsiness for both apnea and controlled subjects in a representative cyclic alternative pattern (CAP) database by implementing our proposed methodology, time distributed convolutional neural network (TD-CNN). Variations in the output of our proposed model are investigated on the basis of their gender, age group. Our specific hypothesis only mentions about the role of TD-CNN for the prediction of drowsiness in SA patients. This technique could be an effective one to understand the neuro connectivity pattern of patients with different degree of SA severity associated with drowsiness.

13.2 Proposed methodology

This section highlights our proposed methodology for the detection of drowsiness both in SA and controlled patients. The methodology begins with collecting the database from an open source. Then, each of the collected data is involved in preprocessing phase. After obtaining a generalized preprocessed image from the signals, we have implemented a TD-CNN for extracting features from the images. Finally, a sequence modeling-based Transformer is introduced to train the training samples and classify the corresponding data. The performance of our classifier is measured through various evaluation parameters and thereby, the drowsiness is detected. Fig. 13.1 shows architecture of our proposed methodology.

13.2.1 Description of dataset

In this work, we have collected samples from CAP sleep database from a publicly available open source, namely PhysioNet. The dataset involves 92 subjects from different sleep disorders and 16 controlled subjects. The subjects in this database are categorized on the basis of their gender, age group. Multiple channels of ECG (for heart rate variability), EEG (for sleep stages), EOG (eyeball movement), EMG (for the movement of submentalis muscle) and respiration signals (for airflow, SaO_2, abdominal and thoracic effect) devices are connected to the patient's body while they are observed under a polysomnography (PSG) test. The signals are recorded from several minutes before sleep and several minutes after getting awake. Sleep experts have labeled the sleep stages as: wake (W), REM, light sleep ($S_1$and S_2), and nonrapid eye movement or deep sleep (S_3 and S_4). The length of total recording was 9–10 hours. The database is available from https://archive.physionet.org/cgi-bin/atm/ATM.

For this research work, we have selected 16 controlled patients who are later on termed as nonapnea subjects and four sleep behavior disorder (SBD) subjects. Patients with SBD syndromes are more prone to SA, therefore the datasets are valid for the detection of sleeping pattern in these patients. Sleep data for each of the patients are categorized into three different modalities such as three-biploar lead EEG (F4-C4, P4-O2, and C4-A1) and a dual-channel EOG (ROC-LOC). The length of each signal without any segmentation is 3 minutes at a sampling frequency of 512 Hz. The details of patients are mentioned in Table 13.1.

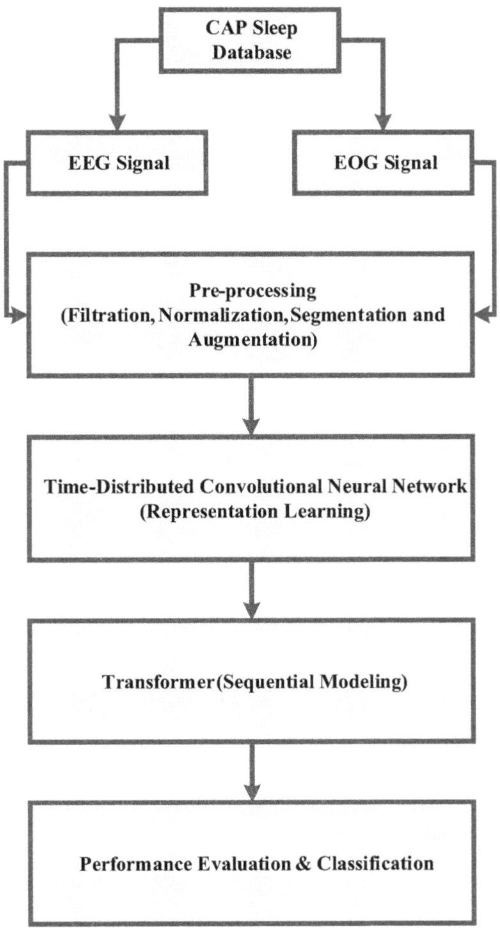

FIGURE 13.1 Framework of proposed methodology.

TABLE 13.1 Demographic characteristics of cyclic alternative pattern (CAP) sleep database.

Types of subjects	No. of subjects	Annotations available at 512 Hz	Gender		Age	Epoch
			M	F	Mean ± standard deviation	
Controlled (non-SA)	16	13	9	4	32 ± 4.89	6063
SBD (SA)	04	03	3	–	69.33 ± 6.12	2879

Non-SA, nonsleep apnea; *SA*, sleep apnea; *SBD*, sleep behavior disorder.

As observed from Table 13.1, we have excluded three controlled subjects (n04, n08, n16) and one SBD patient (sbd02) due to unavailable annotation of sampling frequency. Total 8942 epochs have generated being 13.47% W, 11.65% REM, 74.88% NREM. The sleep stage of NREM is further divided as 5.38% S1, 41.62% S2, and 27.88% S3. Fig. 13.2 shows raw waveform view of multichannel EEG and EOG signals.

13.2.2 Signal preprocessing

This phase involves signal filtration, missing value deletion or imputation, segmentation, augmentation, and Fourier Transform-based conversion.

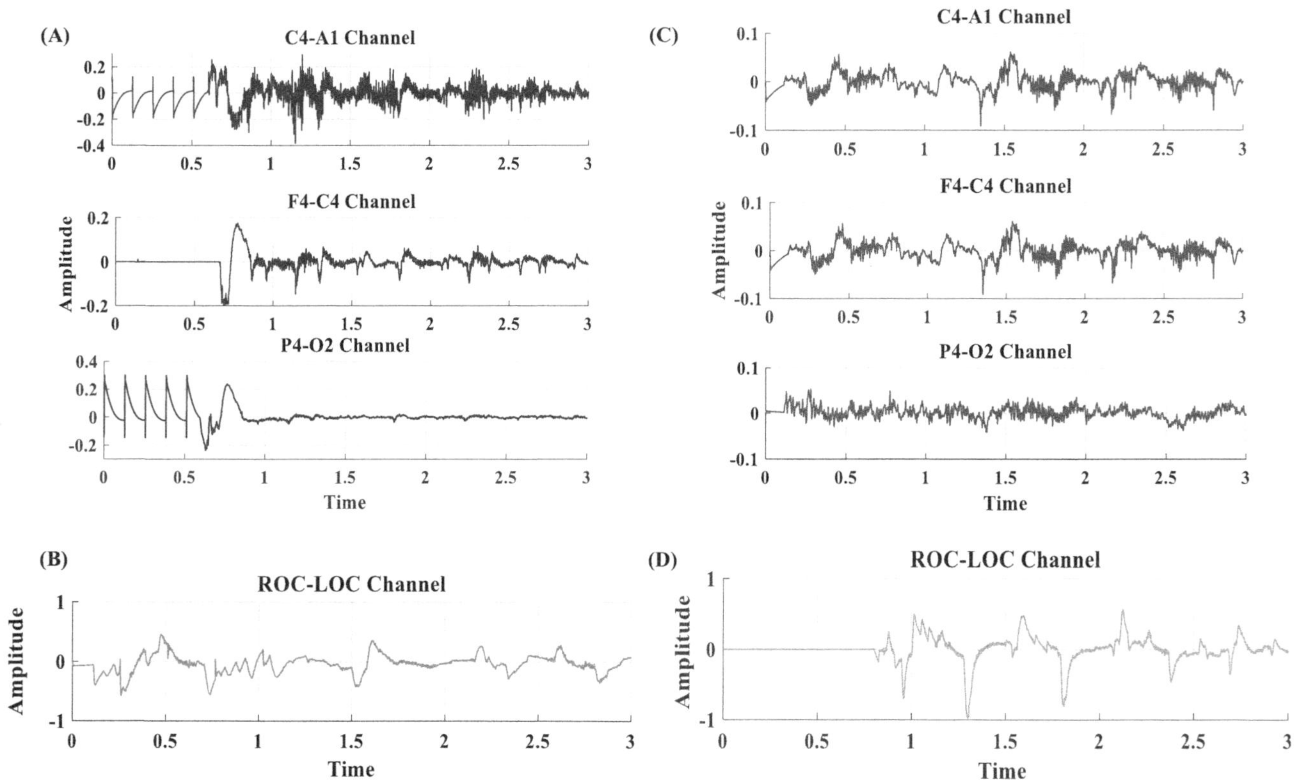

FIGURE 13.2 (A) Original view of EEG channels from one SBD subject. (B) Original view of EOG channel from one SBD subject. (C) Original view of EEG channels from one controlled subject. (D) Original view of EOG channel from one controlled subject.

13.2.2.1 Signal filtration

The raw physiological signals collected from PSG device are involved in preprocessing to remove the unwanted noise from the signal. For EEG signal, a zero-phase high-pass FIR filter of 100th order with a cut-off frequency of 0.25 Hz is used as a preprocessing technique [13]. For ECG signal, the same filter is used at 0.1 Hz cut-off frequency. For EOG signal, the cut-off frequency varies between 0.25 and 0.5 Hz [14].

13.2.2.2 Missing value deletion or imputation

The dataset is further examined to delete any missing values from the data or impute any missing values into the data. For that purpose, a technique called Little's Missing Completely at Random Test (MCRT) [15] is used at null hypothesis ($P > .05$). The test confirms that all the missing values do not put much significance to our collected dataset. Unsurprisingly, there is no requirement for missing value imputation as the existing dataset already have missing values in it. Hence, the variable with numeric value "0" would be deleting from the data dimension. Additionally, categorical and continuous variable are handled separately as they carry different characteristics. Continuous variables are normalized and scaled with a specific standardization technique; thereby the values are represented in terms of "mean ± standard deviation." Meanwhile, the categorical variables are converted into encoded vectors with equal columns of unique categories which are otherwise used as "dummy variables."

13.2.2.3 Signal normalization

This study uses multimodal signals for the detection of day time sleepiness in SA patients. Different physiological signals pursuits multiple data points with a diversified range, thereby affect the trained model. To reduce the affect, the dataset is normalized at a common range.

13.2.2.4 Data segmentation

After signal filtration and normalization of the raw EEG and EOG signals in the previous subsection, the corresponding signals are segmented at 30-second interval to highlight the prior features and obtain a clearer image. For this study, each of the image view is segmented in a 30-second interval of window without any time overlap [16]. This phase of preprocessing is helpful in feature extraction.

13.2.2.5 Augmentation

As large numbers of EEG and EOG channels are associated with different classes, there is a probability of imbalanced dataset. To overcome this issue, we need a data regularization technique, also known as data augmentation [17]. Augmented data help in training the model without any overfitting or underfitting issue. It also counters the impact of class bias due to the regularized image. Hence, this component is highly essential to be implemented as a preprocessing phase. Initially, augmentation starts with incrementing the class samples in the preexisting class with less number of samples. Then, new samples are generated for the same class and reduce the chance of overfitting. For this study, we have tried various augmentation strategies such as image shifting, resizing, cropping, and horizontal flipping.

Finally, Short time Fourier Transform (STFT) is applied to each augmented image and then converts them into a 2-dimensional spectrogram. The spectrogram view represents spectrum of both EEG and EOG signals in terms of frequency over a horizontal time axis [18]. Small frames of time window 30 seconds have taken from the segmented signal and STFT is applied to it (as mentioned in the spectrogam view of EEG) (Fig. 13.3).

13.2.3 Time-distributed convolutional neural network

The Convolutional Neural Network is used for the purpose of identifying spatial features and later on implicates those simple spatial features to generate a large dimensional feature vector for the same data. Other neural network techniques such as Artificial Neural Network (ANN) [19], Recurrent Neural Network (RNN) [20], Long Short Term Memory (LSTM) [21] are proved to be tough competitors, but the remarkable milestone created by CNN in case of image processing, it becomes the popular choice for researchers in the task of image classification. Our propose technique Time Distributed Convolutional Neural Network (TD-CNN) is the integration of multiple CNN [22]. The state-of-art of TD-CNN initiates with multiple convolution layers stacked with Batch Normalization and some Max pooling layers, followed by a drop out layer [23]. The last part of CNN architecture is a flatten layer. As a beginner, we have reviewed several architectures such as Inception, VGG-19, ResNet etc. The architectural representation of CNN layers in depicted in Fig. 13.4.

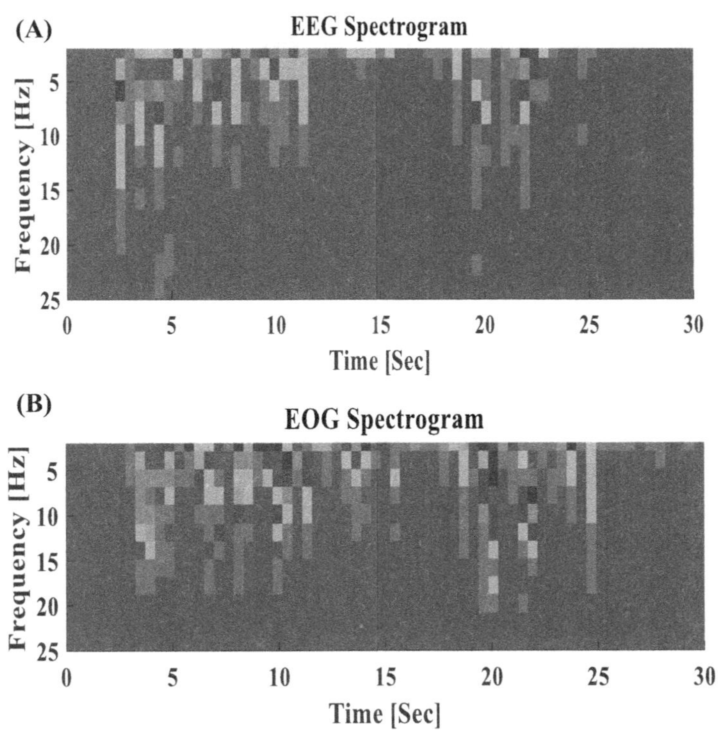

FIGURE 13.3 (A) Spectrogram-based view of EEG. (B) Spectrogram-based view of EOG.

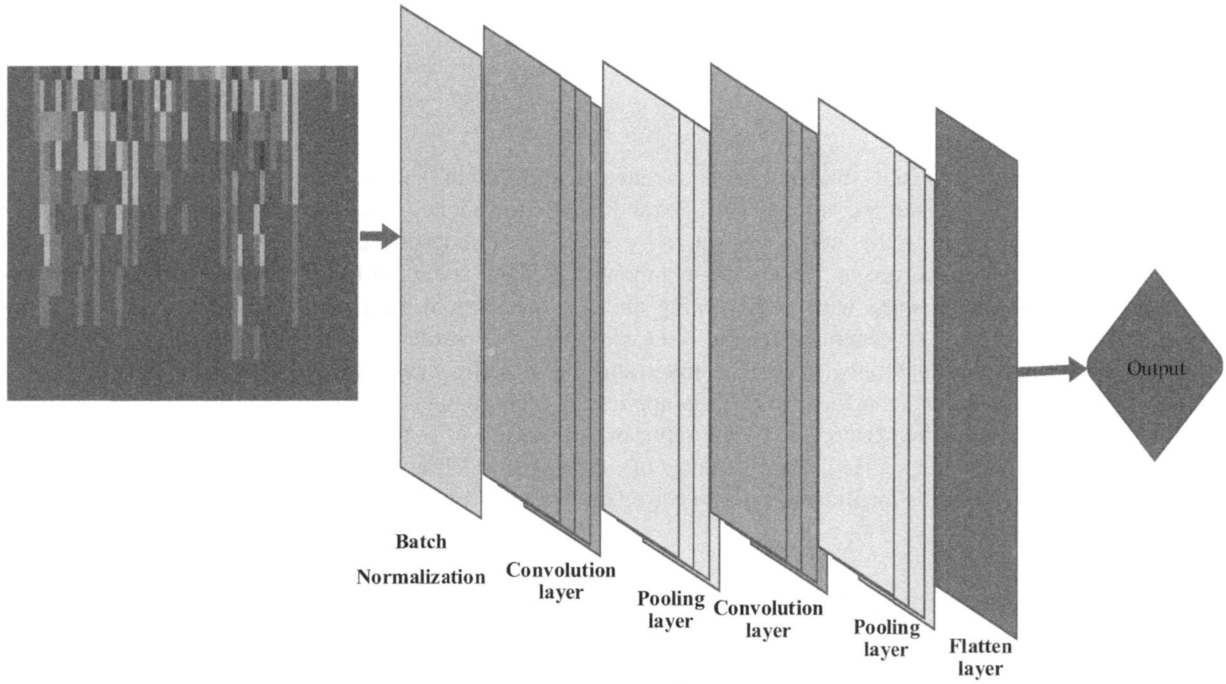

FIGURE 13.4 Architectural representation of CNN layers.

Batch normalization: Before training the convolution layer of our proposed TD-CNN model, both EEG and EOG signals are fed to a batch normalization part. Batch normalization for an input signal S_i is given in Eq. (13.1).

$$b_n = \alpha \times \left(\frac{S_i - m}{\sqrt{\gamma^2 + \chi}} \right) + \delta \tag{13.1}$$

where b_n stands for batch normalization, α is the scale parameter, m is the mini-batch mean, γ mini-batch variance, χ is the random disturbance, and δ is the shift parameter.

Convolution layer: A convolution layer is chosen for the feature mapping of physiological signals like EEG and EOG. It is the most preferable and simpler way for two-dimensional feature transformation.

Pooling layer: A pooling layer is used for reducing the dimensions of feature map derived from the neighboring convolution layer. All the pooling layers use max pooling function with 2×2 kernel.

Drop out layer: Random nodes from the CNN model is dropped out with the help of drop out layer in order to reduce the overfitting, thereby prevents the complex adaptations on all the training samples.

Table 13.2 lists all the layers used in a TD-CNN model.

13.2.4 Tuning of hyperparameters

The process in which a specified value is set for our hyperparameter is known as hyperparameter tuning. Hyperparameters are assigned values for particular neural network architecture. Although we have overlooked several tuning strategies such as random search, semiautomated search, hyperband search that can be used for a neural network model; we finalize random search strategies to select hyperparameters for TD-CNN model. Random search strategy is used to train the optimal parameters meanwhile eradicate aliasing [24]. We have performed hyperparameter tuning by comparing all the reviews strategies, but stick to random search strategy only whose values are mentioned in Table 13.3.

13.2.5 Analytical study

Demographic characteristics of all the SBD and controlled patients' area are analyzed by implementing descriptive statistical analysis. From that, mean and standard deviations of the individual characteristics are calculated. Later on, inferential statistical analysis, namely analysis of variance (ANOVA) test is performed to detect the significant variables. Hence, the

TABLE 13.2 Listed CNN layers with their corresponding values.

Layers	No. of filters used	No. of kernels used	Stride	Output
B.N_1	–	–	–	2510 × 2510
Conv._1	128	3 × 3	1	2080 × 2080 × 128
Maxpool_1	–	2 × 2	5	1040 × 1040 × 128
Dropout_1	–	–	–	–
Conv._2	64	3 × 3	1	1040 × 1040 × 64
Maxpool_2	–	2 × 2	5	520 × 520 × 64
Dropout_2	–	–	–	–
Conv._3	32	3 × 3	1	520 × 520 × 32
Maxpool_3	–	2 × 2	5	260 × 260 × 32
Dropout_3	–	–	–	–
Conv._4	16	3 × 3	1	260 × 260 × 16
Maxpool_4	–	2 × 2	5	130 × 130 × 16
Dropout_4	–	–	–	–
Conv._5	8	3 × 3	1	130 × 130 × 08
Maxpool_5	–	2 × 2	5	65 × 65 × 08
Dropout_5	–	–	–	–
Conv._6	4	3 × 3	1	65 × 65 × 04
Maxpool_6	–	2 × 2	5	32 × 32 × 04
Dropout_6	–	–	–	

TABLE 13.3 Selected hyperparameters through random search.

Hyperparameters	Value
Convolutional layers	6
Kernel size	3
Pooling layer	6
Kernel size	2
Drop out rate	0.25

statistical impacts on the features are determined by testing their categorical potentiality. All the retrieved features are demonstrated by a statistical relevance value "*P*" using ANOVA test. Total 10 features are selected on rank basis from both EEG and EOG signal as mentioned in Table 13.4. For this study, we have not used any correlation-based feature selection algorithms as all the features show their statistical significance by indicating the "*P*" close to zero. Mean and standard deviation of features after ANOVA test considering EEG and EOG features is mentioned in Table 13.5.

13.2.6 A sequence modeling-based transformer for classification

Although, transformers have been used since a long time for the classification of audio and test-based data; the experiments are still going on in the field of computer vision tasks like image processing. Integrating with various CNN

TABLE 13.4 Feature ranks for selected features of EEG and EOG.

Feature rank	EEG features	EOG features
1	Beta frontal (F4)	Amplitude velocity ratio (AVR)
2	Alpha frontal (F4)	Blink duration (BD)
3	Theta frontal (F4)	Peak closing velocity (PCV)
4	Beta central (C4)	
5	Alpha central (C4)	
6	Theta central (C4)	
7	Beta parietal (P4)	
8	Alpha parietal (P4)	
9	Theta parietal (P4)	
10	Beta occipital (O2)	
11	Alpha occipital (O2)	
12	Theta occipital (O2)	

TABLE 13.5 Mean and standard deviation of features after ANOVA test.

Modality	EEG (F4-C4)	EEG (P4-O2)	EEG (C4-A1)	EOG	EEG multimodal	EEG multimodal + EOG
Mean ± standard deviation	0.58 ± 0.05	0.51 ± 0.04	0.46 ± 0.05	0.33 ± 0.05	0.42 ± 0.06	0.48 ± 0.06

model, Raghu et al. [25] have proposed a Vision Transformer (VT) that segregates each images into patches. Later on, each of these patches is fed to the input layer of the transformer and performs the classification task alike text embedding. Despite of several advantageous functionality of VT, the use is limited as it takes one patch at a time for processing. However, a sequence modeling-based Transformer accepts the entire embedding at once as an input [26].

As for the study, we are using spectrogram-based view of EEG and EOG; hence it is not divided into patches. Rather, we inject the series of time steps to the Transformer showing the vertical axis of each frame as frequency and horizontal axis as time. Before sending the frames as an input, each of the spectrogram are convoluted independently. The convolution process is time-consuming here, as each of the frames undergoes the convolution flow. Each of the convolution flow carries different weight, thereby it is difficult to correlate two different features resulted by two different convolution flow. But, it must be ensured that all the convolution flow should detect same set of features.

Mentioning the transformer, the architecture is same as of RNN and LSTM. It involves two major blocks: an encoder block for feature mapping and a decoder block to interpret the outputs of encoder and classify them [27]. The details of each block are explained in Fig. 13.5.

Encoder block: The encoder part of Transformer is a stack of "N" number of encoders. Its internal architecture involves a feed-forward neural network (FFNN) and a multihead attention part. Both the parts have a layered normalization and a residual connection. For this study, we have considered a two stacked encoder block by taking "$N = 2$."

To deal with the multimodal feature sequence, we use multihead attention (MHA) mechanism along with the encoder block to execute the dynamic correlation between different modalities and learn from their sequence pattern. It also promotes the intra-modality dependencies between two different feature locations. MHA is derived from self-attention which plays an important role for successful sequence learning by the Transformer. The mechanism of

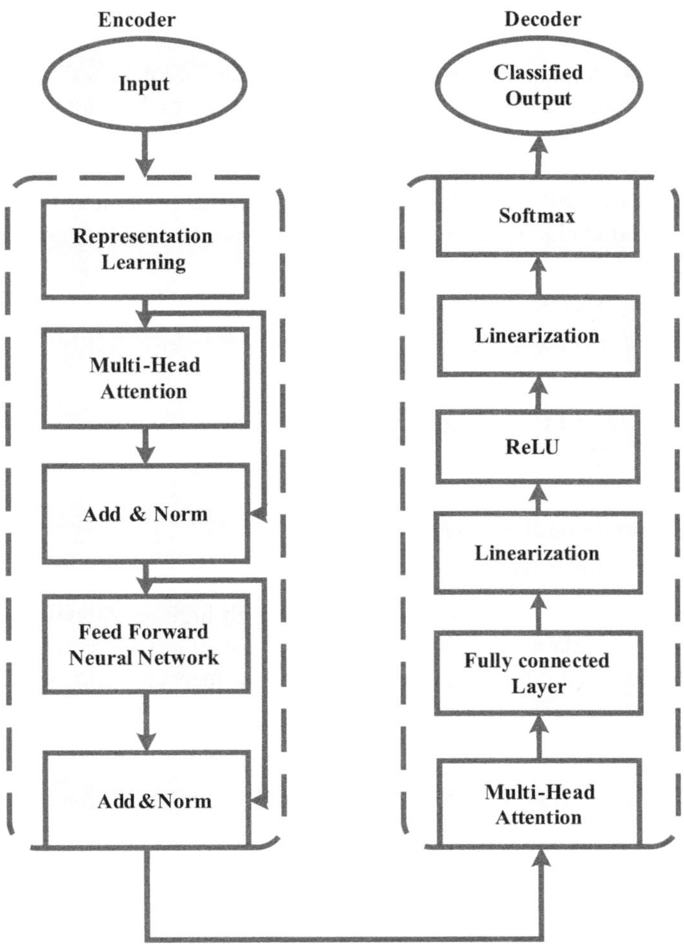

FIGURE 13.5 Overview of sequential modeling-based Transformer.

self-attention is based on correlating the distinct points of a single sequence in order to determine the distinct representation of a similar sequence. However, the MHA is used multiple times on a temporal sequence to get a desirable representation.

To perform the MHA in encoder block, the input is linearly projected "z" times into three different matrices D^Q, D^V, and D^K of dimensions d^Q, d^V, and d^K to determine three dot products: the query (Q), the value (V), and the key (K), respectively. The result is then multiplied by a predefined weight mask before implicating the softmax function on it, thereby the attention source matrix M is obtained by Eq. (13.2) as:

$$soft\ \max\left(\frac{QK^t}{\sqrt{d^K}} \times M_z\right) \tag{13.2}$$

where M_z is the bidirectional attention weight mask being used to ignore the features from the far space head and consider only near space features. The output of MHA is then concatenated and translated linearly into the sequence with desirable dimension. The output sequence from MHA is then normalized and fed to a FFNN block after a residual join and again perform layered normalization.

Decoder block: The decoder block comprises one fully connected layers and one FFNN stacked together. The fully connected layer is proved to be a robust one for spatial translation of input frames without parameter optimization. The second sublayer of FFNN involves two input normalization part with a ReLU activation function. At the last part, a softmax function is implicated to perform the classification task and we obtain the classified output.

The section explains the integration of TD-CNN along with a sequence modeling-based Transformer which is the novelty of this work.

13.2.7 Evaluation matrices

The performance of sequence modeling-based Transformer is evaluated by some standard matrices such as accuracy, precision, recall, and F1-score. All the matrices values are determined from a confusion matrix (CM) which is the visual representation between actual classes and predicted classes. From the CM, four parameters are obtained as: true positive (TP), true negative (TN), false positive (FP), and false negative (FN). Here TP is the SA events predicted correctly, TN is the non-SA or controlled events predicted correctly. Similarly, FP is the non-SA events predicted as SA and FN is the SA events predicted as non-SA. All the performance-based evaluations are done through training and test samples. The formula for accuracy, precision, recall, and F1-score are given in the following equations:

$$\text{Accuracy} = \frac{TP + TN}{TP + FP + FN + TN} \tag{13.3}$$

$$\text{Precision} = \frac{TP}{TP + FP} \tag{13.4}$$

$$\text{Recall} = \frac{TP}{TP + FN} \tag{13.5}$$

$$\text{F1} - \text{Score} = \frac{2 \times \text{precision} \times \text{recall}}{\text{precision} + \text{recall}} \tag{13.6}$$

For the detection of drowsiness, higher number of false alarm will be shown for a model with high recall and low precision. On the other hand, there is a probability of missing the alarm if the model will treat in such a worst classifier with high precision and low recall. Hence, considering the absolute difference between these two measures will detect the drowsiness in both SBD and controlled patients.

13.3 Results

13.3.1 Experimental set-up

After ANOVA test and modality fusion, 80% of the entire data are considered as training samples and 20% are stored as test samples. Later on, from the 80% training samples, 20% are used as validation samples. During hyperparameter tuning and optimization, both the training and validation samples are used to the parameters in each layers of TD-CNN model. The model continuous to train until a callback function is called for termination. In case of early termination halts the training model, hence the model would not make any improvement on validation samples. This makes the model safe from overfitting. The test samples are only meant for accessing the model's performance. On the other hand, training the model and measuring the model performance repeatedly on the same data may produce an overoptimistic result. To avoid this problem, the entire dataset will perform a fivefold cross validation (CV). The detail of CV is mentioned in Fig. 13.6.

13.3.2 Experimental results

Our proposed methodology is tested on two different modalities such as EEG and EOG for the detection of drowsiness in SBD and controlled patients. Each signal is involved with basic preprocessing techniques, namely filtration, missing value deletion or imputation, normalization, segmentation, and augmentation. Then, the augmented portion passes through STFT to convert into 2-D spectrogram. For feature extraction, each spectrogram-based images are fed to TD-CNN to extract the spatial features automatically. Finally, we introduced a sequence modeling where encoder-decoder based Transformer is added for classifying the data and detect the drowsiness of corresponding patients. To validate the model's performance, confusion matrix is visualized from which various standard evaluation matrices are calculated. The CM for hybridized EEG + EOG modality is depicted in Fig. 13.7.

The result of our proposed methodology is validated for both singular modality and multimodality. The performance of model is evaluated on the basis of four measuring parameters such as accuracy, precision, recall, and F1-score (as mentioned in Section 13.2.7). From Table 13.6, we observed that the model shows better classification performance by considering hybridized modality of EEG + EOG signals as compared to other single-modality approach. The accuracy of individual EOG and EEG signal obtain 84.62% and 83.44% accuracy, 77.68% and 79.47 precision, 81.17% and 83.02% recall, 84.75% and 81.8% F1-score. Contrast to that, the multimodal-based approach that is the classification by considering EEG + EOG have achieved 94.90% accuracy, 89.32% precision, 91.64% recall, and 92.71% F1-score.

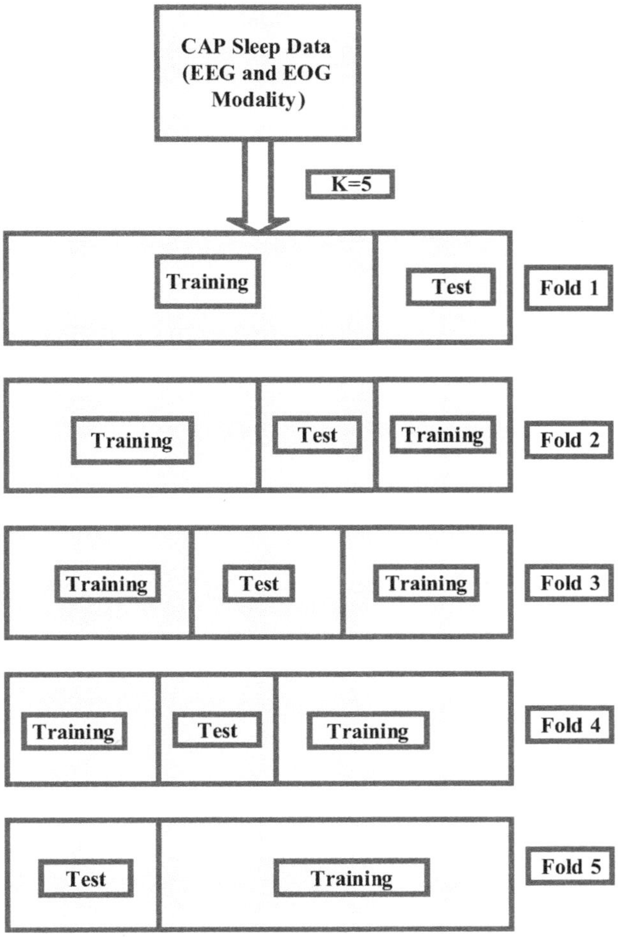

FIGURE 13.6 Fivefold cross validation.

Hence, it is quite clear that for a better classification result, multimodal approach is proved to be a better one compared to single-modal approach. Extending our work, we performed an ablation study by testing individual preprocessing components on multimodal EEG + EOG signals. From Table 13.7, it is shown that the spectrogram without any preprocessing techniques perform the worst during classification with 72.14% accuracy. But, the spectrogram with our proposed methodology is again performed to be the best with 94.90% accuracy.

13.4 Discussion

The purpose of this work is to detect the drowsiness in SBD patients having SA symptoms and controlled subjects by considering multimodal physiological signals. A comparison-based analysis is also performed between uni-modal classification and multimodal classification with TD-CNN integrated Transformer. Table 13.6 shows two major findings. Firstly, we have observed a specific trend in precision and recall when considering both uni-modal approaches, that is, EEG and EOG. However, there is a notable disparity between these precisions and recall values. Secondly, the disparity is minimized in case of multimodal approach (EEG + EOG), thereby improving the performance accuracy. The results also show that the classification performance of Transformer by considering EEG is proved to be better than EOG signal. Hence, uni-modal approach of EEG is enough to detect drowsiness in SBD and controlled subjects.

We have also performed an ablation study on different preprocessing components to know their individual importance on our proposed TD-CNN and sequence modeling-based transformer. Table 13.7 has clearly mentioned the importance of each preprocessing components. When we started adding individual preprocessing components to the signals, the performance of classifier is remarkably increased. However, the implementation of single components such as filtration or segmentation or augmentation is still underperformers with accuracy of 75.61%, 76.55%, and 76.91%, respectively. Additionally, dual approach of preprocessing components such as filtration + augmentation, filtration + augmentation, and segmentation + augmentation shows an average performance with 78.99%, 83.64%, and 82.48%

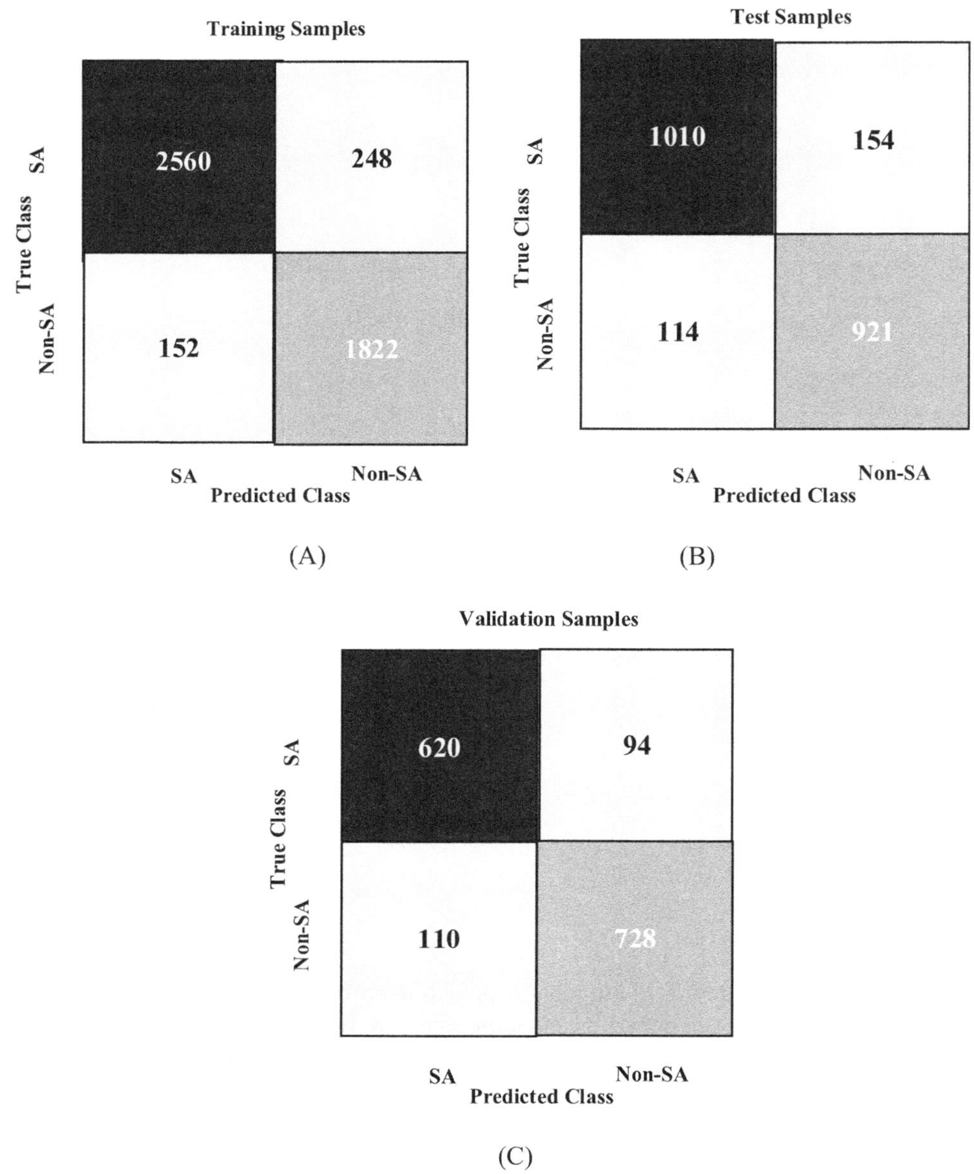

FIGURE 13.7 Confusion matrix for (A) training samples, (B) test samples, (C) validation samples.

TABLE 13.6 Performance of Transformer for the classification of different physiological signals.

Modality	Accuracy (%)	Recall (%)	Precision (%)	F1-score (%)
EOG	83.44	81.17	77.68	84.75
EEG	84.62	83.02	79.47	81.8
EEG + EOG	94.90	93.64	89.32	92.71

accuracy. But our proposed methodology involving all the preprocessing components performs the highest with 94.90% accuracy. The ablation study shows that each of the components has their specific importance in making the signal a generalized one to perform classification task.

Finally, our proposed work is compared with several existing works in order to make it a fair comparison. Table 13.8 shows the comparison analysis of our proposed work with other existing works.

TABLE 13.7 Ablation study by comparing various preprocessing components on multimodal signals.

Tested components on EEG + EOG modality	Accuracy (%)	Recall (%)	Precision (%)	F1-score (%)
Spectrogram without preprocessing + TD-CNN + Transformer	72.14	65.81	67.65	74.92
Spectrogram with filtration + TD-CNN + Transformer	75.61	76.14	71.90	75.17
Spectrogram with segmentation + TD-CNN + Transformer	76.55	78.64	75.25	76.72
Spectrogram with augmentation + TD-CNN + Transformer	76.91	79.98	78.41	77.17
Spectrogram with filtration and augmentation + TD-CNN + Transformer	78.99	80.07	80.97	79.39
Spectrogram with filtration and segmentation + TD-CNN + Transformer	83.64	82.76	81.79	84.91
Spectrogram with segmentation and augmentation + TD-CNN + Transformer	82.48	84.16	81.59	85.14
Proposed methodology	94.90	93.64	89.32	92.71

TABLE 13.8 Comparison of proposed methodology with several contemporary works.

Authors	Modality used	No. of subjects involved	Classifier used	Accuracy (%)
Budak et al. [28]	EEG	16	LSTM	94.31
Fujiwara et al. [29]	EEG and HRV	34	MSPC	90
Belakdhar et al. [30]	Single channel EEG	Not mentioned	ARM-based processor	88.80
Chaabene et al. [31]	EEG	Not mentioned	CNN	90.42
Arefnezhad et al. [32]	ECG	20	Deep CNN	79.33
Proposed methodology	EEG and EOG	16	TD-CNN with sequence modeling-based transformer	94.90

HRV, Heart rate variability; *MSPC*, multivariate statistical process control.

The comparison table is settled on the basis of types of modality used, number of subjects involved, proposed classifier and its corresponding classification accuracy. From the classification results performed by authors Budak et al. [28], Belakdhar et al. [30], and Chaabene et al. [31], it is observed that most of them have used single channel modality (EEG) and still getting a good accuracy. However, Fujiwara et al. [29] have used a dual modality of EEG and heart rate variability (HRV) to detect the drowsiness in OSA patients by using multivariate statistical process control (MSPC) processor. The accuracy obtained in that work is 90%. In Ref. [32], a ECG-based drowsiness detection system is proposed by using deep CNN, but the accuracy is showing poor (79.33%). Moreover, our proposed methodology has used a dual modality approach by considering EEG and EOG signals with the help of a sequence modeling-based Transformer, thereby achieves highest accuracy of 94.90%.

The comparison analysis proves two major points. Firstly, multimodal approach for automated drowsiness detection is definitely a wise decision as it proves the classification result. Secondly, the implication of a sequence modeling-based Transformer gives the most optimal results than other machine learning and deep learning-based classifiers. Hence, our proposed methodology is seemed to be robust and reliable for the detection of drowsiness in SA and controlled subjects.

The recent study also possesses some limitations that must be taken into account when analyzing our obtained results. First, the sample size used in this study is small; hence there is a need of more samples by involving more physiological signals. Second, there is an interindividual difference between subjects that creates a potential issue for generating models for classification and drowsiness detection. Third, there is a variation in features when we deal with multimodal analysis that ultimately creates feature imbalance. Hence, for a robust drowsiness detection system, we must stabilize our features. For this more feature representation techniques are needed.

13.5 Conclusion

In this study, a drowsiness detection technique is proposed by considering TD-CNN with sequence learning-based Transformer. The research work begins with collecting EEG and EOG channels of both controlled subjects and SBD patients having apnea symptoms from CAP sleep database. Then, the dataset is fed to a preprocessing phase including filtration, missing value deletion or imputation, normalization, segmentation, and augmentation. The augmented images are then converted into a 2-D spectrogram view by using STFT. After that, a TD-CNN is implicated on that spectrogram to extract spatial features from that. Finally, the data are classified with the help of sequential modeling-based Transformer. The performance of Transformer is evaluated through various standard performance matrices determined from confusion matrix. Our proposed methodology achieves 94.90% accuracy from the Transformer-based model. The comparison study indicates that our proposed methodology is better compared to other contemporary works. From an ablation study, we have understood the importance of various preprocessing components on the bio-signals. This study draws a conclusion that multimodal approach is more significant compared to uni-modal approach. Therefore our proposed methodology has more potential for detecting drowsiness in both sleep disorder and controlled subjects.

References

[1] D.J. Gottlieb, N.M. Punjabi, Diagnosis and management of obstructive sleep apnea: a review, JAMA 323 (2020) 1389–1400.

[2] D.P. White, Sleep apnea, Proc. Am. Thorac. Soc. 3 (2006) 124–128.

[3] A. Malhotra, I. Ayappa, N. Ayas, N. Collop, D. Kirsch, N. Mcardle, et al., Metrics of sleep apnea severity: beyond the apnea-hypopnea index, Sleep 44 (2021) zsab030.

[4] S. Javaheri, J. Dempsey, Central sleep apnea, Compr. Physiol. 3 (2013) 141–163.

[5] J.V. Rundo, Obstructive sleep apnea basics, Clevel. Clin. J. Med. 86 (2019) 2–9.

[6] M. Ohayon, E.M. Wickwire, M. Hirshkowitz, S.M. Albert, A. Avidan, F.J. Daly, et al., National Sleep Foundation's sleep quality recommendations: first report, Sleep. Health 3 (2017) 6–19.

[7] L. Mei, X. Li, S. Wang, R. Si, T. Ji, Z. Xu, et al., The impacts of obstructive sleep apnea severity on brain white matter integrity and cognitive functions in children: a diffusion tensor imaging study, Nat. Sci. Sleep. (2021) 2125–2135.

[8] H.-J. Li, X.-J. Dai, H.-H. Gong, X. Nie, W. Zhang, D.-C. Peng, Aberrant spontaneous low-frequency brain activity in male patients with severe obstructive sleep apnea revealed by resting-state functional MRI, Neuropsychiatric Dis. Treat. (2015) 207–214.

[9] S.M. Tashjian, D. Goldenberg, M.M. Monti, A. Galván, Sleep quality and adolescent default mode network connectivity, Soc. Cognit. Affect. Neurosci. 13 (2018) 290–299.

[10] J.R. Lunsford-Avery, K.S. Damme, M.M. Engelhard, S.H. Kollins, V.A. Mittal, Sleep/wake regularity associated with default mode network structure among healthy adolescents and young adults, Sci. Rep. 10 (2020) 1–8.

[11] C.W. Wu, P.-Y. Liu, P.-J. Tsai, Y.-C. Wu, C.-S. Hung, Y.-C. Tsai, et al., Variations in connectivity in the sensorimotor and default-mode networks during the first nocturnal sleep cycle, Brain Connect. 2 (2012) 177–190.

[12] A.T. Mulgrew, N. Al Lawati, N.T. Ayas, N. Fox, P. Hamilton, L. Cortes, et al., Residual sleep apnea on polysomnography after 3 months of CPAP therapy: clinical implications, predictors and patterns, Sleep. Med. 11 (2010) 119–125.

[13] H. Wang, X. Cheng, P. Song, L. Yu, W. Hu, L. Zhao, An extrapolated impulse response filter design with sparse coefficients based on a novel linear approximation of matrix, Circuits Syst. Signal. Process. 34 (2015) 2335–2361.

[14] B. Chandrakar, O. Yadav, V. Chandra, A survey of noise removal techniques for ECG signals, Int. J. Adv. Res. Computer Commun. Eng. 2 (2013) 1354–1357.

[15] J.C. Jakobsen, C. Gluud, J. Wetterslev, P. Winkel, When and how should multiple imputation be used for handling missing data in randomised clinical trials—a practical guide with flowcharts, BMC Med. Res. Methodol. 17 (2017) 1–10.

[16] M. Perslev, M. Jensen, S. Darkner, P.J. Jennum, C. Igel, U-time: a fully convolutional network for time series segmentation applied to sleep staging, Adv. Neural Inf. Process. Syst. 32 (2019).

[17] S.C. Wong, A. Gatt, V. Stamatescu, M.D. McDonnell, Understanding data augmentation for classification: when to warp? in: 2016 International Conference on Digital Image Computing: Techniques And Applications (DICTA), 2016, pp. 1–6.

[18] A. Ullah, S.M. Anwar, M. Bilal, R.M. Mehmood, Classification of arrhythmia by using deep learning with 2-D ECG spectral image representation, Remote. Sens. 12 (2020) 1685.

[19] O.I. Abiodun, A. Jantan, A.E. Omolara, K.V. Dada, N.A. Mohamed, H. Arshad, State-of-the-art in artificial neural network applications: a survey, Heliyon 4 (2018) e00938.

[20] A. Graves, Generating sequences with recurrent neural networks, arXiv preprint arXiv:1308.0850, 2013.

[21] T.N. Sainath, O. Vinyals, A. Senior, H. Sak, Convolutional, long short-term memory, fully connected deep neural networks, in: 2015 IEEE International Conference on Acoustics, Speech and Signal Processing (ICASSP), 2015, pp. 4580–4584.

[22] A. Chowanda, Spatiotemporal features learning from song for emotions recognition with time distributed CNN, in: 2021 1st International Conference on Computer Science and Artificial Intelligence (ICCSAI), 2021, pp. 407–412.

[23] C. Cooney, R. Folli, D. Coyle, Optimizing layers improves CNN generalization and transfer learning for imagined speech decoding from EEG, in: 2019 IEEE International Conference on Systems, Man and Cybernetics (SMC), 2019, pp. 1311–1316.

[24] S.R. Young, D.C. Rose, T.P. Karnowski, S.-H. Lim, R.M. Patton, Optimizing deep learning hyper-parameters through an evolutionary algorithm, in: Proceedings of the Workshop on Machine Learning in High-Performance Computing Environments, 2015, pp. 1–5.

[25] M. Raghu, T. Unterthiner, S. Kornblith, C. Zhang, A. Dosovitskiy, Do vision transformers see like convolutional neural networks? Adv. Neural Inf. Process. Syst. 34 (2021) 12116–12128.

[26] H. Liu, S. Cui, X. Zhao, F. Cong, Detection of obstructive sleep apnea from single-channel ECG signals using a CNN-transformer architecture, Biomed. Signal. Process. Control. 82 (2023) 104581.

[27] J. Jiang, J.B. Kim, Y. Luo, K. Zhang, S. Kim, AdaMCT: adaptive mixture of CNN-transformer for sequential recommendation, arXiv preprint arXiv:2205.08776, 2022.

[28] U. Budak, V. Bajaj, Y. Akbulut, O. Atila, A. Sengur, An effective hybrid model for EEG-based drowsiness detection, IEEE Sens. J. 19 (2019) 7624–7631.

[29] K. Fujiwara, E. Abe, K. Kamata, C. Nakayama, Y. Suzuki, T. Yamakawa, et al., Heart rate variability-based driver drowsiness detection and its validation with EEG, IEEE Trans. Biomed. Eng. 66 (2018) 1769–1778.

[30] I. Belakhdar, W. Kaaniche, R. Djemal, B. Ouni, Single-channel-based automatic drowsiness detection architecture with a reduced number of EEG features, Microprocess. Microsyst. 58 (2018) 13–23.

[31] S. Chaabene, B. Bouaziz, A. Boudaya, A. Hökelmann, A. Ammar, L. Chaari, Convolutional neural network for drowsiness detection using EEG signals, Sensors 21 (2021) 1734.

[32] S. Arefnezhad, A. Eichberger, M. Frühwirth, C. Kaufmann, M. Moser, I.V. Koglbauer, Driver monitoring of automated vehicles by classification of driver drowsiness using a deep convolutional neural network trained by scalograms of ECG signals, Energies 15 (2022) 480.

Chapter 14

Comparison of deep transfer learning models to detect obstructive sleep apnea with single-channel electrocardiogram

R.H. Talwekar and Nivedita Singh

Department of Electronics and Telecommunication Engineering, Government Engineering College, Raipur, Chhattisgarh, India

14.1 Introduction

Sleep is rejuvenating physiological process which must be accomplished according to the standard of the Association of American Medical Colleges (AAMC) and as per age. Postpandemic situation drastically changed the lifestyle of people and also disturb the circadian rhythm of most of us irrespective of age. It is well known that a normal adult human being has to take sound sleep of 6–8 hours. Over usage of gadgets, late-night exposure, and longer screen time are various factors that affect sleep behavior and various sleep-related disorder.

14.1.1 Sleep apnea

Obstructive sleep apnea (OSA) is one of a kind of sleep disorder when people suffer from a partial and complete collapse of breathing during sleep. Fig. 14.1 depicts different types of sleep apnea that occur during different stages of sleep. Sleep apnea is categorized as OSA and central sleep apnea (CSA) which is defined in Fig. 14.1.

The two major classes of sleep disorders are dyssomnias and parasomnias. OSA is also categorized as a sleep-breathing disorder and is considered dyssomnia. "Sleep Apnea occurred due to hypopnea and apnea; apnea is defined as a total collapse of the respiratory tract, whereas hypopnea is defined as the reduction of air passage for more than, and equal to 10 sec. Severe sleep apnea may cause high blood pressure, arrhythmia, sudden death, and other cardiac anomalies" [1]. The significant feature to determine the presence of OSA is an apnea-hypopnea index (AHI) [2]. The male–female ratio of OSA slightly differs in both and it was established that men are suffering from OSA more than women [3–5]. "AHI score is one of the most significant features to detect OSA and classify it as well using ECG signal" [6].

14.1.2 Historical background of obstructive sleep apnea

According to Cheng et al., sleep apnea detection using electrocardiogram (ECG) in recurrent neural networks (RNN) performed better than a conventional neural network (CNN). The RNN model performed better when the number of epochs increased from 20 to 50 [7]. Whereas a study [8] proposed that OSA detection and classification performed

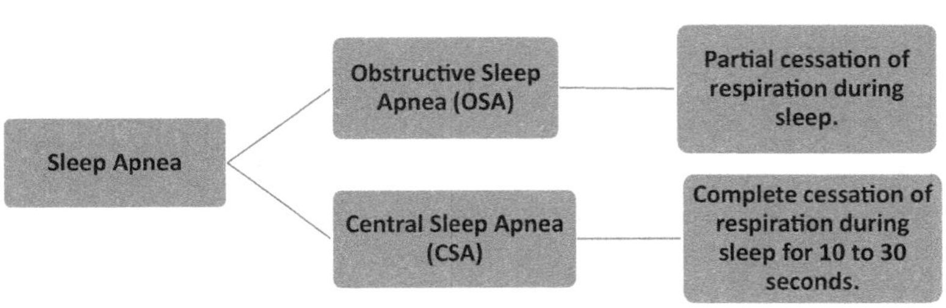

FIGURE 14.1 Types of sleep apnea.

Intelligent Computing Techniques in Biomedical Imaging. DOI: https://doi.org/10.1016/B978-0-443-15999-2.00002-5

better in the CNN model as compared to the RNN long short-term memory (LSTM) and machine learning (ML) classifiers. When the epoch is set to 15, setting the epoch is a tedious task and one should be very cautious while setting it to avoid overfitting or underfitting.

A study [9] reported that the single-channel ECG signal was divided into per segment and recording and then preprocessed to detect RR interval which was fed into 1D CNN. The accuracy obtained was 87.9% and 97.1%, respectively. This shows that per-recording data have performed better than per-segment data. A study [10] investigated different RNN, CNN, and deep hybrid models. The results showed that for the hybrid model and LSTM, the system performance was better than other models. The research [11] published demonstrates sleep apnea detection using CNN for SpO_2 signal is an alternative to polysomnography (PSG) and it is a cost-effective as well as portable method. The study [12] reported that classification based on the severity of sleep apnea using only instantaneous heart rate or only SpO_2 signal in the LSTM-RNN method was performed satisfactorily.

Deep learning has attracted the focus of many researchers in the area of automatic detection of OSA using single-channel ECG signals. A CNN architecture is very efficient in image classification as feature selection is not a criterion [13]. Urtnasan et al. [14] showed that the CNN having six layers for the detection of OSA using ECG has better accuracy. Jiang et al. [15] proposed a model to detect sleep apnea using electroencephalogram (EEG) signals in multiscale parallel CNN. Haidar et al. [16] showed that to detect OSA the modern and advanced classifier using CNN outperforms the traditional machine learning classifier. Van Steenkiste et al. [17] investigated the detection of sleep using chest and abdominal data in RNN LSTM performs better than other conventional classifiers [18]. Some researchers identified OSA detection using RNN architecture performs with average accuracy as compared to the machine learning classifier [13–19]. Many researchers reported the classification of only apnea while neglecting hypopnea [20] but the identification of hypopnea is equally important as apnea [21,22] and its treatment [23,24].

There are various methods and modalities investigated by researchers for the detection of sleep apnea namely PSG, EEG, electronyogam (EMG), SpO_2 (saturation of peripheral oxygen), ECG-derived respiration, ECG, and HRV. Sometimes more than one modality in combination is investigated to enhance performance or to show robustness.

14.1.3 Polysomnography

The most popular gold standard of sleep apnea detection PSG as shown in Fig. 14.2 is widely accepted. In this figure, different channels are used to record different physiological signals namely EEG, ECG, EOG, EMG, and SpO_2 are shown. Mere dependency on PSG is not a wise choice as it requires a well-equipped sleep laboratory and a skilled person to record data, this process is a complex and expensive tool. PSG is inconvenient, time-consuming, and expensive. PSG has a complex and uncomfortable process to record physiological signals and needs a skilled physician to acquire them.

PSG has certain limitations although it motivated to adopt a single-channel ECG modality to investigate the detection of OSA. Using ECG for the detection of OSA has manyfold advantages as it improves the processing time and complexity of the research work. "ECG signals are implemented for study since this signal is the reflection of

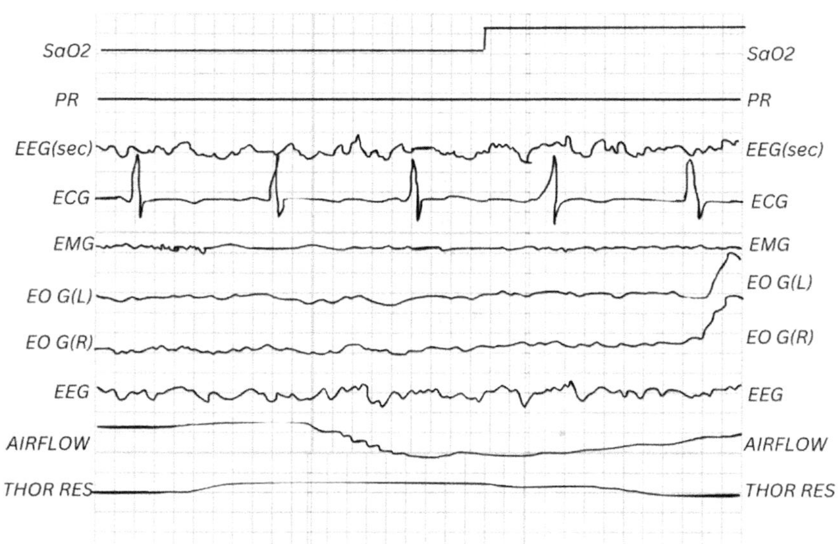

FIGURE 14.2 Polysomnography.

cardiac behavior which may be affected by another physiological signal such as respiratory signals. OSA is another class of respiratory and sleep disorders and hence single channel ECG succeeds as the most appropriate signal for OSA detection" [23].

14.1.4 Electrocardiogram

Electrocardiogram is a very important physiological signal which is helpful to diagnose various cardiovascular diseases. ECG is not only one of the major tools to detect a cardiac problem but also very crucial for the diagnosis of OSA. The single-channel ECG signals are recorded during sleep in the sleep laboratory by experts. The change in the QRS wave or RR interval may signify an anomaly in the health condition which has to be taken care of in advance. The proposed approach in this regard will certainly assist a person suffering from a sleep disorder through advanced and automatic detection using single-channel ECG.

Fig. 14.3 depicts the standard ECG signal and its various waves which have vital importance in the study of OSA. Mainly QRS complex is considered to study the OSA in this chapter.

14.1.5 Process flow diagram of detection of obstructive sleep apnea

The flow diagram of the detection is shown in Fig. 14.4 to comprehend the complete approach to the detection of OSA using a single-channel ECG signal. The study was performed on open-source online data available on physionet apnea data. Data collected are then preprocessed to remove artifacts and provide crisp data. In the next stage of postprocessing, the time series data are converted into equivalent images of size [224 224 3].

Further, the image data were classified using various deep CNN transfer learning algorithms to check the robustness of the model. This chapter includes comparisons of four transfer learning CNN models: ShuffleNet, GoogLeNet, ResNet50, and ResNet101 for the detection of OSA.

14.1.6 Objectives of chapter

- This chapter deals with the detection and characterization of sleep apnea in humans using ECG and HRV.
- The main goal of this chapter is the early detection of sleep apnea to avoid any comorbidity specifically cardiac issues.
- Denoising of ECG signal and determining QRS complex to get RR interval using Pan Tompkins's algorithm is implemented.
- Image classification using deep learning classifiers is employed while converting the time series ECG signals into image signals in continuous wavelet transform.

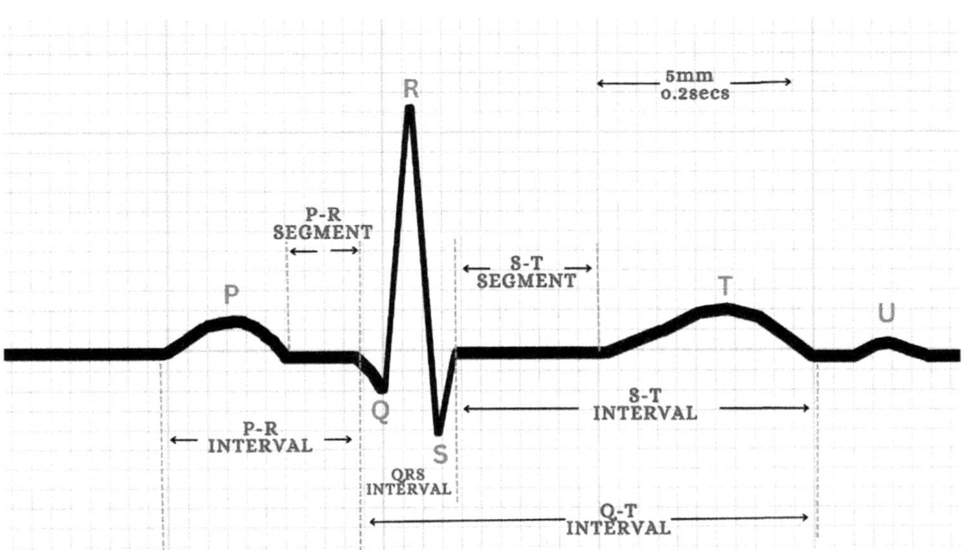

FIGURE 14.3 Typical diagram of ECG.

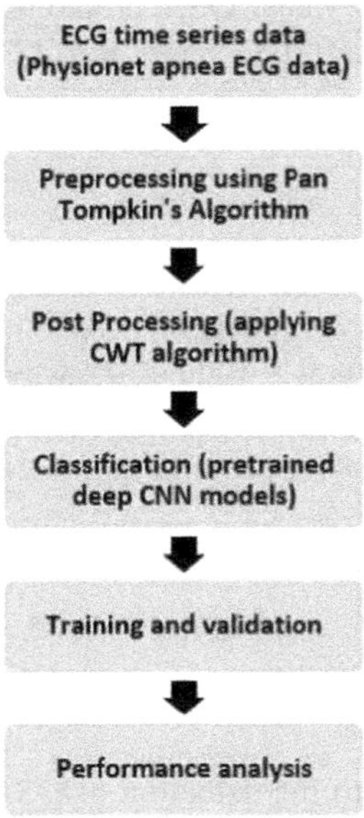

FIGURE 14.4 Process flow diagram.

- Selection of a suitable pretrained model for the specific task and the size of data followed by transfer learning.
- Tuning of hyperparameters is done to optimize the performance of different classifiers, cross validation is done using 10-K fold validation processes.
- Comparison and analysis of different deep learning pretrained models GoogLeNet, ShuffleNet, ResNet50, and ResNet101 are accomplished.
- This chapter is demonstrating the detection of sleep apnea efficiently with high accuracy in an easy way and with low computational tasks.
- To achieve this, deep neural network classifiers are the most promising and robust. The coding of deep learning in Matlab is easier and takes less time.

14.2 Methods and materials

14.2.1 Description of datasets

"The Physionet Apnea-ECG database is downloaded from its official website. This database contains a total of 70 records from 57 men and 13 women divided into 35 training data and 35 testing data. The sampling frequency of the single-channel ECG is 100 Hz. The total duration of sleep recording is 7–8 hours approximately. According to the AHI label, the dataset consists of three groups A (apnea), B (borderline), and C (control). To record the ECG the electrode position was connected at (modified lead V2) in the standard sleep laboratory" [25].

14.2.2 Preprocessing

The online single-channel ECG signal was preprocessed by the Pan Tompkins' algorithm to remove noise and detect QRS complex for both apnea and normal records which are depicted in Fig. 14.5.

"Raw ECG data for the complete night is preprocessed by the Pan Tompkins algorithm to detect QRS complex. Single-channel ECG is preprocessed using a low pass filter, high pass filter, and derivative filter which is followed by squaring and moving window integration" [26]. The data are segmented in 1 minute from each hour of data per record. It is seen that the heart rate of apnea is 54.48 while the normal subject has a 62.30 heart rate.

14.2.3 Data postprocessing

The time series data are converted into equivalent image data using continuous wavelet transform (CWT). CWT method is used to compress the data and also to convert the time series data into the equivalent time-frequency domain in the form of a spectrogram which could be easily implemented in DCN models.

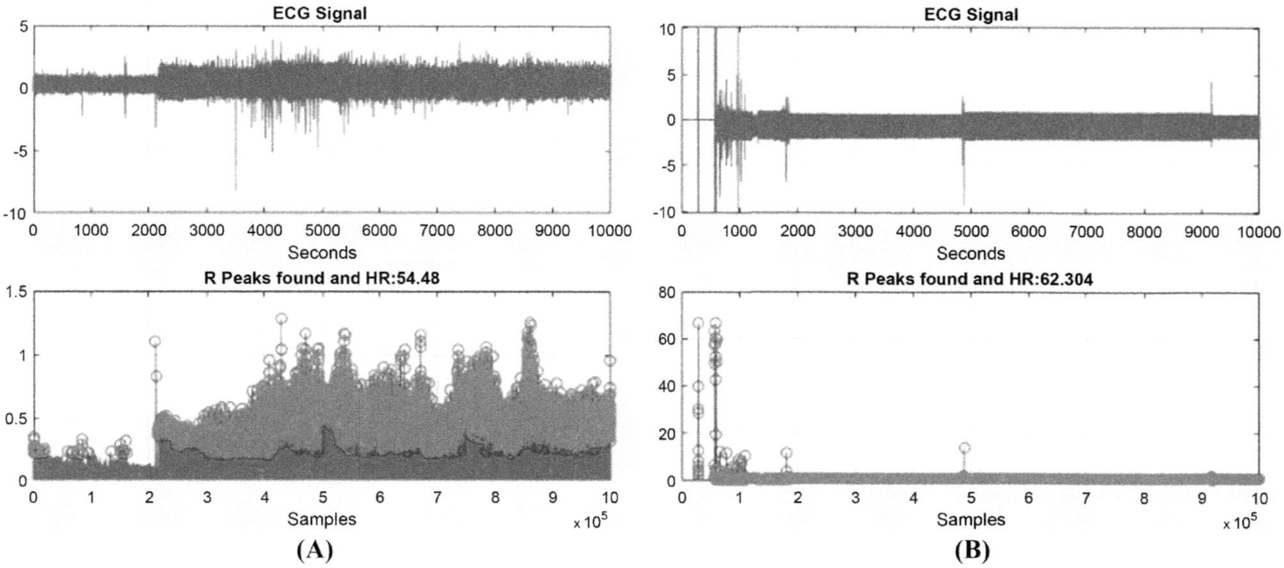

FIGURE 14.5 (A) ECG of apnea. (B) ECG of normal.

14.3 Transfer learning models

Transfer learning is a most advantageous technique when time is a major concern as it requires less processing time. It does not require large data as compared to other deep learning architectures prepared from scratch which is why it is popular to implement. With transfer learning, a proposed model can be built easily as it is already trained for extremally large data with a greater number of classes. It takes days, weeks, or even months to train the network if it is implemented from scratch with the same large amount of data.

Transfer learning performs feature extractions automatically unlike it has to be extracted manually in machine learning classifiers. Although the domain knowledge and dominant features and their significance must be well acknowledged by the researchers. Since the neural network has unique characteristics to select dominant and significant features. For any classification problem, it determines the certain set of features that archives better performance. It could be implemented for any kind of problem for large data size efficiently and with less processing and training time. Before implementing transfer learning modifications have to be performed for some layers of output. For using transfer learning techniques, pretrained models are employed to leverage the deep learning techniques in biomedical image processing. Selection of a pretrained model from a wide variety of models is an important task and has to be done carefully. Following are the eligibility criteria for selecting pretrained models first what is the problem, second data size, and finally hyperparameter tuning.

Henceforth these proposed CNN transfer learning models are implemented on an MSI single GPU desktop with inbuilt 16 GB RAM and an Intel i7 processor 64 bits. Data preprocessing and postprocessing for training and testing are implemented in MATLAB 2022b. The dataset of ECG signals is preprocessed and then converted to the image size of [224 224 3] before being fed to different models.

14.3.1 Convolutional neural network

Fig. 14.6 shows the basic CNN architecture. It consists of the input layer, output layer, and hidden layers.

The very first layer after the input layer is the convolution layer which is intended to extract features. The convolution layer is followed by Relu and then the reduction of features is achieved through the max pooling layer. Generally, cascading of these layers leads to an increased number of hidden layers. The same set of hidden layers is cascaded 5ive times before the output layer to generate the final feature map. The output layer consists of three consecutive layers, respectively, fully connected layer, a softmax layer, and classification which produces the final output of the CNN.

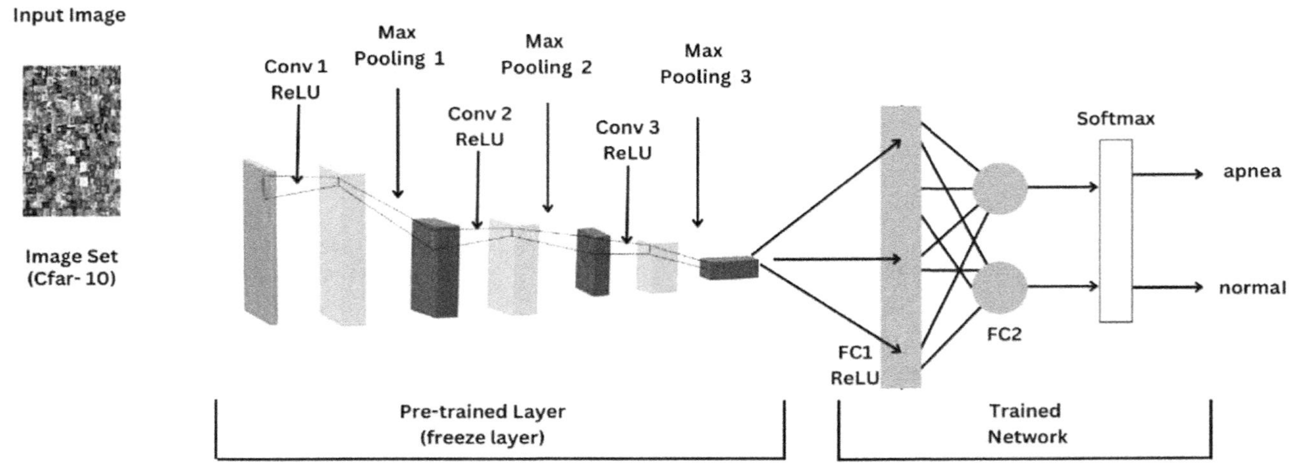

FIGURE 14.6 The basic architecture of CNN.

14.3.2 ShuffleNet architecture

ShuffleNet is a CNN that is trained on more than a million images from the ImageNet database. "The network can classify images into 1000 object categories, such as keyboard, mouse, pencil, and many animals. As a result, the network has learned rich feature representations for a wide range of images." The network has an image input with an image size of 224-by-224 [27]. There are two classes defined in this classification, that is, apnea and normal, and the number of images are 1000 where 500 images are from each class, out of which 600 images are used for training and 400 images are used for validation. This network is designed using convolution and grouped convolution and further no channel shuffle is implemented as shown in Fig. 14.7.

14.3.3 GoogLeNet architecture

"GoogLeNet architecture processes RGB images of size [224 224 3]. Fig. 14.8 represents the sequence of layers followed in this architecture enables a reduction in error rate as compared to the earlier model ShuffleNet. The architecture consists of convolution layers, pooling layers, and inception layers" [28]. The activation function is the Relu layer as shown in Fig. 14.8. Inception layers are playing a significant role in this architecture as it provides us alternatives to choose either a stack of 3X3 filters or 5X5 filters or a max pooling layer.

14.3.4 ResNet50

"ResNet50 is typically designed as a residue network in deep neural network architecture (He & Sun, n.d.) This Residual Network consists layer by layer and is made of a stack of following layers with convolution, pooling, activation, and fully-connected layers. These residual functions are mapped to identify the variation in output to the input from a residual block as shown in Fig. 14.9" [29].

14.3.5 ResNet101

"ResNet-101 is a convolutional neural network that is 101 layers deep as shown in Fig. 14.10. The pretrained network can classify images into 1000 object categories, viz pen, torch, mouse, pencil, and many animals. As a result, the network has learned rich feature representations for a wide range of images" [29].

14.4 Result and discussion

14.4.1 Performance comparison of ShuffleNet, GoogLeNet, ResNet50, and ResNet101

Table 14.1 shows the architecture of different pretrained DCNN models. Training options are fixed for all the pretrained models such as optimizer—"adam," learning rate—"0.001," mini batch size—128, number of epochs for training and testing—30, and finally, the environment is set as single GPU. A single CPU takes more processing time as compared

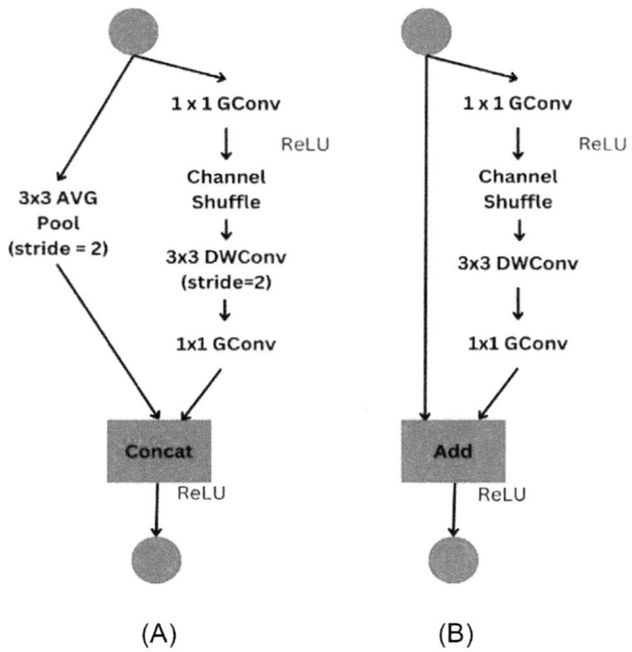

FIGURE 14.7 ShuffleNet architecture.

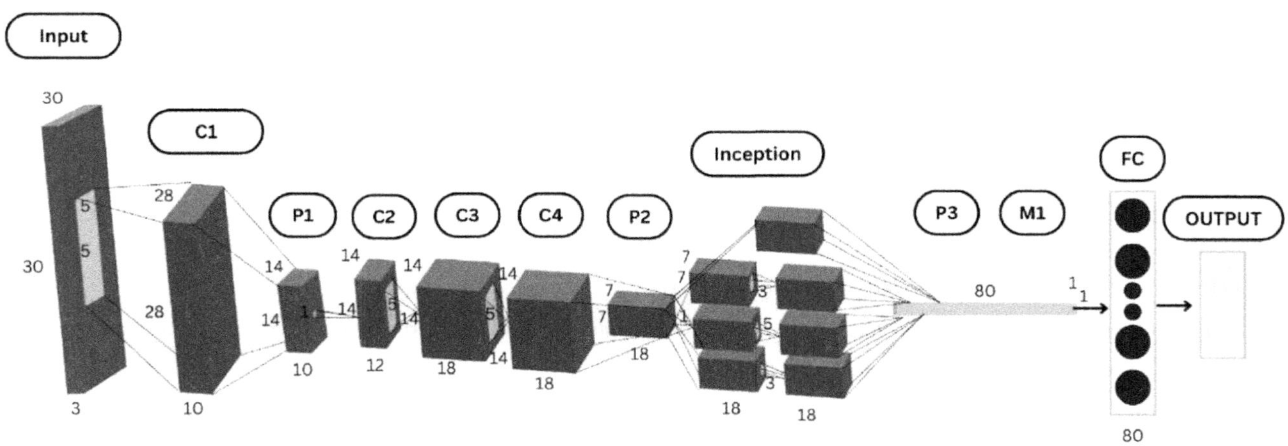

FIGURE 14.8 The architecture of the GoogLeNet pretrained model.

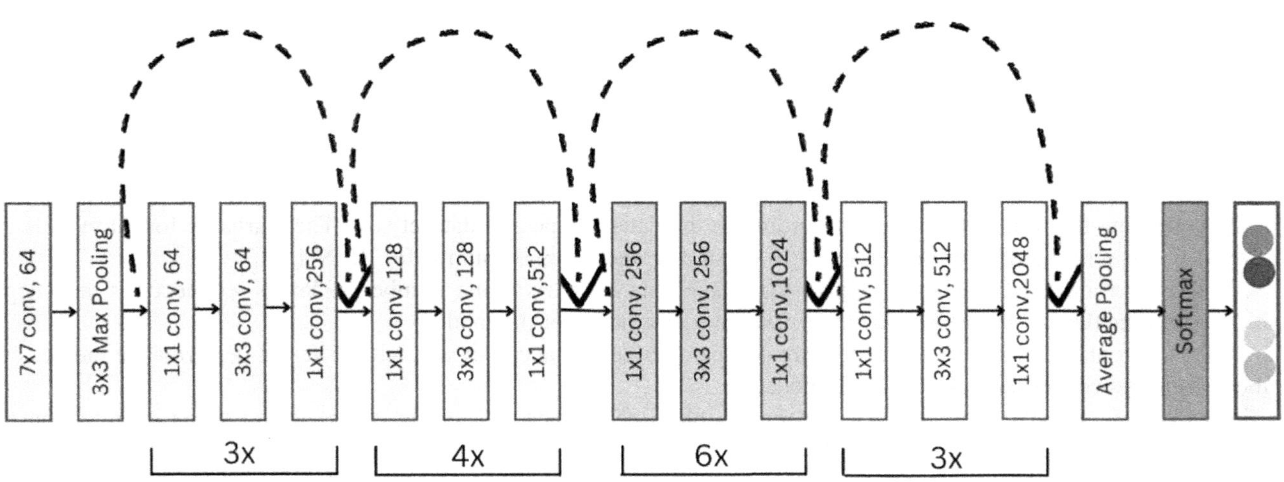

FIGURE 14.9 ResNet50 architecture.

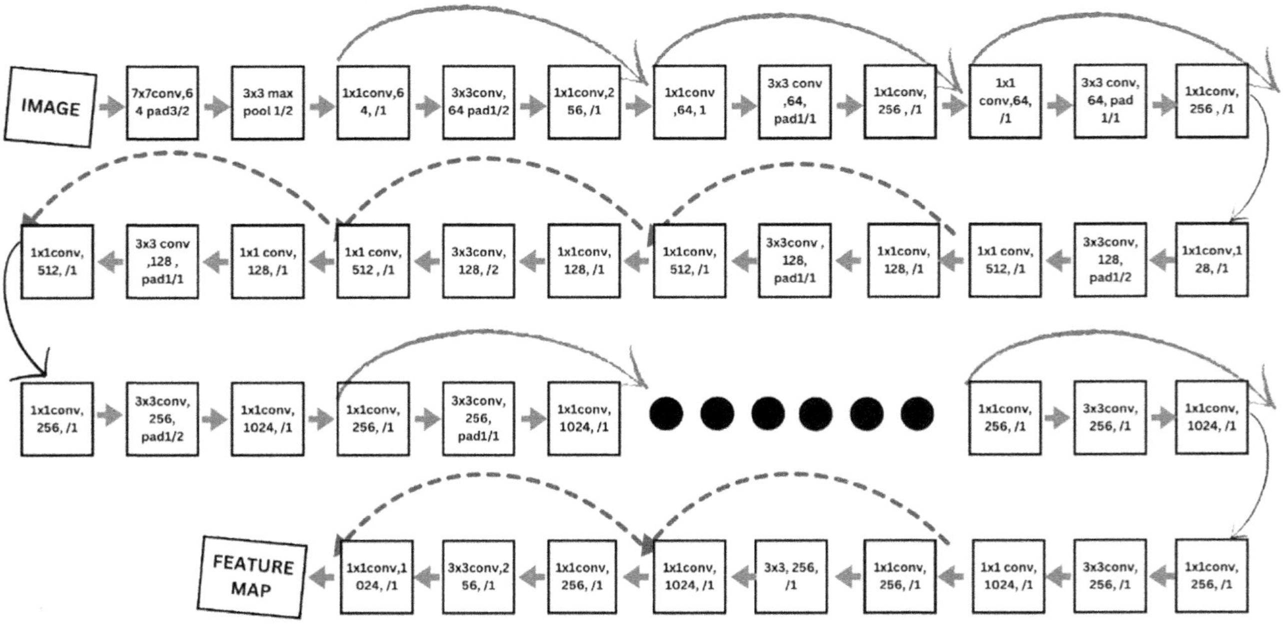

FIGURE 14.10 ResNet101 architecture.

TABLE 14.1 Architecture of different pretrained models.

Type of CNN	ShuffleNet	GoogLeNet	ResNet50	ResNet101
Optimizer	Adam	Adam	Adam	Adam
Total no. of layers	172	144	177	347
Total no. of connections	187	170	192	379
InitialLearnRate	0.001	0.001	0.001	0.001
Validation frequency	50 iterations	50 iterations	50 iterations	50 iterations
Max epochs	30	30	30	30
Mini batch size	128	128	128	128
Evolution environment	GPU	GPU	GPU	GPU
Learnable	1.4 million	6.9 million	25.5 million	44.6 million
Elapsed time	1 min 28 s	1 min 54 s	6 min 19 s	40 min 18 s

to a single GPU. According to the architecture of the pretrained models, the performance varies. This also plays a vital role to determine the best-suited model according to the dataset type and dataset size. The learnable for all models are different that is least number of the learnable is for shuffle and the maximum is for ResNet101.

All four pretrained models are designed for image data with a varying number of layers, connections, and depth which quantifies the process and hence improves the performance of the model.

The results obtained for all four models are compared based on accuracy and with different performance indices calculated from the confusion matrix of each model.

Fig. 14.11 depicts the confusion matrix for the ShuffleNet transfer learning model which was performed on the Physionet Apnea-ECG database. The validation accuracy obtained is 98.21% for ShuffleNet as shown in Fig. 14.12.

Fig. 14.13 depicts the confusion matrix for the GoogLeNet transfer learning model with values of $a = 392$, $b = 0$, $c = 0$, and $d = 392$. Fig. 14.14 represents a validation accuracy of 93.45% for GoogLeNet.

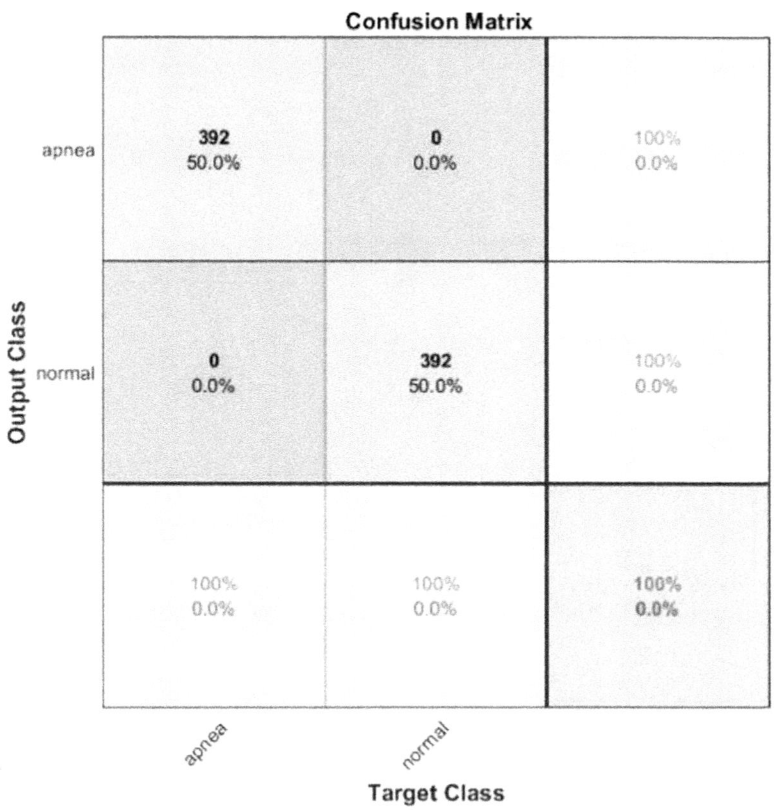

FIGURE 14.11 Confusion matrix for ShuffleNet.

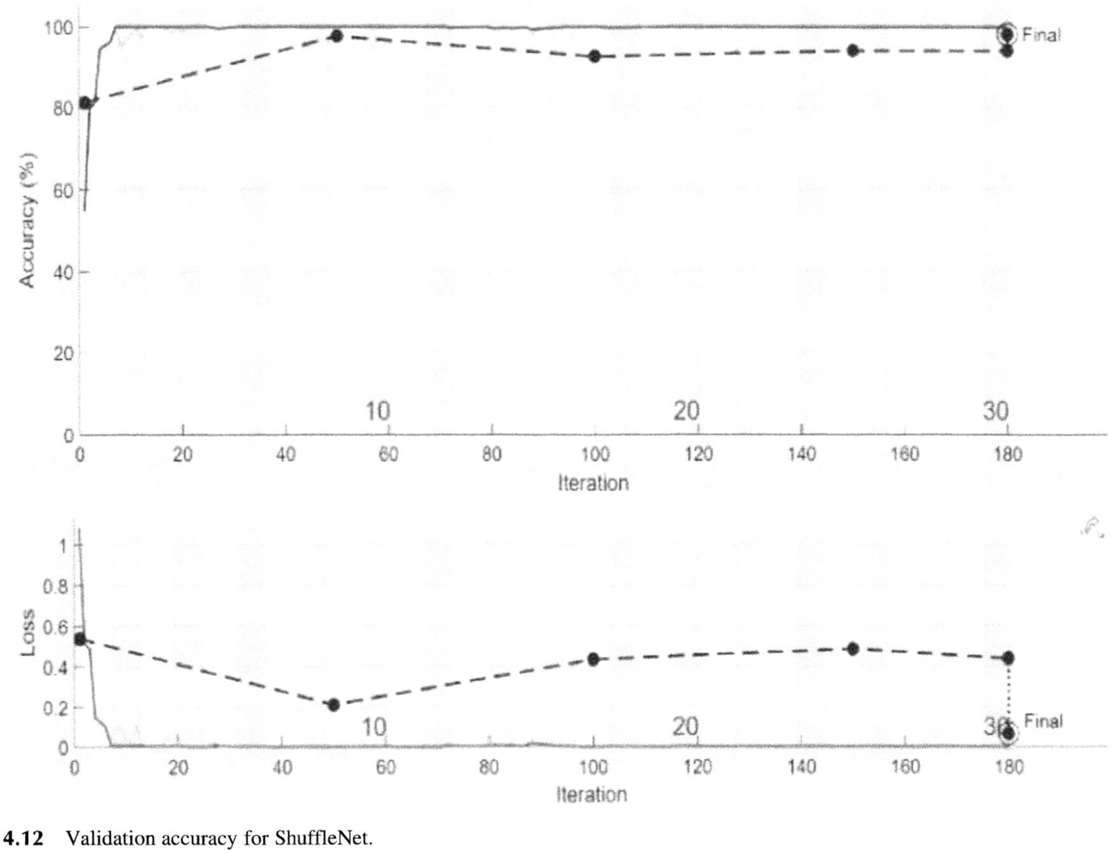

FIGURE 14.12 Validation accuracy for ShuffleNet.

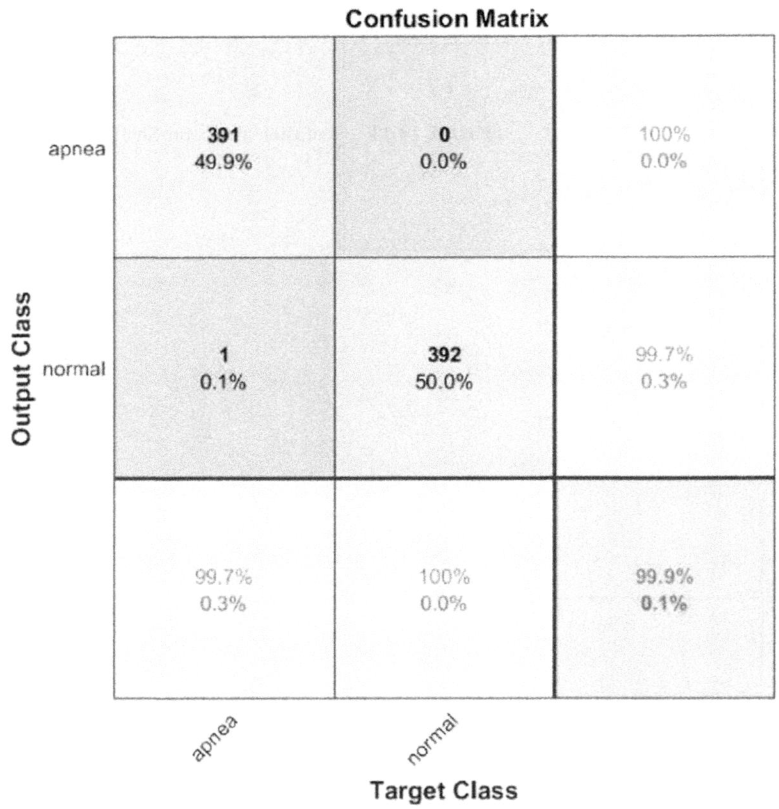

FIGURE 14.13 Confusion matrix for GoogLeNet.

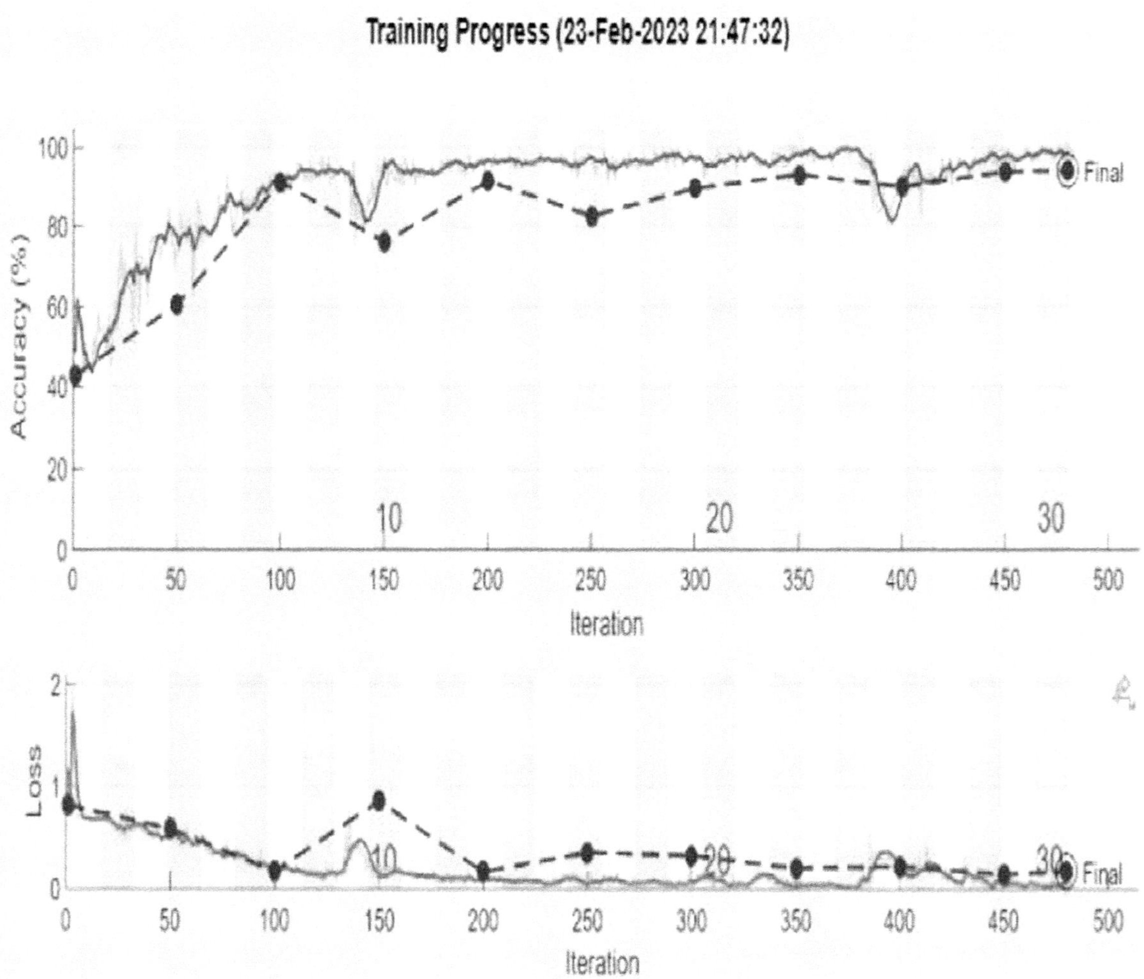

FIGURE 14.14 Validation accuracy for GoogLeNet.

Fig. 14.15 depicts the confusion matrix for the ResNet50 pretrained model and the validation accuracy obtained is 95.54% as shown in Fig. 14.16.

Fig. 14.17 depicts the confusion matrix for the ResNet101 transfer learning model which was performed on the Physionet Apnea-ECG database. The validation accuracy obtained is 96.43% for ResNet101 as is represented in Fig. 14.18.

Table 14.2 shows a comparison of the performance indices for all four pretrained models GoogLeNet, ShuffleNet, ResNet50, and ResNet101. The highest accuracy achieved 100% is by the ShuffleNet transfer learning model and the least is by the ResNet50 model. Variations in the performance of all the models are small and the overall performance of the models signifies that the accuracy is more than 99% which indicates the overfitting of these models.

Fig. 14.19 depicts a comparison of performance indices of all four models and investigated that almost all the transfer learning models are very close, no significant difference was found. This is another limitation of the research, although the layers and complexity of each architecture are very much different, the performance matrix is almost the same.

Both the limitation of overfitting and redundant performances could be removed by increasing the dataset size or using clinical data for testing. This work can be further extended using large-size clinical image data.

Table 14.3 enlists the performances of different research methods and is compared with existing results for the detection of OSA using single channel ECG signal. It is concluded from the table that the accuracy achieved by CNN and the hybrid model has better performance than RNN and conventional classifiers. This chapter investigated that different DCNN classifiers were implemented for OSA detection and the most suitable classifier is ShuffleNet for the type and size of dataset chosen.

Table 14.4 shows the overall performance of the proposed models and concludes that ShuffleNet performed outstanding and that is 100% accuracy.

Table 14.4 represents the result of this research which is very important for further research in the study of OSA detection as it is depicted in Fig. 14.20.

The outcome of the proposed work is as follows: (1) Comparison of the four different transfer learning architectures: ShuffleNet, GoogLeNet, ResNet50, and ResNet101 to detect OSA. (2) Comparison of the performance of all the models was evaluated using only single-channel ECG signals. (3) The performance is better than the existing approach for

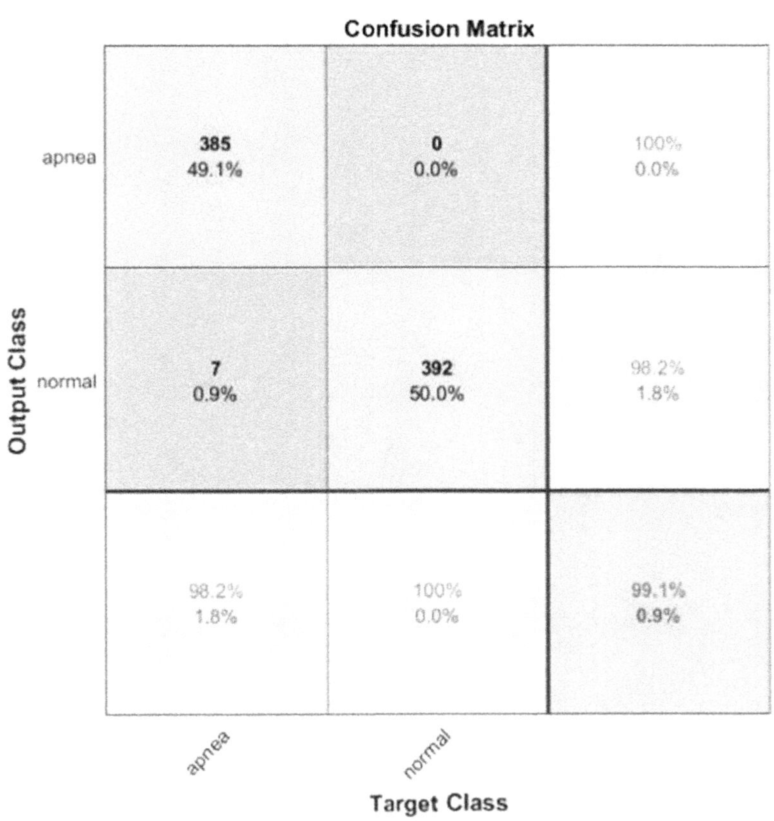

FIGURE 14.15 Confusion matrix for ResNet50.

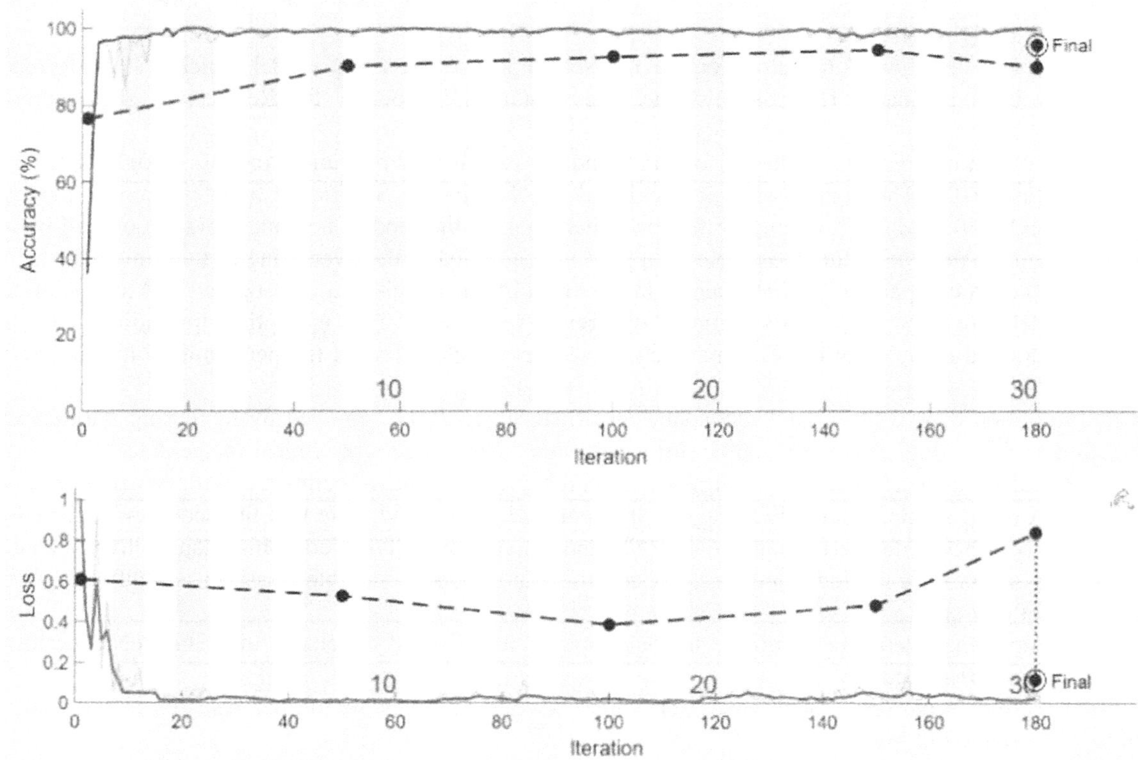

FIGURE 14.16 Validation accuracy for ResNet50.

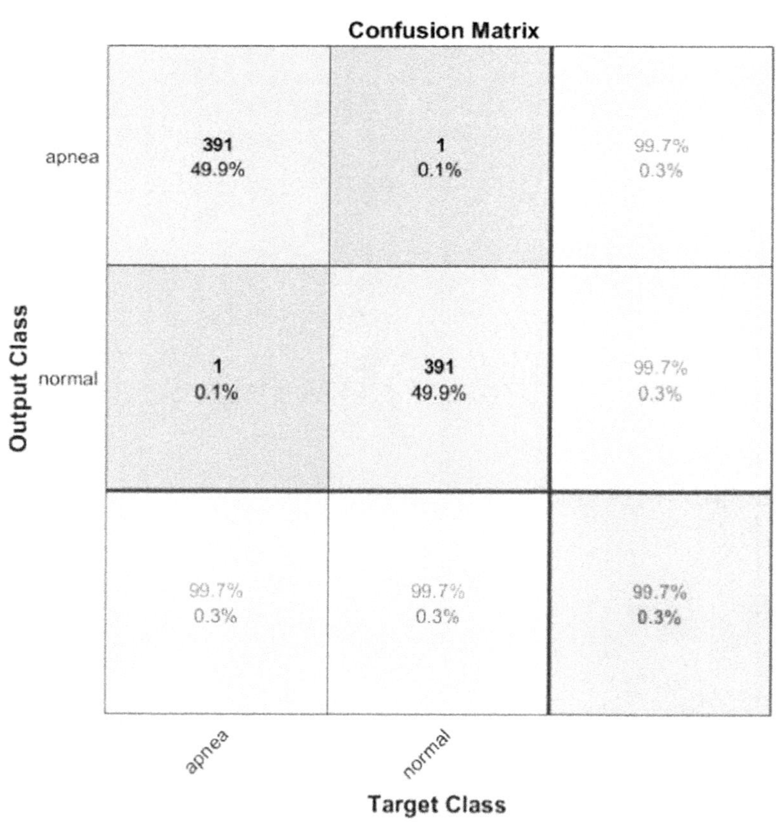

FIGURE 14.17 Confusion matrix for ResNet101.

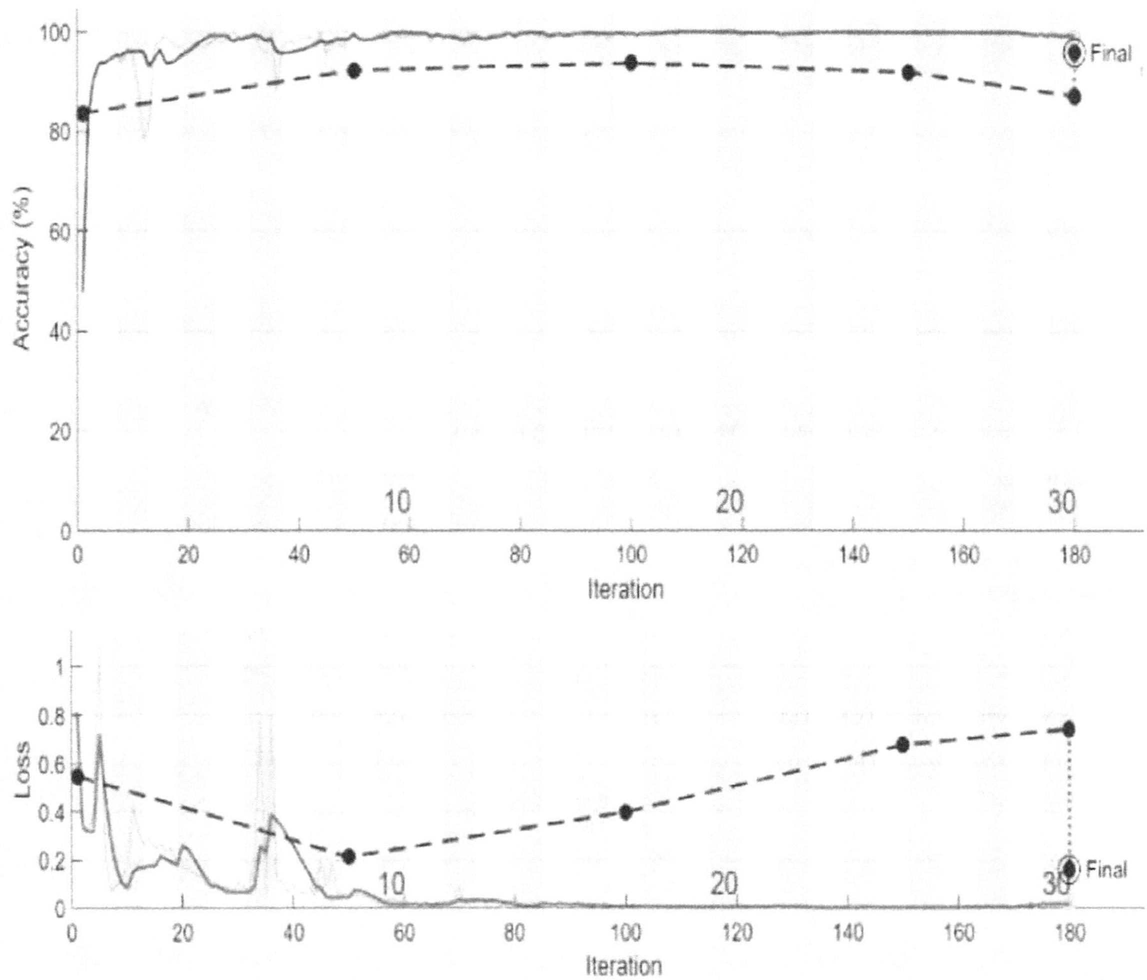

FIGURE 14.18 Validation accuracy for ResNet101.

TABLE 14.2 Comparison of performance indices.

Classifier	a (TN)	b (FP)	c (FN)	d (TP)	TPR (sensitivity)	TNR (specificity)	PPV	NPV	TA (accuracy)
GoogLeNet	391	0	1	392	0.997455471	1	1	0.99744898	0.99872449
ShuffleNet	392	0	0	392	1	1	1	1	1
ResNet50	385	0	7	392	0.98245614	1	1	0.98214286	0.991071429
ResNet101	391	1	1	391	0.99744898	0.99744898	0.99744898	0.99744898	0.99744898

OSA detection using single-channel ECG signal in ShuffleNet and GoogLeNet transfer learning models. All the methods performed better than the traditional method and also require less computing time which leads to the utility of the proposed method.

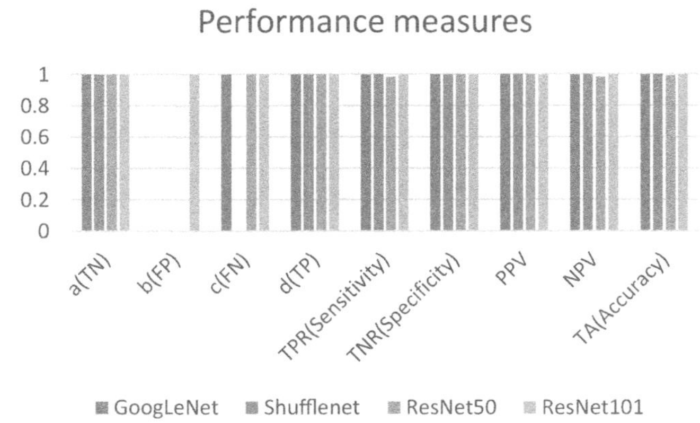

FIGURE 14.19 Comparison chart of different pretrained models.

TABLE 14.3 Comparison of performance measures of existing research.

Paper	Classifier	Signal	Result (in %)		
			Accuracy	Sensitivity	Specificity
[24]	DNN, 1D CNN, 2D CNN, RNN, LSTM, and GRU	ECG	99.0	99.0	NA
[30]	CNN	ECG	99.0	98.0 (f1-score)	NA
[31]	1 D CNN	ECG	97.79	92.23	NA
[32]	RNN, CNN, LSTM, and CNN LSTM	ECG	86.25	NA	NA
[10]	LSTM-RNN GRU, deep hybrid model	ECG	80.67	75.04	84.13
[33]	LSTM-RNN	Instantaneous heart rate (IHR)	85	NA	NA

TABLE 14.4 Results achieved in the proposed work.

Models\Optimizer	Accuracy
GoogLeNet	0.99872449
ShuffleNet	1
ResNet50	0.99107143
ResNet101	0.99744898

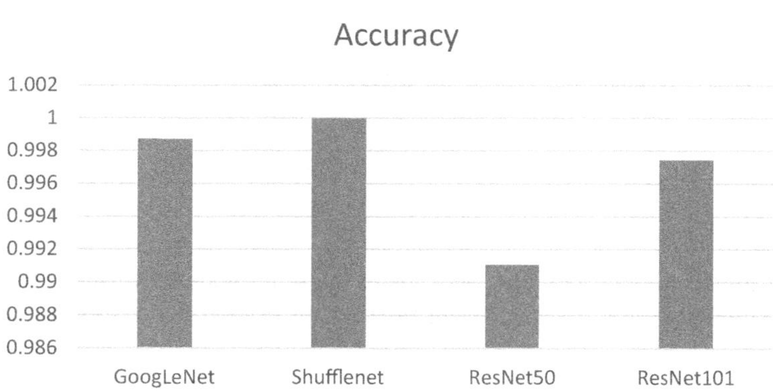

FIGURE 14.20 Comparison of accuracy of a pretrained model.

References

[1] M.J. Sateia, International classification of sleep disorders-third edition (ICSD-3), Chest 146 (2014) 1387–1394.

[2] J. Zhang, Q. Zhang, Y. Wang, C. Qiu, A real-time auto-adjustable smart pillow system for sleep apnea detection and treatment, in: Proceedings of the 12th International Conference on Information Processing in Sensor Networks (IPSN), Philadelphia, PA, USA, 8–11 April 2013; pp. 179–190.

[3] T. Young, M. Palta, J. Dempsey, J. Skatrud, S. Weber, S. Badr, The occurrence of sleep-disordered breathing among middle-aged adults, N. Engl. J. Med. 328 (1993) 1230–1235.

[4] T. Young, L. Evans, L. Finn, M. Palta, Estimation of the clinically diagnosed proportion of sleep apnea syndrome in middle-aged men and women, Sleep 20 (1997) 705–706.

[5] T. Gislason, B. Benediktsdóttir, Snoring, apneic episodes, and nocturnal hypoxemia among children 6 months to 6 years old, Chest 107 (1995) 963–966.

[6] T. Penzel, J. McNames, P. de Chazal, et al., Systematic comparison of different algorithms for apnoea detection based on electrocardiogram recordings, Med. Biol. Eng. Comput. 40 (2002) 402–407. Available from: https://doi.org/10.1007/BF02345072.

[7] M. Cheng, W.J. Sori, F. Jiang, A. Khan, S. Liu. Recurrent neural network based classification of ECG signal features for obstruction of sleep apnea detection (2017) 199–202. Available from: https://doi.org/10.1109/CSE-EUC.2017.220.

[8] Y. Bai, Detection of sleep apnea syndrome by CNN based on ECG. J. Phys.

[9] M. Cheng, W.J. Sori, F. Jiang, A. Khan, Recurrent neural network-based classification of ECG signal features for obstruction of sleep apnea detection, RICAI (2020) 250–254.

[10] M. Bahrami, M. Forouzanfar, Detection of sleep apnea from single-lead ECG: Comparison of deep learning algorithms, 2021 IEEE International Symposium on Medical Measurements and Applications (MeMeA), Lausanne, Switzerland (2021), 1–5. Available from: https://doi.org/10.1109/MeMeA52024.2021.9478745.

[11] S.S. Mostafa, F. Mendonça, A.G. Ravelo-García, F. Morgado-Dias, A systematic review of detecting sleep apnea using deep learning, Sensors (Basel) 19 (22) (2019) 4934. Available from: https://doi.org/10.3390/s19224934.

[12] R.K. Pathinarupothi, E.S. Rangan, E.A. Gopalakrishnan, R. Vinaykumar, K.P. Soman, Single sensor techniques for sleep apnea diagnosis using deep learning, 2017 IEEE International Conference on Healthcare Informatics (ICHI), Park City, UT, USA (2017), 524–529. Available from: https://doi.org/10.1109/ICHI.2017.37.

[13] Y. LeCun, Y. Bengi, G. Hinton, Deep learning, Nature 521 (7553) (2015) 436–444. Available from: https://doi.org/10.1038/nature14539.

[14] E. Urtnasan, J.U. Park, E.Y. Joo, K.J. Lee, Automated detection of obstructive sleep apnea events from a single-lead electrocardiogram using a convolutional neural network, J. Med. Syst. 42 (6) (2018) 104. Available from: https://doi.org/10.1007/s10916-018-0963-0.

[15] D. Jiang, Y. Ma, Y. Wang, A multi-scale parallel convolutional neural network for automatic sleep apnea detection using single-channel EEG signals, in: 11th International Congress on Image and Signal Processing, BioMedical Engineering and Informatics (CISP-BMEI); 2018 Oct 13–15; Beijing, China; New Jersey: IEEE; 2018. Available from: https://doi.org/10.1109/CISP-BMEI.2018.8633132.

[16] R. Haidar, I. Koprinska, B. Jeffries, Sleep apnea event detection from nasal airflow using convolutional neural networks, in: D. Liu, S. Xie, Y. Li, D. Zhao, E.S. El-Alfy (Eds.), Neural Information Processing. ICONIP 2017. Lecture Notes in Computer Science. Cham: Springer; 2017:10638. Available from: https://doi.org/10.1007/978-3-319-70139-4_83.

[17] T. Van Steenkiste, W. Groenendaal, D. Deschrijver, T. Dhaene, Automated sleep apnea detection in raw respiratory signals using long short-term memory neural networks, IEEE J. Biomed. Health Inform. 23 (6) (2019) 2354–2364. Available from: https://doi.org/10.1109/JBHI.2018.2886064.

[18] S. Hochreiter, J. Schmidhuber, Long short-term memory, Neural Comput. 9 (8) (1997) 1735–1780. Available from: https://doi.org/10.1162/neco.1997.9.8.1735.

[19] Y. Yu, X. Si, C. Hu, J.A. Zhang, A review of recurrent neural networks: LSTM cells and network architectures, Neural Comput. 31 (7) (2019) 1235–1270. Available from: https://doi.org/10.1162/neco_a_01199.

[20] R.B. Berry, R. Brooks, C.E. Gamaldo, S.M. Harding, C. Marcus, B. Vaughn, The AASM manual for the scoring of sleep and associated events, Rules, Terminology, and Technical Specifications, American Academy of Sleep Medicine, Illinois, Darien, 2012.

[21] B. Duce, J. Milosavljevic, C. Hukins, The 2012 AASM respiratory event criteria increase the incidence of hypopneas in an adult sleep center population, J. Clin. Sleep. Med. 11 (12) (2015) 1425–1431. Available from: https://doi.org/10.5664/jcsm.5280.

[22] K.R. Casey, R. Tiwari, The (still) elusive definition of hypopnea, J. Clin. Sleep. Med. 14 (12) (2018) 1971–1972. Available from: https://doi.org/10.5664/jcsm.7512.

[23] A. Korotinsky, S.Z. Assefa, M. Diaz-Abad, E.M. Wickwire, S.M. Scharf, Comparison of American Academy of Sleep Medicine (AASM) versus Center for Medicare and Medicaid Services (CMS) polysomnography (PSG) scoring rules on AHI and eligibility for continuous positive airway pressure (CPAP) treatment, Sleep. Breath. 20 (4) (2016) 1169–1174. Available from: https://doi.org/10.1007/s11325-016-1327-y.

[24] S. Arora, J.F. Collen, What are we missing? Hypertension and hypopneas, J. Clin. Sleep. Med. 16 (10) (2020) 1653–1654. Available from: https://doi.org/10.5664/jcsm.8764.

[25] T. Penzel, G.B. Moody, R.G. Mark, A.L. Goldberger, J.H. Peter, The apnea-ECG database, Computers Cardiol. (2000) 255–258.

[26] J. Pan, W.J. Tompking, A real-time QRS detection algorithm, IEEE Trans. Biomed. Eng. 32 (1985) 230–236.

[27] ImageNet. http://www.image-net.org.

[28] C. Szegedy, W. Liu, Y. Jia, P. Sermanet, S. Reed, D. Anguelov, et al., Going deeper with convolutions, in: Proceedings of the IEEE Conference on Computer Vision and Pattern Recognition, pp. 1–9. 2015.

[29] K. He, X. Zhang, S. Ren, J. Sun, Deep residual learning for image recognition, in: Proceedings of the IEEE Conference on Computer Vision and Pattern Recognition, pp. 770–778. 2016.
[30] E. Urtnasan, et al., Identification of sleep apnea severity based on DL from a short-term normal ECG, J. Korean Med. Sci. 35 (2020).
[31] A. John, et al., A 1D-CNN based DL technique for sleep apnea detection in IoT, Sensors (2021).
[32] A. Sheta, H. Turabieh, T. Thaher, J. Too, M. Mafarja, M.S. Hossain, et al., Diagnosis of obstructive sleep apnea from ECG signals using ML and DL classifiers, Appl. Sci. 11 (2021).
[33] R.K. Pathinarupothi, R. Vinay Kumar, E. Rangan, E. Gopalakrishnan, K.P. Soman, Instantaneous heart rate as a robust feature for sleep apnea severity detection using deep learning (2017). Available from: https://doi.org/10.1109/BHI.2017.7897263.

Chapter 15

Anxiety brain topographic maps and its classification using quantitative electroencephalography

Ching Wei Ong, Humaira Nisar and Danyal Mahmood
Department of Electronic Engineering, Faculty of Engineering and Green Technology, Universiti Tunku Abdul Rahman, Kampar, Malaysia

15.1 Introduction

Anxiety disorder is an illness that can cause continuous fear, worry, and stress when one suffers from it. The feeling of being anxious will grow from time to time and it may interfere with one's ability to carry out normal daily routine tasks [1]. An anxiety disorder may have different causes [2]. Research suggests that genetics may be one of the direct causes of mental illness and disorders. The effect of anxiety disorder may worsen if one is genetically inherited and frequently faces difficult situations. Research has also proven that the lack of differential diagnosis during check-ups may lead to a person experiencing signs and symptoms of anxiety as it can be rooted in any other medical problems. In some instances, a bad suit in the environment can also affect a person's mental health and behavior. For example, financial instability, poor relationship with friends and family, and social judgments.

A person with an anxiety disorder may give up easily and is normally pessimistic all the time. Due to the overwhelming anxiety, the individual might turn to drugs or alcohol to relieve stress. As anxiety continues to overpower the person, he or she may lose self-control and end up committing a crime or attempting suicide.

There are a few types of anxiety disorders: panic disorder, social anxiety disorder (SAD), generalized anxiety disorder (GAD), various phobia-related disorders, etc. People that suffer from panic disorder often portray symptoms such as having a fast heartbeat and shortness of breath. Whereas people who suffer from SAD have an extreme fear of perceiving negative judgments from other people. Whereas in GAD, the subject experiences excessive and unrealistic worry about daily tasks. The causes of GAD include being in a stressful environment or traumatic events in life.

Children or youth nowadays also suffer from anxiety disorders. A total of 20.5% of the 80,879 youths globally are affected by anxiety and the result shows a tremendous increase in the statistics of adolescents portraying anxiety symptoms, especially during the COVID-19 pandemic [3]. This results in changes in a child's social behavior. As a result, the cognitive ability of the child is affected, and this can cause a drop in the child's academic performance.

There are two methods to diagnose anxiety disorders. The first method is through screening or lab tests. However, this process is time-consuming, and the result obtained only diagnoses its presence or absence. Therefore the data obtained from screening are insufficient for the diagnosis of anxiety disorder. Hence, the second approach which involves assessment is highly recommended as questions are asked specifically based on the problem faced by the patient to help develop a relevant diagnosis method, and recommendations for tests to be taken are done subsequently. Thus assessments are widely employed in the current practice for anxiety detection.

Different types of data such as medical history and questionnaires are used to diagnose anxiety disorders. Machine learning algorithms are widely employed for anxiety classification. Some studies used functional magnetic resonance imaging (fMRI) or electroencephalogram (EEG) to record brain activity which acts as the data source to classify anxiety [4]. Besides that, measurements of electrocardiogram (ECG) and DNA methylation were also used to evaluate the presence of anxiety disorders. Facial movements and verbal features also contribute to the study of anxiety detection.

The use of EEG has increased rapidly in recent years due to its noninvasive, low-cost, and effective nature. Even further than that, the use of neural networks and machine learning techniques to analyze the EEG data, and the

Intelligent Computing Techniques in Biomedical Imaging. DOI: https://doi.org/10.1016/B978-0-443-15999-2.00003-7

assessment of different brain conditions and illnesses such as bipolar disorders or attention-deficit/hyperactivity disorder (ADHD) has become much easier. While a subject is having anxiety, the brain produces signals which changes neural communication and causes EEG signals to behave differently. This variation in the brain signals can be detected by machine learning techniques and it might help to detect anxiety in the subjects. Due to the complexity of EEG data and significant changes in the EEG data due to the age and other mental conditions of the subject, it is very difficult to formulate a vigorous system of detecting anxiety. The sensitivity of EEG sensors and the incorporation of different types of artifacts in the EEG data such as ocular and muscular artifacts, make it more challenging to detect anxiety. Hence, a robust machine learning technique is required to classify different EEG patterns from EEG signals. In this chapter, we will discuss the brain signals acquired using EEG from the Database for Anxious States based on a Psychological Stimulation (DASPS) dataset [5] to detect and classify the anxiety states experienced by the participants using a machine learning algorithm.

15.2 Treatment of anxiety disorder

15.2.1 Early history of treatments used

There are several ways to treat anxiety [6]. In the 1990s, the psychological and medical approach was developed to treat anxiety. A drug known as selective serotonin reuptake inhibitor (SSRI) was viewed as one of the top choices of medication for treating depression and SAD. However, there are certain side effects of SSRIs such as an increase in the risk of suicidality, syndrome disease, etc. Due to these side effects, psychological treatment is preferred for anxiety disorders.

One of the psychological treatments used is cognitive-behavioral therapy (CBT). CBT focuses on using various techniques to help people overcome thoughts of fear and anxiety. Cognitive restructuring methods are implemented to help reduce overwhelming thoughts so that the patients are more relieved to confront their anxiety. CBT technique is normally used to treat patients with panic disorder [6]. During CBT sessions, the patients are taught to relax and breathe differently. In addition, they are also taught to confront their fear and emotions through repeated exposure to averted situations in a face-to-face setting. Furthermore, CBT is used because it is directive and collaborative. It has been proven to have a decreasing effect on anxiety by the end of each CBT session even though one may feel emotionally uncomfortable in the beginning. The result is long term; hence it is concluded that CBT has a positive effect on the treatment of anxiety disorders.

15.2.2 Neurofeedback and biofeedback treatment

Aside from medication and CBT, neurofeedback or biofeedback training can also be used to treat anxiety. There are a total of four steps in the neurofeedback treatment for anxiety: EEG recording, EEG signal processing, mining vision, and production of feedback signals [7]. EEG signals can be segmented into five frequency bands: delta (1–4 Hz), theta (4–8 Hz), alpha (8–12 Hz), beta (13–30 Hz), and gamma (30–100 Hz).

During the EEG recording, the electrodes are connected to the central, frontal, and parietal regions. Six frequency bands of interest such as delta, theta, alpha, and sensorimotor rhythm (SMR) (12–15 Hz), beta (15–18 Hz), and beta2 (20–33 Hz) were used [8]. In this research, it was found that patients with severe anxiety have low alpha activity in the parietal region and low beta activity in the frontal region. Besides that, high beta2 activity was also recorded in the anxious subjects. As a result, the protocol was designed by increasing the alpha and beta activity at the same time so that beta2 activity can be slowed. The feedback signals were generated using audio-visual tracks. When the alpha wave amplitude is low, it indicates the occurrence of anxiety. Therefore alpha reinforcement was carried out, resulting in a state whereby the subject is calm and relaxed and there is a low level of concentration. On the other hand, a high level of beta2 frequency indicates an alert and panic state. Hence, beta2 amplitude is reduced so that the patient can be in a focused state without panicking.

Biofeedback training is also effective in increasing the alpha asymmetry index of the frontal region (e.g., F4 - F3) [9]. The increase in the frontal asymmetry index can be associated with the increase in alpha power in the right frontal hemisphere compared to the left hemisphere. This indicates an opposing trait to anxious subjects who have low alpha power. Hence, it is believed that frontal alpha asymmetry neurofeedback could be useful for treating anxiety and certain phobias.

There is another approach in which the theta frequency is elevated instead of the beta frequency. In this approach, the patient is subjected to "deep meditation" in an extremely calm state hence stress and anxiety were removed significantly [8].

15.3 Background

In literature, different techniques are used by researchers to conduct experiments to study anxiety. Some researchers designed their experimental protocols and collected the EEG data of the participants. Whereas other researchers used publicly available datasets. The DASPS EEG dataset (https://ieee-dataport.org/open-access/dasps-database) is a publicly available dataset available in the IEEE Data port collected by Ref. [5]. The data are gathered while conducting a behavioral therapy session known as in vivo exposure therapy. Twenty-three healthy subjects were physically exposed to their feared stimuli of six different situations to induce anxiety in them. In this work, Hamilton anxiety rating scale (HAM-A) and self-assessment manikin (SAM) were used to rate the severity of anxiety faced by the participants. Results show that high anxiety level is detected during low valence and high arousal (LVHA). The EEG signals of the participants were captured before and after the anxiety stimulation using a wireless EMOTIV EPOC+ headset. The blind source separation (BSS) algorithm was implemented in the preprocessing step for artifact removal [5]. The BSS method is used to separate a set of mixed signals into the respective source signals which include the original signals and noises. There are a few categories of BSS techniques such as principal component analysis (PCA), independent component analysis (ICA), and canonical correlation analysis (CCA). CCA was used as it can maximize the correlations across the consecutive observed EEG datasets [5]. The use of FastICA algorithm was proposed to maximize statistical independence between the resulting components in a single EEG dataset [10]. A bandpass filter is used to divide the EEG source signal into its respective subbands. Ref. [5] has obtained the best accuracy result of 67% obtained for two-level anxiety classification using root mean square (RMS) features in k-Nearest Neighbor (k-NN) classifier [5] and 83.5% for all time, frequency, and time-frequency features in stacked sparse autoencoder (SSAE). The function of an autoencoder is to compress the input data by selecting only the most important features [11]. It is done through the introduction of multiple hidden layers in the encoder network. After that, the network is learned to reconstruct the input EEG data.

Decision tree (DT) and random forest were implemented to classify two-level anxiety states during the 15-second trial using the same DASPS dataset [10]. After the completion of feature extraction from EEG signals, a recursive feature elimination with cross-validation (RFECV) technique was applied in the stage of feature selection in which the irrelevant features were discarded. The DT classifier can produce a higher classification accuracy of 70.25% using Hjorth parameters than the Random Forest classifier as well as the other classifiers [5]. Besides that, SSAE with RFEVC was proposed to improve the performance of the existing neural network system such that the highest accuracy of 83.93% was obtained.

Another commonly used dataset to study anxiety is the DEAP dataset (https://www.eecs.qmul.ac.uk/mmv/datasets/deap/) [12]. A previous study has used the publicly available DEAP dataset coupled with their homemade dataset for validation purposes [13]. The DEAP data experiment comprised 32 participants whereas there were 20 participants involved in the homemade data experiment. For the EEG data preprocessing in this study, a filtering method was used to remove the artifacts. Then, linear discriminant analysis (LDA), CNN, and PCA algorithms were studied and compared for feature extraction. In one of the previous studies, CNN showed the highest accuracy among the three algorithms, followed by PCA and LDA [13]. Since EEG signals are generally nonlinear and non-Gaussian, in terms of dimensionality reduction it is harder to separate the different classes from the EEG data using PCA and LDA as there are still some correlations between the classes. Whereas CNN is useful in extracting representational abstract features from the raw EEG data. In terms of anxiety classification, Takagi–Sugeno–Kang (TSK) fuzzy classifier had given the best results of 76.68% and 75.23% in identifying anxiety from the DEAP and homemade datasets, respectively.

EEG signals from the DEAP database were used to develop an emotion detection system [14]. Features such as differential entropy (D_fE), rational asymmetry (RASM), and linear formulation of differential entropy (LF-D_fE) were extracted from the EEG data. Then, artificial neural network (ANN)—MLP and recurrent neural network (RNN)—bidirectional LSTM (BiLSTM) classifiers were implemented. The highest accuracies of 76% and 75.5% were obtained in classifying both levels of arousal and valence using LF-D_fE features in the BiLSTM network.

A study was proposed using stop signal task (SST) for anxiety induction [15]. Eighty-one participants were involved in the experiment and their EEG signals were recorded when conflict processes were stimulated in the brain. State-trait anxiety inventory (STAI) score for each participant was evaluated. Generally, the STAI score is evaluated based on a 4-point scale with a total of 20 questions portraying a set of stressful situations. The total score of this report is 80 and high anxiety is often indicated by a higher STAI score ranging from 45 to 80. EEG signals were collected, and CNN architectures were used to predict the occurrence of anxiety. In their work, 3D CNN was proposed to improve the performance of the system using 2D CNN. Goal-conflict-specific-rhythmicity (GCSR) is used as an early biomarker used for anxiety detection [15]. It is measured during SST at the right frontal cortex of the brain. From the results, participants with high anxiety and high STAI score exhibited more positive GCSR values whereas smaller or negative GCSR values were found for participants with no or low anxiety.

One of the previous studies used a dataset obtained from Changhai Hospital in Shanghai [16]. They proposed using the CNN algorithm for EEG classification to study anxiety and depression states. The EEG data were preprocessed using EEGLAB to remove artifacts and then passed through Butterworth's sixth-order filter for further filtering. Next, phase lag index (PLI) was applied to calculate the functional connectivity matrix of the preprocessed data. After that, this matrix was used to train different deep learning models such as deep belief network (DBN), LDA, and CNN for anxiety classification and the results were compared. The highest accuracy of 67.67% was obtained using functional connectivity features with the CNN classifier.

Another study used healthy brain network (HBN) public dataset [13]. It has 92 participants mainly children and adolescents. The screen for child anxiety-related disorders (SCARED) scale was used to rate their anxiety levels in them. This study proposed using the correlation of eye movements of the subjects and EEG signals for anxiety detection during resting state EEG. In this work, the PCA algorithm and the Augmented Lagrange Multipliers method (ALM) were used in the preprocessing step. Power spectrum features (PSD) were extracted from the EEG data using the TEAP toolbox available in MATLAB or Octave software whereas features from the eye movements were captured using an interval algorithm [13]. From the analytical results, the fusion of both EEG and eye movement features using group sparse canonical correlation analysis (GSCCA) had proved to increase the accuracy of classifying anxiety compared to unimodal EEG. Besides that, the gamma band in EEG was found to have the highest accuracy of 78.57% using k-NN and 82.70% using SVM classifiers in detecting anxiety compared to other frequency bands in the fusion model.

Another study used the social interaction anxiety scale (SIAS) dataset that has 89 participants to evaluate the risk of SAD [17]. They proposed deep learning models to identify and classify the severity of SAD based on the neural brain information extracted in the default mode network (DMN). EEG data of the participants were recorded during resting state as DMN is believed to be active at this condition. For signal filtering, FIR band-pass filter was applied to define the frequency range whereas the removal of the artifacts was carried out using BESA software. Effective connectivity features were extracted from the EEG signals and quantified using partial directed coherence (PDC) algorithm to predict SADly. The highest accuracy was obtained for all levels of severity with an average percentage of 93% using the combination of both CNN and long short-term memory (LSTM) models.

Another research [18] proposed a trait anxiety detection framework using MUSE EEG Headband with a built-in noise removal circuit. Sixty-five subjects were involved, and their brain signals were recorded during the resting condition. Trait anxiety is a type of anxiety that shows part of a person's personality, and it does not need to be intentionally induced by certain stressful conditions or dangers. For instance, trait anxiety can be described as more of a personal thought or feeling that arises due to some situations faced in daily routines. In their work, additional steps of channel selection and feature selection were incorporated to improve the accuracy result. Analysis of variance (ANOVA) was applied to the raw EEG data by studying the strength of the energy at different intervals of frequencies (power spectral density) to select a significant set of channels before proceeding to feature extraction. The features extracted from the EEG signals include maximum absolute value, signal energy, and signal sum [18]. The wrapper method was then applied for feature selection. Features that can produce the highest classification accuracy based on the classifiers implemented were chosen. The training of the models was done using Weka 3.8 tool. Random Forest classifier yielded the highest accuracy of 87.69% and 83.07% for both two-class and three-class anxiety classifications, respectively, followed by multilayer perceptron (MLP) and logistic regression (LR). Besides that, the smaller feature vector length proposed also contributed to better anxiety classification accuracy.

Table 15.1 shows the overall comparison of accuracy results based on EEG-based anxiety detection across the different datasets discussed in this section.

15.4 Anxiety classification system

In this chapter, the EEG data from the DASP dataset was used for the classification of anxiety disorders. The EEG data were divided into light and severe anxiety for two-class classifications and normal, light, moderate, and severe for four-class classifications. The EEG data were preprocessed using artifact removal techniques to remove any artifacts or noise from it. Then for the feature extraction, EEG data from all channels were used. The features selected for this study were divided into three categories, that is, band power, differential asymmetry, and statistical features—RMS. Once all the features were extracted, different machine-learning models were applied to them for both two-class and four-class classifications. The machine learning models used for the classifications of the data include DTs, LDA, LR, Naïve Bayes (NB) classifiers, support vector machines (SVM), nearest neighbor classifiers, and ensemble classifiers. Fig. 15.1 shows the block diagram of this anxiety classification system. The classification results for each classifier were recorded and tabulated. A comparison was made between the results. Finally, the results were compared with previous studies to find the significance of our findings.

15.4.1 EEG data collection

In this study, the DASPS public dataset which comprises EEG signals of 23 healthy subjects collected after the anxiety stimulation is used. Prior to the data collection, all participants signed a consent form for volunteering in the study. During the setup, exposure therapy was used to induce a bearable anxious feeling through the recitation of different stressful situations by the psychotherapist. There was a total of six exposure therapy sessions. The first session lasted for 30 seconds with the recitation of a stressful situation for 15 seconds, and the recall of the situation by the participant for 15 seconds. Next, 30 seconds were given to the respective subject to fill in the SAM questionnaire after the exposure therapy session ended. The same procedure was repeated for the remaining five sessions, followed by filling in the SAM each time. In addition, the anxiety level of each participant was recorded manually at the start and end of the experiment based on the HAM-A scale. Fig. 15.2 shows the sequential steps involved in the anxiety stimulation process.

TABLE 15.1 Summary of background research on EEG-based anxiety detection using different datasets.

Researcher	Type of anxiety study/ dataset	Feature extraction	Classification method	Best accuracy result (%)
Baghdadi et al. [5]	DASPS	RMS	k-NN	67.00
		All features	SSAE	83.50
Shikha et al. [10]	DASPS	Time domain features	Decision tree	70.25
		All features	SSAE + RFECV	83.93
Zhang et al. [13]	DEAP	All features	TSK fuzzy system	76.68
Joshi and Ghongade [14]	DEAP	LF-D_fE	BiLSTM	76.00
Wang et al. [15]	SST	Spatial and temporal features	3D-CNN	–
Xie et al. [16]	Changhai Hospital	Functional connectivity	CNN	67.67
Zhang et al. [13]	HBN	PSD + eye movement features	SVM	82.70
Al-Ezzi et al. [17]	SAD	Effective connectivity	CNN + LSTM	93.00
Arsalan and Majid [18]	Trait anxiety	Selected features using the wrapper method	Random forest	87.69

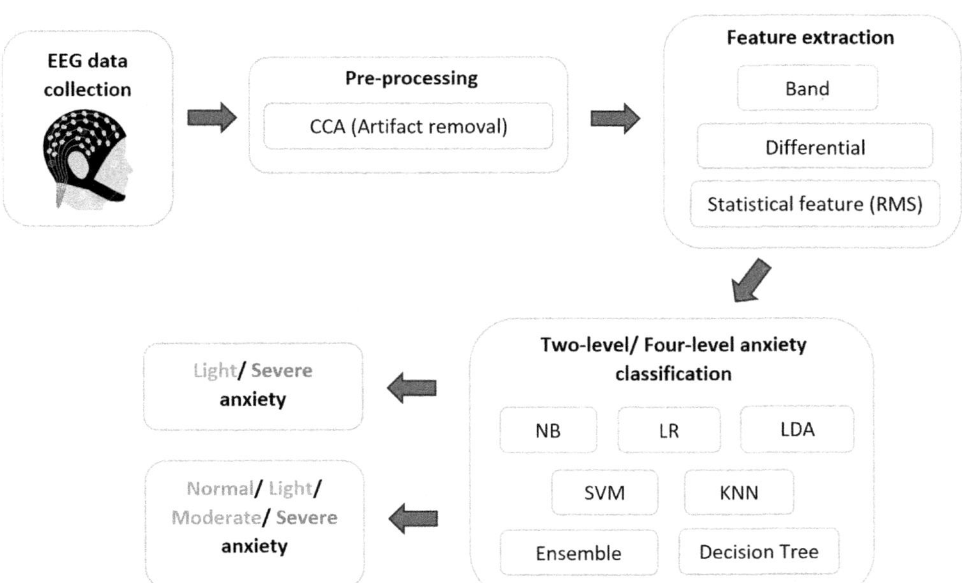

FIGURE 15.1 Block diagram of the anxiety classification system.

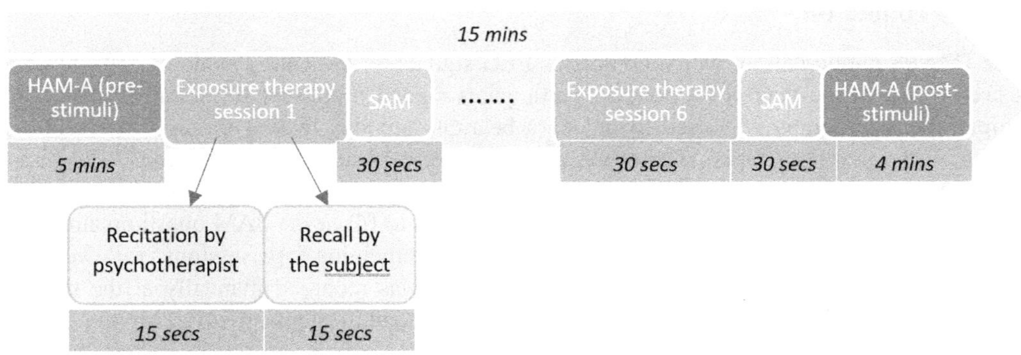

FIGURE 15.2 Anxiety stimulation process presented in DASPS EEG acquisition.

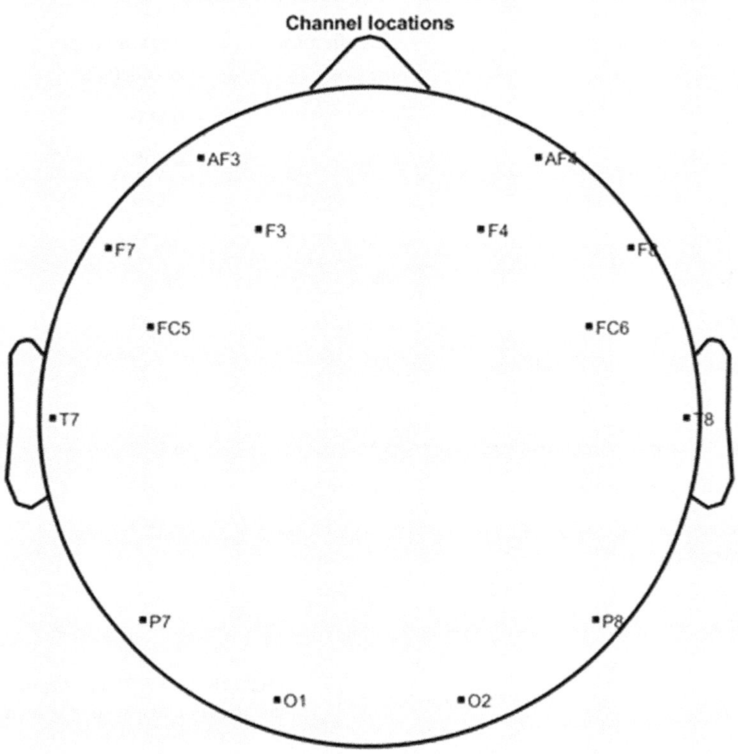

FIGURE 15.3 Fourteen electrode-channel locations of the Emotiv EPOC EEG headset.

The EEG data of each participant was recorded during the exposure therapy sessions. The EEG recording process was stopped before the SAM fill-in. Hence, the recording lasted for 3 minutes (30 seconds × six sessions) and was carried out using an Emotiv EPOC + EEG headset with 14 channels and two reference electrodes as shown in Fig. 15.3. The electrodes were divided into frontal lobe (F3, F4, F7, F8, AF3, AF4, FC5, FC6), temporal lobe (T7, T8), parietal lobe (P7, P8), and occipital lobe (O1, O2). The sampling rate was set to 128 Hz. Hence, in 3 minutes (180 seconds), there will be a total of 23,040 (128 seconds-1 × 180 seconds) EEG data points/samples. These data points were divided into 12 epochs with 1920 data points each and stored in a single file. Both raw EEG data as well as preprocessed EEG of the 23 participants are publicly available in the DASPS database in ".edf" and ".mat" formats, respectively.

15.4.2 Data analysis

Before the preprocessing step, the SAM and HAM-A scores were re-evaluated in the work of previous researchers [5], to discard those with major differences compared to the real rating. Besides that, the comparison of data from all participants was made by calculating the coefficient of variation (CV) for their ratings in the six exposure therapy sessions.

As a result, a mean CV of 0.58 and 0.42 was obtained for valence and arousal, respectively, indicating a low variability between the participants' ratings. Besides that, the increase in the number of subjects with severe anxiety after the experiment was also recorded by comparing the scores obtained in Hamilton 1 (HAM-A prestimuli) and Hamilton 2 (HAM-A poststimuli). Hence, this shows that the results obtained using the exposure method are reliable and it is effective for anxiety induction purposes.

15.4.3 Preprocessing

The recorded brain waves are contaminated with noises or artifacts that may affect the study of EEG signals. Thus it is important to utilize a suitable method to preprocess the EEG data so that a clean signal can be obtained before proceeding to feature extraction in the next stage. Data preprocessing is commonly known as a vital step in machine learning such that it reformats the raw data into modified data to enhance learning ability. As mentioned earlier, physiological artifacts such as EOG and EMG artifacts are usually found in recorded EEG signals [19]. EOG artifacts which are also known as ocular artifacts originate from eye movement and eye blinks. EOG artifacts can also be detected by the EEG, and they are usually found in the frontal region of the brain. On the other hand, EMG artifacts originate from the muscle movement caused by different types of muscle groups in the subject. These movements can take place when the subject is talking, swallowing, or slightly moving his or her head during the experiment. Generally, EMG artifacts usually have a large amplitude with wide frequency domain distributions. Hence, the subjects were asked to minimize their gestures during the EEG recording session to maintain the quality of the EEG data.

To remove the physiological artifacts [20], the CCA method was chosen because it can maximize correlations between successively observed EEG datasets. The MATLAB EEGLAB package was used to perform this. The MATLAB autocorrelation function is used because it enables the identification of the smaller periodic EEG signal that is obscured by skeletal and ocular disturbances. Additionally, a finite impulse response (FIR) bandpass filter was used to separate the EEG source signal into each of its associated subbands.

15.4.4 Feature extraction

Various EEG features had been studied in the field of biomedical research for the diagnosis and classification of epileptic syndromes, anxiety, depression, autism, emotion, etc. [21]. In this study, a frequency-domain feature, an electrode combination feature, and a statistical feature are used for anxiety classification.

15.4.4.1 Frequency-domain feature—band power

The power spectral density (PSD) was determined using Welch's method with an overlapping window of 75%. In other words, the EEG signals were divided into several overlapping segments to allow periodogram averaging. This helps to reduce the variance between the EEG segments effectively. The frequency range of each band was defined to locate the frequency bins that intersect the frequency range of interest. The relative power is calculated using Eq. (15.1).

$$\text{Relative power} = \frac{\text{Absolute Power of Band}}{\text{Total Power}} \tag{15.1}$$

This step is repeated by modifying the frequency range to calculate the relative power of individual bands such as in delta (1–4 Hz), theta (4–8 Hz), alpha (8–13 Hz), beta (13–30 Hz), and gamma (30–64 Hz) bands, respectively. The average power of each band for all 14 channels was calculated. Welch's method was used for feature extraction because it is computationally efficient and effective in reducing the steps involved in calculations to analyze the EEG signals.

15.4.4.2 Electrode combination feature—differential asymmetry

Electrode combination feature known as differential asymmetry is defined as the relative difference of power between two impulses coming from opposite hemispheres of the brain. Frontal alpha asymmetry is one of the EEG features used in emotion classification [22]. Usually, a greater right frontal activity relates to negative emotions such as anxiety. On top of that, a high level of anxiety is indicated by a low differential alpha power. We extracted the differential asymmetry features from all five frequency bands for anxiety classification. Eq. (15.2) is used to compute the asymmetry in this work as shown below:

$$Asymmetry\ Index = \ln(power)|_{L\ Channel} - \ln(power)|_{R\ Channel} \tag{15.2}$$

The alpha asymmetry was computed using the band power features generated previously for all 23 participants using custom MATLAB scripts. For better call-out functions, these features were stored in multiple arrays. For example, the natural logarithm of channel 1 (AF3) power divided by channel 14 (AF4) power is stored as a variable representing the asymmetry index for the specific frontal region. The next variable is used to perform the division of the natural logarithm of channel 2 (F7) power and channel 13 (F8) power, followed by the division of other left and right channels' powers. Differential asymmetry is known as the electrode combination feature because it simplifies the data of two channels (electrodes) into only one. Hence, there are a total of 7 asymmetry features instead of 14 features for every frequency band. The arrays were then concatenated horizontally to form a table stored in MATLAB. The table is then exported to an Excel file. The same step is repeated for the remaining four frequencies using the recorded band power features of the respective band.

15.4.4.3 Statistical feature—root mean square

The next feature proposed in this work is the statistical feature—RMS. Statistics are used to determine the size of a range of quantities. When there are both positive and negative variations as seen in an EEG signal, RMS can be useful. RMS can be utilized in many different domains, although it is most frequently used in the field of signals. Eq. (15.3) is used to calculate RMS as shown below:

$$\mathrm{RMS} = \sqrt{\frac{1}{n}\sum_{i} x_i^2} \tag{15.3}$$

EEGLAB is used to compute the RMS value for each of the 14 channels across 12 epochs. The preprocessed data with extension.mat were converted to multiple datasets with the extension.set which is readable by EEGLAB. A total of 23 datasets were loaded one at a time and the combination of lowpass and highpass FIR filters was implemented to create a bandpass filter that cuts off the EEG signal at a certain frequency range. A slight buffer is given to allow EEGLAB to perform filtering at the frequency as intended. For instance, the beta range is defined as 12.6–32 Hz in the algorithm whereas the output after running the command is computed from 13 to 30 Hz. The RMS score for each channel was computed at the end of each epoch. The mean RMS value per channel was calculated and the same steps above were repeated for delta, alpha, theta, and gamma bands. Table 15.2 shows the summary of the features extracted from EEG signals in the present work.

15.4.5 Classification of human anxiety

After the features were extracted from all channels, a holdout validation was carried out such that the data of the first feature—band power of all frequencies, were divided into two partitions: training data and test data. A standard separation way has been carried out with 80% of the data allocated for training whereas the remaining 20% is allocated for testing. The same procedure was repeated for differential asymmetry and mean RMS features. This method is known as holdout validation.

Next, the Classification Learner app in MATLAB was used to classify two-level and four-level anxiety using the training data of band power. In the literature review, valence and arousal were used as indications to determine the

TABLE 15.2 Summary of the total number of features extracted from different EEG frequency bands.

	Type of features		
	Band power	Differential asymmetry	RMS
Delta (1–4 Hz)	14	7	14
Theta (4–8 Hz)	14	7	14
Alpha (8–13 Hz)	14	7	14
Beta (13–30 Hz)	14	7	14
Gamma (30–64 Hz)	14	7	14
Total:	70	35	70

anxiety classes. However, in this work, the HAM-A scores obtained during the EEG recording were utilized as the responding variable for anxiety classification. Both Hamilton 1 and Hamilton 2 were used to compare the accuracy result and the effect of applying different datasets was observed.

In this study, k-fold cross-validation was applied to the training data before the classification process to ensure a better estimation done by the models on the unseen dataset. A given dataset is divided into k sections or folds, and each fold serves as a testing set at some point in the k-fold cross-validation [23]. The first fold is utilized in the last iteration to test the model, while the remaining folds are used to train the model. The second iteration uses the second fold from the right as the testing set and the remaining data as the training set. Up until the k-number of folds has been employed as the testing set, this process is repeated. Finally, an average accuracy will be yielded, which serves as the final accuracy result. Several trials of $k = 5$, 6, 7, 8, 9, and 10 were implemented to compare the accuracy results obtained. Anxiety classification was then performed with all the machine-learning models available in MATLAB.

Different machine learning-based feature classification models were employed in this work. As the dataset was labeled and categorical, supervised learning methods such as DTs, discriminant analysis, LR, NB classifiers, SVM, nearest neighbor classifiers, kernel approximation classifiers, and ensemble classifiers were used for classification. DTs can be used if the number of classes is known. The model classifies the data into two classes at each step on the bases of some condition creating a tree-like structure. The LDA separates the features by projecting the data into different dimensions within the feature space and finds the maximum division between the classes. LR can be used if the data are categorical. It uses mathematical equations to find a relation between two data factors and uses this relationship to predict classes in features. On the other hand, a NB classifier uses the Bayes theorem to calculate conditional probability based on which it classifies the data. SVM uses kernels to project the data in feature space into higher dimensions and then creates a decision boundary that best separates the data into classes. The boundary is created in such a way that the classes can have a maximum margin. Similarly, the k-NN algorithm divides items into clusters based on feature similarity. It works on the assumption that data with similar characteristics cluster together. Ensemble classifiers are another method of feature classification in which the data are passed to different classifiers and each classifier provides its predictions. Another classifier uses the predictions from all those classifiers and provides a final decision and classifies the data into classes. After applying all these methods, the training accuracies obtained using the models were recorded in a table.

15.4.5.1 Decision trees

A DT is a visual representation of every scenario that could result in a decision, depending on specific parameters. A condition is created on the features for each step or node of a DT that is used for classification to fully separate all of the labels or classes in the dataset [24]. Since a DT generally makes up the structure of a tree and performs decision-making that is like a human being, hence it is easy to understand and interpret. After the tree classifiers were trained, these models were then exported, and anxiety classification was carried out using the same models.

15.4.5.2 Linear discriminant

LDA is a technique used to maximize the dimensionality reduction in machine learning algorithms. In other words, LDA can reduce the features implemented in the classification process, hence making the whole process less complex. The model was trained using band power features, asymmetry, and mean RMS, respectively. The predicted anxiety classes were compared with the actual classes to yield test accuracy.

15.4.5.3 Logistic regression

LR usually applies to only two-level classification scenarios. A logistic function was used to design the modeling of the responding variable. For instance, it will compute the predictions in terms of binary data whereby the answer will be either a yes or no by referring to a threshold value. Another key property of LR is that the predictions of final output are usually carried out based on the highest probability. Therefore it is important to keep the data as less biased as possible to prevent the occurrence of misclassification.

15.4.5.4 Naïve Bayes

The operation of a NB classifier follows a theorem which is known as the Bayes theorem [25]. Based on this theorem, it can be understood that each feature in the extracted feature list is independent of one another. In other words, there will be more probabilities of a certain event (e.g., severe anxiety) as all features have a fair amount of effect on the output. No priorities were given over one another. The probability function in Bayes theorem can be expressed as shown in Eq. (15.4).

$$P(A|,B) = \frac{P(B|,A)P(A)}{P(B)} \tag{15.4}$$

15.4.5.5 *Support vector machines*

As mentioned in the literature review, EEG signals are nonlinear and non-Gaussian. Hence, in SVM, it will search for a boundary or so-called hyperplane so that the features in EEG data can be separated into distinct classes. The margin is expected to be maximized optimally as with a large margin; the problem of overfitting can be reduced.

15.4.5.6 *k-Nearest neighbors*

The k-NN algorithm divides items into clusters based on feature similarity. k-NN falls under the category of nonparametric models because it does not use the training dataset to infer parameters for a discriminative function that predicts previously unobserved data. It works by remembering the training dataset. Based on how similar fresh data (testing data) is to previously known data (training data), the k-NN algorithm predicts responses for those samples. It bases its main operations on distance measures and assumes that data with similar characteristics cluster together.

15.4.5.7 *Ensemble learning*

Through the process of ensemble learning, a new classifier that outperforms all its constituent classifiers is created from a variety of base classifiers. The algorithm employed, the hyperparameters, the representation, or the training data may vary among these base classifiers. Reducing bias and variance is the ensemble methods' main goal [26]. Hence, this method is also proposed to perform anxiety classification using EEG signals since the dataset used in this work may be of high bias due to the unbalanced dataset.

15.5 Results and discussion

To obtain classification accuracy, MATLAB R2019a software is used. Different machine learning models are implemented for different features to classify anxiety levels. The training and test accuracies of the models were compared with the state of the art.

15.5.1 Brain topographic maps for different features

Brain topographic maps are useful for visualizing areas of the brain that are responsible for anxiety detection. Fig. 15.4 shows the anxiety brain topographic map of the 23 participants using beta band power features. The maps are color

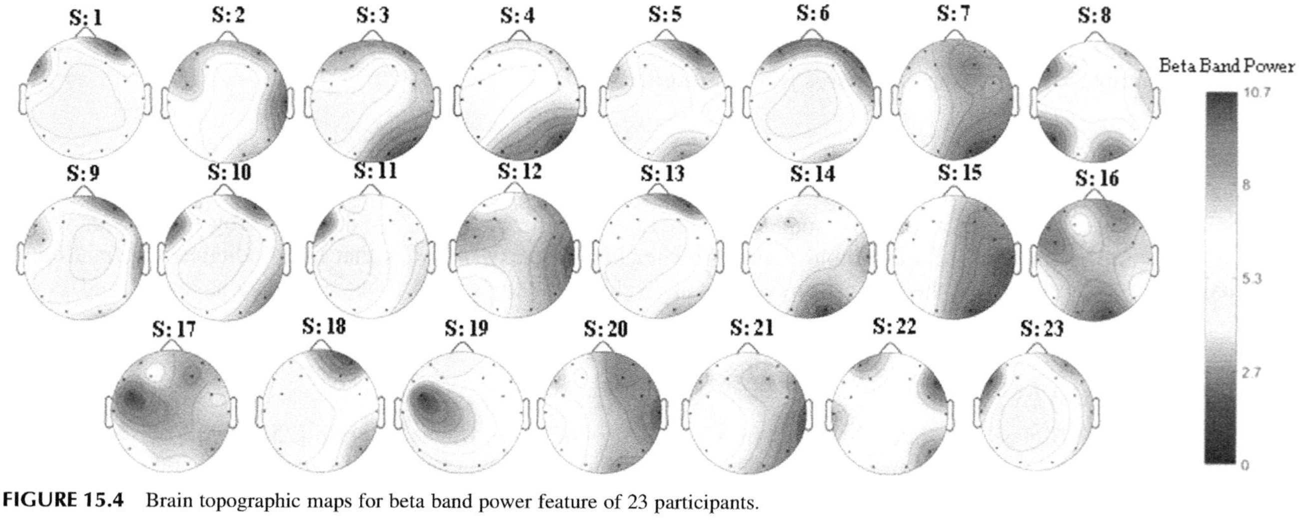

FIGURE 15.4 Brain topographic maps for beta band power feature of 23 participants.

coded which means that the concentration of red color shows that the activities in this portion of the brain is higher than the other. The brain is generally divided into two vertical halves known as left and right hemispheres. Participants with level-4 anxiety have higher beta band power in the right hemisphere of the brain. This is because the right hemisphere of the brain is usually more concerned with emotion and feelings compared to the left hemisphere which is responsible for speech and logical thinking. Besides that, the results show that heightened activity in the right occipital (electrode O2) and temporal lobe (electrode T8) contributes to higher anxiety. This is because the temporal lobe consists of an important area such as the amygdala which is responsible for emotional processing such as anxiety. Ironically, lower activity is observed in the right occipital region of participants with level-1 anxiety. Therefore the relationship between band power features and anxiety level is established.

Fig. 15.5 shows the anxiety brain topographic map of the 23 participants using differential asymmetry features extracted from alpha frequency. Alpha frequency is chosen as a sample for asymmetry features as it functions at a resting state and research has proven that resting EEG asymmetry is strongly related to anxiety response and cortisol level. Most of the participants with level-4 anxiety have high negative alpha asymmetry at the parietal region of the brain and slightly negative alpha asymmetry at the frontal lobe. The alpha asymmetry observed in the frontal lobe such as FP3 and FP4 corresponds to a higher activation at the right hemisphere according to the method of calculation as shown in Eq. (15.2). On the other hand, a lower resting parietal EEG activity is also an indication of high anxiety and negative emotions.

Fig. 15.6 shows the anxiety brain topographic map of the 23 participants using average RMS features extracted from beta frequency. A higher average RMS value is observed at the occipital region and frontal region of participants

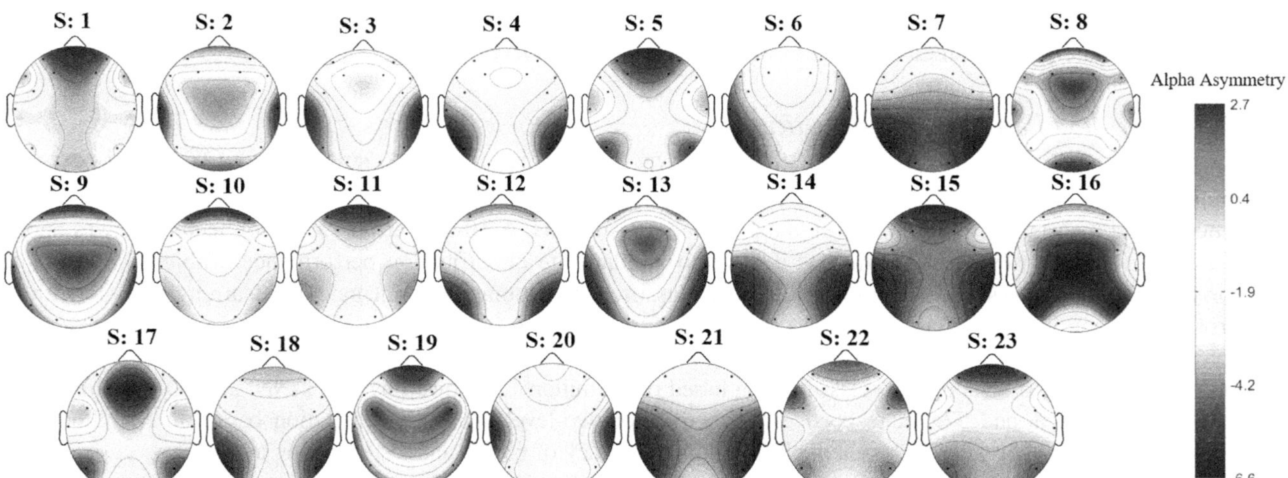

FIGURE 15.5 Brain topographic maps for alpha asymmetry features of 23 participants.

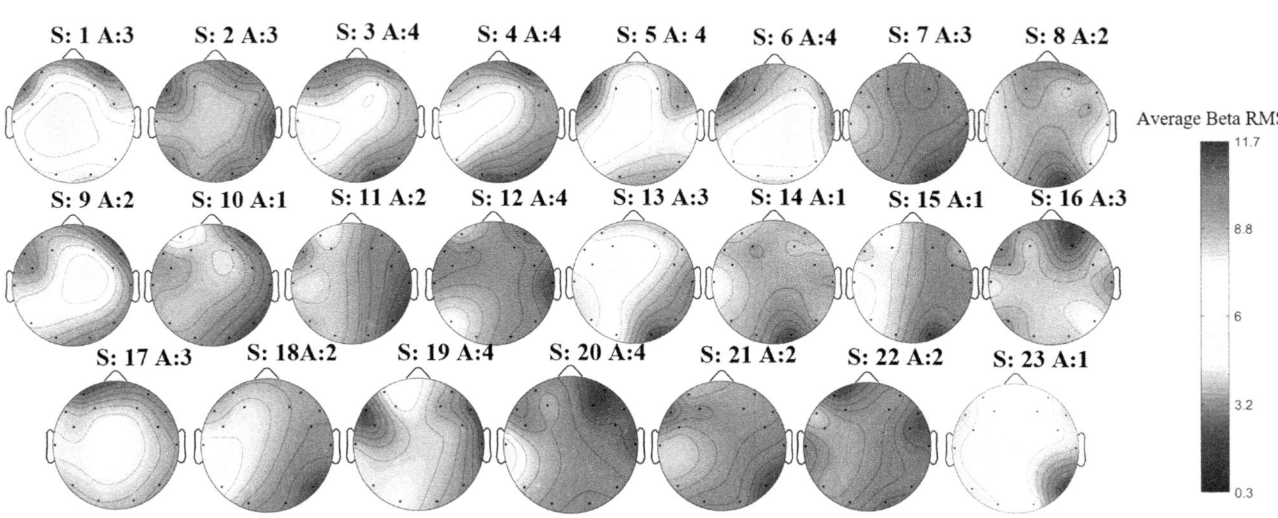

FIGURE 15.6 Brain topographic maps for average beta RMS features of 23 participants (S) and their anxiety (A) levels.

with level-4 anxiety. The anxiety level for all participants is also shown in Fig. 15.6, which also stands for Figs. 15.4 and 15.5. This result shows a good correlation between RMS features and the brain regions that are responsible for emotion processing.

15.5.2 Classification using band power features

The band power features for all frequencies, namely delta (1–4 Hz), theta (4–8 Hz), alpha (8–13 Hz), beta (13–30 Hz), and gamma (30–64 Hz) were extracted from all 14 channels. The features were classified using a built-in algorithm based on the total score recorded in Hamilton 1 and Hamilton 2 to detect anxiety at two and four levels, respectively. In this study, a total of 23 subjects were involved in the anxiety classification process whereby the data of random 19 subjects were allocated for training the models, and the data of the remaining four subjects were reserved for validation or testing. This method of separation is based on the 80:20 ratio. Fig. 15.7 shows the distribution of the 19 subjects involved in training data. This distribution is done using a random separation algorithm written in MATLAB whereby 80% of the subjects were allocated for training data. The distribution was then portrayed in the form of a bar chart with the number of subjects on the y-axis and anxiety levels based on HAM-A scores on the x-axis.

From Fig. 15.7, it can be observed that the training dataset is imbalanced in Hamilton 2 due to more subjects with severe anxiety than light anxiety or normal. On the other hand, the training dataset in Hamilton 1 is balanced as it is distributed fairly across all anxiety levels despite two or four levels. Therefore it is expected that better classification results will be yielded based on Hamilton 1 compared to that of Hamilton 2. Besides we hypothesize that two-level anxiety would provide a better classification accuracy compared to four-level anxiety as the overall distribution of two-level anxiety is more proportionate compared to that of four-level anxiety.

15.5.2.1 Two-level anxiety classification

Classification of anxiety levels was carried out using DTs, LDA, LR classifiers, NB classifiers, SVMs, k-NNs, and ensemble classifiers. These models were trained with the train data consisting of 70 power features (14 channels × five frequencies) and then tested with the unseen data of four test subjects. A 10-fold cross-validation technique was applied to the training data to prevent overfitting as well as to validate the classification performance. Table 15.3 shows the comparison of accuracy results for two-level anxiety classification among different machine learning models with 10-fold cross-validation based on Hamilton 1 and Hamilton 2.

From Table 15.3, it can be observed that the DT classifier computes the highest training accuracy of 73.7% and test accuracy of 75% each for two-level anxiety classification based on Hamilton 1, followed by fine k-NN and LDA, yielding training accuracies of 63.2% and 57.9%, respectively. Whereas for two-level anxiety classification based on Hamilton 2, the ensemble (bagged trees) classifier yields the highest training accuracy of 63.2% and test accuracy of 75%.

An ensemble classifier is the evolution of a normal DT classifier [27]. However, in terms of compatibility, the bagging method of performing classification is more suitable to be used for the case of Hamilton 2 due to its ability to

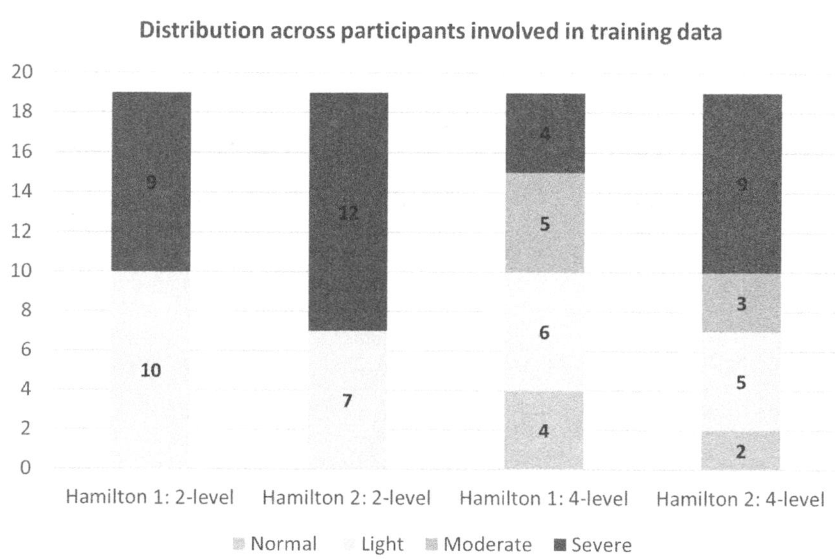

FIGURE 15.7 Distribution of anxiety levels for the 19 subjects involved in training data.

TABLE 15.3 Accuracy results for two-level anxiety classification using band power features based on Hamilton 1 and Hamilton 2.

Machine learning model	Hamilton 1		Hamilton 2	
	Training accuracy	Test accuracy	Training accuracy	Test accuracy
Decision tree	**73.7%**	**75%**	36.8%	25%
Linear discriminant	57.9%	50%	42.1% (biased)	–
Logistic regression	47.4% (biased)	–	42.1%	50%
Gaussian Naïve Bayes	52.6% (biased)	25%	47.4%	50%
Kernel Naïve Bayes	52.6%	–	47.4% (biased)	–
Linear SVM	52.6% (biased)	25%	47.4% (biased)	–
Quadratic SVM	47.4% (biased)	50%	31.6%	–
Cubic SVM	47.4%	–	52.6% (biased)	–
Fine Gaussian SVM	47.4% (biased)	25%	57.9% (biased)	–
Medium Gaussian SVM	42.1% (biased)	–	63.2% (biased)	–
Coarse Gaussian SVM	52.6% (biased)	–	63.2% (biased)	–
Fine k-NN	63.2%	50%	36.8%	–
Medium k-NN	63.2% (biased)	–	52.6% (biased)	–
Coarse k-NN	52.6% (biased)	–	63.2% (biased)	–
Cosine k-NN	52.6% (biased)	25%	47.4% (biased)	–
Cubic k-NN	47.4% (biased)	25%	52.6% (biased)	–
Weighted k-NN	57.9%	50%	57.9% (biased)	–
Ensemble (boosted trees)	52.6% (biased)	–	57.9% (biased)	–
Ensemble (bagged trees)	57.9%	–	**63.2%**	**75%**
Ensemble (subspace discriminant)	42.1%	25%	52.6%	–
Ensemble (subspace k-NN)	47.4%	–	36.8%	–
Ensemble (RUSBoosted trees)	47.4%	–	36.8%	50%

The symbol "–" indicates that the test accuracy is not computed.

minimize the variance of predictions on the imbalanced training dataset as illustrated in Fig. 15.7 (12 subjects with class 2 anxiety whereas 7 subjects with class 1 anxiety).

Multiple levels of predictions were carried out by implementing a few models with the same learning algorithm using different subsets of the training data each time and the average of the predicted results was evaluated as the final output of the two-level anxiety classification system for Hamilton 2.

Furthermore, the test accuracies for two-level anxiety classification based on Hamilton 1 and Hamilton 2 were also compared. For instance, the test accuracy under Hamilton 1 is not computed for some of the models listed in Table 15.3 which include the LR model, kernel NB, and ensemble (bagged trees). This is because the models were not trained well such that the predictions made by them were biased toward class 1 anxiety.

On the other hand, the test accuracies under Hamilton 2 were not computed for the majority of the machine learning models as they made biased predictions due to the imbalanced training dataset as mentioned earlier. For instance, even though SVM and k-NN classifiers obtained a training accuracy of 63.2% each as seen in Table 15.3, this result is not accurate because it is biased toward class 2 anxiety. This problem may be due to the higher number of subjects experiencing anxiety elevation after the video-playing, hence the data used are not sufficient to train the model to understand low-level anxiety (class 1 anxiety). Besides that, high variance in the test data may also affect the test accuracy result, making it unreliable, especially for models that were not trained well in the first place.

Overall, the machine learning models were trained better based on Hamilton 1 to detect two-level anxiety compared to Hamilton 2. This implies that a fair distribution of response variables across the training dataset plays an important role in maintaining accuracy as well to prevent overfitting. However, since the dataset used in this work is generally small, a better approach would be to start by using a larger dataset that consists of more participants to increase the number of sample data for each of the anxiety classes.

15.5.2.2 Four-level anxiety classification

Table 15.4 compares the training and test accuracies for each of the models using all 70 power features for four-level anxiety classification based on Hamilton 1 and Hamilton 2.

From Table 15.4, it can be observed that the ensemble (RUSBoosted trees) classifier yields the highest training accuracy of 47.4% and test accuracy of 50% for four-level anxiety classification based on Hamilton 1 or Hamilton 2, SVM, coarse k-NN, and ensemble (boosted trees) classifiers yield the highest training accuracy of 47.4% each but the result shown in confusion matrix is biased toward class 4 anxiety. Hence, it is not a valid result since there is more sample data of class 4 anxiety in the training dataset compared to other classes. Thus the test accuracies were not computed for these classifiers of Hamilton 2 due to the imbalanced dataset used for four-level anxiety classification as shown in Fig. 15.7. The figure shows that out of 19 subjects in column Hamilton 2: four-level nine participants were experiencing severe (class 4) anxiety after the anxiety induction experiment whereas there are only two subjects categorized under normal (class 1) anxiety. This results in a lack of sufficient training data for class 1 anxiety, causing the training model to recognize only patterns of the band power features extracted from class 4 anxiety data.

TABLE 15.4 Accuracy results for 4-level anxiety classification using band power features based on Hamilton 1 and Hamilton 2.

ML model	Hamilton 1		Hamilton 2	
	Training accuracy	Test accuracy	Training accuracy	Test accuracy
Decision trees	21.1%	25%	31.6%	–
Linear discriminant	10.5%	–	31.6%	50%
Kernel Naïve Bayes	36.8%	25%	42.1% (biased)	–
Linear SVM	10.5%	25%	26.3% (biased)	–
Quadratic SVM	21.1%	25%	15.8% (biased)	–
Cubic SVM	10.5%	25%	21.1%	–
Fine Gaussian SVM	15.8% (biased)	–	**47.4% (biased)**	–
Medium Gaussian SVM	15.8% (biased)	–	**47.4% (biased)**	–
Coarse Gaussian SVM	26.3% (biased)	–	**47.4% (biased)**	–
Fine k-NN	21.1%	50%	31.6%	50%
Medium k-NN	5.3%	25%	26.3%	–
Coarse k-NN	31.6% (biased)	–	**47.4% (biased)**	–
Cosine k-NN	5.3%	25%	26.3%	–
Weighted k-NN	10.5%	25%	42.1% (biased)	–
Ensemble (boosted trees)	31.6% (biased)	–	47.4% (biased)	–
Ensemble (bagged trees)	47.4%	25%	36.8%	50%
Ensemble (subspace discriminant)	10.5%	25%	26.3%	25%
Ensemble (subspace k-NN)	15.8%	–	26.3%	–
Ensemble (RUSBoosted trees)	**47.4%**	**50%**	42.1%	–

The symbol "–" indicates that the test accuracy is not computed.

15.5.3 Classification using differential asymmetry

The asymmetry index of all brain regions, namely frontal, parietal, temporal, and occipital were extracted for all five frequencies each. The features were classified based on the total score recorded in Hamilton 1 and Hamilton 2 to detect anxiety at two and four levels, respectively. The same 19 subjects as in the anxiety classification process using band power features were allocated for training the models and the data of the remaining four subjects were reserved for testing. Hence, there will be a total of 35 asymmetry features (seven features × five frequencies) that function as predictors for the anxiety classes. The 10-fold cross-validation technique was applied to the training data to prevent overfitting problems.

15.5.3.1 Two-level anxiety classification

Table 15.5 shows the comparison of accuracy results among different machine learning models for two-level anxiety classification using asymmetry features based on Hamilton 1 and Hamilton 2.

From Table 15.5, it can be observed that the highest training accuracy of 68% is by ensemble classifier using asymmetry features for two-level anxiety detection based on Hamilton 1, and testing accuracy of 100% is achieved, whereas the highest training accuracy of 63.2% and test accuracy of 75% are obtained by k-NN classifier based on Hamilton 2.

TABLE 15.5 Accuracy results for two-level anxiety classification using differential asymmetry features based on Hamilton 1 and Hamilton 2.

ML model	Hamilton 1		Hamilton 2	
	Training accuracy	Test accuracy	Training accuracy	Test accuracy
Decision tree	42.1%	50%	36.8%	75%
Linear discriminant	31.6%	50%	26.3%	75%
Logistic regression	36.8%	50%	31.6%	25%
Gaussian Naïve Bayes	52.6%	25%	47.4%	50%
Kernel Naïve Bayes	47.4%	50%	57.9% (biased)	–
Linear SVM	36.8%	50%	52.6% (biased)	–
Quadratic SVM	36.8%	50%	63.2% (biased)	–
Cubic SVM	47.4%	25%	57.9%	75%
Fine Gaussian SVM	47.4% (biased)	–	63.2% (biased)	–
Medium Gaussian SVM	36.8%	25%	63.2% (biased)	–
Coarse Gaussian SVM	47.4% (biased)	–	63.2% (biased)	–
Fine k-NN	47.4%	25%	**63.2%**	**75%**
Medium k-NN	42.1% (biased)	50%	47.4%	75%
Coarse k-NN	52.6% (biased)	–	63.2% (biased)	–
Cosine k-NN	47.4%	50%	52.6%	75%
Cubic k-NN	36.8%	50%	47.4%	75%
Weighted k-NN	47.4%	50%	52.6% (biased)	–
Ensemble (boosted trees)	52.6% (biased)	–	63.2% (biased)	–
Ensemble (bagged trees)	47.4%	25%	52.6%	75%
Ensemble (subspace discriminant)	31.6%	50%	36.8%	–
Ensemble (subspace k-NN)	52.6%	50%	52.6%	75%
Ensemble (RUSBoosted trees)	**68.4%**	**100%**	42.1%	50%

The symbol "–" indicates that the test accuracy is not computed.

Generally, the models trained based on Hamilton 1 have fewer problems of biased classification compared to that based on Hamilton 2. This can be justified by comparing the test accuracies in both Hamiltons. Although the anxiety levels for all four test cases in both Hamiltons are of level 2, most of the test accuracies generated in Hamilton 1 are lower or closer to their respective training accuracy. However, for Hamilton 2, the majority achieved the test accuracy of 75% despite having low training accuracy. This is due to the imbalanced data in the training dataset as well as the test dataset, resulting in more predictions that are biased toward class 2 anxiety. Besides that, since all classes in the test dataset under Hamilton 1 and Hamilton 2 for two-level anxiety classification were referring to severe anxiety, therefore the probability of the models detecting class 2 anxiety would be higher.

15.5.3.2 Four-level anxiety classification

Table 15.6 shows the comparison of accuracy results among different machine learning models for four-level anxiety classification using asymmetry features based on Hamilton 1 and Hamilton 2.

From Table 15.6, it can be observed that the highest training accuracy of 42.1% and test accuracy of 75% were obtained using ensemble (RUSBoosted trees) classifier for anxiety classification based on Hamilton 1. Whereas for Hamilton 2, the training accuracy of 52.6% was obtained using the Cosine k-NN classifier. Overall, the training accuracies for the four-level anxiety classification based on Hamilton 1 are lower than the ones based on Hamilton 2. However, fewer biasing issues were found in classification using Hamilton 1 than Hamilton 2. On the other hand, the result obtained in Hamilton 2 is biased toward both class 2 and class 4 anxiety. Hence, it can be concluded that the models were not trained well due to the imbalanced dataset based on Hamilton 2.

TABLE 15.6 Accuracy results for 4-level anxiety classification using differential asymmetry features based on Hamilton 1 and Hamilton 2.

ML model	Hamilton 1		Hamilton 2	
	Training accuracy	Test accuracy	Training accuracy	Test accuracy
Decision tree	15.8%	50%	26.3%	75%
Linear discriminant	10.5%	–	10.5%	50%
Kernel Naïve Bayes	36.8%	25%	47.4% (biased)	–
Linear SVM	5.3%	25%	42.1% (biased)	–
Quadratic SVM	26.3%	25%	47.4%	50%
Cubic SVM	10.5%	25%	42.1%	50%
Fine Gaussian SVM	26.3% (biased)	–	47.4% (biased)	–
Medium Gaussian SVM	10.5%	25%	42.1% (biased)	–
Coarse Gaussian SVM	21.1%	–	47.4% (biased)	–
Fine k-NN	31.6%	25%	36.8%	25%
Medium k-NN	5.3%	–	36.8%	75%
Coarse k-NN	31.6% (biased)	–	47.4% (biased)	–
Cosine k-NN	10.5%	–	**52.6% (biased)**	–
Cubic k-NN	15.8%	–	31.6%	75%
Weighted k-NN	10.5%	25%	31.6%	–
Ensemble (boosted trees)	31.6% (biased)	–	47.4% (biased)	–
Ensemble (bagged trees)	5.3%	–	31.6%	75%
Ensemble (subspace discriminant)	10.5%	–	10.5%	25%
Ensemble (subspace k-NN)	31.6%	25%	36.8%	25%
Ensemble (RUSBoosted Trees)	**42.1%**	**75%**	21.1%	25%

The symbol "–" indicates that the test accuracy is not computed.

15.5.4 Classification using statistical feature—RMS

The RMS values for each channel were extracted for all 12 epochs. These values were extracted from five frequencies, delta, theta, alpha, beta, and gamma, respectively. The mean RMS was also calculated for each epoch in each channel. The features were classified based on the total score recorded in Hamilton 1 and Hamilton 2 to detect anxiety at two and four levels, respectively. The same 19 subjects as in the previous anxiety classification process were allocated for training and the data of the remaining four subjects were reserved for testing. Hence, there will be a total of 70 statistical features (14 mean RMS × five frequencies) that function as predictors for the anxiety classes. The 10-fold cross-validation technique was applied to the training data to validate the classification performance.

15.5.4.1 Two-level anxiety classification

Table 15.7 shows the comparison of accuracy results among different machine learning models for two-level anxiety classification using mean RMS features based on Hamilton 1 and Hamilton 2.

From Table 15.7, it can be observed that the highest training accuracy of 73.7% was obtained using ensemble (subspace discriminant) classifier for two-level anxiety classification based on Hamilton 1. Under Hamilton 2, the highest training accuracy obtained is 73.7%, using ensemble (RUSBoosted trees) classifier. Overall, the models were trained

TABLE 15.7 Accuracy results for two-level anxiety classification using mean RMS features based on Hamilton 1 and Hamilton 2.

ML model	Hamilton 1		Hamilton 2	
	Training accuracy	Test accuracy	Training accuracy	Test accuracy
Decision tree	68.4%	50%	52.6%	75%
Linear discriminant	63.2%	25%	52.6%	25%
Logistic regression	63.2%	25%	42.1%	–
Gaussian Naïve Bayes	57.9% (biased)	25%	57.9%	75%
Kernel Naïve Bayes	52.6%	50%	42.1%	50%
Linear SVM	68.4%	25%	68.4%	50%
Quadratic SVM	68.4%	–	47.4%	50%
Cubic SVM	68.4%	–	57.9%	50%
Fine Gaussian SVM	57.9% (biased)	–	63.2% (biased)	–
Medium Gaussian SVM	68.4%	25%	63.2% (biased)	–
Coarse Gaussian SVM	52.6% (biased)	–	63.2% (biased)	–
Fine k-NN	68.4%	25%	52.6%	25%
Medium k-NN	52.6% (biased)	–	57.9% (biased)	–
Coarse k-NN	52.6% (biased)	–	63.2% (biased)	–
Cosine k-NN	63.2% (biased)	25%	52.6%	75%
Cubic k-NN	52.6% (biased)	–	68.4%	75%
Weighted k-NN	57.9% (biased)	–	52.6% (biased)	75%
Ensemble (boosted trees)	52.6% (biased)	–	63.2% (biased)	–
Ensemble (bagged trees)	42.1%	50%	63.2%	–
Ensemble (subspace discriminant)	**73.7%**	**25%**	47.4%	25%
Ensemble (subspace k-NN)	68.4%	25%	47.4%	50%
Ensemble (RUSBoosted trees)	63.2%	–	**73.7%**	**50%**

The symbol "–" indicates that the test accuracy is not computed.

well using mean RMS features. However, the test accuracies obtained using the ensemble classifiers based on both Hamilton 1 and Hamilton 2 were relatively low, such as 25% and 50%, respectively. This is due to the high variance in the training data. In other words, the machine learning models have trained accurately but they did not perform the classification well on test data since the features in the test data were not consistent (scattered). Furthermore, some of the test accuracies were not computed as shown in Table 15.7. This is because the imbalanced dataset used for training in some models may produce inaccurate results.

15.5.4.2 Four-level anxiety classification

Table 15.8 shows the comparison of accuracy results among different machine learning models for four-level anxiety classification using mean RMS features based on Hamilton 1 and Hamilton 2.

From Table 15.8, it can be observed that the highest training accuracy of 47.4% and test accuracy of 50% were achieved by ensemble (RUSBoosted trees) classifier in four-level anxiety detection using mean RMS features based on Hamilton 1. On the other hand, the highest training accuracy of 68.4% and test accuracy of 50% were obtained using tree classifiers based on Hamilton 2. Overall, better results were obtained based on Hamilton 2 compared to that Hamilton 1. However, the result obtained based on Hamilton 2 is biased toward class 2 and class 4 anxiety due to more subjects involved in these two classes.

TABLE 15.8 Accuracy results for four-level anxiety classification using mean RMS features based on Hamilton 1 and Hamilton 2.

ML model	Hamilton 1		Hamilton 2	
	Training accuracy	Test accuracy	Training accuracy	Test accuracy
Decision tree	10.5%	25%	**68.4% (biased)**	**50%**
Linear discriminant	21.1%	25%	42.1%	25%
Kernel Naïve Bayes	26.3%	25%	52.6% (biased)	–
Linear SVM	10.5%	25%	47.4%	50%
Quadratic SVM	5.3%	25%	42.1%	50%
Cubic SVM	–	–	31.6%	75%
Fine Gaussian SVM	21.1% (biased)	–	47.4% (biased)	–
Medium Gaussian SVM	10.5% (biased)	–	47.4% (biased)	–
Coarse Gaussian SVM	26.3% (biased)	–	47.4% (biased)	–
Fine k-NN	15.8%	25%	47.4%	25%
Medium k-NN	15.8%	–	57.9%	75%
Coarse k-NN	31.6% (biased)	–	47.4% (biased)	–
Cosine k-NN	10.5%	–	42.1%	75%
Cubic k-NN	15.8%	25%	57.9% (biased)	75%
Weighted k-NN	5.3%	25%	42.1%	50%
Ensemble (boosted trees)	31.6% (biased)	–	47.4% (biased)	–
Ensemble (bagged trees)	26.3%	25%	47.4%	75%
Ensemble (subspace discriminant)	10.5%	25%	31.6%	25%
Ensemble (subspace KNN)	10.5%	25%	47.4%	25%
Ensemble (RUSBoosted trees)	**47.4%**	**50%**	47.4%	25%

The symbol "–" indicates that the test accuracy is not computed.

15.5.5 Comparison of classification results for single feature and multiple features based on Hamilton 1 and Hamilton 2

The classification results using band power features are included in Table 15.9 and Table 15.10 for comparison. A total of 105 band power and asymmetry features were used to predict the anxiety classes based on Hamilton 1 and Hamilton 2. Finally, all 175 features including band power, asymmetry, and statistical features (mean RMS values) were used to perform the two-level and four-level anxiety classification based on both Hamiltons.

From Table 15.9, it can be observed that the training accuracy for the two-level anxiety classification obtained based on Hamilton 2 increases from 63.2% to 68.4% after adding differential asymmetry and RMS features to the existing predictor group. For four-level anxiety classification, the training accuracy has increased from 47.4% for only band power features to 68.4% for all features. This shows a big improvement because, with more input features, the ML model can reduce the training error caused by high bias in the training dataset.

TABLE 15.9 Training accuracy results for two-level and four-level anxiety classification using multiple features based on Hamilton 1 and Hamilton 2.

		Band power (70)		Band power + asymmetry (105)		Band power + asymmetry + RMS (175)	
	No. of classes	Best model	Training accuracy	Best model	Training accuracy	Best model	Training accuracy
Hamilton 1	2	Decision tree	73.7%	Logistic regression	68.4%	Ensemble (bagged trees)	68.4%
Hamilton 2	2	Ensemble (bagged trees)	63.2%	Ensemble (boosted trees)	63.2% (biased)	**Ensemble (bagged trees)**	**68.4%**
Hamilton 1	4	Ensemble (RUSBoosted trees)	47.4%	Ensemble (RUSBoosted trees)	47.4%	Ensemble (RUSBoosted trees)	36.8%
Hamilton 2	4	Ensemble (boosted trees)	47.4% (biased)	Kernel Naïve Bayes	47.4% (biased)	**Decision tree**	**68.4% (biased)**

TABLE 15.10 Test accuracy results obtained for two-level and four-level anxiety classification using multiple features based on Hamilton 1 and Hamilton 2.

		Band power (70)		Band power + asymmetry (105)		Band power + asymmetry + RMS (175)	
	No. of classes	Best model	Test accuracy	Best model	Test accuracy	Best model	Test accuracy
Hamilton 1	2	Decision tree	75%	Logistic regression	50%	Ensemble (bagged trees)	50%
Hamilton 2	2	Ensemble (bagged trees)	75%	Ensemble (boosted trees)	–	Ensemble (bagged trees)	–
Hamilton 1	4	Ensemble (RUSBoosted trees)	50%	Ensemble (RUSBoosted trees)	50%	Ensemble (RUSBoosted trees)	25%
Hamilton 2	4	Ensemble (boosted trees)	–	Kernel Naïve Bayes	–	**Decision tree**	**50%**

TABLE 15.11 Training accuracies comparison for two-level and four-level anxiety classifications.

Feature	Type of anxiety classification	This work		Baghdadi et al. [5]	
		Best classifier	Training accuracy	Best classifier	Training accuracy
Band power	2-level	Ensemble (bagged trees)	63.2%	SVM	**66.3%**
	4-level	Ensemble (boosted trees)	47.4%	k-NN	**57.6%**
RMS	2-level	Ensemble (RUSboosted trees)	73.7%	k-NN	67.0%
	4-level	Decision tree	**68.4%**	SVM	59.1%

However, for the two-level anxiety classification based on Hamilton 1, there is a slight decrease in the training accuracy from 73.3% for band power features to 68.4% for all features. This may be due to the increase in complexity of the training model. This is because, with more input features, the model will attempt to fit the majority of the data points in the training dataset, hence increasing the misclassification probability.

From Table 15.10, it can be observed that the testing accuracy for four-level anxiety classification based on Hamilton 2 has improved as well such that a test accuracy of 50% is achieved using all features. Although the test dataset extracted based on Hamilton 2 was imbalanced, the test accuracy has improved in the sense that by using all features, the dataset has a lower variance. Hence, the model trained with all features is more consistent on average compared to only using band power features.

Hence, it can also be concluded that in this work, the ensemble and DT classifiers outperformed other classifiers, that is, k-NN, SVM, LDA, etc. for anxiety classification using EEG signals. In other words, EEG-based anxiety classes in this work are best classified using ensemble classifiers, rather than individual distinct classifiers.

15.5.6 Comparison of classification results with the state-of-the-art

Table 15.11 shows the comparison between the training accuracies obtained in this work with a previous study [5] for two-level and four-level anxiety classifications. The features used for comparison include band power and RMS.

From Table 15.11, it can be observed that higher classification accuracies of 73.7% and 68.4% from ensemble classifiers using RMS features for both two-class and four-class anxiety, respectively, are obtained in this work as compared to the previous study [5]. However, for band power features, the training accuracies obtained by the previous study [5] using SVM and k-NN classifiers were slightly higher compared to our results. This may be due to the difference between the feature extraction methods used in both works. Besides that, the separation of training data and testing data also plays an important role because the subjects allocated for training and testing may be different for both works. Hence, the distribution of anxiety classes across the subjects chosen will be different as well.

In addition, the different features used will also affect the accuracy result. For instance, the features used in this work were extracted from all five frequencies whereas only four frequencies (delta is excluded) were implemented by Ref. [5]. Furthermore, a different number of k-folds used in the cross-validation can also result in different training accuracy. For instance, 10-fold cross-validation has produced the best accuracy result in this work whereas fivefold cross-validation work was utilized by Ref. [5].

Hence, we may say that many factors contribute to achieving an accurate classification result. Therefore analysis of the EEG data is an important step and various approaches or alternatives can be implemented to study the classification performance. Optimization can also be carried out when it is necessary to improve anxiety classification accuracy.

15.6 Conclusion

Anxiety disorder is becoming a common problem globally, especially in children and youth. Aforetime, drugs were used as a medication for treating anxiety and depression but there were side effects of these drugs. CBT provided a slightly better solution for the treatment of anxiety disorder. Neurofeedback training (NFT), on the other hand, is deemed to have the best results in curing anxiety disorder by increasing the alpha and beta activity at the same time using audio-visual tracks as feedback signals. In this chapter, two-level and four-level anxiety classifications based on

Hamilton 1 and Hamilton 2 were carried out using single and multiple features with different machine learning algorithms. The EEG data from DASPS public dataset were used which had artifacts in it. The artifacts from EEG data were removed using bandpass filtering and CCA methods. The features were then extracted from the EEG signal including band power, differential asymmetry, and statistical feature—mean RMS. Several classification methods were used to classify the features. Furthermore, single and multiple feature classification has been tested and compared. Ten-fold cross-validation has been implemented for training. By doing so, the training accuracy and test accuracy based on Hamilton 2 have increased significantly by using multiple features. In this work, the statistical, that is, RMS features performed better than the band power features in both two-class and four-class anxiety classifications. The results were compared with the previously conducted studies and the proposed method provided better results in terms of classification accuracies.

The main problem encountered is the imbalanced dataset used in this work. One of the recommendations is to increase the size of the training data using data augmentation. Deep learning models can be used with data augmentation to artificially generate new data based on the existing EEG data to increase the size of training data. When the size of training data is larger, better predictions of anxiety classes can be made. In addition, feature selection can also be implemented in the future so that redundant features can be eliminated to improve the accuracy of the classifiers. For instance, Pearson correlation coefficient can be calculated to study the relationship between different features, hence using only the important features for classification.

Acknowledgments

This work is financially supported by the Universiti Tunku Abdul Rahman Research Fund (UTARRF) (Grant No. IPSR/RMC/UTARRF/2022-C2/H03), (Grant No. IPSR/RMC/UTARRF/2023-C2/H01), and Excellent Research Center Award Fund 2023, Centre for Healthcare Science and Technology, UTAR, Malaysia.

References

[1] M.G. Craske, et al., What is an anxiety disorder? Focus 9 (3) (2011) 369–388.

[2] J. Driskell, Introduction to anxiety disorders, All About Anxiety: An Introductory Guide to Neuroscience, Assessment, and Intervention, Pressbooks, 2022.

[3] N. Racine, et al., Global prevalence of depressive and anxiety symptoms in children and adolescents during COVID-19: a meta-analysis, JAMA Pediatr. 175 (11) (2021) 1142–1150.

[4] M. Arif, et al., Classification of anxiety disorders using machine learning methods: a literature review, Insights Biomed. Res. 4 (1) (2020) 95–110.

[5] A. Baghdadi, et al., Dasps: a database for anxious states based on a psychological stimulation. arXiv preprint arXiv:1901.02942, 2019.

[6] A. Bystritsky, et al., Current diagnosis and treatment of anxiety disorders, P & T 38 (1) (2013) 30–57.

[7] F. Abbasi, K. Shariati, F. Tajikzadeh, Comparison of the effectiveness of cognitive behavioral therapy and neurofeedback: reducing anxiety symptoms, Arch. Neurosci. 5 (3) (2018).

[8] A. Moradi, et al., Treatment of anxiety disorder with neurofeedback: a case study, Procedia-Social Behav. Sci. 30 (2011) 103–107.

[9] R. Mennella, E. Patron, D. Palomba, Frontal alpha asymmetry neurofeedback for the reduction of negative affect and anxiety, Behav. Res. Ther. 92 (2017) 32–40.

[10] Shikha, M. Agrawal, M.A. Anwar, D. Sethia, Stacked sparse autoencoder and machine learning based anxiety classification using EEG signals, in: The First International Conference on AI-ML-Systems. 2021.

[11] A. Ng, Cs294a lecture notes: sparse autoencoder. URL: https://web.stanford.edu/class/cs294a/sparseAutoencoder2011new.pdf, 2010.

[12] S. Koelstra, et al., Deap: a database for emotion analysis; using physiological signals, IEEE Trans. Affect. Comput. 3 (1) (2011) 18–31.

[13] X. Zhang, et al., Fusing of electroencephalogram and eye movement with group sparse canonical correlation analysis for anxiety detection, IEEE Trans. Affect. Comput. 13 (2) (2020) 958–971.

[14] V.M. Joshi, R.B. Ghongade, EEG based emotion detection using fourth order spectral moment and deep learning, Biomed. Signal. Process. Control. 68 (2021) 102755.

[15] Y. Wang, et al., AnxietyDecoder: an EEG-based anxiety predictor using a 3-D convolutional neural network, in: 2019 International Joint Conference on Neural Networks (IJCNN). 2019. IEEE.

[16] Y. Xie, et al., Anxiety and depression diagnosis method based on brain networks and convolutional neural networks, in: 2020 42nd Annual International Conference of the IEEE Engineering in Medicine & Biology Society (EMBC). 2020. IEEE.

[17] A. Al-Ezzi, et al., Severity assessment of social anxiety disorder using deep learning models on brain effective connectivity, IEEE Access. 9 (2021) 86899–86913.

[18] A. Arsalan, M. Majid, A study on multi-class anxiety detection using wearable EEG headband, J. Ambient. Intell. Humanized Comput. 13 (12) (2022) 5739–5749.

[19] D. Mahmood, H.N. Riaz, H. Nisar, Introduction to non-invasive biomedical signals for healthcare, Advances in Non-Invasive Biomedical Signal Sensing and Processing with Machine Learning, Springer, 2023, pp. 1–24.

[20] D. Mahmood, H. Nisar, Y.V. Voon, Removal of physiological artifacts from electroencephalogram signals: a review and case study, in: 2021 IEEE 9th Conference on Systems, Process and Control (ICSPC 2021). 2021. IEEE.

[21] R. Nawaz, et al., Comparison of different feature extraction methods for EEG-based emotion recognition, Biocybern. Biomed. Eng. 40 (3) (2020) 910–926.

[22] R. Nawaz, H. Nisar, Y.V. Voon, The effect of music on human brain; frequency domain and time series analysis using electroencephalogram, Ieee Access. 6 (2018) 45191–45205.

[23] D. Anguita, et al., The 'K' in K-fold cross validation, in: ESANN. 2012.

[24] A. Roy, A dive into decision trees. Ha mtad April, 2020. 21: p. 2022.

[25] D. Berrar, Bayes' theorem and naive Bayes classifier, Encycl. Bioinforma. Comput. Biol. 403 (2018) 412.

[26] T. Acharya, Advanced ensemble classifiers. June 14th, 2019.

[27] S.A.A. Shah, et al., Decision trees based classification of cardiotocograms using bagging approach, in: 2015 13th International Conference on Frontiers of Information Technology (FIT). 2015. IEEE.

Chapter 16

Smart computing in brain-computer interface and neuroscientific research: opportunities, methods, and challenges

Harish Velingkar[1], Roopa R. Kulkarni[2] and Prashant P. Patavardhan[3]

[1]*Department of Electronics and Communication Engineering, Agnel Institute of Technology & Design, Assagao, Goa, India,* [2]*Department of Electronics and Communication Engineering, Dayananda Sagar Academy of Technology & Management, Bengaluru, Karnataka, India,* [3]*Department of Electronics and Communication Engineering, RV Institute of Technology & Management, Bengaluru, Karnataka, India*

16.1 Introduction

The technology behind brain-machine interfacing (BMI) or brain-computer interfacing (BCI) involves use of sophisticated sensors that can detect and interpret signals from the brain and translate them into commands that machines can interpret. This technology has enormous potential in the fields of healthcare, entertainment, and defense, and it could significantly enhance the quality of life of individuals with physical disabilities or limited mobility.

Intelligent computing in the field of BMIs and neuroscientific research opens up significant opportunities for advancement. These opportunities include:

1. *Advanced Pattern Recognition*: Intelligent computing techniques, such as machine learning (ML) and deep learning, have the potential to analyze and classify complex brain signals. By identifying patterns, recognizing mental states, and decoding neural activity, these methods enhance the accuracy and intuitiveness of BMI control.
2. *Adaptive and Learning Systems*: Intelligent computing enables the development of adaptive BMIs that can learn and adapt to changes in users' brain signals over time. These systems continuously improve their performance by adjusting to individual preferences, compensating for noise or signal variations, and enhancing user experience and control.
3. *Improved Communication and Control*: BMIs offer individuals with limited mobility or communication abilities to interact with the external world. Intelligent computing can facilitate effective communication and control by translating brain signals into commands for prosthetic devices, robotics, or virtual environments, providing greater independence and functionality.
4. *Assistive Technologies*: Intelligent computing empowers the development of assistive technologies that enhance the quality of life for individuals with neurological disorders or disabilities. By combining BMIs with intelligent algorithms, these technologies can assist in motor rehabilitation, cognitive enhancement, or the restoration of lost sensory functions.

Before delving into the various ML techniques employed in the study of BMIs and neuroscientific research, let us explore the different types of BCIs and gain an understanding of the physiology of brain waves.

16.1.1 Brain-computer interface and its subtypes

BCI can be categorized into distinct branches based on various factors, such as the nature of recorded signals, sensor placement, and the specific application domain. Some common branches of BCI include:

1. *Invasive BCI:* This type of BCI involves implanting electrodes directly into the brain, typically for individuals with severe motor impairments. Invasive BCI can provide high-quality signals, but it also carries risks associated with invasive procedures.

Intelligent Computing Techniques in Biomedical Imaging. DOI: https://doi.org/10.1016/B978-0-443-15999-2.00005-0

2. *Noninvasive BCI*: This type of BCI involves using sensors placed on the scalp or other body parts to record brain signals. Noninvasive BCI is generally safer and more accessible than invasive BCI, but the quality of the signals can be lower.
3. *Motor imagery BCI*: This type of BCI involves detecting patterns in brain signals that correspond to motor imagery, such as imagining moving a limb. Motor imagery BCIs offer the potential to exert control over external devices, such as prosthetics or robotic arms, through the power of mental imagery.
4. *P300 BCI*: This type of BCI relies on detecting the P300 waveform, an electrical signal that occurs in response to a stimulus, such as a flashing light. P300 BCI can be used to control computer interfaces, such as spellers or virtual keyboards.
5. *Hybrid BCI*: This type of BCI combines two or more types of signals or methods, such as combining EEG and fMRI or invasive and noninvasive methods, to improve the quality and reliability of the signals.

These mentioned branches of BCI represent only a glimpse into the diverse landscape of this field, which is continually evolving and witnessing the emergence of novel technologies and applications.

The use of BCI technology is gaining traction in predicting diseases, especially in neurodegenerative diseases such as Parkinson's, Alzheimer's, and epilepsy [1]. BCI technology can detect subtle changes in brain activity that may be indicative of the early stages of these diseases. Researchers have been using ML algorithms to analyze brain signals recorded through BCI, which allows them to identify specific patterns that may be early signs of the disease [2]. BCI has also been used to monitor changes in the brain over time, allowing researchers to track the progression of diseases and potentially predict the rate at which the disease will progress [3]. The potential for BCI to predict diseases is enormous, as early diagnosis of these diseases can help individuals receive timely treatment, potentially delaying the onset of symptoms and improving overall outcomes. Nevertheless, it is essential to acknowledge that BCI technology is still in its nascent phases, and further research is required to ascertain its precision and efficacy in disease prediction.

16.1.2 The physiology of brain waves

The human brain generates different types of electrical waves that can be measured using electroencephalography (EEG) technology. These waves are categorized into several types based on their frequencies and amplitudes, and each type is associated with different mental states and conditions.

The first wave is characterized as Delta waves, encompassing the lowest frequency range of 0.5–4 Hz and exhibiting the highest amplitude. Delta waves occur during deep sleep or in conditions such as brain damage or brain tumors.

The second type is Theta waves. These waves fall within a frequency range of 4–8 Hz and have a lower amplitude than Delta waves. Theta waves occur during sleep, meditation, and deep relaxation.

The third type is Alpha waves. These waves have a range of 8–13 Hz and a lower amplitude than Theta waves. Alpha waves occur when an individual is awake but relaxed, such as during meditation or creative thinking.

The fourth type is Beta waves, with a frequency range of 13–30 Hz and a higher amplitude than Alpha waves. Beta waves occur when an individual is awake, alert, and focused, such as during problem-solving or mental tasks.

The fifth type is Gamma waves, which have the highest frequency, ranging from 30–100 Hz, and a lower amplitude than Beta waves. Gamma waves occur during high mental activity, such as during problem-solving, perception, and consciousness is shown in Fig. 16.1.

16.1.3 How AI can be used for intelligent computing of EEG signals and brain image analysis

ML algorithms are predominantly data-oriented, that is, they need a large dataset having some quantifiable features from which the said model could extract patterns and infer the accuracy of the model. Therefore the first steps toward analyzing a biosignal such as an EEG or a brain image would be preparing a large set of relevant and operable dataset. The dataset could be in the form of a graph derived from EEG sensors, corresponding potential values or brain/nerve images obtained through imaging techniques such as magnetic resonance imaging (MRI) or positron emission tomography (PET).

The next step would be preprocessing of the dataset. EEG signals are generally very fragile in nature and tend to include a lot of artifacts. After filtering the artifacts, the essential and nonredundant information could be extracted.

Deep learning serves as an exemplar among ML algorithms employed in brain imaging. It harnesses the power of neural networks, a collection of algorithms tailored to identify patterns in data, for the purpose of analyzing brain images. Researchers have used deep learning algorithms to analyze brain images of individuals with Alzheimer's disease and have been able to predict with a high level of accuracy, which individuals will go on to develop the disease [5].

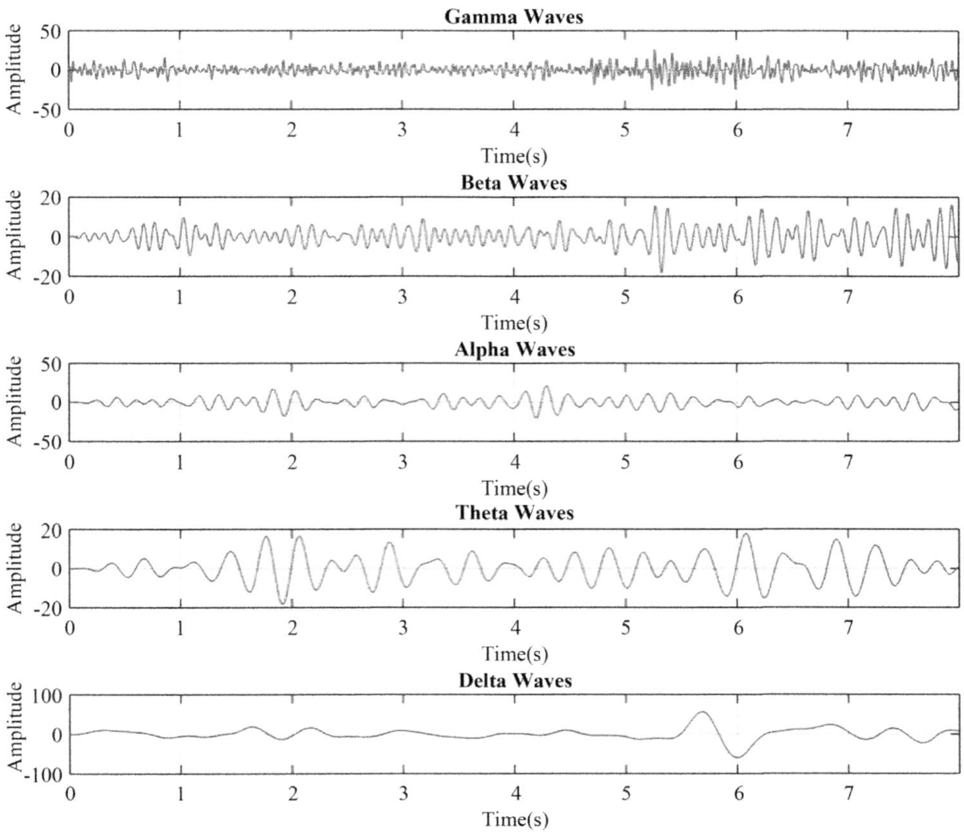

FIGURE 16.1 Brain wave patterns of normal adult [4].

One of the limitations of using AI algorithms in brain imaging is the need for large datasets to train these algorithms effectively. Brain imaging datasets can be challenging to obtain due to the cost and time required to acquire high-quality images. Additionally, as with any medical imaging technology, there are concerns around data privacy and security, as brain images contain sensitive personal information [6].

Despite these challenges, the use of AI algorithms in brain imaging to predict diseases shows great promise in improving the accuracy and speed of diagnosis, allowing for earlier interventions and potentially better outcomes for patients. As the technology continues to advance and more data becomes available, the potential for AI algorithms in medical imaging to transform healthcare is enormous.

16.2 The architecture of brain-machine interface

Fig. 16.2 depicts the typical architecture of a BMI process. Let us have a look at the individual components of the block diagram in detail.

16.2.1 The signal acquisition

The signal acquisition block is a crucial component in the architecture of a BMI. It is responsible for capturing neural signals from the user's brain, which serve as the input for the BMI system. During this stage, noninvasive or invasive techniques are employed to obtain the neural activity. Noninvasive methods typically involve the use of EEG electrodes placed on the scalp as per the 10–20 electrode system [7] which measures electrical activity on the scalp, while invasive approaches may involve implanting electrodes directly into the brain to capture neural signals at a higher resolution [8]. The signal acquisition block plays a vital role in ensuring the quality and integrity of the acquired signals, as they serve as the foundation for subsequent processing and interpretation stages within the BMI architecture. Additionally, this block must address challenges such as signal noise, artifacts, and individual variability to obtain reliable neural data for accurate and effective control of the BMI system.

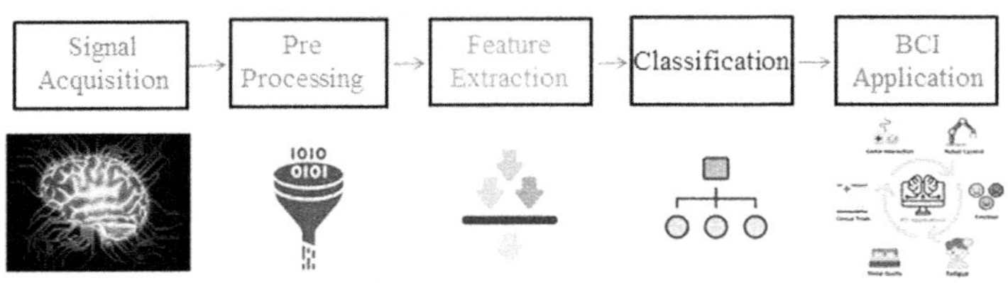

FIGURE 16.2 Architecture of brain-machine interface.

16.2.2 Preprocessing of the acquired signals

The artifacts can either be external, that is, due to sources such as electronic gadgets, surrounding noise, and power line fluctuations, or it could be due to physiological noise due to eye blinks, movements of head, skin resistance, etc. In the preprocessing step, different band filters (such as low pass filter [LPF], band pass filter [BPF] and high pass filter [HPF]) are used to eradicate such artifacts provided; their frequency bands do not overlap. However, the frequency bands of EEG signals typically tend to overlap in which case techniques such as regression, common average referencing, empirical mode decomposition, blind source separation, and nonlinear mode decomposition are used [9].

Further the data size may be reduced by eliminating irrelevant features not suitable for the specific research purpose. Preprocessing helps in eliminating a principle called "error in – error out," typical in AI algorithms, which means that data fed with errors into a model tends to produce biased prediction of the new data.

16.2.3 Feature extraction by training of the AI model

Most ML algorithms are Model Skeptic, that is, their structure does not depend on the data that would be fed but rather are chosen based on their empirical performance.

Fig. 16.3 shows the different ML algorithms that could be implemented by the researcher for training of the AI model. Depending on the application, the research objectives and the type of dataset available, a suitable ML algorithm may be implemented. These algorithms are briefly explained below.

16.2.3.1 Supervised learning

A "Supervised Learning" model makes use of labeled training data in order to classify or predict the outcome effectively. Generally two types of algorithms make use of Supervised Learning model, that is, classification algorithms, which are used to segregate test data into different classes, and regression algorithms, which are used to predict the occurrence of a continuous variable. For example, we can have a labeled dataset of EEG signals with eye-blinking artifacts and detect noise spikes in the live EEG recording of patients [10] or a labeled dataset of early occurrence of post-traumatic gliosis, predicting an occurrence of seizure [11]. Following are some of the most popular supervised learning techniques. Each of the techniques is explained from the perspective of an application in the study of bio medical engineering (BME) and neuroscientific research.

16.2.3.1.1 Support vector machine

Support vector machine (SVM) is an influential and adaptable ML algorithm, extensively utilized for classification and regression tasks. Its operation revolves around identifying the hyperplane that optimally separates two classes within a high-dimensional space, thereby enabling effective pattern recognition. In ML, a hyperplane is a multidimensional geometric plane that acts as a decision boundary, helping to separate data points into distinct classes or categories. It is defined by a set of parameters and represents the dividing surface between different regions in the feature space, enabling classification or regression tasks to be performed. The position and orientation of the hyperplane are determined by the ML algorithm based on the input data and the desired objective.

Fig. 16.4 explains the following concepts:

1. *Support Vectors*: Support vectors refer to the data points that are in closest proximity to the hyperplane. These specific points play a crucial role in defining the separating line or decision boundary. By utilizing these support vectors, the ML algorithm determines the optimal position and orientation of the hyperplane, effectively delineating distinct classes or categories within the dataset.
2. *Hyperplane*: In the aforementioned diagram, the hyperplane refers to the decision plane or space that acts as a boundary, segregating a set of objects into distinct classes based on their characteristics or features.

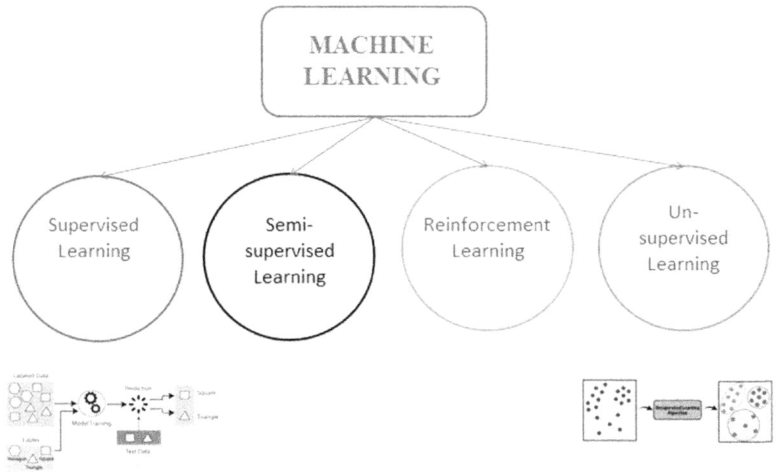

FIGURE 16.3 Types of machine learning algorithm.

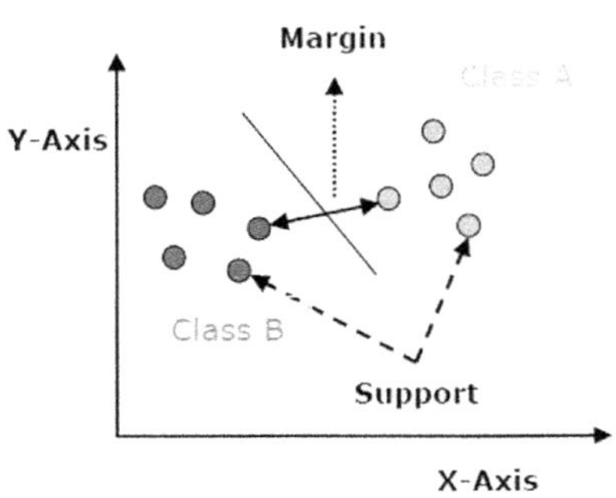

FIGURE 16.4 Concept of support vector machine [12].

3. *Margin*: The margin can be defined as the separation or gap between two lines formed by the closest data points of different classes. It can be calculated as the perpendicular distance from the line to the support vectors. A large margin is generally regarded as desirable, indicating a greater degree of separation between classes, while a small margin is considered less favorable, implying a potential overlap or ambiguity in classification.

SVMs exhibit notable usefulness in handling intricate, nonlinear datasets by virtue of their ability to map the data into a higher-dimensional space, where it can be linearly separated. This characteristic enables SVMs to effectively tackle complex data patterns and enhance classification performance. This ability to handle nonlinear data makes SVMs a popular choice in many different fields, including computer vision, natural language processing, and bioinformatics [13]. Despite their power and flexibility, SVMs can be computationally expensive to train on large datasets and can sometimes suffer from overfitting if the model is too complex.

In the analysis of neuroimaging data, including MRI, researchers have leveraged SVM to study and compare datasets obtained from individuals with Alzheimer's disease and healthy control subjects [14].

By extracting features from the MRI data, such as volume measurements of different brain regions, the SVM algorithm can learn to distinguish between the two groups with high accuracy. In one such study, researchers used SVM to analyze MRI data from 30 patients with Alzheimer's disease and the SVM model was able to accurately classify patients with Alzheimer's disease and healthy controls with 91.6% accuracy [15].

16.2.3.1.2 Neural network algorithms

A neural network, inspired by the structure and functioning of the human brain, represents a form of ML algorithm. Comprising interconnected nodes or neurons organized in layers, neural networks possess the capability to acquire the ability to identify patterns within datasets. They are often used for tasks such as image recognition, natural language processing, and speech recognition, among others [16]. Neural networks can undergo training through diverse approaches, encompassing supervised learning, unsupervised learning, and reinforcement learning techniques. They are capable of handling complex and nonlinear data, making them a powerful tool in many different fields. However, they can also be computationally expensive and require a large amount of data to train effectively.

Resembling the structure of the human brain, a neural network comprises artificial neurons, commonly referred to as nodes, arranged in a layered configuration. These nodes are organized into three layers: the input layer, the hidden layer(s), and the output layer as shown in Fig. 16.5.

Within a neural network, data serve as the informational input for each node. These nodes, in turn, multiply the inputs by random weights, perform calculations, and incorporate a bias. Subsequently, activation functions, which are nonlinear in nature, are employed to determine the firing of specific neurons. Through this iterative process, neural networks are able to extract meaningful patterns and make informed decisions based on the given inputs.

Neural networks have been used in brain sleep analysis to classify sleep stages from EEG signals. EEG signals provide a measure of brain activity and are used to diagnose sleep disorders, such as sleep apnea and narcolepsy. The neural network is able to identify sleep stage transitions, which is important in the diagnosis of sleep disorders. This type of analysis can be used to identify abnormalities in sleep patterns and aid in the diagnosis of sleep disorders, which can improve patient outcomes and inform treatment decisions [18].

16.2.3.1.3 Decision tree algorithms

Decision tree algorithms, widely adopted in supervised ML, present a prominent method for classification and regression tasks. The functioning of decision trees involves segregating data into subsets by applying a series of rules, typically derived from data features. By establishing rules that give rise to a tree-like structure, decision tree algorithms operate with individual nodes representing decisions based on specific features. Through training on labeled data, where each input is associated with the correct classification or regression output, the tree can be honed and subsequently employed to make predictions on new, unlabeled data.

Fig. 16.6 resembles the structure of a decision tree.

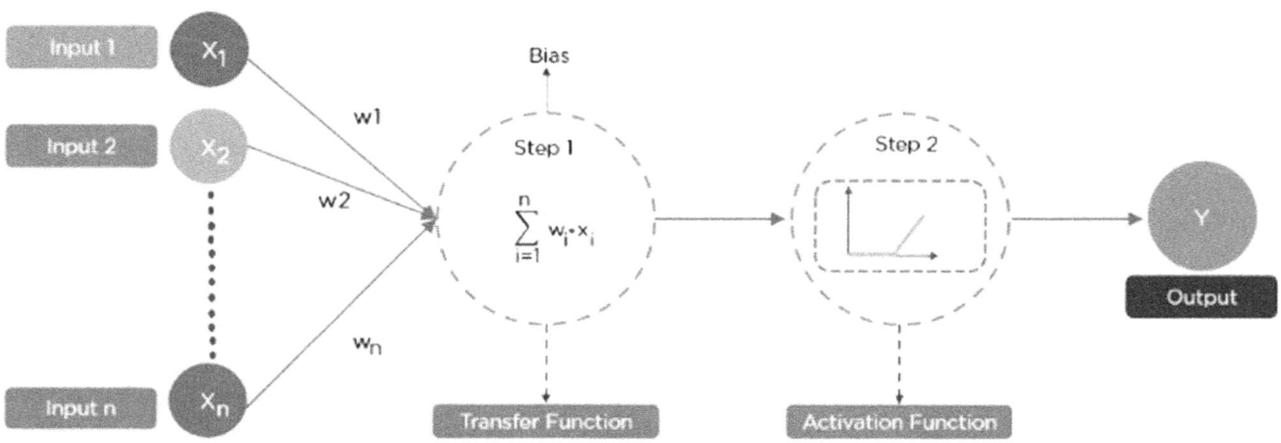

FIGURE 16.5 Concept of neural network algorithms [17].

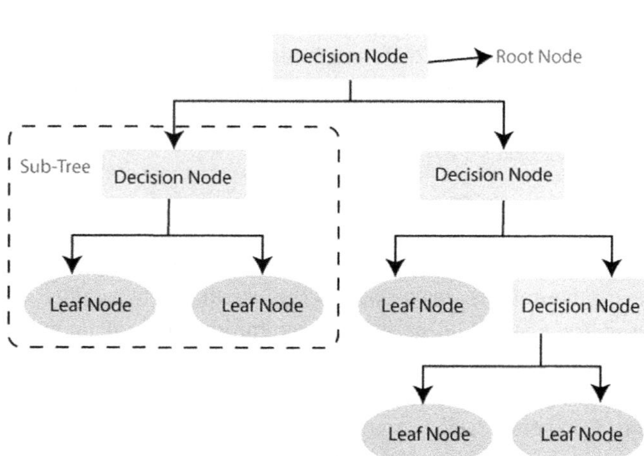

FIGURE 16.6 The concept of decision tree [19].

A decision tree encompasses two key nodes: the decision node and the leaf node. Decision nodes are utilized for making decisions and contain multiple branches, while leaf nodes represent the output of those decisions and do not possess any additional branches.

Decisions or tests within a decision tree are performed based on the features present in the given dataset. This tree serves as a graphical representation, offering a comprehensive overview of all possible solutions to a problem or decision based on the specified conditions. Its nomenclature as a decision tree arises from its resemblance to an actual tree, starting with a root node and subsequently expanding into branches to construct a tree-like structure. To construct a decision tree, the CART algorithm (Classification and Regression Tree) is employed. The algorithm facilitates the creation of a tree by posing questions or queries and, based on the response (Yes/No), further splits the tree into subtrees, progressively refining the decision-making process.

Decision trees have a number of advantages, including their simplicity and interpretability, which make them particularly useful in domains where the results of the analysis need to be explained to nonexperts. They are also capable of handling both numerical and categorical data and can handle missing data and outliers.

16.2.3.2 Semisupervised learning

It is a Machine Learning (ML) algorithm in which the data are divided into labeled and unlabeled type.

Semisupervised learning represents a ML approach that amalgamates labeled and unlabeled data during the model training process. Diverging from conventional supervised learning that relies solely on labeled data, semisupervised learning leverages both labeled and unlabeled data to augment learning outcomes. In the field of BCI, semisupervised learning can be highly valuable. Collecting labeled brain signals for training BCI models is often challenging due to the time and resource-intensive nature of data annotation. However, unlabeled data, such as raw neural activity, are more readily available. By incorporating this unlabeled data, the algorithms can capture underlying patterns and structures, enabling the discovery of hidden relationships and enhancing the generalization capabilities of the BCI system [20].

16.2.3.3 Reinforcement learning

Reinforcement learning is an ML technique that enables an agent to learn by interacting with an environment to maximize a cumulative reward signal. For example, a robot may learn to navigate a maze by exploring its environment and receiving a reward signal for finding the correct path to the end as explained in Ref. [21]. The reinforcement learning algorithm will use this feedback to adjust its actions and learn to make better decisions in the future. Reinforcement learning has numerous applications, such as game playing, robotics, and autonomous driving. It has also been used to develop conversational agents, such as chatbots, that can learn from human interactions to improve their responses over time [22]. This model, however, is seldom used for biomedical applications.

16.2.3.4 Unsupervised learning

It pertains to a category of ML where algorithms are trained on datasets lacking labeled examples. In this approach, the algorithm is not provided with specific instructions but instead discovers patterns and structures within the data independently. Presented below are a few notable examples of unsupervised learning algorithms.

16.2.3.4.1 Clustering algorithms

Clustering algorithms are a type of unsupervised ML algorithm that can be used to group similar data points together based on their characteristics. Unlike supervised learning, where the data are labeled, in unsupervised learning, the data are unlabeled and the goal is to identify patterns or structure within the data. Clustering algorithms work by assigning data points to clusters based on their similarity to other data points in the same cluster, while maximizing the dissimilarity between data points in different clusters.

Various clustering approaches have been developed, each tailored to specific data distributions. Below, we provide a concise overview of three commonly employed approaches [23]:

1. *Centroid-Based Clustering*: Centroid-based clustering involves organizing data into nonhierarchical clusters, distinguishing it from hierarchical clustering, as described later. The widely utilized algorithm for centroid-based clustering is k-means. While these algorithms are efficient, they can be influenced by initial conditions and outliers. In this course, the emphasis is on k-means due to its efficiency, effectiveness, and simplicity as a clustering algorithm.
2. *Density-Based Clustering:* Density-based clustering is a method that establishes connections between regions of significant data density to form clusters. This approach enables the identification of clusters with arbitrary shapes, as

long as densely populated areas can be linked. However, density-based clustering algorithms encounter challenges when dealing with datasets that exhibit varying densities and high dimensions. Additionally, it is important to note that these algorithms, by design, do not assign outliers to any specific clusters.

3. *Distribution-Based Clustering:* Distribution-based clustering operates under the assumption that data are composed of distributions, such as Gaussian distributions. In Fig. 16.7, the algorithm based on distribution clusters the data into three distinct Gaussian distributions. As the distance from the center of a distribution increases, the probability of a point belonging to that distribution decreases, which is visually represented by the diminishing bands. It is important to note that when the type of distribution in the data is unknown, alternative algorithms should be considered for clustering purposes.

Clustering algorithms have shown promise in the field of neuroscience for identifying patterns of brain abnormalities in medical imaging data. For example, researchers have used clustering algorithms to identify groups of patients with similar patterns of brain activity or structure, which can then be used to predict clinical outcomes or aid in the diagnosis of neurological disorders.

One study used clustering algorithms to identify subgroups of patients with schizophrenia based on patterns of brain activity measured using functional magnetic resonance imaging (fMRI). The researchers found that patients in different subgroups had distinct patterns of brain activity that were associated with different symptom profiles and levels of cognitive impairment [24].

16.2.3.4.2 Principle component analysis

Principal component analysis (PCA) is an unsupervised learning algorithm that is widely used in data analysis and pattern recognition. PCA is particularly useful in reducing the dimensionality of high-dimensional datasets, where the number of features or variables is much larger than the number of observations or samples. The main objective of PCA is to identify the most important features or components that capture the most variation in the data. By identifying these key features, PCA can effectively summarize and represent the data in a lower-dimensional space while preserving the most important information. PCA is a powerful tool for exploratory data analysis, data visualization, and feature engineering and is commonly used in applications such as image processing, signal processing, finance, genetics, and many others.

PCA can be a powerful tool in predicting diseases related to the brain, such as Alzheimer's disease as described in Ref. [25]. In this study, the percentage of subjects correctly classified was 95, 70, and 81 for cardiothoracic ratio (CTR), mildAD, and modAD groups, respectively.

16.2.4 Importance of cross-validation in neuroscientific data

Cross-validation serves as a crucial technique in ML for evaluating a model's performance on unseen data and mitigating the risk of overfitting. Overfitting often arises when a model becomes overly complex, exhibiting strong performance on the training data but faltering when faced with the test data. Through cross-validation, this challenge is

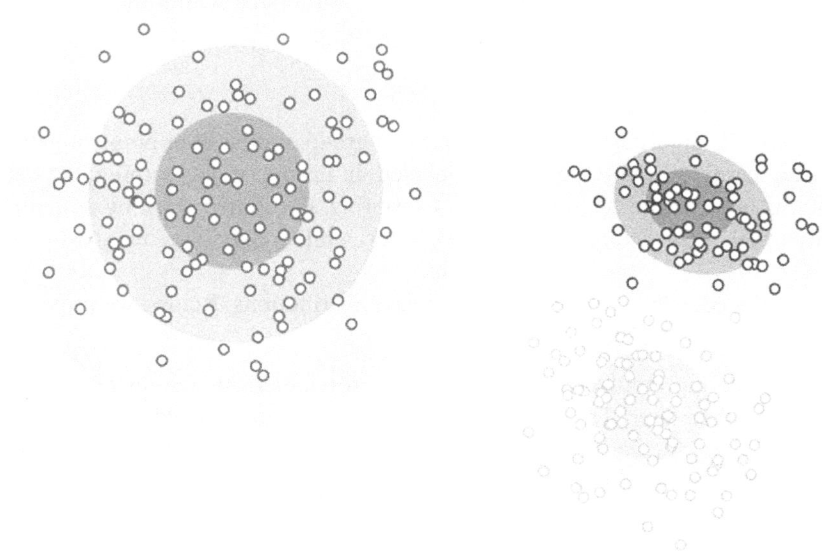

FIGURE 16.7 Distribution-based clustering [23].

tackled by assessing the model's performance across multiple data subsets, thereby providing an estimate of its ability to generalize and perform well on new, unseen data.

There are several types of cross-validation techniques, including:

1. *K-Fold Cross-Validation*: This approach involves partitioning the data into k subsets or folds. During each iteration, the model is trained on k-1 folds, while the remaining fold is utilized for testing. This process is repeated k times, ensuring that each fold serves as the test set once. The final model performance is determined by averaging the performance across all k-folds.
2. *Leave-One-Out Cross-Validation*: In this methodology, every individual observation within the dataset is consecutively held out as a test set, while the model is trained on the remaining data. This iterative process is repeated for each observation, and the final model performance is determined by averaging the performance across all the observations.
3. *Stratified Cross-Validation*: This technique is used when the data are imbalanced and the classes are not represented equally. In stratified cross-validation, the data are divided into folds such that each fold contains an equal proportion of the classes.
4. *Time Series Cross-Validation*: This technique is used when the data are a time series, and the order of the data points is important. In time series cross-validation, the data are split into training and test sets such that the test set contains only future data points.

Cross-validation is a critical technique in analyzing neuroscientific data, particularly in studies that involve brain imaging analysis [26]. By employing these techniques, we can mitigate the risk of the model overfitting to the training data and enhance its ability to generalize effectively to new, unseen data. This can help us identify the brain regions that are consistently activated in response to visual stimuli across different participants and avoid spurious activations due to noise or individual variability.

16.2.5 Emerging trends and potential research areas

Outlined below are several real-life applications in the mental health field that demonstrate the potential and diverse domains where biosignals and smart computing techniques can be effectively deployed.

16.2.5.1 Precision medicine

Precision medicine is an approach to healthcare that considers individual variability in genes, environment, and lifestyle to develop tailored treatments for patients. It aims to provide the right treatment to the right patient at the right time. In precision medicine, the focus is on understanding the unique characteristics of each patient's disease and developing treatments that are specific to those characteristics.

ML has the potential to revolutionize the field of precision medicine in neurophysiological treatments. Neurological disorders, such as epilepsy, Parkinson's disease, and Alzheimer's disease, are complex and multifactorial, with multiple genes and environmental factors contributing to the disease [27]. ML can help to identify patterns and relationships within large and complex datasets, such as genetic and imaging data, to identify patient-specific disease characteristics and predict treatment responses. For example, ML algorithms could be used to analyze brain imaging data to identify patterns of brain activity that are associated with different types of neurological disorders. This information could be used to develop personalized treatment plans for patients with these disorders [28].

16.2.5.2 Study of emotions and physiological modifications

The study of emotions and physiological modifications has been an area of great interest to researchers in fields such as psychology, neuroscience, and physiology. Emotions represent intricate psychological states that frequently entail alterations in physiological responses, including fluctuations in heart rate, respiration, and skin conductance.

One of the most interesting aspects of this research is the discovery that these physiological modifications can be intentionally controlled through techniques such as meditation, deep breathing, and visualization. For example, studies have shown that the practice of meditation can lead to reduced heart rate, lower levels of the stress hormone cortisol, and increased activity in brain regions associated with positive emotions. Similarly, visualization techniques have been shown to elicit changes in physiological responses, such as decreased blood pressure and heart rate [29].

Understanding the relationship between emotions and physiological modifications can have important implications for mental health and well-being. For instance, identifying specific physiological responses that are associated with certain emotions could help in the development of more targeted treatments for anxiety and depression [30].

16.2.5.3 Brain-computer interface

BCI is a technology that enables communication between the brain and a computer. It allows individuals to interact with technology using their thoughts, bypassing the need for traditional physical interfaces such as keyboards or mice. AI algorithms could be used to enhance the capabilities of BCI by improving the accuracy and speed of communication between the brain and computer [31].

AI algorithms could be trained to interpret EEG signals more accurately, allowing individuals to control a computer or prosthetic device more precisely using their thoughts [32]. This could be particularly useful for individuals with motor disabilities, allowing them to regain some of their lost functionality. Furthermore, AI could be used to develop predictive models for BCI, allowing the system to anticipate the user's intentions and respond accordingly. For example, if an individual is typing on a virtual keyboard using a BCI, AI could predict the next letter they intend to type based on their previous inputs, allowing for faster and more efficient communication.

16.2.6 Challenges of intelligent computing in BCI and neuroscientific research

Intelligent computing in the realm of BCIs and neuroscientific research introduces specific challenges that are distinctive to this field. These challenges encompass:

1. *Data Interpretation*: A key hurdle lies in accurately interpreting and comprehending the intricate signals generated by the brain. BCIs generate substantial amounts of raw data, including neural signals and physiological parameters, necessitating advanced algorithms and models for effective analysis and interpretation.
2. *Signal Processing*: BCIs often come across as weak and noisy signals, making it difficult to extract meaningful information. Intelligent computing techniques must overcome signal processing challenges like noise reduction, artifact elimination, and feature extraction to enhance the accuracy and dependability of BCI systems.
3. *Ethical and Privacy Concerns*: The utilization of intelligent computing in BCIs gives rise to ethical considerations concerning privacy, consent, and potential misuse of brain data. Safeguarding the confidentiality and security of neural information, while ensuring informed consent from users, is paramount.
4. *Calibration and Adaptation*: BCIs often necessitate calibration for each user, which can be time-consuming and burdensome. Intelligent computing approaches need to streamline the calibration process and develop adaptive systems that can learn and adapt to changes in the user's brain signals over time.
5. *Validation and Generalization*: Validating and generalizing intelligent computing models and algorithms across diverse populations, experimental conditions, and real-world scenarios pose challenges. Ensuring the robustness and reliability of BCI systems beyond controlled laboratory settings is vital for their widespread adoption.

16.3 Conclusion

The chapter presented the benefits of using computational intelligence, specifically ML, in mental health disciplines like BMI and neuroscientific research. The biosignals commonly employed in the fields of psychology and mental health were elucidated, along with their corresponding psychological associations. The distinct stages of ML, encompassing preprocessing, training, validation, and testing, were delineated, emphasizing the significance of meticulously executing each step to facilitate the model's ability to generate reliable forecasts. The article also outlined various learning algorithms such as supervised, unsupervised, semisupervised, and reinforcement learning. Special focus was given to the validation and test phases, underscoring their significance in fostering dependable generalization and mitigating the risk of overfitting. While acknowledging certain limitations, the utilization of ML in mental health domains and the interpretation of biosignals hold the potential to yield more reproducible and widely applicable outcomes in the realm of neuropsychological research.

References

[1] Z. Li, K. Hwang, K. Li, J. Wu, T. Ji, Graph-generative neural network for EEG-based epileptic seizure detection via discovery of dynamic brain functional connectivity, Sci Rep. 12 (1) (2022) 18998. Available from: https://doi.org/10.1038/s41598-022-23656-1. PMID: 36348082.

[2] S.K. Mudgal, S.K. Sharma, J. Chaturvedi, A. Sharma, Brain computer interface advancement in neurosciences: applications and issues, Interdiscip. Neurosurg. 20 (2020) 100694. ISSN 2214-7519.

[3] A.N. Belkacem, N. Jamil, J.A. Palmer, S. Ouhbi, C. Chen, Brain computer interfaces for improving the quality of life of older adults and elderly patients, Front Neurosci. 14 (2020) 692. Available from: https://doi.org/10.3389/fnins.2020.00692. PMID: 32694979.

[4] Real Time EEG Based Automatic Brainwave Regulation by Music – Scientific Figure on ResearchGate. Available from: https://www.researchgate.net/figure/Brain-waves-pattern-on-a-normul-adult_fig1_305656782.

[5] C.Y. Cheung, A.R. Ran, S. Wang, V.T.T. Chan, Kaiser, A deep learning model for detection of Alzheimer's disease based on retinal photographs: a retrospective, multicentre case-control study, Lancet Digit Health 4 (2022) e806−15. Available from: https://doi.org/10.1016/S2589-7500(22)00169-8. Published OnlineSeptember 30, 2022.

[6] D. Eke, I.E.J. Aasebø, S. Akintoye, W. Knight, A. Karakasidis, E. Mikulan, et al., Pseudonymisation of neuroimages and data protection: increasing access to data while retaining scientific utility, Neuroimage: Reports 1 (4) (2021) 100053. ISSN 2666-9560.

[7] U. Herwig, P. Satrapi, C. Schönfeldt-Lecuona, Using the international 10−20 EEG system for positioning of transcranial magnetic stimulation, Brain Topogr. 16 (2) (2003) 95−99. Available from: https://doi.org/10.1023/b:brat.0000006333.93597.9d. PMID: 14977202.

[8] A.K. Shah, S. Mittal, Invasive electroencephalography monitoring: indications and presurgical planning, Ann Indian Acad Neurol. 17 (Suppl 1) (2014) S89−94. Available from: https://doi.org/10.4103/0972-2327.128668. PMID: 24791095.

[9] J. Metsomaa, J. Sarvas, R.J. Ilmoniemi, Blind source separation of event-related EEG/MEG, IEEE Trans. Biomed. Eng. 64 (9) (2017) 2054−2064. Available from: https://doi.org/10.1109/TBME.2016.2616389. Epub 2016 Oct 12. PMID: 28113203.

[10] M. Rashida, M.A. Habib, Quantitative EEG features and machine learning classifiers for eye-blink artifact detection: a comparative study, Neurosci. Inform. 3 (1) (2023) 100115. Available from: https://doi.org/10.1016/j.neuri.2022.100115. ISSN 2772-5286.

[11] S.M. Usman, M. Usman, S. Fong, Epileptic seizures prediction using machine learning methods, Comput. Math Methods Med. (2017) 9074759. Available from: https://doi.org/10.1155/2017/9074759. Epub 2017 Dec 19. PMID: 29410700.

[12] Tutorials Point, "Support Vector Machine". Available from: https://www.tutorialspoint.com/machine_learning_with_python/classification_algorithms_support_vector_machine.htm.

[13] E. Byvatov, G. Schneider, Support vector machine applications in bioinformatics, Appl Bioinformatics. 2 (2) (2003) 67−77. PMID: 15130823.

[14] Y. Vichianin, A. Khummongkol, P. Chiewvit, A. Raksthaput, S. Chaichanettee, N. Aoonkaew, et al., Accuracy of support-vector machines for diagnosis of Alzheimer's disease, using volume of brain obtained by structural MRI at Siriraj Hospital, Front. Neurol. 12 (2021) 640696. Available from: https://doi.org/10.3389/fneur.2021.640696.

[15] A. Sharma, S. Kaur, N. Memon, A. Jainul Fathima, S. Ray, M.W. Bhatt, Alzheimer's patients detection using support vector machine (SVM) with quantitative analysis, Neurosci. Inform. 1 (3) (2021) 100012ISSN 2772-5286. Available from: https://doi.org/10.1016/j.neuri.2021.100012.

[16] S. Shah, H. Dudhrejia, Speech recognition using neural networks, Int. J. Eng. Res. Technol. 07 (10) (2018).

[17] Biswal Avijeet, "Top 10 Deep Learning Algorithms You Should Know in 2024", Feb 15 2024. Available from: https://www.simplilearn.com/tutorials/deep-learning-tutorial/deep-learning-algorithm.

[18] J.B. Stephansen, A.N. Olesen, M. Olsen, et al., Neural network analysis of sleep stages enables efficient diagnosis of narcolepsy, Nat. Commun. 9 (2018) 5229. Available from: https://doi.org/10.1038/s41467-018-07229-3.

[19] Decision Tree Classification Algorithm. Available from: https://www.javatpoint.com/machine-learning-decision-tree-classification-algorithm.

[20] Y. Wei, J. Li, H. Ji, L. Jin, L. Liu, Z. Bai, et al., A semi-supervised progressive learning algorithm for brain-computer interface, IEEE Trans. Neural. Syst. Rehabil. Eng. 30 (2022) 2067−2076. Available from: https://doi.org/10.1109/TNSRE.2022.3192448. Epub2022. PMID: 35853068.

[21] D. Venkata Vara Prasad, Knowledge based reinforcement learning robot in maze environment, Int. J. Comput. Appl. 14 (7) (2011).

[22] E. Ricciardelli, D. Biswas, Self-improving chatbots based on reinforcement learning, Conference: 4th Multidisciplinary Conference on Reinforcement Learning and Decision Making (RLDM2019), May 2019.

[23] D. Xu, Y. Tian, A comprehensive survey of clustering algorithms, Ann. Data. Sci. 2 (2015) 165−193. Available from: https://doi.org/10.1007/s40745-015-0040-1.

[24] M.L. Seghier, Clustering of fMRI data: the elusive optimal number of clusters, PeerJ. 6 (2018) e5416. Available from: https://doi.org/10.7717/peerj.5416. PMID: 30310731.

[25] M. Pagani, D. Salmaso, G. Rodriguez, D. Nardo, F. Nobili, Principal component analysis in mild and moderate Alzheimer''s disease − a novel approach to clinical diagnosis, Psychiatry Res. 173 (1) (2009) 8−14. Available from: https://doi.org/10.1016/j.pscychresns.2008.07.016. Epub 2009 May 14. PMID: 19443186.

[26] N. Kriegeskorte, Crossvalidation in brain imaging analysis, bioRxiv017418; https://doi.org/10.1101/017418.

[27] S.J. MacEachern, N.D. Forkert, Machine learning for precision medicine, Genome 64 (4) (2021) 416−425. Available from: https://doi.org/10.1139/gen-2020-0131.

[28] S. Quazi, Artificial intelligence and machine learning in precision and genomic medicine, Med. Oncol. 39 (2022) 120. Available from: https://doi.org/10.1007/s12032-022-01711-1.

[29] R. Ward, P. Marsden, Physiological responses to different WEB page designs, Int. J. Human Comput. Studies 59 (1-2) (2003) 199−212.

[30] S.M.A. Shah, D. Mohammad, M.F.H. Qureshi, M.Z. Abbas, S. Aleem, Prevalence, psychological responses and associated correlates of depression, anxiety and stress in a global population, during the coronavirus disease (COVID-19) pandemic, Community Ment. Health J. 57 (1) (2021) 101−110. Available from: https://doi.org/10.1007/s10597-020-00728-y. Epub 2020 Oct 27. PMID: 33108569.

[31] X. Zhang, Z. Ma, H. Zheng, T. Li, K. Chen, X. Wang, et al., The combination of brain-computer interfaces and artificial intelligence: applications and challenges, Ann. Transl. Med. 8 (11) (2020) 712. Available from: https://doi.org/10.21037/atm.2019.11.109. PMID: 32617332.

[32] S. Nayak, R.K. Das, Application of artificial intelligence (AI) in prosthetic and orthotic rehabilitation, Service Robotics (2020). Available from: https://doi.org/10.5772/intechopen.93903.

Chapter 17

Automated detection of diabetic retinopathy from gray-scale fundus images using GLCM and GLRLM-based textural features—a comparative study

Subhanshu Gupta and Dilip Singh Sisodia

Department of Computer Science and Engineering, National Institute of Technology Raipur, Raipur, Chhattisgarh, India

17.1 Introduction

The prevalence of diabetic eye disease is rapidly rising and is projected to become a global pandemic in the coming decades. The International Diabetes Federation reports that diabetes will impact around 537 million individuals aged 20–79 in 2021, which is equivalent to 1 in every 10 people [1]. This number is expected to rise to 643 million by 2030 and 784 million by 2045 [1]. Among the complications of diabetes, diabetic retinopathy (DR) is the most common and often goes unnoticed until it leads to sudden blindness. DR is the primary cause of preventable blindness or vision loss in developing nations and represents a significant disabling complication of diabetes mellitus. Early detection of DR enables laser therapy [2] to be performed, which prevents patients from DR. The methods which are used in today's time period are very costly, nonefficient, and time-consuming processes. There are five levels (0, 1, 2, 3, 4), 0 is the minimum and 4 is the most severe stage in DR. These levels help the ophthalmologist (eye specialist) decide which level the diabetic fundus exists. The detection of DR became much easier with the help of an automated tool that uses texture features for image analysis [3]. The dataset used from Kaggle contains images of the colored fundus. This paper focuses on the extraction of different feature from colored fundus images using gray level cooccurrence matrix (GLCM) [4] and gray level run-length matrix (GLRLM) [5] techniques. These technique takes gray-scale images as the input and create a matrix as the output. Therefore the conversion of the images needs to be done using the image library used in Python. DR [6] has features like mean, dissimilarity, correlation, energy, entropy, skewness, homogeneity, dissimilarity, short-run emphasis (SRE), long-run emphasis (LRE), run entropy (RE), and run percentage (RP). These features are applied in machine learning models as input which is of different kinds like decision tree, AdaBoost, random forest, SVM, naïve Bayes, XGBoost, and decision tree. The stacking method increases the accuracy using multiple classifiers and regression models. The model we have used on the stacking method is random forest, which is a machine learning classifier.

These classification models use all of the features extracted from the fundus images to find out if the person has DR or not [7]. Energy, mean, and entropy are the main features from our extracted features list, which helps to detect diabetic eyes using textures. Energy shows if the image is uniform or not, where entropy is used to find the randomness in the digital image, and mean reveals the general brightness of the image.

17.2 Literature review

For DR detection, many researchers have worked on various techniques, such as detecting blood vessels and various lesions. These researchers used two ways of automating blood vessel segmentation [8,9] and of identifying the lesion [10,11]. Saleck et al. proposed a method for the detection of the tumor using GLCM texture [12]. Radhakrishnan and

Intelligent Computing Techniques in Biomedical Imaging. DOI: https://doi.org/10.1016/B978-0-443-15999-2.00011-6

Kuttiannan have proposed a method for extracting features using GLCM and GLRLM [13]. Nayak et al. proposed a method to extract the biological features using image preprocessing, morphological processing techniques, and texture analysis methods applied to the fundus images [2]. Agurto et al. proposed using amplitude-modulation-frequency-modulation methods to find the difference between normal and diabetic images [14]. Researchers proposed a method to find out DR using textural features with the help of the local binary pattern [15]. Ashraf et al. focused on the textual analysis of micro-patterns of the region of interest (ROI), which is a dangerous region in the fundus image for detecting microaneurysms and hemorrhages [6]. Akara et al. proposed an exudates detection method based on mathematical morphology on retinal images of nondilated pupils for low-quality images.

In this study, a comparative approach is used to detect DR using various machine learning algorithms. The images were first converted to gray-scale and then subjected to the GLCM and GLRLM techniques, which generated matrices containing extracted features. Various models were employed to analyze and identify DR patients.

17.3 Methodology

This section provides a concise overview of the proposed work's workflow, accompanied by a visual representation in Fig. 17.1. Initially, the data samples were subjected to processing using the Python library. Subsequently, the processed images were fed into two primary functions, namely GLCM and GLRLM, to extract features. These extracted features were then inputted into two distinct learning models, namely individual and ensemble models, to generate an output that determines the presence or absence of DR in a person. The subsequent sections provide a brief description of each step incorporated into the framework.

17.3.1 Dataset

The Kaggle platform serves as a widely utilized data repository, hosting various datasets, including one commonly employed for studying DR disease. Dataset [16], sourced from Kaggle, comprises 17,000 colored fundus images. However, for the purposes of this study, a subset of 8000 images was selected. It is important to note that all images in the dataset represent different ROI.

17.3.2 Data preprocessing

Data preprocessing involves the balancing of noisy data and the application of various filters tailored to the research objectives. In this particular study, the dataset was organized into large and small ROI. Additionally, the "RGB to gray-scale" function was utilized to convert the colored images into gray-scale format.

After completing the preprocessing step, the extraction of highly relevant features becomes the key element in the framework. This study specifically focuses on computing textural features using GLCM and GLRLM. These approaches enable the derivation of features from digital images, and each approach is elaborated below.

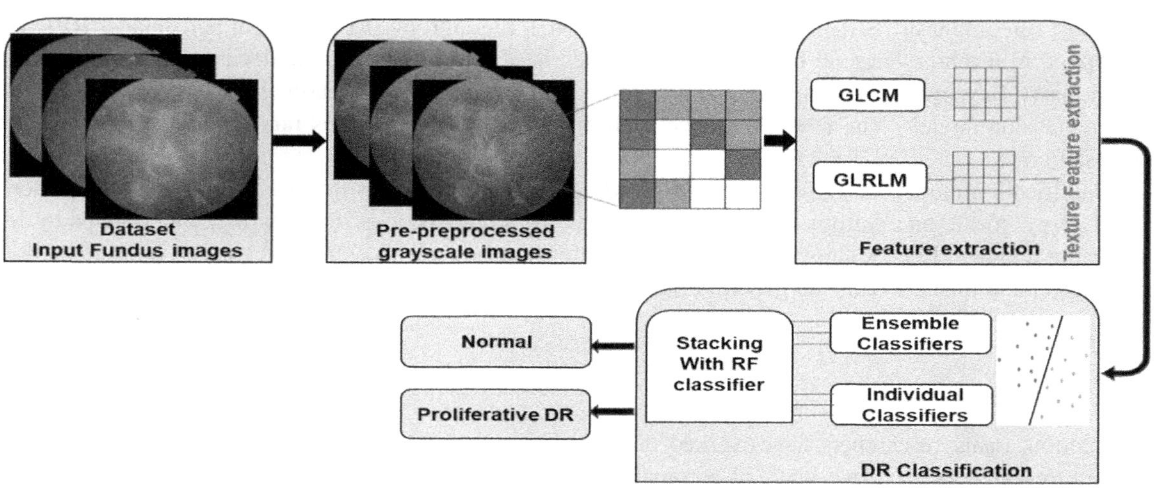

FIGURE 17.1 Workflow diagram of the proposed automated detection of diabetic retinopathy.

17.3.3 Gray level cooccurrence matrix

The utilization of the GLCM in analyzing textural features plays a vital role in detecting DR disorder [17]. These features provide valuable insights into the arrangement and connections among gray-level intensities within an image. Examining cooccurrence patterns in gray-scale images, we have calculated various textural features, including mean, dissimilarity, contrast, homogeneity, variance, correlation, energy, entropy, and skewness [13]. Each feature holds unique significance, highlighting different aspects within a specific image. These GLCM-based textural features significantly contribute to distinguishing between normal and pathological retinal tissues.

The GLCM processes the pixels of the digital images using the function $p(i,j|d,\Theta)$ to create the matrix, which helps to extract the features from digital images. It works for different values of d and θ, where d denotes the distance and θ denotes the angle for d. In θ, we choose different angles, which are 0, 45, 90, and 135 degrees [13]. The feature extraction consists of two different parts depending on the features to be extracted. Table 17.1 provides a comprehensive list of mathematical functions utilized to calculate specific features. It encompasses the functions the features employ and their corresponding descriptions within the GLCM context.

17.3.4 Gray level run-length matrix

This work examines additional textural features in gray-scale images using the GLRLM technique. GLRLM analyzes consecutive pixels with the same gray-level intensity along a given direction, capturing important textural characteristics of retinal images. By identifying pixels with the same gray level intensity and examining their neighboring pixels [18], GLRLM-based features are able to extract textural information from digitalized images effectively. The study focuses on several GLRLM-based features, including SRE, LRE, run length nonuniformity (RLN), RP, and RE. Table 17.2, in this

TABLE 17.1 List of all textural-based features extracted using the GLCM approach, along with their respective mathematical formulas and concise explanations.

Features	Formula	Description
Skewness	$\sum_{i=1}^{Ng} (i-\mu)^3 p(i)$	Measure how unbalances the distribution of the gray level in the image.
Variance	$\sum_{i=1}^{Ng}\sum_{j=1}^{Ng} (i\mu)^2 p(i,j)$	Measure the heterogeneity.
Entropy	$\sum_{i=1}^{Ng}\sum_{j=1}^{Ng} p(i,j)\log[p(i,j)]$	The measure of randomness which is used to characterize the texture of an image.
Mean	$\sum_{i=1}^{Ng} i * p(i)$	Reveals the general brightness of the image.
Correlation	$\frac{\sum_{i=1}^{Ng}\sum_{j=1}^{Ng} (i*j)*p(i,j) - \mu_x\mu_y}{\sigma_x\sigma_y}$	The measure of gray tone linear dependencies in the image.
Homogeneity	$\sum_{i=1}^{Ng}\sum_{j=1}^{Ng} \frac{1}{1+(i-j)^2} p(i,j)$	Measure image homogeneity as it assumes larger values for smaller gray tones.
Energy	$\sum_{i=1}^{Ng}\sum_{j=1}^{Ng} (p(i,j))^2$	The measure of the extent of pixel pair repetitions.
Dissimilarity	$\sum_{i,j=0}^{Ng} p(i,j) * \lvert i-j \rvert$	Have higher values when the local region has high contrast.
Contrast	$\sum_{n=0}^{Ng-1} (i-j)^2 \left\{ \sum_{i=1}^{Ng}\sum_{j=1}^{Ng} p(i,j) \right\}$	Measure the intensity of the pixels. It determines the difference in color and brightness of images.

Note: Ng is the number of distinct gray levels in the images.

TABLE 17.2 List of all textural-based features extracted using the GLRLM approach, along with their respective mathematical formulas and concise explanations.

Features	Formula	Description
SRE	$\frac{1}{n}\sum_{i,j}\frac{p_\theta(i,j)}{j^2}$	It is the distribution of the short or the long homogenous in the image.
LRE	$\frac{1}{n}\sum_{i,j} j^2 * p_\theta(i,j)$	It is the distribution of the short or the long homogenous in the image.
RLN	$\frac{1}{n}\sum_{j}\left(\sum_{i} p_\theta(i,j)\right)^2$	Measure the nonuniformity of the gray level in the images.
RP	$\sum_{i,j}\frac{p_\theta(i,j)}{p_\theta(i,j)*j}$	Measure coarseness of the texture.
RE	$\sum_{i,j} p_\theta\left(i,j\right)\log_2(p_\theta(i,j))$	Measure coarseness of the texture.

study, presents the function, mathematical formulations, and description utilized for extracting the GLRLM. The function employed for determining pixel intensity and creating the matrix in GLRLM [13] is referenced in Eq. (17.1).

$$(p(i,j)|\Theta) \tag{17.1}$$

17.3.5 Classifiers

Following the feature extraction stage, the next step is classification, which involves predicting the class of given data points. Classification is a part of supervised learning, where the input data are accompanied by corresponding target labels. In order to classify distinct features in normal and diabetic fundus images, the extracted features are inputted into two different categories of classifiers: ensemble and traditional machine learning. These classifiers include SVM, naïve Bayes, decision tree, random forest, XGBoost, and AdaBoost, which are utilized for training purposes. The performance of the classifiers is evaluated using various assessment metrics, such as accuracy, precision, recall, and F-score. Below is a brief description of each classifier that was studied.

17.3.5.1 Ensemble learners

Ensemble learning, a machine learning technique that combines multiple models to create an optimized model [19], has proven to be effective in various applications [20]. These models are employed to enhance prediction accuracy and reduce variance. Drawing inspiration from previous studies, our research focuses on exploring models like XGBoost, AdaBoost, and random forest for the detection of DR. We assess their performance using various evaluation metrics.

17.3.5.1.1 Gradient boosting

XGBoost is a powerful gradient-boosting library renowned for its efficiency, versatility, and portability. It operates within a machine learning framework and places a strong emphasis on computational speed and model performance. The library incorporates various features, including model features, system features, and algorithm features. XGBoost utilizes a gradient-boosting algorithm, an ensemble technique that combines new models in a way that corrects errors and yields satisfactory results [21].

17.3.5.1.2 Adaptive boosting

AdaBoost is an ensemble classifier that harnesses the power of multiple classifiers by aggregating their outputs as input. By combining several weak classifiers, it constructs strong classifiers that achieve higher accuracy scores when compared to using a single weak classifier alone [22].

17.3.5.1.3 Random forest

Random forest is a supervised learning algorithm that utilizes an ensemble of decision trees. It is predominantly trained using the bagging method, which combines multiple models to improve accuracy. The random forest can be employed for both classification and regression tasks. It shares similar parameters with a decision tree. Unlike the decision tree,

TABLE 17.3 The hyperparameters of the classifiers utilized in this experiment.

Ensemble learning	Parameters
XGBoost	Booster (default = gbtree), max_depth (default = 6), gamma (default = 0), max_delta_step (default = 0)
AdaBoost	n_estimators (default = 50), learning (default = 1)
Random forest	n_estimators (default = 10), max_depth (default = none)

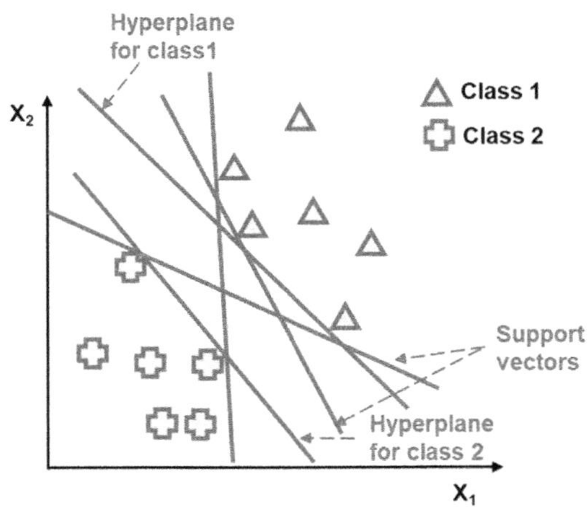

FIGURE 17.2 Illustration of the working of support vector machine.

however, random forest focuses on searching for the best features rather than identifying important features. For a more comprehensive understanding, readers can refer to the detailed description provided in Ref. [23].

The performance of the aforementioned classifiers is fine-tuned through extensive experimentation by adjusting the hyperparameters of each classifier. The optimal set of hyperparameters utilized in this study is listed in Table 17.3.

17.3.5.2 Individual learners

Ensemble learning classifiers generally exhibit superior predictive accuracy and generalization compared to individual classifiers. However, individual traditional classifiers hold importance in various aspects, such as interpretability, simplicity, robustness, utilization of prior knowledge, and serving as baseline models for performance evaluation. Consequently, the performance of the proposed approach is compared not only to ensemble learning classifiers but also to traditional individual classifiers. This study explores SVM, naïve Bayes, and decision trees as individual classifiers, with their brief descriptions provided in the subsequent sections.

17.3.5.2.1 Support vector machine

Support vector machine, commonly referred to as SVM, is a supervised learning algorithm utilized for both classification and regression tasks. SVM possesses a unique characteristic of simultaneously minimizing classification errors while maximizing the geometric margin [24]. Fig. 17.2 visually represents the fundamental concept of a SVM classifier. SVM is based on the principle of Structural Risk Minimization, aiming to strike a balance between accurate classification and optimal margin. The provided images offer an illustration of SVM, showcasing the generation of hyperplanes between two distinct classes [24].

17.3.5.2.2 Naïve Bayes

Naïve Bayes is a simple yet powerful supervised learning algorithm used for classification tasks. It is based on Bayes' theorem and assumes that the features are conditionally independent of each other given the class label. The algorithm calculates the probability of each class given a set of input features and selects the class with the highest probability as

TABLE 17.4 Individual classifiers and parameters.

Individual classifier	Parameter
Support vector machine	Kernel = (default = "rbf"), Degree (default = 3), gamma (default = "auto")
Naïve Bayes	Array-like, shape (n_classes), var_smoothing (default = 1e-9)
Decision tree	Splitter (default = best), max_depth (default = none)

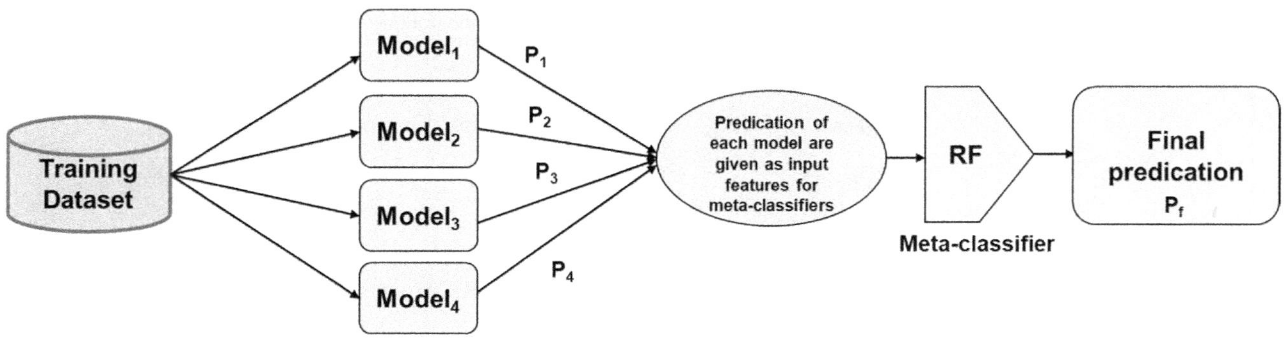

FIGURE 17.3 An illustration of the stacking approach in ensemble learning.

the predicted class. Naïve Bayes is computationally efficient and can handle high-dimensional data. It is particularly effective when there is a large amount of data available for training. However, the assumption of feature independence can limit its accuracy when dealing with correlated features [25].

17.3.5.2.3 Decision tree

The decision tree is a widely used supervised learning algorithm capable of handling classification and regression tasks. It constructs a tree-like model representing a sequence of decisions and their potential outcomes [26]. By recursively splitting the data based on input feature values, it creates a hierarchy of decision rules. Decision trees offer interpretability as the resulting tree structure can be easily comprehended and visualized. They are versatile in handling both categorical and numerical data and exhibit resilience to missing values. However, overfitting can occur if the tree becomes overly complex. To mitigate this, techniques such as pruning and setting depth limits can be employed. Overall, decision trees serve as a powerful tool for data analysis and decision-making in diverse domains. Extensive experimentation was conducted to fine-tune the performance of the individual classifiers mentioned above. This involved adjusting the hyperparameters of each classifier to optimize their performance. The optimal set of hyperparameters used in this study can be found in Table 17.4.

17.3.5.3 Stacking approach

Another method studied in this work is stacking, which is an ensemble technique that enhances prediction accuracy by combining multiple models. It entails training several base models on the same dataset and using a metamodel to generate predictions based on the base models' outputs. By leveraging the strengths of diverse models, stacking can capture intricate relationships between features and target variables, making it suitable for various data types and problem domains. In this study, stacking is employed by utilizing the aforementioned classifiers, with random forest serving as the metaclassifier responsible for generating the final predictions shown in Fig. 17.3.

17.3.6 Evaluation measures

The abovementioned classifiers are evaluated using well-known performance assessment metrics such as accuracy, precision, recall, and F-score computed from the confusion matrix. In the confusion matrix, the entries such as *TP*, *TN*, *FP*, and *FN* provide valuable information. The true-positive (*TP*) rate represents the proportion of correctly predicted

TABLE 17.5 Mathematical equations of each performance assessment matrix utilized in the study.

Measures	Formula	Description
Accuracy	$\frac{TP+TN}{TP+TN+FP+FN}$	Accuracy shows the detail you need to better understand the performance of your classification model.
Precision	$\frac{TP}{TP+FP}$	It is defined as the ratio of correctly predicted DR cases to the total number of healthy and diseased patients.
Recall	$\frac{TP}{P}$	The recall measures the proportion of correctly predicted disease presence cases out of the total number of cases of DR.
F-Score	$2*\frac{P*R}{P+R}$	It is defined as the number of true positives over the number of true positives plus the number of false positives.

TABLE 17.6 Obtained classification performance from ensemble learning classifiers.

Ensemble classifier	Accuracies	Precision	Recall	F-1 Score
XGBoost	74.2	0.6874	0.6532	0.7285
AdaBoost	68.1	0.6542	0.6014	0.6647
Random forest	**83.3**	**0.7041**	**0.6412**	**0.7945**

diseased cases. The false-positive (*FP*) rate indicates the ratio of mispredicted healthy cases. The true-negative (*TN*) rate represents the total number of correctly predicted healthy cases. On the other hand, the false-negative (*FN*) rate indicates the total number of DR cases that were misclassified as healthy. Table 17.5 displays the mathematical formulations and brief descriptions of all these measures.

17.4 Experiment results

The following sections showcase the outcomes of the proposed work, as depicted in Fig. 17.1. The study involved the extraction of textural features using GLCM and GLRLM from gray-scale images of normal and DR patients. These features were then utilized for classification using three different types of machine learning classifiers: ensemble learning, individual classifiers, and stacking approaches. The performance results of these classifiers are presented in the subsequent sections for comprehensive analysis.

17.4.1 Ensemble classifier results

In this chapter, the performance of ensemble learners, namely XGBoost, AdaBoost, and random forest, was evaluated. The classifiers were assessed based on accuracy, precision, recall, and F-score, and the results are presented in Table 17.6. Upon analyzing the table, it can be concluded that the random forest classifier outperformed the other classifiers, achieving the highest classification accuracy, precision, recall, and F-score of 83.3%, 0.7041, 0.6412, and 0.7945, respectively. This finding is highlighted in Table 17.6 using bold font. On the other hand, the AdaBoost classifier exhibited the lowest performance among all the classifiers utilized in this experiment.

17.4.2 Individual classifier results

In the subsequent experiment, we also assessed the effectiveness of traditional individual machine learning classifiers. Table 17.7 presents the performance assessment metrics of the achieved classifier. By comparing Tables 17.6 and 17.7, it becomes evident that ensemble learning classifiers outperformed individual classifiers. Specifically, when comparing

TABLE 17.7 Obtained classification performance from traditional individual classifiers.

Individual classifier	Accuracies	Precision	Recall	F-1 Score
Support vector machine	71.8	0.7024	0.7198	0.7125
Naïve Bayes	73.09	0.6987	0.7045	0.7015
Decision tree	79.6	0.7154	0.7514	0.7324

TABLE 17.8 Obtained classification performance of stacking ensemble learning classifier.

Classifier used	Accuracy	Precision	Recall	F-1 score
Random forest	93.6	0.8041	0.8547	0.8286

the individual classifiers solely, the decision tree exhibited superior performance, surpassing naïve Bayes and SVM with the highest classification accuracy, precision, recall, and F-score of 79.6%, 0.7154, 0.7514, and 0.7324, respectively. This significant finding is highlighted in Table 17.7, while naïve Bayes stands as the second-highest classifier.

17.4.3 Stacking method result

To further enhance the prediction of DR disease, we implemented a stacked ensemble classifier in this study. The stacked ensemble consists of other classifiers acting as base classifiers, with the random forest serving as the metaclassifier. The results of this approach are presented in Table 17.8, which displays the classification accuracy, precision, recall, and F-score as 93.6%, 0.8041, 0.8547, and 0.8286, respectively. By examining Tables 17.6–17.8, we can conclude that the stacking approach proves to be more effective in automatically classifying DR from gray-scale fundus images based on textural features.

17.5 Conclusion and future work

This chapter aimed to discuss automated classification of DR from fundus images using various classifiers, such as SVM, decision tree, naïve Bayes, AdaBoost, XGBoost, and random forest, while employing textural feature extraction. Additionally, different preprocessing techniques, including gray-scale conversion, were explored. The evaluation of the proposed approach utilized a Kaggle dataset, where RGB images were converted to gray-scale, and two textural feature extraction methods (GLCM and GLRLM) were applied. These extracted features were utilized to differentiate between normal and diabetic fundus images. The classifiers' performance was assessed regarding accuracy, precision, recall, and F-score. Among the individual classifiers, the decision tree achieved the highest accuracy of 79.6%, while the random forest achieved the highest accuracy of 83.3% among ensemble learning classifiers. Furthermore, the stacking approach was employed to evaluate the performance of multiple classifiers, with the random forest achieving the highest accuracy of 93.6%. These results demonstrate that the stacking method significantly improved the accuracy of the random forest approach.

In the future, the aim is to extract biological features, such as exudates, red lesion detection, and optic disk, from the dataset of diabetic eye images. Additionally, we plan to incorporate different models, such as fuzzy C-means, in conjunction with machine learning classifiers.

References

[1] H. Sun, et al., IDF diabetes atlas: global, regional and country-level diabetes prevalence estimates for 2021 and projections for 2045, Diabetes Res. Clin. Pract. 183 (2022) 109119.

[2] J. Nayak, P.S. Bhat, R. Acharya U, C.M. Lim, M. Kagathi, Automated identification of diabetic retinopathy stages using digital fundus images, J. Med. Syst. 32 (2) (2008) 107–115. Available from: https://doi.org/10.1007/s10916-007-9113-9.

[3] D.S. Sisodia, R.B. Pachori, L. Garg, Handbook of Research on Advancements of Artificial Intelligence in Healthcare Engineering, IGI Global, 2020.

[4] R.M. Haralick, I. Dinstein, K. Shanmugam, Textural features for image classification, IEEE Trans. Syst. Man. Cybern., SMC-3 6 (1973). Available from: https://doi.org/10.1109/TSMC.1973.4309314.

[5] X. Tang, Texture information in run-length matrices, IEEE Trans. Image Process. 7 (11) (1998) 1602–1609. Available from: https://doi.org/10.1109/83.725367.

[6] M.N. Ashraf, Z. Habib, M. Hussain, Texture feature analysis of digital fundus images for early detection of diabetic retinopathy, in: Proc. 11th International Conference on Computer Graphics, Imaging and Visualization, CGiV 2014, 2014, pp. 57–62. doi: 10.1109/CGiV.2014.29.

[7] D.S. Sisodia, S. Nair, P. Khobragade, Diabetic retinal fundus images: preprocessing and feature extraction for early detection of diabetic retinopathy, Biomed. Pharmacol. J. 10 (2) (2017) 615–626. Available from: https://doi.org/10.13005/bpj/1148.

[8] P. Liskowski, K. Krawiec, Segmenting retinal blood vessels with deep neural networks, IEEE Trans. Med. Imaging 35 (11) (2016) 2369–2380. Available from: https://doi.org/10.1109/TMI.2016.2546227.

[9] R. Kolar, et al., Retinal vessel segmentation by improved matched filtering: evaluation on a new high-resolution fundus image database, IET Image Process. 7 (4) (2013) 373–383. Available from: https://doi.org/10.1049/iet-ipr.2012.0455.

[10] V. Gulshan, et al., Development and validation of a deep learning algorithm for detection of diabetic retinopathy in retinal fundus photographs, JAMA - J. Am. Med. Assoc. 316 (22) (2016) 2402–2410. Available from: https://doi.org/10.1001/jama.2016.17216.

[11] M.D. Abràmoff, et al., Improved automated detection of diabetic retinopathy on a publicly available dataset through integration of deep learning, Investig. Ophthalmol. Vis. Sci. 57 (13) (2016) 5200–5206. Available from: https://doi.org/10.1167/iovs.16-19964.

[12] M.M. Saleck, A. Elmoutaouakkil, M. Moucouf, Tumor detection in mammography images using fuzzy C-means and GLCM texture features, in: Proc. 14th International Conference on Computer Graphics, Imaging and Visualization, CGiV 2017, 2018, pp. 122–125. doi: 10.1109/CGiV.2017.22.

[13] M. Radhakrishnan, T. Kuttiannan, Comparative analysis of feature extraction methods for the classification of prostate cancer from TRUS medical images, IJCSI Int. J. Comput. Sci. Issues 9 (1) (2012) 171–179.

[14] C. Agurto, et al., Multiscale AM-FM methods for diabetic retinopathy lesion detection, IEEE Trans. Med. Imaging 29 (2) (2010) 502–512. Available from: https://doi.org/10.1109/TMI.2009.2037146.

[15] B. Abdillah, A. Bustamam, D. Sarwinda, Classification of diabetic retinopathy through texture features analysis, International Conference on Advanced Computer Science and Information Systems, ICACSIS 2017, 2018, pp. 333–337. doi: 10.1109/ICACSIS.2017.8355055.

[16] Laila, BMI_6334 Sample Diab Ret (Diabetic Retinopathy Samples), Kaggle [Online]. Available: http://www.kaggle.com/lrasmy/sample-diab-retn. (accessed 05.09.18).

[17] K. Cao, J. Xu, W.Q. Zhao, Artificial intelligence on diabetic retinopathy diagnosis: an automatic classification method based on grey level co-occurrence matrix and naive Bayesian model, Int. J. Ophthalmol. 12 (7) (2019). Available from: https://doi.org/10.18240/ijo.2019.07.17.

[18] S. Das, U.R. Jena, Texture classification using combination of LBP and GLRLM features along with KNN and multiclass SVM classification, 2nd International Conference on Communication, Control & Intelligent Systems, CCIS 2016, 2017, pp. 115–119. doi: 10.1109/CCIntelS.2016.7878212.

[19] T.G. Dietterich, Ensemble learning, Handbook of Brain Theory and Neural Networks 2 (1) (2002) 110–125.

[20] A. Anuragi, D.S. Sisodia, R.B. Pachori, Epileptic-seizure classification using phase-space representation of FBSE-EWT based EEG sub-band signals and ensemble learners, Biomed. Signal. Process. Control. 71 (2022) 103138.

[21] T. Chen, C. Guestrin, XGBoost : a scalable tree boosting system, Xgboost: A Scalable Tree Boosting System, 2016, pp. 785–794.

[22] R.E. Schapire, Explaining adaboost, Empirical. Inference: Festschrift in Honor of Vladimir N. Vapnik (2013) 37–52. Available from: https://doi.org/10.1007/978-3-642-41136-6_5.

[23] L. Breiman, Random forests, Mach. Learn. 45 (1) (2001) 5–32.

[24] M.M. Adankon, M. Cheriet, Support vector machine, Encyclopedia of Biometrics, Springer, 2009, pp. 1303–1308.

[25] R. Webb, Geoffrey I and Keogh, Eamonn and Miikkulainen, Naive bayes, Encycl. Mach. Learn. Data Sci. 15 (2010) 713–714. Available from: https://doi.org/10.1007/978-1-4899-7502-7.

[26] A.J. Myles, R.N. Feudale, Y. Liu, N.A. Woody, S.D. Brown, An introduction to decision tree modeling, J. Chemom. 18 (6) (2004) 275–285. Available from: https://doi.org/10.1002/cem.873.

Chapter 18

Leg postural identification framework based on one-dimensional force data using machine learning models

K. Adalarasu[1], K. Ghousiya Begum[1], M. Jagannath[2], K. Aravindh[1] and S.J. Shivasundar[1]

[1]*School of Electrical and Electronics Engineering, SASTRA Deemed to be University, Thanjavur, Tamil Nadu, India,* [2]*School of Electronics Engineering, Vellore Institute of Technology (VIT), Chennai, Tamil Nadu, India*

18.1 Introduction

Postural classification is an important concept in today's world. To better comprehend a patient's body position and movement patterns, healthcare providers need postural categorization. It offers a standardized manner of describing postural abnormalities, which can help with diagnosis, arranging treatments, and keeping track of patient's progress. The patient's postural classification might assist physicians in creating focused interventions to address certain postural disorders and prevent complications linked to bad postures [1]. Researchers looking into the biomechanics and health implications of various postures can also benefit from postural classification. In order to keep the body posture balanced and to avoid putting too much strain on the muscles, ligaments, and joints, proper posture is essential. Finding postural patterns can help in the early detection of conditions like scoliosis, kyphosis, and lordosis. In order to enhance posture, mobility, and general health, postural categorization can also help in choosing the right treatments and exercise routines. Correcting bad postures may help people feel better by lowering their chance of getting hurt, reducing their pain and musculoskeletal disorders (MSD) [2]. Also, wearable sensors have been suggested to obtain the human body's postural data [3,4]. This postural classification is essential for preserving people's physical health and quality of life. Early identification of such issues allows for timely interventions and preventive measures to improve overall well-being and reduce the chances of accidents.

People who participate in physical activities and sports can benefit from better posture in terms of their overall performance. It enhances coordination, stability, and balance, which can result in better performance [5]. By squeezing the lungs and limiting airflow, poor posture can have an impact on breathing. Breathing and oxygen intake can be enhanced by recognizing and correcting the postures. Positive psychological effects of good posture include increased self-confidence and self-esteem. Additionally, it fosters a sense of control and well-being, which enhances the mental health. Posture classification plays a major role in robotics too. The postural bracket enables a robot to understand mortal gestures, thereby establishing a perfect communication between humans and robots to have an effective robot performance [6]. The postural bracket enables robots to respond to different postures and movements of humans, making them more flexible in controlling their movements. There are various 3D reconstruction techniques available for estimating human pose. One such method involves using a 3D multiview sports (basketball) dataset along with the VGG11 architecture [7]. Another approach uses 3D human body data points captured from two cameras, wherein SVM-based segmentation methods are applied [8]. Also, 2D images reveal the postural coordinates for which VGG-19 and support vector machine (SVM) achieved the best performance [9]. But, there are a few glitches while processing 3D and 2D image datasets before training the model. The input is of high dimensionality and has vast variations in expected postures. The primary goal of the proposed work is to identify and classify different human postures using 1D force data.

- To acquire/record force data for different standing postures using a force platform.
- Extracting the features such as vertical ground reaction force (VGRF) and center of pressure (COP) from the data obtained from SENSIX force platform.
- Classification of different postures using the extracted features from VGRF using machine learning algorithms.

Fig. 18.1 depicts the steps followed for implementing our proposed strategy.

Intelligent Computing Techniques in Biomedical Imaging. DOI: https://doi.org/10.1016/B978-0-443-15999-2.00018-9

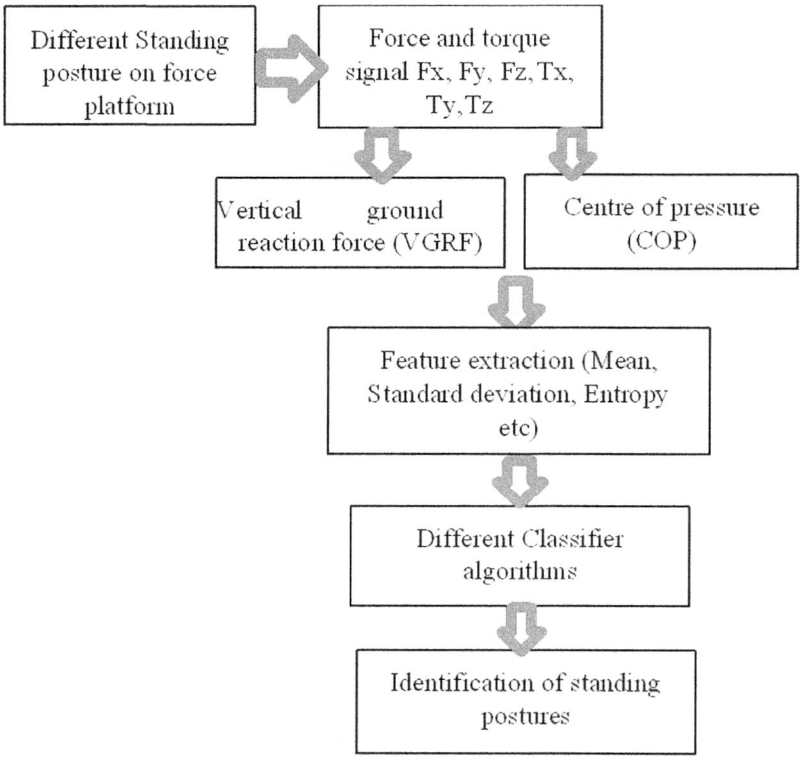

FIGURE 18.1 Block diagram of the proposed approach.

18.2 Literature review

Artificial intelligence approaches have been widely employed in the classification of body postures using data acquired from wearable sensors. A study has been made on sensor fusion basketball shooting posture as there is no specific system to identify different basketball shooting positions. A sensor fusion framework has been proposed to acquire data on shooting position for basketball using an accelerometer and gyroscope attached to the player's foot and hand. A convolutional neural networks model has been deployed for posture identification and recognition [10].

In a research study [11], the researchers conducted an experiment on single-image depth map estimation that refers to the process of inferring the depth information of a scene or object from a single 2D image. In the context of posture recognition, the use of single-image depth map estimation can be beneficial in improving the accuracy and robustness of posture recognition systems. Image-based pose detection can indeed be challenging because a colored image alone does not directly provide 3D information. While a colored image captures the appearance and visual features of a person or object, it lacks explicit depth information. Hence, the researchers included a two-step approach. The first step is used to evaluate the depth map of black and white image with the advanced generation module Pix2Pix. In order to improve the performance of color image recognition in the identification stage (second step) of the posture recognition system, an approach has been proposed that augments the estimated depth map. This involves developing a two-stream convolutional neural network (CNN) architecture that processes both the color image and the estimated depth image. The two-stream CNN architecture is designed to leverage the complementary information provided by the color image and the estimated depth map. By incorporating the depth information into the recognition process, the system can benefit from additional cues related to spatial relationships and 3D structure, which are not readily available from the color image alone.

In previous works, various approaches have been used to detect sitting postures, including camera-based methods and wearable sensor-based techniques. However, these approaches may have limitations such as privacy concerns with video-based approaches or user discomfort with wearable sensors. To address these issues, a technique called SitR has been proposed, which aims to detect sitting positions using only RF (radio frequency) signals. The SitR technique utilizes RF signals, specifically RFID (radio frequency identification) technology, to recognize different sitting postures. RFID tags are attached to the user's back in a light and low-cost manner. By analyzing the RF signals reflected or scattered by the tags, the SitR system can infer the user's sitting posture. The effective features were extracted from measured step sequences and using a suitable machine learning algorithm was deployed to make SitR achieve robust and high performance [12].

A fuzzy classifier-based soft margin minimization (FC-SMM) has been proposed for classifying the body postures. The segmented images of the human body obtained using the Gaussian mixture model (GMM) were employed to construct a three-dimensional body model from which features were extracted. Both the hierarchical fuzzy classifier and 3D fuzzy body voxel features were used to classify the bending, standing, lying, and sitting postures [13].

To recognize the sitting posture, multimodal data from 20 participants have been obtained from infrared and pressure array sensors. It is used to train attention and multitask deep learning model in recognizing the various postures that acquired a 93.08% F1 score [14]. A RIPPER rule ensembled learning algorithm based on multiple features has been proposed for human posture recognition. This employed a random subspace method for feature selection and a bagging approach for sampling the training data [15].

A machine learning-based pressure sensor internet of things (IoT) cushion has been suggested for 15 different sitting postures recognition and stretch identification. It is impressive that the random forest algorithm achieved such a high accuracy of 98.82% in recognizing the postures in the study. Based on the authors' conclusion, it appears that they found a correlation between body mass index (BMI) and the performance of the machine learning models used for posture recognition [16]. For surveillance tasks, human posture recognition and tracking using unmanned aerial vehicles have been developed with the application of you only look once (YOLOV3) algorithm for human detection, and simple online real-time tracking (SORT) is for human tracking. The posture identification system also uses fuzzy logic to understand the poses and provided an accuracy of 95.2% [17]. Most of the study involves the use of high dimensional 3D and 2D image datasets which will increase the computational burden while processing. So, the main objective of the preferred study is to recognize the different human postures using 1D force data.

18.3 Materials and methods

18.3.1 Participants and study protocol

Twenty-five persons including 17 males and 8 females were chosen as subjects for this classification. Each subject was of different heights and weights. Each subject was asked to do all three postures (sitting, double support standing, squat standing) for 15 seconds each with 5 seconds gap in between. The summary of the subjects includes the average age of 19.96 ± 0.2 years, average height of 166.6 ± 9.7 cm, and an average mass of 60.4 ± 9.5 kg. Two trials were taken for each subject in a randomized manner. Subjects were debriefed and their observations and comments were recorded for future reference. Subjects who could not complete the study for any reason were also debriefed and reasons for non-completion had been recorded. All subjects were orally explained about the study and informed consent to participate in the study was obtained.

18.3.2 Data acquisition

The SENSIX force platform and the entire setup are shown in Fig. 18.2A and B. The force plate is made up of a mechanical structure that is put under strain by an effort given to its upper side. The force platform is equipped with an integrated device that is mounted on a dedicated platform. The platform has specific dimensions, measuring 600 mm in length and 400 mm in width. Additionally, the integrated device is positioned at a height of 2.4 mm vertically above

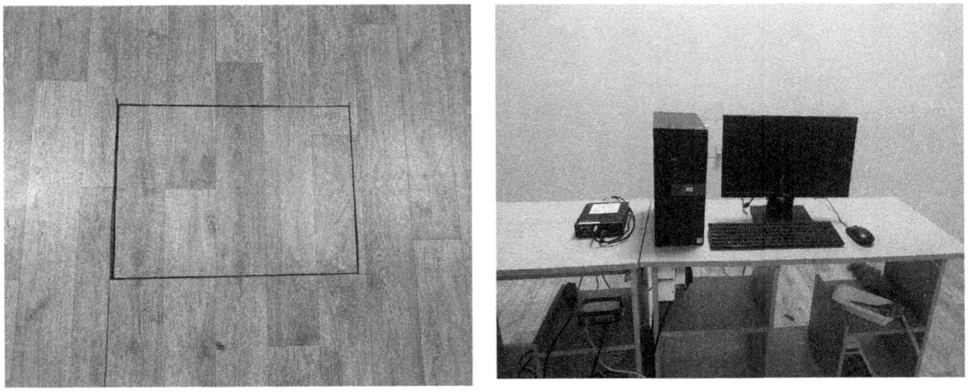
(A) (B)

FIGURE 18.2 (A) SENSIX force platform, (B) force platform setup.

the top of the plate. Extensometry is the technique used to quantify this strain and determine the effort components from a linear relationship. Eight bending beams connect the rigid upper and lower components of the mechanical structure. The beams have strain gauges. To create eight (complete) Wheatstone full-bridges, the number of strain gauges and their placement have been selected in an optimal manner. Each bridge supplies a measuring voltage that is amplified by an instrumentation amplifier and is supplied with a stabilized reference voltage. The 8 V acquired allow for the following calculation (Eq. (18.1)) of the six effort components:

$$F = GU \tag{18.1}$$

where F is the torque effort which is expressed in Newton, U is the vector gathering the eight voltage measurements expressed in Volts, G is the 6×8 calibration matrix.

18.3.3 Experimental protocol

The experiment is performed on different standing postures of humans using 1D data. All the subjects were asked to warm up for 5 seconds before starting the experiment to get familiarize themselves with the different standing postures as well as to avoid any sort of cramps. The force platform was utilized and subjects were allowed to do three different standing postures two trials each ($25 \times 3 \times 2 = 150$). Sitting (as shown in Fig. 18.3A), double support standing (as shown in Fig. 18.3B), and squat standing (as shown in Fig. 18.3C) are the three postures taken. Subjects are asked to do the given three postures in a randomized trial manner. Each pose has to be done for 15 seconds. The sampling frequency chosen was 1000 Hz and the cut-off frequency chosen was 10 Hz for normalized weights. A low-pass filter allows frequencies below the cutoff frequency to pass through while attenuating higher frequencies. In this case, the purpose of the filter is likely to remove high-frequency noise or unwanted fluctuations in the data, while preserving the lower frequency components relevant to the analysis of human posture. Standing postures are performed in the force platform, and the corresponding force and torque data (F_x, F_y, F_z and T_x, T_y, T_z) are obtained. The statistical features extracted from VGRF are given to the classifier in MATLAB and analyzed. From the classifier algorithm, the identification of standing postures is done. Upon successful completion of the tests, subjects were debriefed and their observations and comments were recorded for future reference. Subjects who cannot complete the study for any reason were debriefed and reasons for noncompletion were recorded. Recorded signals will be analyzed in the lab to assess the postures of subjects through 1D data. All experiments were conducted in accordance with the regulations established by SASTRA Deemed University's Institutional Ethics Committee for Human Volunteer Research. All participants signed an informed consent form before commencing the experiment. The experiments were conducted in the Biomechanics Lab of the School of Chemical and Biotechnology at SASTRA Deemed to be University, Thanjavur.

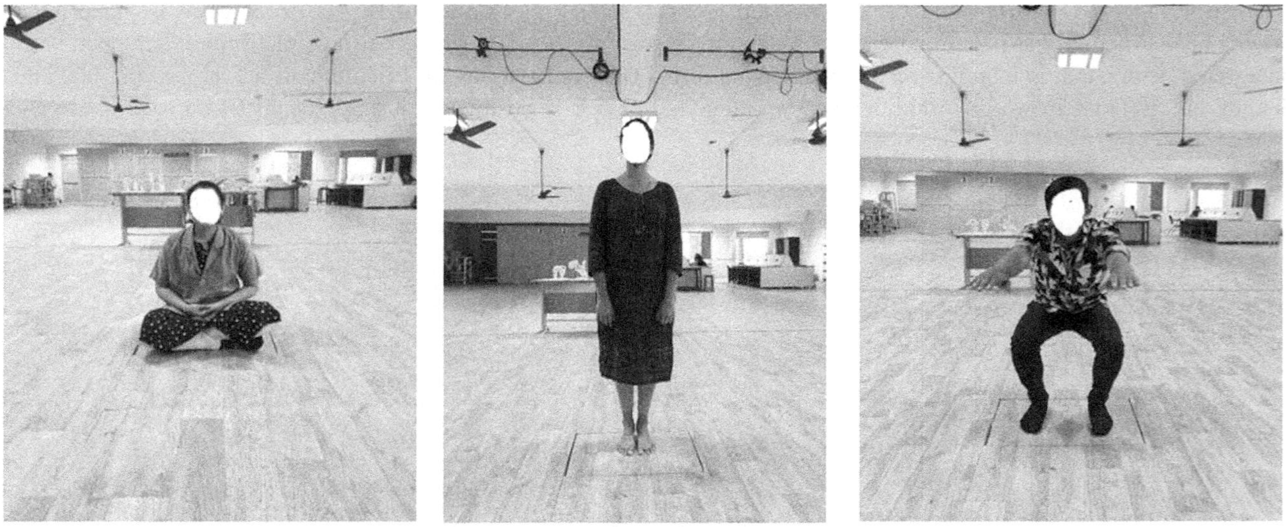

(A) (B) (C)

FIGURE 18.3 (A) Sitting pose, (B) double support standing pose, (C) squat standing pose.

18.3.4 Data processing and feature extraction

Five features have been selected to classify three postures namely three statistical feature such as mean, standard deviation, skewness and two nonlinear time series features such as fuzzy entropy and sample entropy by appropriate formulae as shown in Table 18.1. The statistical features explicate the general pattern in a dataset that aids in classifying. The mean explains the average value in a dataset, the standard deviation gives an understanding of the spread or dispersion of data, whereas the skewness expresses the direction of outliers and the asymmetricity of distribution around the mean. To estimate the randomness in a series of data, sample entropy is used. Sample entropy is a complexity measure of time series without counter of self-matches. It is not sensitive to data corrupted by noise [18] and can be used to quantify the stride-to-stride fluctuation of COP. Sample entropy is the negative natural logarithm of the conditional probability that a subseries of length m that matches pointwise within a tolerance r will also match at the next point. Sample entropy is largely independent of record length and demonstrates relative consistency under circumstances where approximate entropy does not. Fuzzy entropy uses the fuzzy membership function to calculate the similarity of two vectors and replaces the threshold criterion so that fuzzy entropy changes smoothly with the changes of parameters, reduces the dependence of the entropy on parameters, and makes the statistics more stable [19]. Fuzzy entropy considers the imprecise environments from which biological datasets arose and uses an exponential function instead of the traditional Heaviside function. Fuzzy entropy shows its superiority in robustness to noise, input parameter selections, and in eliminating the bias toward regularity. To achieve better classification accuracy, we use a combination of statistical and nonlinear features.

Algorithm is developed for the feature extraction and data points exported from the force platform are processed based on the developed algorithm. Thus the features are extracted from VGRF as shown in Fig. 18.4A. Fig. 18.4B depicts the COP trajectory obtained using force platforms tracks the point of application of the resulting ground reaction forces under the feet that can also be used for the determination of three different postures. The COP trajectory is very important as it reproduces the dynamic nature of the postural control of an individual subject [20]. The mathematical expression for all the features is shown in Table 18.1.

18.3.5 Classifier algorithms

In many different applications, including data mining, pattern recognition, natural language processing, speech, and picture identification, classifiers have been frequently utilized. The input data are analyzed and ML classified based on their properties or attributes are used to identify and categorize the input data into one of the many classes [21]. Here, in the proposed study, the extracted features are fed into the MATLAB classification learner app and the performance of the models are analyzed for all three postures that includes five features. Classifiers namely, decision tree (DT), ensemble (EN), SVM, K-nearest neighbors (KNN), and logistic regression (LR) are used in this study for leg posture recognition. The following states the significance of the employed machine learning models.

1. EN: It classifies by combining the predictions of many models into account to produce the outcome from one final model with an improved decision. It leverages collective knowledge by aggregating the opinions of different classifiers, often resulting in improved overall performance compared to using a single classifier. They can mitigate the weaknesses and biases of individual classifiers, leading to more accurate and reliable predictions.

TABLE 18.1 Feature extraction formulae.

Feature	Mathematical expression
Mean	$\mu = \frac{\sum_{i=1}^{N} x_i}{N}$
Standard deviation	$\sigma = \sqrt{\frac{\sum_{i=1}^{N} (x_i - \mu)^2}{N}}$
Skewness	$S = \sqrt{\frac{\sum_{i=1}^{N} (x_i - \mu)^3}{(N-1)\sigma^3}}$
Fuzzy entropy	$H(X) = \sum_x - p(x)\log_2 \pi(x)$
Sample entropy	$H(X) = - \sum x \in X\ p(x)\log p(x)$ where p is the probability distribution on X

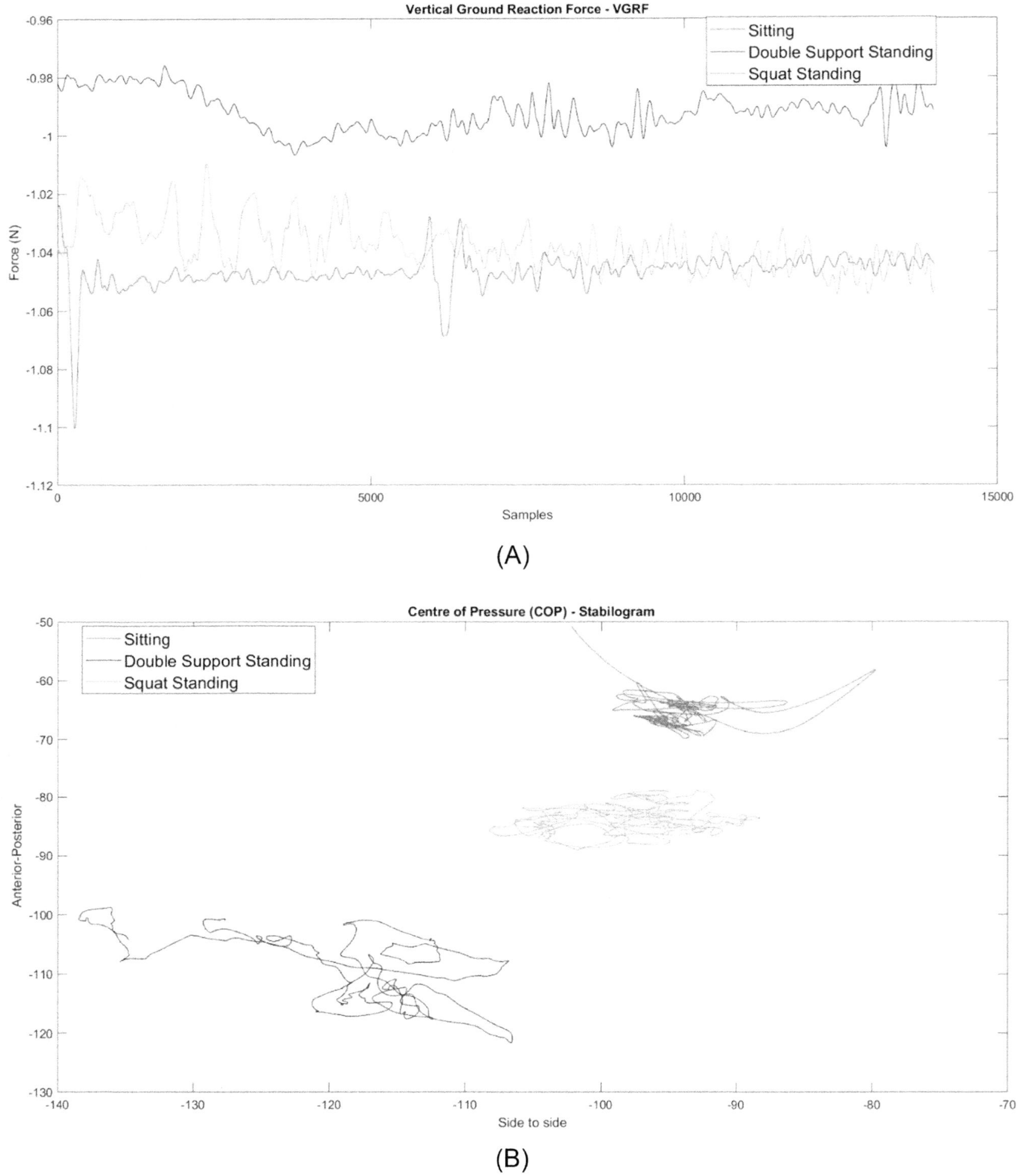

FIGURE 18.4 (A) Vertical ground reaction force plot (VGRF), (B) center of pressure (COP).

2. DT: The classifier has multiple levels branching from the root node to child nodes and repeats this process until it reaches the leaf node, yielding the decision results. It creates a flowchart-like structure to make decisions or predictions. Each internal node represents a feature or attribute, and each leaf node represents a class label or a predicted value.
3. KNN: It determines the similarity among the data point and classifies accordingly by calculating the Euclidean distance of K number of neighbors. It classifies instances based on their similarity to nearby instances in the feature space. It assigns the class label based on the majority vote of the KNN.

4. LR: It gives a probable outcome lying between 0 and 1 and uses a threshold to differentiate between classes by predicting the probability of an instance belonging to a particular class. It models the relationship between the input features and the probability of an event occurring using the logistic function.
5. SVM: It maps the data in n-dimensional space and uses a hyperplane to differentiate the classes by maximizing the marginal distance among the classes and minimizing the errors. The classifier can handle linearly separable and nonlinearly separable data by using different kernel functions. The performance metrics are analyzed for every model.

18.4 Results

Based on three statistical features and nonlinear time domain features, the three poses (sitting, double support standing, squat standing) are classified for posture identification. The dataset is splitted as training dataset (75%) and testing dataset (25%) and different classifiers are deployed. Among which, the EN model has given the maximum testing accuracy of 88% compared to other models as shown in Fig. 18.5. The other performance characteristics of the models such as prediction speed, training time, and model size are also displayed in Table 18.2. The confusion matrices of all the models have been displayed in Fig. 18.6.

Table 18.3 displays the false negative rates in percentage during prediction of ML models. The area under curves (AUC) are shown in Fig. 18.7.

From the simulation results, one can strongly believe that the EN model performs best in categorizing the three different postures with highest accuracy of 88%, lowest FNR of 36.1%, and highest AUC value of 91.5%.

18.5 Discussion

In the study utilizing 2D and 3D techniques, the researchers combined various image processing methods with depth images captured by the Kinect sensor to achieve successful recognition of different human postures. The Kinect sensor is a device that can capture depth information along with red, green and blue (RGB) images, enabling the extraction of 3D data. By leveraging depth images obtained from the Kinect sensor, the researchers were able to enhance the posture recognition process. Depth information provides additional cues about the spatial layout and distances between different body parts, aiding in the accurate identification of postures.

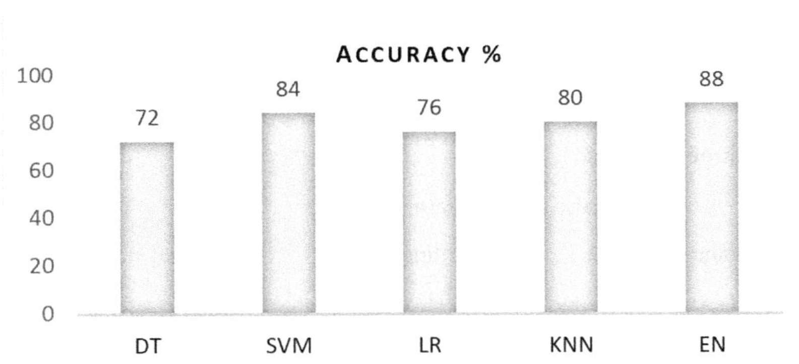

FIGURE 18.5 Percentage of accuracy for ML models. *DT*, Decision tree; *SVM*, support vector machine; *LR*, logistic regression; *KNN*, K-nearest neighbors; *EN*, ensemble.

TABLE 18.2 Performance characteristics and model size.

Model	Prediction speed (Obs/s)	Training time (s)	Model size (kB)
Decision tree	170	12.8	7
Support vector machine	200	4.2	17
Logistic regression	190	5.3	38
K-nearest neighbors	130	1.5	12
Ensemble	260	5.2	227

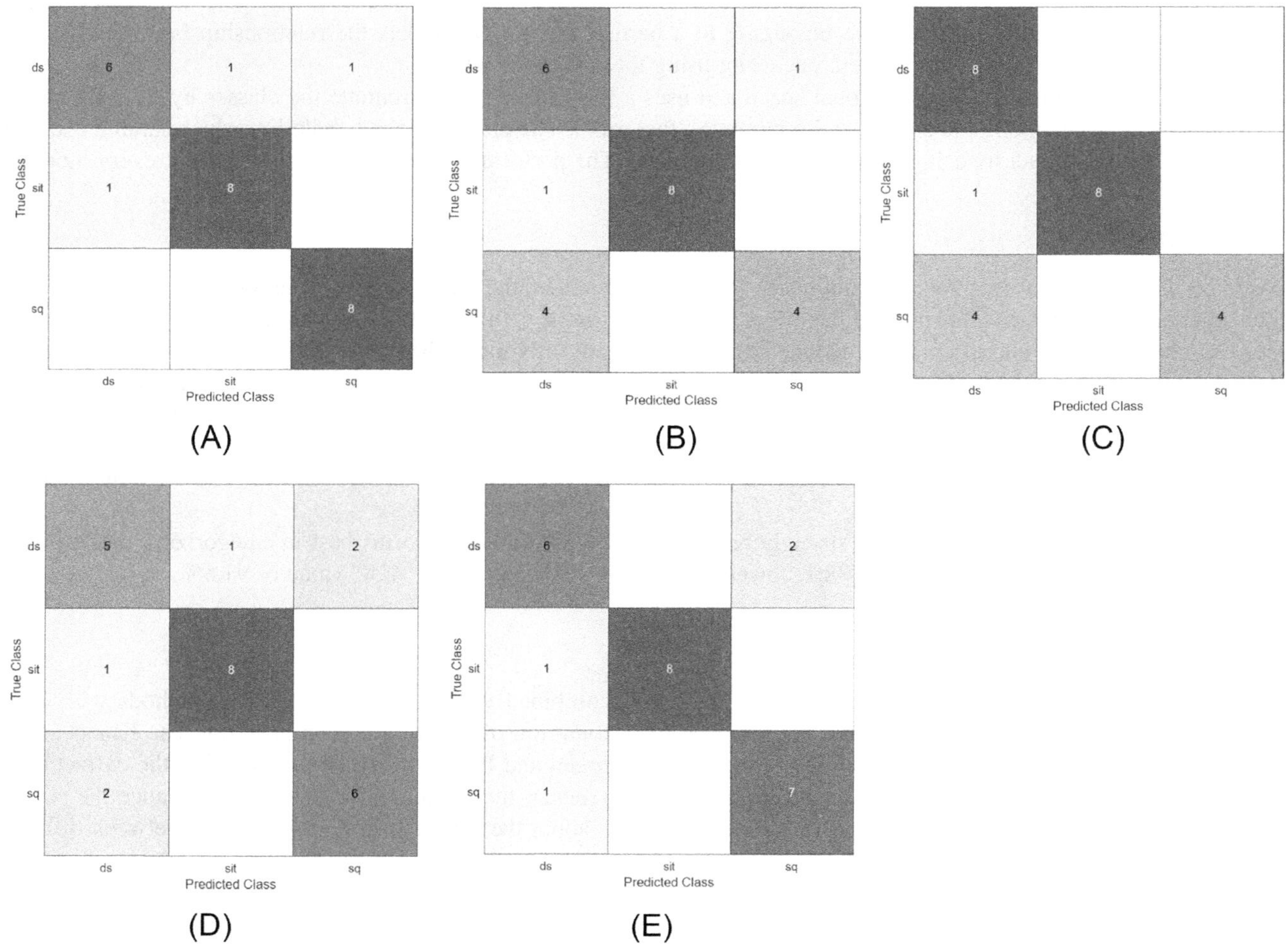

FIGURE 18.6 Confusion matrices for (A) ensemble, (B) decision tree, (C) K-nearest neighbors, (D) logistic regression, (E) support vector machine.

TABLE 18.3 False negative rate for ML models for three postures.

Model	False negative rate (FNR) %		
	Double support standing (ds)	**Sitting (sit)**	**Squat standing (sq)**
Decision tree	25	11.1	50
Support vector machine	25	11.1	12.5
Logistic regression	37.5	11.1	25
K-nearest neighbors	0	11.1	50
Ensemble	25	11.1	0

The combination of image processing techniques with depth images likely involved steps such as image segmentation, feature extraction, and classification. These methods analyze the RGB and depth data to identify specific postural features and patterns, which are then used to classify different postures like sitting, standing, and bending. In the proposed method, we used a force platform and force sensor data instead of a camera and wearable sensors. From VGRF, one can obtain torque and force vectors in x, y, and z directions from which statistical features were extracted. The statistical and nonlinear time domain features can be calculated using the VGRF data. Here, we used a 1D force sensor and MATLAB classifier learner app software to classify three different postures. Postures namely double support

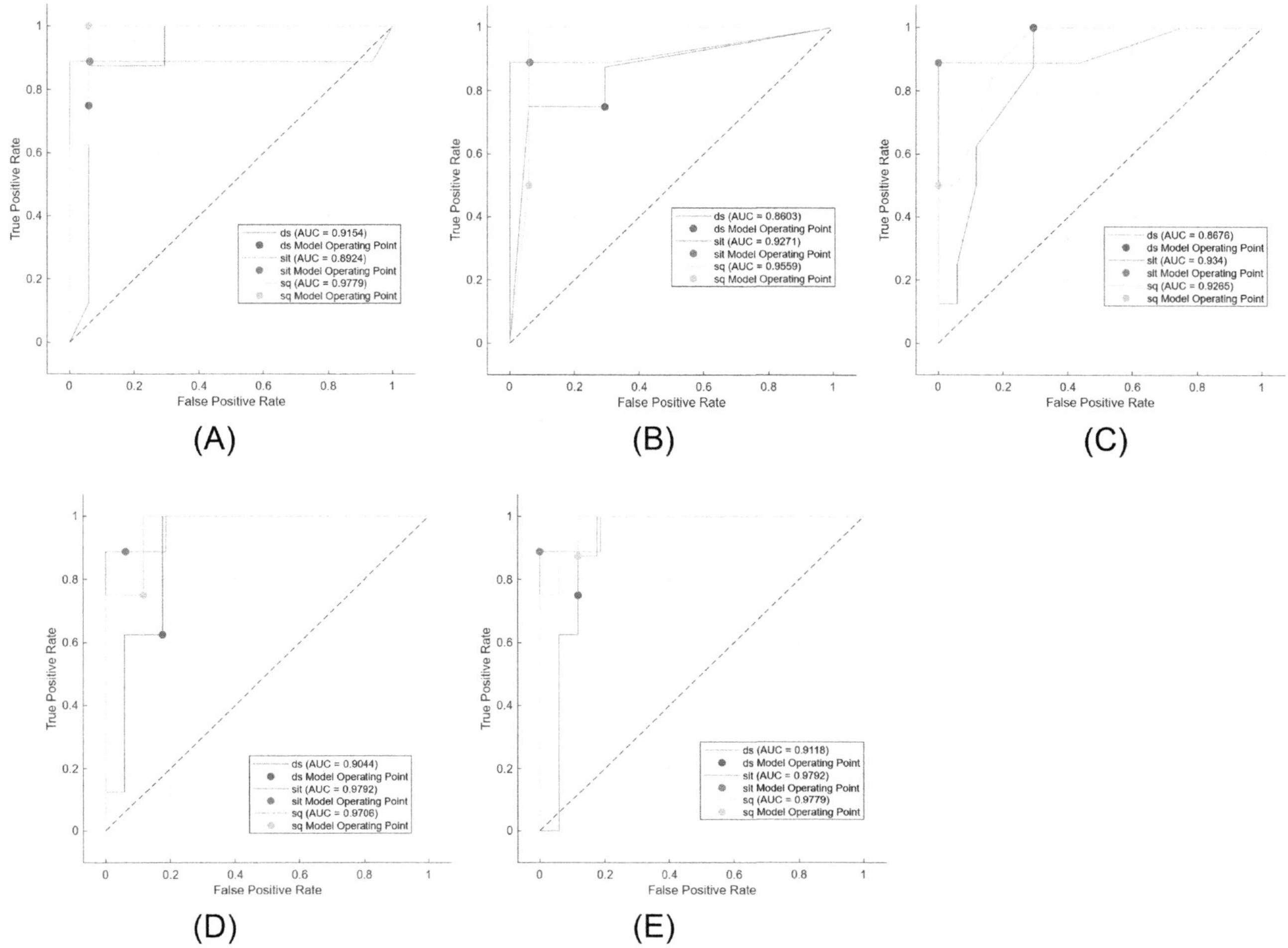

FIGURE 18.7 AUC curves for (A) ensemble, (B) decision tree, (C) K-nearest neighbors, (D) logistic regression, (E) support vector machine.

standing, sitting, and squat standing were chosen. We selected 25 subjects and made them do three different poses in a randomized trial manner and fed them into a classification leaner app. Three statistical parameters such as mean, standard deviation, and skewness along with two nonlinear time domain features sample entropy, fuzzy entropy, were extracted in identifying the postures. All these values were used to train the classifier and the accuracy of the models in classifying the three postures was determined. LR, SVM, DT, EN, and KNN are some of the traditional ML algorithms used. And, the EN model was found to be the one with the highest performance based on the evaluation results and obtained an accuracy of 88%.

The use of skeletal data for recognizing human posture is indeed an important area of study in the field of human–computer interaction. One approach to analyzing skeletal data for posture recognition is to consider both the local relationships between joints and the global spatial location of joints. The local relationship between joints involves analyzing the angles and orientations between adjacent joints, which provides information about the joint flexion, extension, or rotation. These local relationships capture the relative positions and movements of neighboring joints and contribute to understanding the configuration of the body. On the other hand, the global spatial location of joints takes into account the absolute positions of joints in 3D space. This includes the distance properties between different joints, which can be used to capture the overall body posture and spatial relationships between body parts.

In the proposed study, the recognition of human postures was not based on the analysis of skeletal data or the relationships between joints. Instead, the subjects were asked to stand on a force platform, and features were extracted from the collected data. The features extracted from the force platform data were likely related to the distribution of forces and pressure patterns exerted by the subjects while standing in different postures. These features could include parameters such as COP trajectory, force distribution, or other relevant metrics. The experimental findings of the proposed methodology demonstrated that the system was able to successfully recognize the three postures (sitting, double support standing, squat standing) based on the extracted features from the force platform data. This suggests that the features extracted from the force platform data contained valuable information that allowed for accurate classification of the postures. Here, in the present study, the features are extracted manually before applying to machine learning models which is time-consuming. Our future work deals with the use of automatic feature extraction based on deep-learning models with many leg postures. To enhance the accuracy, the number of subjects will be increased, more features will be extracted automatically from force and COP data and feature selection methods for choosing appropriate features will be applied for the posture classification. The study will be extended to real-time applications namely robotics control, rehabilitation, remote monitoring, and human–computer interaction.

18.6 Conclusion

In the study, posture classification for three different postures was performed using a 1D technique. The proposed method aimed to overcome limitations commonly associated with 2D and 3D techniques, such as frame rate, video resolution, background lighting, and image blurring. The features extracted from VGRF have been fed to classifier models and a maximum accuracy of 88% was obtained for the EN model. Thus EN model is used to classify the leg postures of 25 subjects (two trials). The accuracy of the posture classification system could be improved further by increasing the number of subjects, number of features, or choosing appropriate features for the posture classification. Overall, human postural identification has a wide range of applications and can provide valuable insights into health, movement, and behavior. As technology advances, there is potential for further developments in this field, enabling more accurate and real-time analysis of posture for various purposes.

Acknowledgments

The authors of the study express their gratitude to SASTRA Deemed to be University for providing the Biomechanics Laboratory facilities (SASTRA Research & Development Fund), which played a crucial role in conducting the experiments smoothly. The authors also extend their appreciation to the volunteers who participated in the study.

References

[1] K. Sarkar, S. Dev, T. Das, S. Chakrabarty, S. Gangopadhyay, Examination of postures and frequency of musculoskeletal disorders among manual workers in Calcutta, India, Int. J. Occup. Environ. Health 22 (2) (2016) 151–158. Available from: https://doi.org/10.1080/10773525.2016.1189682.

[2] Z. Li, R. Zhang, C.H. Lee, Y.C. Lee, An evaluation of posture recognition based on intelligent rapid entire body assessment system for determining musculoskeletal disorders, Sensors 20 (16) (2020) 1–21. Available from: https://doi.org/10.3390/s20164414.

[3] J. Zhao, E. Obonyo, S.G. Bilén, Wearable inertial measurement unit sensing system for musculoskeletal disorders prevention in construction, Sensors 21 (4) (2021) 1–28. Available from: https://doi.org/10.3390/s21041324.

[4] M. Ghislieri, L. Gastaldi, S. Pastorelli, S. Tadano, V. Agostini, Wearable inertial sensors to assess standing balance: a systematic review, Sensors 19 (19) (2019) 1–25. Available from: https://doi.org/10.3390/s19194075.

[5] D. Singla, Z. Veqar, Methods of postural assessment used for sports persons, J. Clin. Diagnostic Res. 8 (4) (2014) 2–5. Available from: https://doi.org/10.7860/JCDR/2014/6836.4266.

[6] B. Busch, G. Maeda, Y. Mollard, M. Demangeat, M. Lopes, Postural optimization for an ergonomic human-robot interaction, in: 2017 IEEE/RSJ International Conference on Intelligent Robots and Systems (IROS), 2017, pp. 2778–2785. Available from: https://doi.org/10.1109/IROS.2017.8206107.

[7] X. Song, L. Fan, Human posture recognition and estimation method based on 3D multiview basketball sports dataset, Complexity 2021 (2021). Available from: https://doi.org/10.1155/2021/6697697.

[8] C.F. Juang, T.C. Chen, W.C. Du, Human body 3D posture estimation using significant points and two cameras, Sci. World J. 2014 (2014) 23–25. Available from: https://doi.org/10.1155/2014/670953.

[9] P. Lee, T.B. Chen, C.Y. Wang, S.Y. Hsu, C.H. Liu, Detection of postural control in young and elderly adults using deep and machine learning methods with joint–node plots, Sensors 21 (9) (2021). Available from: https://doi.org/10.3390/s21093212.

[10] J. Fan, S. Bi, G. Wang, L. Zhang, S. Sun, Sensor fusion basketball shooting posture recognition system based on CNN, J. Sens. 2021 (2021) 6664776. Available from: https://doi.org/10.1155/2021/6664776.

[11] J. Liu, et al., Single image depth map estimation for improving posture recognition, IEEE Sens. J. 21 (23) (2021) 26997–27004. Available from: https://doi.org/10.1109/JSEN.2021.3122128.

[12] L. Feng, Z. Li, C. Liu, X. Chen, X. Yin, D. Fang, SitR: sitting posture recognition using RF signals, IEEE Internet Things J. 7 (12) (2020) 11492–11504. Available from: https://doi.org/10.1109/JIOT.2020.3019280.

[13] C.-F. Juang, W.-E. Ni, Human posture classification using interpretable 3-D fuzzy body voxel features and hierarchical fuzzy classifiers, IEEE Trans. Fuzzy Syst. 30 (12) (2022) 5405–5418. Available from: https://doi.org/10.1109/TFUZZ.2022.3177028.

[14] X. Zhang, J. Fan, T. Peng, P. Zheng, X. Zhang, R. Tang, Multimodal data-based deep learning model for sitting posture recognition toward office workers' health promotion, Sens. Actuat. A Phys. 350 (2023) 114150. Available from: https://doi.org/10.1016/j.sna.2022.114150.

[15] W. Ding, B. Hu, H. Liu, X. Wang, X. Huang, Human posture recognition based on multiple features and rule learning, Int. J. Mach. Learn. Cybern. 11 (11) (2020) 2529–2540. Available from: https://doi.org/10.1007/s13042-020-01138-y.

[16] K. Bourahmoune, K. Ishac, T. Amagasa, Intelligent posture training: machine-learning-powered human sitting posture recognition based on a pressure-sensing IoT cushion, Sensors 22 (14) (2022). Available from: https://doi.org/10.3390/s22145337.

[17] M.F.R. Lee, Y.C. Chen, C.Y. Tsai, Deep learning-based human body posture recognition and tracking for unmanned aerial vehicles, Processes 10 (11) (2022) 1–22. Available from: https://doi.org/10.3390/pr10112295.

[18] S. Ramdani, F. Bouchara, J. Lagarde, Influence of noise on the sample entropy algorithm, Chaos 19 (1) (2009) 031231–031236. Available from: https://doi.org/10.1063/1.3081406.

[19] W. Chen, J. Zhuang, W. Yu, Z. Wang, Measuring complexity using FuzzyEn, ApEn, and SampEn, Med. Eng. Phys. 31 (1) (2009) 61–68. Available from: https://doi.org/10.1016/j.medengphy.2008.04.005.

[20] T.E. Prieto, J.B. Myklebust, R.G. Hoffmann, E.G. Lovett, B.M. Myklebust, Measures of postural steadiness: differences between healthy young and elderly adults, IEEE Trans. Biomed. Eng. 43 (9) (1996) 956–966. Available from: https://doi.org/10.1109/10.532130.

[21] S. Uddin, A. Khan, M.E. Hossain, M.A. Moni, Comparing different supervised machine learning algorithms for disease prediction, BMC Med. Inf. Decis. Mak. 19 (2019) 281. Available from: https://doi.org/10.1186/s12911-019-1004-8.

Chapter 19

Artificial intelligence for biomedical application associated with fluid-material dynamics

Shashi Kant Verma[1], S.L. Sinha[2] and D.K. Chandraker[3]

[1]Department of Mechanical Engineering, National Institute of Technology Durgapur, Durgapur, West Bengal, India, [2]Department of Mechanical Engineering, National Institute of Technology Raipur, Raipur, Chhattisgarh, India, [3]Reactor Engineering Division, Bhabha Atomic Research Centre, Mumbai, Maharashtra, India

19.1 Introduction

The augmentation of computers and their hardware with higher storage and power capability places a concrete basis for artificial intelligence (AI) applications. AI algorithms can handle huge numbers of produced big data. AI algorithms have become an emerging mechanism to solve computational fluid dynamics (CFD) problems and attract scientists, researchers, and engineers. Wang and Wang [1] discussed the CFD modeling difficulties and solutions by AI technique. Finally, they suggest how to choose the suitable AI algorithm for the CFD problem. Pawar et al. [2] used two different data-driven approaches: a neural network model and a least-squares regression-based model. Babanezhad et al. [3] studied square-shape cavities by using the machine learning (ML) technique. They conclude that AI and CFD cause faster computational interactions to predict the flow field in the square cavity. In this chapter, some good papers have been discussed with selected AI models. Table 19.1 shows the details of previous work performed by different researchers and their coworkers in the field of fluid dynamics by using AI techniques.

TABLE 19.1 Work performed by different researchers and their coworkers in the field of biomedical by using AI techniques.

References	Computing techniques	Remarks
Kumar et al. [4]	Convolutional neural networks (CNN)	All data are focused on breast cancer.
Yoon et al. [5]	The convolutional neural network' (CNN) is the most effective model for medical image analysis	Researchers have discussed the usage of AI for medical image investigation.
Golan et al. [6]	Deep convolutional neural network (CNN)	Researchers have trained a deep convolutional neural network (CNN) for the analysis of computed tomography (CT) scans.
Li et al. [7]	Three-dimensional convolutional neural network	Researchers found that the present 3D CNN has outperformed as compared to prior methods of disease diagnosis.
He et al. [8]	ML, CFD, FSI, and FEA	Researchers have predicted vascular health by using a medical image to develop a CFD and FSI model combine with AI.
Wang and Wang [1]	ANN	The artificial neural network under neural network is widely used in artificial intelligence algorithms.
Baghel et al. [9]	CFD	Researchers have done a CFD analysis of the baby incubator using a turbulence model.

Intelligent Computing Techniques in Biomedical Imaging. DOI: https://doi.org/10.1016/B978-0-443-15999-2.00020-7

CFD and AI are two powerful tools that can be used for biomedical image processing. CFD is a technique used to simulate the movement of fluids through a system, such as blood flow through the human body. AI, on the other hand, involves the usage of ML algorithms to analyze and deduce huge volumes of data. Combining CFD and AI can provide a more comprehensive understanding of intricate biological systems, such as the cardiovascular system. By using CFD to simulate blood flow and AI to analyze the resulting data, researchers can gain insights into the underlying physiological processes and identify potential areas of concern. For example, in the field of cardiology, CFD and AI can be used to analyze blood flow through the heart and identify areas of turbulence or blockages that may lead to cardiovascular disease. This information can be used to guide treatment decisions and improve patient outcomes. In addition to cardiovascular applications, CFD and AI can also be used in other areas of biomedical image processing, such as analyzing brain function and identifying abnormalities in medical imaging scans. Overall, the integration of CFD and AI can provide a powerful tool for biomedical image processing, allowing researchers to gain insights into complex biological systems and improve patient outcomes.

Biomedical image processing has been revolutionized by the introduction of artificial intelligence techniques. Fig. 19.1 shows the interaction of AI algorithms with CFD [1]. With the advancements in machine learning algorithms, AI has proved to be highly effective in analyzing, interpreting, and diagnosing medical images. This chapter will explore the role of AI in biomedical image processing, including its applications, benefits, and challenges.

19.2 Artificial intelligence techniques and their application in fluid dynamics

The term "artificial intelligence" comes from two words "Artificial" and "Intelligence" and was first proposed in 1956. AI is the theory and development of computer systems, which include both machines and software. It enables machines that perform tasks that normally require human intelligence. Fig. 19.2 denotes AI and its domains.

Depending on the training approaches, AI algorithms could be separated into three key categories, such as unsupervised, semisupervised, and supervised learning. The mass transfer process involves different operations such as extraction, absorption, distillation, rectification, adsorption, and drying in chemical industries. Fig. 19.3 represents the perspectives and challenges of AI in fluid-material dynamics [10].

Duraisamy and his coworkers [10] explain the future perspectives and challenges of AI in turbulence modeling. Recently, Mesejo et al. [11] have done a broad review of the application of AI techniques in the field of biomedical. They have discussed the opportunities and limitations of AI in biomedical applications. Fig. 19.4 shows some of the applications of AI techniques in the biomedical field.

AI has a wide range of applications in biomedical image processing, which is shown in Fig. 19.5. AI algorithms can segment medical images to detect regions of interest, such as tumors, blood vessels, or other anatomical structures. This is useful for diagnosis and treatment planning. AI can detect early signs of disease in medical images. For example, AI algorithms can identify patterns in medical images that are indicative of cancer, Alzheimer's disease, or other conditions. AI can reconstruct medical images from noisy or low-quality data, enhancing the clarity and resolution of the images. AI algorithms can align and register medical images from different sources, allowing doctors to compare

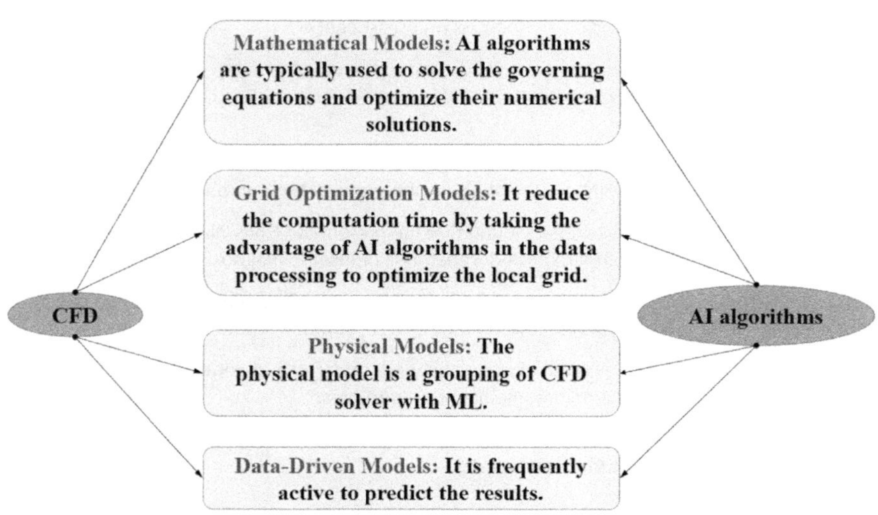

FIGURE 19.1 Interaction of artificial intelligence algorithms with computational fluid dynamics.

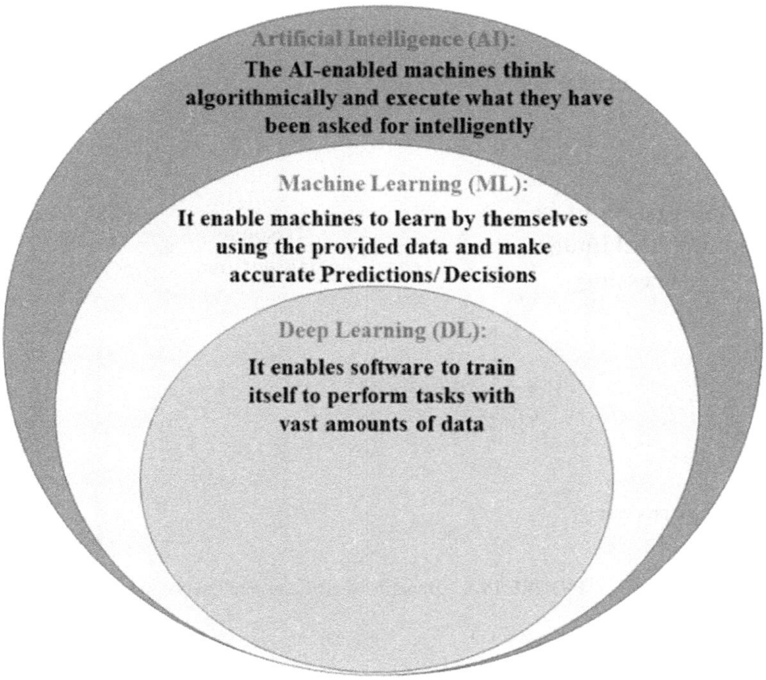

FIGURE 19.2 Artificial intelligence and its domains.

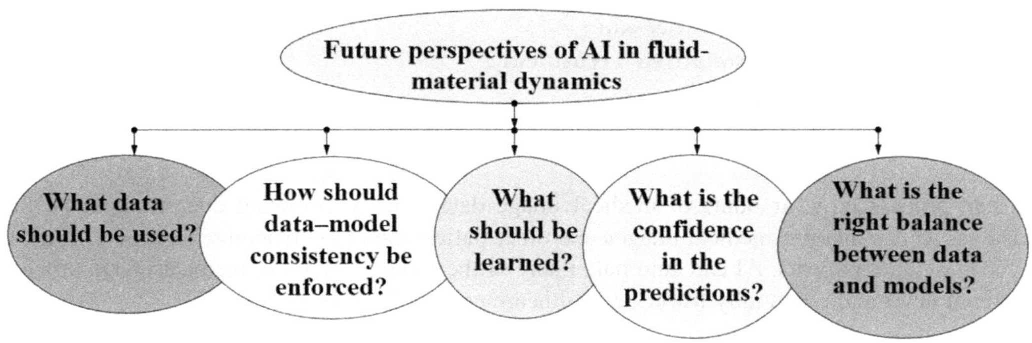

FIGURE 19.3 Perspectives and challenges of artificial intelligence in fluid-material dynamics [10].

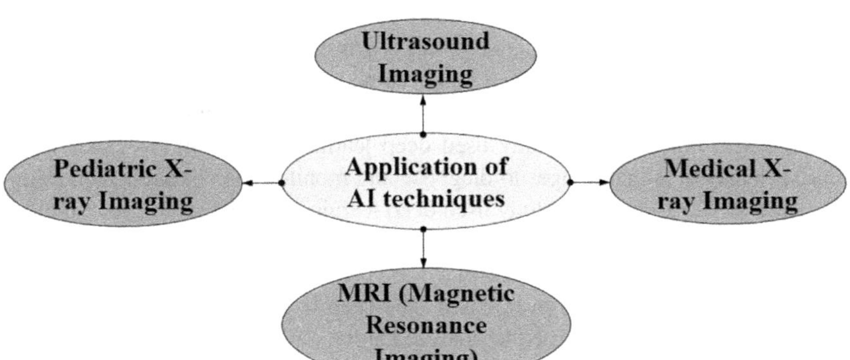

FIGURE 19.4 Application of artificial intelligence techniques in the biomedical field.

images from different time points or imaging modalities. AI can analyze medical images and other patient data to personalize treatment plans, taking into account individual variations in anatomy and physiology.

There are numerous benefits of using AI. Some of those key advantages are shown in Fig. 19.6. AI algorithms can detect subtle changes in medical images that may be missed by human observers, leading to more accurate diagnoses.

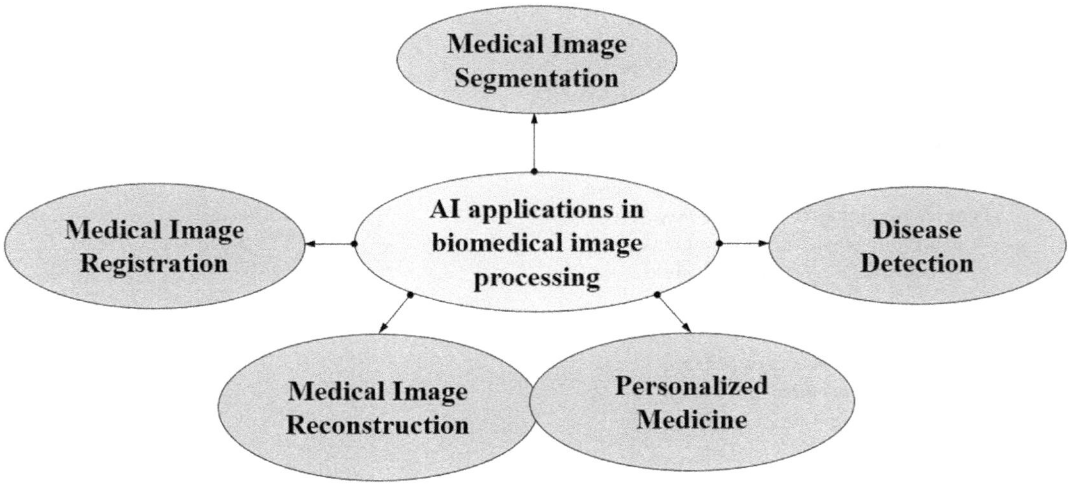

FIGURE 19.5 Artificial intelligence applications in biomedical image processing.

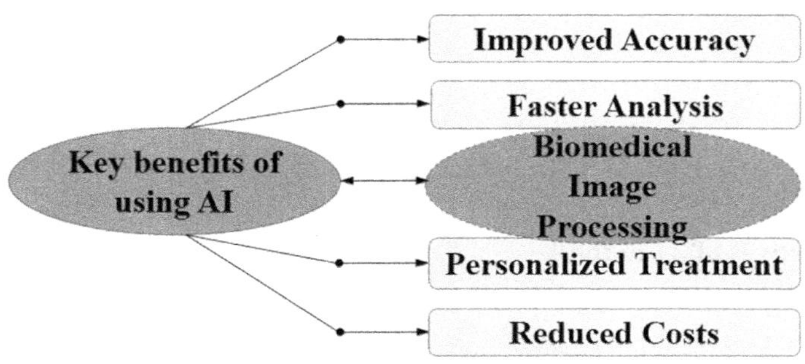

FIGURE 19.6 Benefits of artificial intelligence applications.

AI algorithms can process large amounts of medical image data quickly, allowing doctors to make faster and more informed decisions. AI can analyze medical images and other patient data to personalize treatment plans, leading to better outcomes for individual patients. AI can automate many of the tasks involved in medical image processing, reducing the need for manual labor and potentially lowering healthcare costs.

19.3 Case study examples for artificial intelligence in medical applications

AI algorithms are being used to analyze medical images, such as mammograms and CT scans, to detect and diagnose cancer. One case is the procedure of deep learning algorithms to identify and classify breast cancer cells in mammograms. AI is being used to analyze MRI and CT scans, to identify brain abnormalities and diagnose neurological disorders. For example, a study used a deep learning algorithm to identify Alzheimer's disease in brain MRI scans. AI is being used to analyze images of skin lesions to help diagnose skin diseases. For model, a study used deep learning algorithms to classify skin lesions as benign or malignant. AI is being used to analyze retinal images to diagnose and monitor eye diseases, such as diabetic retinopathy and age-related macular degeneration. For example, a study used deep learning algorithms to detect diabetic retinopathy in retinal images. AI is being use to analyze tissue samples to aid in the identification of cancer. The deep learning algorithms are used to classify breast cancer tumors based on their molecular subtypes. AI is being used to analyze X-rays, CT scans, and other medical imaging data to aid in the diagnosis of diseases. The deep learning algorithms are used to classify lung nodules as benign or malignant in CT scans. AI is being used to analyze cardiac imaging data, such as echocardiograms and cardiac MRI scans, to identify heart disease and monitor cardiac function. The deep learning algorithms are used to predict heart failure based on echocardiogram images.

19.4 Steps to use artificial intelligence for baby incubator's image processing

Using AI for baby incubator image processing can be a helpful tool to monitor the health and well-being of premature infants. Here are some steps we can take to use AI for baby incubator image processing as shown in Fig. 19.7.

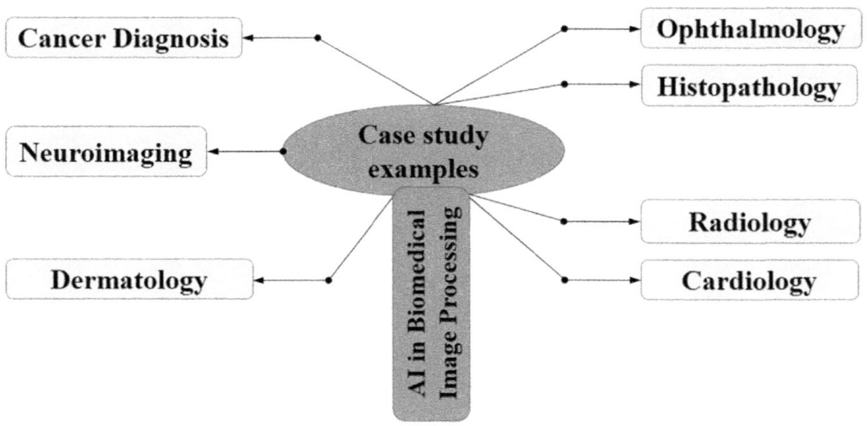

FIGURE 19.7 Case study examples in biomedical image processing.

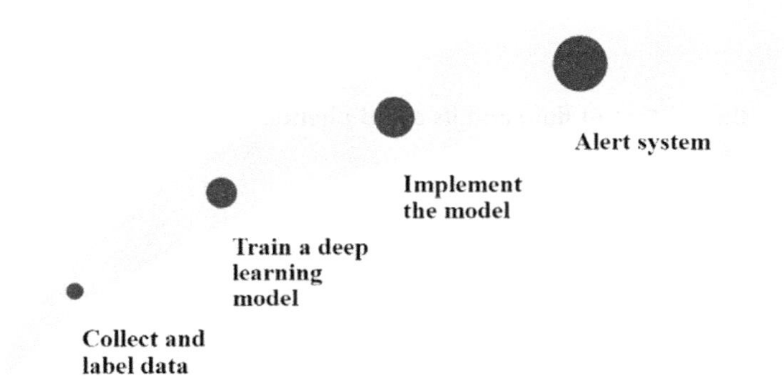

FIGURE 19.8 Use of artificial intelligence for baby incubator image processing.

We will need a dataset of images of premature infants inside the incubator. These images should be labeled with information about the infant's health status, such as heart rate, breathing rate, oxygen saturation, and other vital signs. Fig. 19.8 shows the use of AI for baby incubator image processing. Use a deep learning framework like TensorFlow or PyTorch to train a model on the labeled dataset. The model should be able to detect and analyze vital signs in the images. Once the model is trained, it can be implemented in the incubator system to continuously monitor the infant's health status. The images captured by the incubator camera can be analyzed in real-time using the trained model. The AI system can be set up to alert healthcare providers if the infant's vital signs fall outside of acceptable ranges. This can allow for early intervention and potentially prevent serious health issues. Some basic features that a baby incubator might need are temperature control, humidity control, and oxygen supply. The incubator should have a temperature control system that can maintain a stable temperature range between 32°C and 37°C. The incubator should have a humidification system that can maintain a stable humidity level of 40%–60%. The incubator should have a system for delivering oxygen to the baby if necessary. These are just some basic features that a baby incubator might need, and the actual design and code would depend on the specific requirements and constraints of the project. Fig. 19.9 shows the use of a neural network. It is important to consult with medical professionals and experts in this field before designing or implementing any kind of medical device. An artificial neural network (ANN) model can be a powerful tool for biomedical image processing of baby incubators. By accurately identifying and analyzing different parts of the incubator, the model can help to ensure that the incubator is functioning properly and that the baby is receiving the necessary care.

The neural network can also be used to examine data from heart rate and respiratory rate monitors, to perceive signs of distress or abnormalities in the baby's vital signs. This can help the aware medical staff of potential problems and allow them to take suitable action to address the issue. Furthermore, a neural network can be trained to predict outcomes based on many factors, such as the baby's gestational age, weight, and medical history. This can aid the medical workforce in making informed decisions about treatment and care for the baby. Overall, AI tools can help to advance

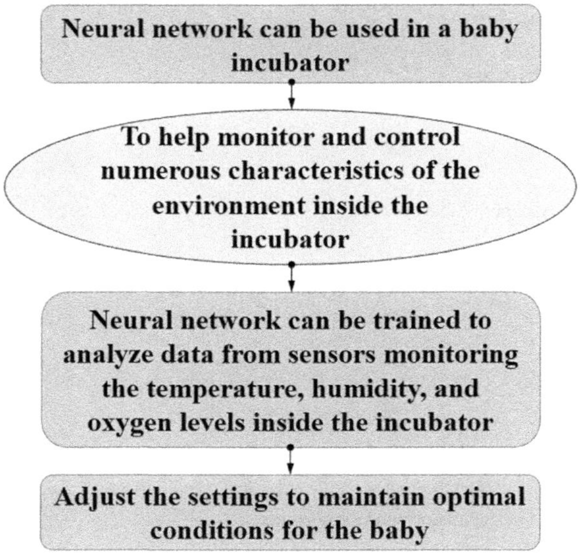

FIGURE 19.9 Use of neural network.

TABLE 19.2 Comparison of different AI techniques in the context of fluid and its allied phenomena.

References	Computing techniques	Remarks
Galan et al. [12]	ML	Use of ML to higher volumetric flow rate experiment for new material discovery.
Brunton et al. [13]	ML	Researchers have discussed the limitations and strengths of ML techniques to the fluid data.
Karniadakis et al. [14]	Deep neural networks (DNNs)	Authors have investigated bubble progression dynamics by DNN techniques, which save the simulation time.
Sofos et al. [15]	AI, ML	DNN has been widely used for mechanics and dynamics of fluid-related phenomena and images.
Zhou et al. [16]	AI	Authors concluded that an AI has a huge potential to control and optimization of turbulence. It will be helpful in near future to identify and visualize novel fluid flow structure.
Zhu et al. [17]	ML (Xtreme gradient boosting (Xgboost models) and ANN)	Authors concluded that the performance of Xgboost models on small data points are better than ANN model. The ANN model is the best choice for large data points of multiphase flow analysis.
Agostini [18]	ML	The dimensionality problem in fluid flow and its associated phenomena can be reduced by ML algorithms with autoencoder.
Gao et al. [19]	DNNs, CNN	The authors have discussed denoising the fluid data for a two-dimensional fluid flow field. They have explained the importance of the treatment of noisy data in the field of biomedical, mechanical, and aerospace engineering.

the worth of care for premature or sick infants by providing more accurate and timely information to medical staff and allowing for more precise control of the environment inside the incubator. Galan et al. [12] have broadly reviewed about smart microfluidics. Table 19.2 shows the comparison of different AI techniques in the context of fluid and its allied phenomena. They had discussed about the use of AI techniques for design and development of smart microfluidics system. They have also discussed the status of ML techniques on smart microfluidics system and associated challenges.

Reeder and Samimy [20] have experimentally visualized the real-time image of the fluid flow used through water tunnels and dual cameras.

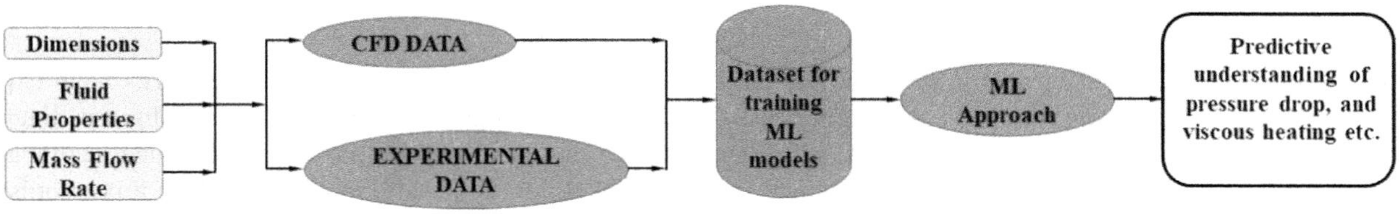

FIGURE 19.10 Overview of machine learning methodologies in die performance prediction [21].

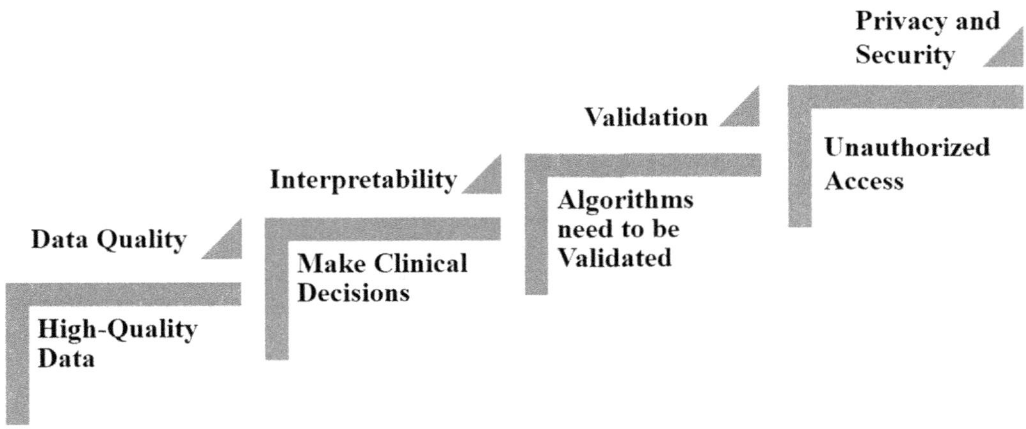

FIGURE 19.11 Challenges that need to be addressed.

McIntyre et al. [21] have done a critical review of the ML approach for the design and control of microfluidics systems. Fig. 19.10 depicts the overview of ML methodologies in die performance prediction.

It is important to note that AI should not replace human caregivers and medical professionals. Rather, it can be used as a tool to assist healthcare providers in monitoring premature infants and providing the best possible care.

19.5 Prospects and challenges of artificial intelligence in biomedical application

Despite the many benefits of AI in biomedical, some challenges need to be addressed. Some of the key challenges are shown in Fig. 19.5.

AI algorithms require high-quality data to learn from, and medical image data can be noisy, incomplete, or subject to artifacts. Fig. 19.11 shows the challenges that need to be addressed. AI algorithms can be difficult to interpret, making it hard to understand how they arrive at their conclusions. This can be a concern for doctors who need to make clinical decisions based on the results of AI analysis. AI algorithms need to be validated on large datasets to ensure their accuracy and reliability. This can be time-consuming and expensive. Medical image data are sensitive and must be protected from unauthorized access. AI algorithms must be designed with security and privacy in mind.

This chapter discusses the implementation of AI algorithms in the fields of fluid-material dynamics. Some good papers have been reported with their artificial intelligence models. Wang and Wang [1] have published a good review in this field. Interested readers can refer to more details. In the last 12 years, the number of publications has increased rapidly in artificial intelligence. Recently, Sofos et al. [9] discussed opportunities and advances in AI algorithms for high-fidelity fluid research. The author has found the gap between AI and fluid–solid interaction and might be the emerging area in this direction. The progress of AI in fluid-material dynamics forms a new research example along with imposing new challenges. The rapid development in the field of AI creates novel research, which is called intelligent fluid-material dynamics. Still, there is less work reported in the direction of the behavior of fluid and its interaction with material using different AI techniques at cryogenics temperature. The algorithms should be capable to captured complex 3D turbulent mixing phenomena and unsteady flow physics in the context of fluid-material dynamics. The challenges are not a barrier, and they motivate us to develop some effective AI techniques in the future. These challenges and research gaps will help us to effectively understand fluid mechanics and dynamics fundamentals.

19.6 Conclusion

AI-based methods such as deep learning, convolutional neural networks, and recurrent neural networks have exposed favorable results in numerous biomedical imaging tasks such as image segmentation, registration, disease diagnosis, and image reconstruction. The application of AI in biomedical imaging has the potential to revolutionize healthcare by improving accuracy, speed, and efficiency in medical imaging modalities. Still, additional research is a prerequisite to validate the reliability and efficiency of AI-based techniques in clinical practice. AI has the prospective to transform biomedical image processing, providing faster and more accurate analysis of medical images and enabling personalized treatment plans. While there are challenges to overcome, the benefits of AI are clear, and the field is likely to see continued growth and development in the ages to come.

References

[1] B. Wang, J. Wang, Application of artificial intelligence in computational fluid dynamics, Ind. Eng. Chem. Res. 60 (7) (2021) 2772–2790. Available from: https://doi.org/10.1021/acs.iecr.0c05045.

[2] S. Pawar, S.E. Ahmed, O. San, A. Rasheed, An evolve-then-correct reduced order model for hidden fluid dynamics, Mathematics 8 (2020) 570. Available from: https://doi.org/10.48550/arXiv.1911.02049.

[3] M. Babanezhad, I. Behroyan, A. Taghvaie Nakhjiri, A. Marjani, S. Shirazian, Simulation of liquid flow with a combination artificial intelligence flow field and Adams–Bashforth method, Sci. Rep 10 (2020) 16719. Available from: https://doi.org/10.1038/s41598-020-72602-6.

[4] Y. Kumar, S. Gupta, R. Singla, Y.-C. Hu, A systematic review of artificial intelligence techniques in cancer prediction and diagnosis, Arch. Comput. Methods Eng. 29 (2022) 2043–2070. Available from: https://doi.org/10.1007/s11831-021-09648-w.

[5] H.J. Yoon, Y. Jin Jeong, H. Kang, J.E. Jeong, D.-Y. Kang, Medical image analysis using artificial intelligence, Prog. Med. Phys. 30 (2) (2019). Available from: https://doi.org/10.14316/pmp.2019.30.2.49.

[6] R. Golan, C. Jacob, J. Denzinger, Lung nodule detection in CT images using deep convolutional neural networks, International Joint Conference on Neural Networks (IJCNN), 2016. https://doi.org/10.1109/IJCNN.2016.7727205.

[7] R. Li, W. Zhang, H.-I. Suk, L. Wang, J. Li, D. Shen, et al., Deep learning based imaging data completion for improved brain disease diagnosis, Med. Image Comput. Comput Assist. Interv. 17 (0 3) (2014) 305–312. doi; 10.1007%2F978-3-319-10443-0-39.

[8] Y. He, H. Northrup, H. Le, A.K. Cheung, S.A. Berceli, Y.T. Shiu, Medical image-based computational fluid dynamics and fluid-structure interaction analysis in vascular diseases, Front. Bioeng. Biotechnol. 27 (2022). Available from: https://doi.org/10.3389/fbioe.2022.855791.

[9] D.K. Baghel, S.L. Sinha, S.K. Dewangan, SST K -ω based air flow and heat transfer assessment in an infant incubator, J. Comp. Appl. Res. Mech. Eng. 12 (2) (2023) 161–175. Available from: https://doi.org/10.22061/jcarme.2022.7590.2010.

[10] K. Duraisamy, G. Iaccarino, H. Xiao, Turbulence modeling in the age of data, Annu. Rev. Fluid Mechanics. 51 (2019) 357–377. Available from: https://doi.org/10.1146/annurev-fluid-010518-040547.

[11] P. Mesejo, R. Martos, Ó. Ibáñez, J. Novo, M. Ortega, A survey on artificial intelligence techniques for biomedical image analysis in skeleton-based forensic human identification, Appl. Sci. 10 (2020) 4703. Available from: https://doi.org/10.3390/app10144703.

[12] E.A. Galan, H. Zhao, X. Wang, Q. Dai, W.T.S. Huck, S. Ma, Intelligent microfluidics: the convergence of machine learning and microfluidics in materials science and biomedicine, Matter 3 (2020) 1893–1922. Available from: https://doi.org/10.1016/j.matt.2020.08.034.

[13] S.L. Brunton, B.R. Noack, P. Koumoutsakos, Machine learning for fluid mechanics, Annu. Rev. Fluid Mech. 52 (2020) 477–508. Available from: https://doi.org/10.1146/annurev-fluid-010719-060214.

[14] C. Lin, Z. Li, L. Lu, S. Cai, M. Maxey, G.E. Karniadakis, Operator learning for predicting multiscale bubble growth dynamics, J. Chem. Phys. 154 (2021) 104118. Available from: https://doi.org/10.1063/5.0041203.

[15] F. Sofos, C. Stavrogiannis, K.K. Exarchou-Kouveli, D. Akabua, G. Charilas, T.E. Karakasidis, Current trends in fluid research in the era of artificial intelligence: a review, Fluids 7 (2022) 116. Available from: https://doi.org/10.3390/fluids7030116.

[16] Y. Zhou, D. Fan, B. Zhang, R. Li, B.R. Noack, Artificial intelligence control of a turbulent jet, J. Fluid Mech. 897 (2020) A27. Available from: https://doi.org/10.1017/jfm.2020.392.

[17] L.-T. Zhu, J.-X. Tang, Z.-H. Luo, Machine learning to assist filtered two-fluid model development for dense gas–particle flows, AIChE J. 66 (2020) e16973. Available from: https://doi.org/10.1002/aic.16973.

[18] L. Agostini, Exploration and prediction of fluid dynamical systems using auto-encoder technology, Phys. Fluids 32 (2020) 067103. Available from: https://doi.org/10.1063/5.0012906.

[19] H. Gao, L. Sun, J.-X. Wang, Super-resolution and denoising of fluid flow using physics-informed convolutional neural networks without high-resolution labels, Phys. Fluids 33 (2021) 073603. Available from: https://doi.org/10.1063/5.0054312.

[20] M.F. Reeder, M. Samimy, The evolution of a jet with vortex-generating tabs: real-time visualization and quantitative measurements, J. Fluid Mech. 311 (1996) 73–118. Available from: https://doi.org/10.1017/S0022112096002510.

[21] D. McIntyre, A. Lashkaripour, P. Fordyce, D. Densmore, Machine learning for microfluidic design and control, Lab. Chip 22 (2022) 2925. Available from: https://doi.org/10.1039/D2LC00254J.

Chapter 20

MRI and SPECT image fusion based on adaptive fusion rule in transform domain

A.D. Prasad[1] and Padma Ganasala[2]

[1]Civil Engineering Department, National Institute of Technology Raipur, Raipur, Chhattisgarh, India, [2]Electronics and Communications Engineering, Gayatri Vidya Parishad College of Engineering, Visakhapatnam, Andhra Pradesh, India

20.1 Introduction

MRI is an anatomical imaging modality. MRI image has high spatial resolution and gives the clear picture of the anatomy. The lesions or abnormalities can be identified in MRI image. However, the physiological status of the identified lesion is not present in MRI, which is very essential for treatment planning. This functional information is available in the SPECT image. This has poor spatial resolution and hence it is difficult to localize the identified abnormality. Image fusion is used to fill this gap and make use of the complementary information given by different imaging modalities. The main purpose of MRI and SPECT fusion is to get a composite image in which functional information given by SPECT image is overlaid on anatomical details given by MRI image without distortion. Thereby, functional image enriched with anatomical details is achieved. This is useful for precise diagnosis and treatment planning in many specialties. Through image fusion, it is possible to reveal crucial diagnostic information present in the individual images which may not be perceived when they are viewed independently.

Multimodality medical image fusion is characterized as hardware-based fusion and software-based fusion. The dual imaging modalities like positron emission tomography and computed tomography (PET-CT) scanner provide the fused image at the acquisition stage itself and is an example for hardware-based fusion. However, designing dual scanners for combinations like MRI and SPECT is a challenging task because of the strong interference of signals. Currently, those are under developing stage. In case of software based image fusion, input images are acquired separately by means of individual imaging systems. These acquired source images are registered to fix the spatial orientation and resolution imbalances. The registered images are fused using a software algorithm. Software-based fusion methods have received a lot of research attention over the past decade. Though, there is a tremendous progress in medical image fusion methods over the past decade, many of the methods fail to provide optimal fusion results in terms of fusion accuracy and computational efficiency. The existing methods suffer from spectral or spatial distortion in certain cases of source images. This issue has been addressed in the present work.

Transparency techniques were used for MRI and SPECT image fusion [1]. In this method, MRI is made as background and SPECT as foreground. In this method, using a fixed value of transparency factor in combining SPECT and MRI resulted in a dark look for the fused image. Later, this is improved by using variable weighting factor through simplex optimization [2]. Though variable weight is used here, it is still a weighted averaging method. Therefore the fused image is deprived of contrast.

MRI and positron emission tomography (PET) images are fused in intensity hue saturation (IHS) color model [3]. In this methodology, SPECT or PET image (pseudocolor image) is converted to IHS color space, then its intensity component is replaced with MRI image and then converted back to red, green and blue (RGB) color space. In this process, there is a possibility for loss of information as the intensity component of SPECT image is completely discarded. He et al. [4] used principle component analysis to combine MRI and intensity component.

The important task in image fusion is to transfer the source image details into the fused image. Hence, there is necessity for the tool that extract the details present in the source images accurately. Multiresolution transform can do this task precisely. Transforms like discrete wavelet transform (DWT), nonsubsampled contourlet transform

Intelligent Computing Techniques in Biomedical Imaging. DOI: https://doi.org/10.1016/B978-0-443-15999-2.00001-3

(NSCT), and nonsubsampled shearlet transform (NSST) have been applied for image fusion in the literature. The process involved in transform domain image fusion is to apply the transform on input images and obtain approximate and detail information. Approximate and detail information of the source images is combined by the corresponding fusion strategies. The quality of the fusion output depends on the transform employed and the fusion strategy followed. Gaussian filtering-based preprocessing followed by DWT fusion in IHS color space is employed in Ref. [5]. NSCT is used to combine intensity component (I) of SPECT with MRI image [6]. In this method, I and MRI are fragmented into low frequency (LF) and high frequency (HF) components by NSCT. Then, LF components are combined using energy rule and HF components are combined using variance rule. However, NSCT is computationally expensive. Combination of DWT and Contourlet transform is also used for SPECT and MRI fusion [7]. Both DWT and contourlet transform are shift variant transforms. As a result, the fused image would have ringing artifacts.

NSST is applied for image fusion [8,9]. In this methodology, an important thing is how subbands are combined after NSST decomposition. The framework used for combining subband is called as fusion rule. Wider range of fusion rules suggested by the researchers are to improve fusion quality. These span from simple averaging to neural network-based fusion rules and, recently, deep convolution neural network-based fusion rules [10]. Maximum selection and spatial frequency rules are employed in transform domain for MRI and SPECT image fusion [11]. Pulse coupled neural network (PCNN)-based fusion approach is applied in NSST domain for medical image fusion in Ref. [12]. PCNN-based fusion rules in NSCT domain are adapted for image fusion [13]. Parameters of the dual channel PCNN are estimated using fractal dimension in Ref. [14]. Selection of suitable values for PCNN parameters is one of the main issues in these methods. Also, the computational complexity of these methods is relatively high. Li et al [15] proposed Laplacian re-decomposition for medical image fusion. Laplacian pyramid-based decomposition and convolutional neural network (CNN) reconstruction are employed for image fusion [16]. These methods have high computational complexity. Seal et al. [17] proposed PET-CT image fusion method based on random forest and shift invariant wavelet transform. Recently, optimization techniques such as Grasshopper optimization [18] and Adolescent identity search [19] algorithms have been adapted for fusion in NSST domain. However, the efficiency of medical image fusion method can be further improved by implementing pre- or postenhancing methods. Another factor which influences the SPECT and MRI image fusion is color space in which fusion is performed. Performance evaluation of color spaces for SPECT and MRI fusion is performed in Ref. [20]. It is inferred that the luminance, in-phase, quadrature (YIQ) color model is suitable for image fusion. Coupled neural P systems-based fusion rules are employed in NSST domain [21]. Bilateral filter-based double layer decomposition and gradient energy-based fusion rules are employed in Ref. [22]. However, in these methods, due to unavoidable loss of information, suboptimal fusion results have been obtained.

Recently, deep learning methodology is being applied for fusion of various categories of images. CNN and sparse representation-based methodology is adapted for fusion of medical images [23]. Encoder-decoder-based network has been utilized for fusion in Ref. [24]. Dense convolutional network with dual attention mechanism is applied for multimodality medical image fusion [25]. All these CNN-based models require extensive training and fail to generalize in certain cases.

In the proposed method, MRI and SPECT image fusion is performed in YIQ color space. Conversion of RGB color space to YIQ color space is described in the following section. Based on the radio pharmaceutical used and the organ/tissue being imaged, the information content of SPECT image may be different. If the SPECT image consists of less information, then most of its area appears dark. In those regions, the anatomical details given by MRI image should turn up. If the SPECT image has high information content, that should get blended with MRI details. The proposed methodology solves these issues by using two different fusion rules. Further, the fused image is enhanced through post-enhancement method based on morphological features.

20.2 Materials

20.2.1 Nonsubsampled shearlet transform

NSST is a shift invariant multiscale transform. It can extract detail information at different scales through Laplacian pyramid decomposition in nonsubsampled form and directional decomposition is obtained through nonsubsampled shear filters. Shearlet filters make the transform less computationally complex. NSST's shift invariance property and high directional sensitivity makes it suitable for image fusion.

20.2.2 YIQ color model

YIQ color model is a de-correlative color model. In this color model, intensity information is extracted into Y channel and chromatic information is extracted into I channel and Q channel. The RGB into YIQ conversion formula is given in Eq. (20.1).

$$\begin{bmatrix} Y \\ I \\ Q \end{bmatrix} = \begin{bmatrix} 0.299 & 0.587 & 0.114 \\ 0.596 & -0.275 & -0.321 \\ 0.212 & -0.523 & 0.311 \end{bmatrix} \begin{bmatrix} R \\ G \\ B \end{bmatrix} \tag{20.1}$$

In YIQ color model, the in-phase channel I and quadrature channel Q carries chromatic information. The luminance channel (Y) contains intensity information.

20.3 Proposed method

The proposed methodology is shown in Fig. 20.1. The SPECT image is converted from RGB into YIQ color space by using the transformation given in Eq. (20.1). The luminance channel (Y) is fused with MRI image by adaptive fusion rule-based NSST fusion scheme to get a new luminance channel (Y_{NEW}) containing entire anatomical information without any distortion. Further, this Y_{NEW} is enhanced through morphological image enhancement method to highlight the vital anatomical details in Y_{NEW} and is denoted as Y_{EN}. This Y_{EN} along with original chromatic channels I and Q are transformed again to RGB color space that gives the fused functional and anatomical image.

20.3.1 Nonsubsampled shearlet transform-based fusion method with adaptive fusion rule

Based on the organ being imaged in functional and anatomical image fusion, two different cases arise. In one case, the functional image would consist of very less information, and in other case, the functional image would consists of more information. If the functional image consists of less information, all the energy of the anatomical image must be transferred to the fused image. If the functional image consists of more information, then details of both functional and anatomical images are to be transferred to the fused image. Hence, these two conditions are handled differently in the proposed method.

The approximate and detail information of Y and MRI image are extracted into LF and HF subbands using NSST. Then LF information is fused using adaptive LF fusion rule discussed in the following section. HF information is fused by means of absolute maximum fusion rule. Fig. 20.2 shows that inverse NSST of fused LF and HF subbands gives the new Y component (Y_{New}) with all anatomical details.

20.3.2 Adaptive low frequency fusion rule

Depending on the amount of information present in the functional image, fusion rules are to be adapted. Based on the amount of information present in the source images, suitable saliency matrix must be employed in the fusion rule. The amount of information present in an image can be estimated by its entropy value. Entropy of an image is calculated by using Eq. (20.2).

$$E = -\sum_{i=1}^{n} p_i \log_2(p_i) \tag{20.2}$$

If the entropy of the luminance component Y of SPECT is greater than the fixed threshold, which implies that functional image has more information, then texture energy measure-based fusion rule is employed. If the entropy of Y is less than the threshold (T), which implies that the functional image has less information, then the local energy centered fusion strategy is employed to combine the LF subbands.

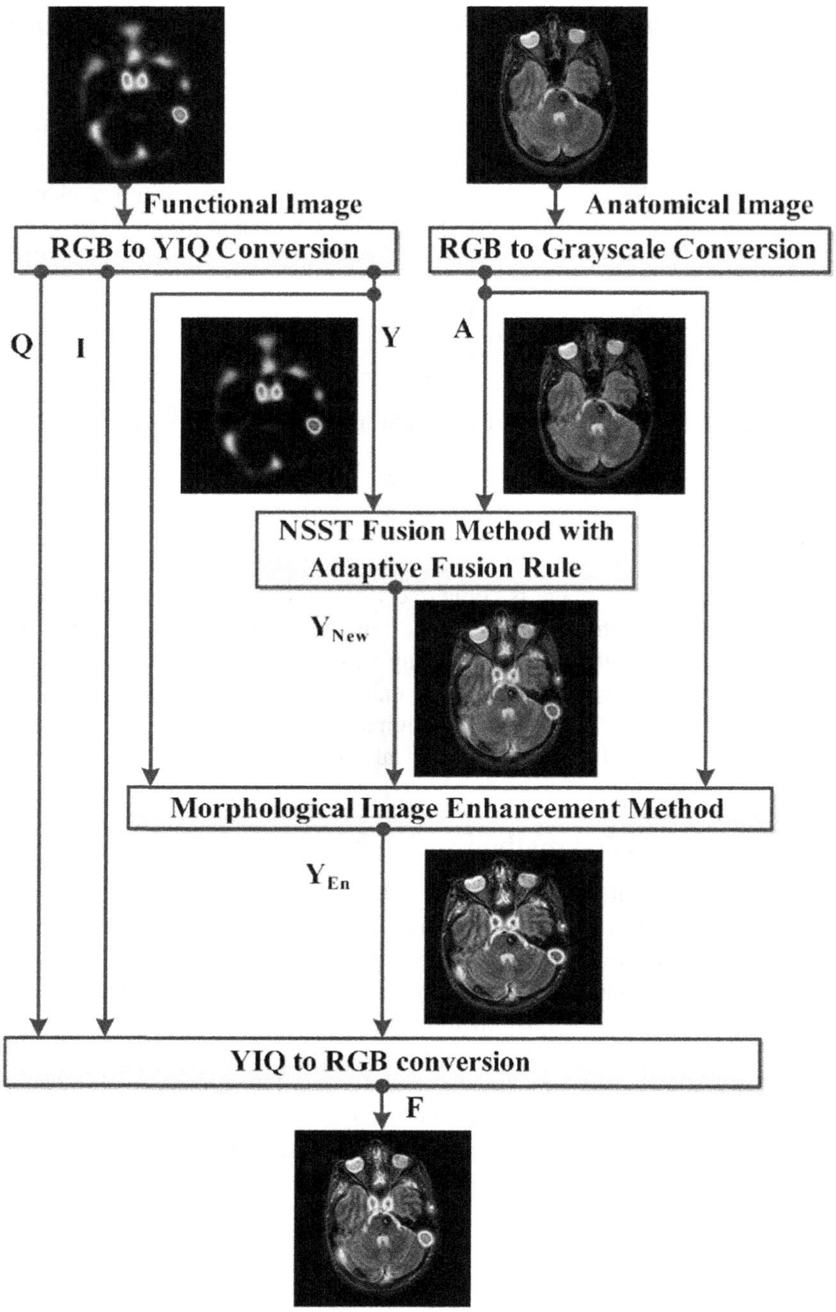

FIGURE 20.1 The proposed MRI and SPECT image fusion framework.

20.3.2.1 Texture-based low frequency fusion rule

The texture energy measure is obtained from three basic vectors: [1,2], [-1, 0, 1], and [-1, 2, -1]. These are equivalent to weighted averaging, edge recognition, and spot recognition, respectively. Each of these vectors are convolved with itself or the other ones. Through this process, nine 3 × 3 filters are molded. These are listed as follows:

$$m_1 = \begin{bmatrix} 1 & 2 & 1 \\ 2 & 4 & 2 \\ 1 & 2 & 1 \end{bmatrix} \tag{20.3}$$

$$m_2 = \begin{bmatrix} -1 & 0 & 1 \\ -2 & 0 & 2 \\ -1 & 0 & 1 \end{bmatrix} \tag{20.4}$$

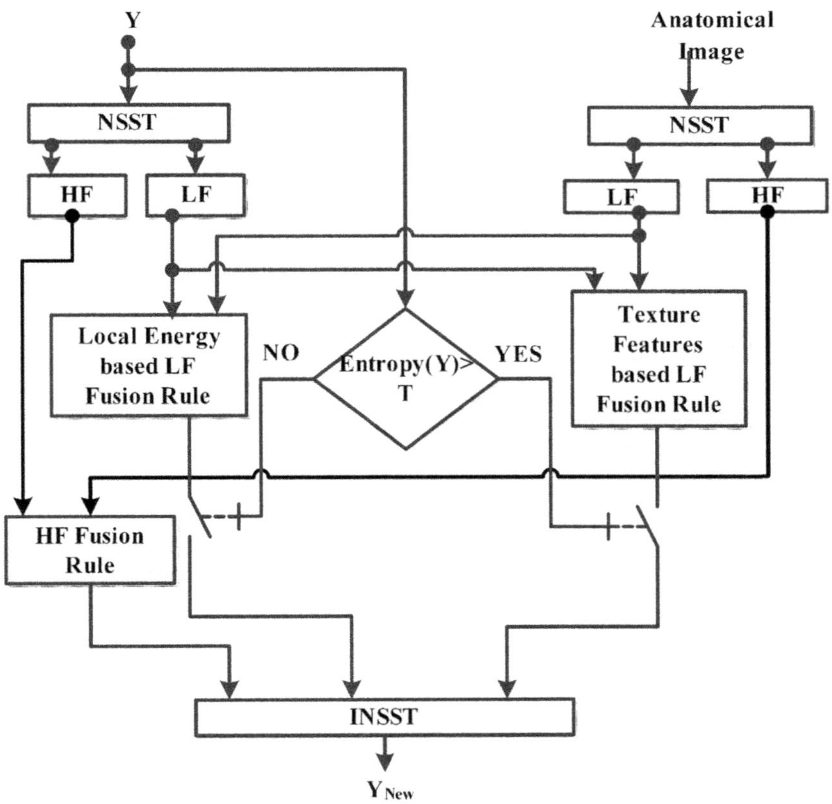

FIGURE 20.2 The proposed adaptive fusion rule-based fusion method in nonsubsampled shearlet transform domain.

$$m_3 = \begin{bmatrix} -1 & 2 & -1 \\ -2 & 4 & -2 \\ -1 & 2 & -1 \end{bmatrix} \tag{20.5}$$

$$m_4 = \begin{bmatrix} 1 & 0 & -1 \\ 0 & 0 & 0 \\ -1 & 0 & 1 \end{bmatrix} \tag{20.6}$$

$$m_5 = \begin{bmatrix} 1 & -2 & 1 \\ 0 & 0 & 0 \\ -1 & 2 & -1 \end{bmatrix} \tag{20.7}$$

$$m_6 = \begin{bmatrix} -1 & -2 & -1 \\ 0 & 0 & 0 \\ 1 & 2 & 1 \end{bmatrix} \tag{20.8}$$

$$m_7 = \begin{bmatrix} 1 & -2 & 1 \\ -2 & 4 & -2 \\ 1 & -2 & 1 \end{bmatrix} \tag{20.9}$$

$$m_8 = \begin{bmatrix} -1 & -2 & -1 \\ 2 & 4 & 2 \\ -1 & -2 & -1 \end{bmatrix} \tag{20.10}$$

$$m_1 = \begin{bmatrix} 1 & 0 & -1 \\ -2 & 0 & 2 \\ 1 & 0 & -1 \end{bmatrix} \tag{20.11}$$

The mask m_1 gives the average energy of the image. The masks m_2 and m_3 extract edge details. The masks m_6 and m_8 provide horizontal edges. The mask m_4 gives the waves. Masks m_5 and m_9 extract ripples in the image. The activity of LF subband coefficient is formulated by using these 3 × 3 texture masks. Let LY be the LF subband of the Y component. The texture energy measures TLY_1, TLY_2, TLY_3, TLY_4, TLY_5, TLY_6, TLY_7, TLY_8, and TLY_9 corresponding to the filters m_1, m_2, m_3, m_4, m_5, m_6, m_7, m_8, and m_9 are given by Eq. (20.12).

$$TLY_i = \sum_{p=-1}^{1}\sum_{q=-1}^{1} LY(i+p,j+q)^2 m_i(p,q) \tag{20.12}$$

Let the normalized texture energy measures be $NTLY_i$.

$$NTLY_i = \text{normalized}(|TLY_i|) \tag{20.13}$$

Finally, the saliency ALY of the LF subband LY is given by Eq. (20.14).

$$ALY = \sum_{i=1}^{9} NTLY_i \tag{20.14}$$

Similarly, the saliency measure ALM for LF subband LM of the MRI image M is calculated. Fused LF subband LF is obtained by comparing these activity measures.

$$LF(i,j) = \begin{cases} LY(i,j) & \text{if} ALY(i,j) \geq ALM(i,j) \\ LM(i,j) & \text{otherwise} \end{cases} \tag{20.15}$$

20.3.2.2 Energy-based low frequency fusion rule

Local energy at each coefficient of LF subband is calculated by using mask m.

$$m = \frac{1}{16}\begin{bmatrix} 1 & 2 & 1 \\ 2 & 4 & 2 \\ 1 & 2 & 1 \end{bmatrix} \tag{20.16}$$

$$ALY(i,j) = \sum_{r=-1}^{1}\sum_{s=-1}^{1} LY(i+r,j+s)^2 m(r+2,s+2) \tag{20.17}$$

$$ALM(i,j) = \sum_{r=-1}^{1}\sum_{s=-1}^{1} LM(i+r,j+s)^2 m(r+2,s+2) \tag{20.18}$$

$$LF(i,j) = \begin{cases} LY(i,j) & \text{if} ALY(i,j) \geq ALM(i,j) \\ LM(i,j) & \text{otherwise} \end{cases} \tag{20.19}$$

20.3.3 High frequency fusion rule

HF subband coefficients hold edge features of input images. HF subband coefficients where edges are present would have high absolute value related to those of the even sections. Therefore the activity HF subband coefficient is formulated by way of the sum of absolute value of coefficients within a 3 × 3 window. At each location of the HF subband HY of luminance component Y, the saliency AHY is calculated by using Eq. (20.20). Similarly, the saliency AHM of HF subband HM of MRI is calculated by using Eq. (20.21). The fused HF subband coefficients are obtained by using Eq. (20.22).

$$AHY(i,j) = \sum_{p=-1}^{1}\sum_{q=-1}^{1} abs(AHY(i+p,j+q)) \tag{20.20}$$

$$AHM(i,j) = \sum_{p=-1}^{1}\sum_{q=-1}^{1} abs(AHM(i+p,j+q)) \tag{20.21}$$

$$HF(i,j) = \begin{cases} HY(i,j) & \text{if } AHY(i,j) \geq AHM(i,j) \\ HM(i,j) & \text{otherwise} \end{cases} \tag{20.22}$$

20.3.4 Morphological features-based fused image enhancement

Anatomical and functional details revealed by MRI and SPECT image are highlighted in the fused image through morphological feature-based enhancement method. The bright features of luminance component Y of SPECT image and MRI image are extracted through white top hat transform (WTHT). The WTHT extracts the bright features by subtracting the morphological opened image corresponding to a structuring element (SE) from the source image. Bright features of different sizes are extracted using SE of different sizes. The bright features corresponding to SE_i to be used for enhancement of fused image are obtained by taking the maximum of bright features of Y and MR. The final bright features BF^F are obtained by adding bright features corresponding to all SEs.

$$BF_i^Y = WTHT(Y, SE_i) \tag{20.23}$$

$$BF_i^{MR} = WTHT(MR, SE_i) \tag{20.24}$$

$$BF_i^F = \max(BF_i^Y, BF_i^{MR}) \tag{20.25}$$

$$BF^F = \sum_{i=1}^{n} BF_i^F \tag{20.26}$$

The dark features are extracted through black top hat transform (BTHT). BTHT extracts dark features by subtracting the original image from the morphological closed image. Again, to get the dark features having different sizes and shapes, the structuring elements having different sizes are used. The dark features of Y corresponding to SE_i are denoted as $DF^Y{}_i$ and are obtained by Eq. (20.27). Similarly, the dark features of MR corresponding to SE_i are denoted as $DF^{MR}{}_i$ and are obtained by Eq. (20.28). The dark features $DF^F{}_i$ corresponding to SE_i are taken as the maximum of $DF^Y{}_i$ and $DF^{MR}{}_i$. Finally, the dark features used for enhancement are given by addition of $DF^F{}_i$ corresponding to all SEs.

$$DF_i^Y = BTHT(Y, SE_i) \tag{20.27}$$

$$DF_i^{MR} = BTHT(MR, SE_i) \tag{20.28}$$

$$DF_i^Y = \max(DF_i^Y, DF_i^{MR}) \tag{20.29}$$

$$DF^F = \sum_{i=1}^{n} DF_i^F \tag{20.30}$$

The enhanced Y component Y_{En} with all anatomical and functional details is obtained by mixing these bright and dark features with the fused image Y_{New} obtained by fusion of Y and MR.

$$Y_{EN} = Y_{New} + 0.5BF^F - 0.5DF^F \tag{20.31}$$

20.4 Results and discussions

Key emphasis of the presented research is to design an effective MRI and SPECT image fusion method. The spatial resolution of the fused SPECT and MRI image should be as high as that of the MRI. Its chromatic or color content must be like that of the SPECT image because different levels of functional activity are embodied with diverse colors in SPECT. In the evaluation of the fused images, these two aspects are to be considered.

The proposed adaptive fusion rule-based MRI and SPECT image fusion methodology is authorized on nine MRI and SPECT images [26]. MRI is a gray-scale image and SPECT is color image. Each image is of size 256×256. These are the images of the patients who suffered from various neurodegenerative diseases like brain attack, brain tumor, degenerative disease, and infectious disease. These are divided into two groups. The first group consists of five datasets as presented in Fig. 20.3. Second group contains three datasets (datasets 6–9) as shown in Fig. 20.4. The first column in Figs. 20.3 and 20.4 shows MRI images and the second column shows SPECT images. The effectiveness of the anticipated adaptive fusion rule is established over comparison with five existing approaches. The first technique is IHS color model. In this method, the fused image is obtained by replacing the intensity component of the SPECT image with MRI image. This method is referred as IHS in the further discussion. The second, third, and fourth methods use YIQ color model. In these methods, the intensity component of the SPECT and MRI are fused in NSST domain. The second technique utilizes energy-centered fusion rule aimed at LF subband fusion then the absolute maximum-based rule for HF subband fusion. This approach is represented as energy-based method in the further discussion. In case of the third

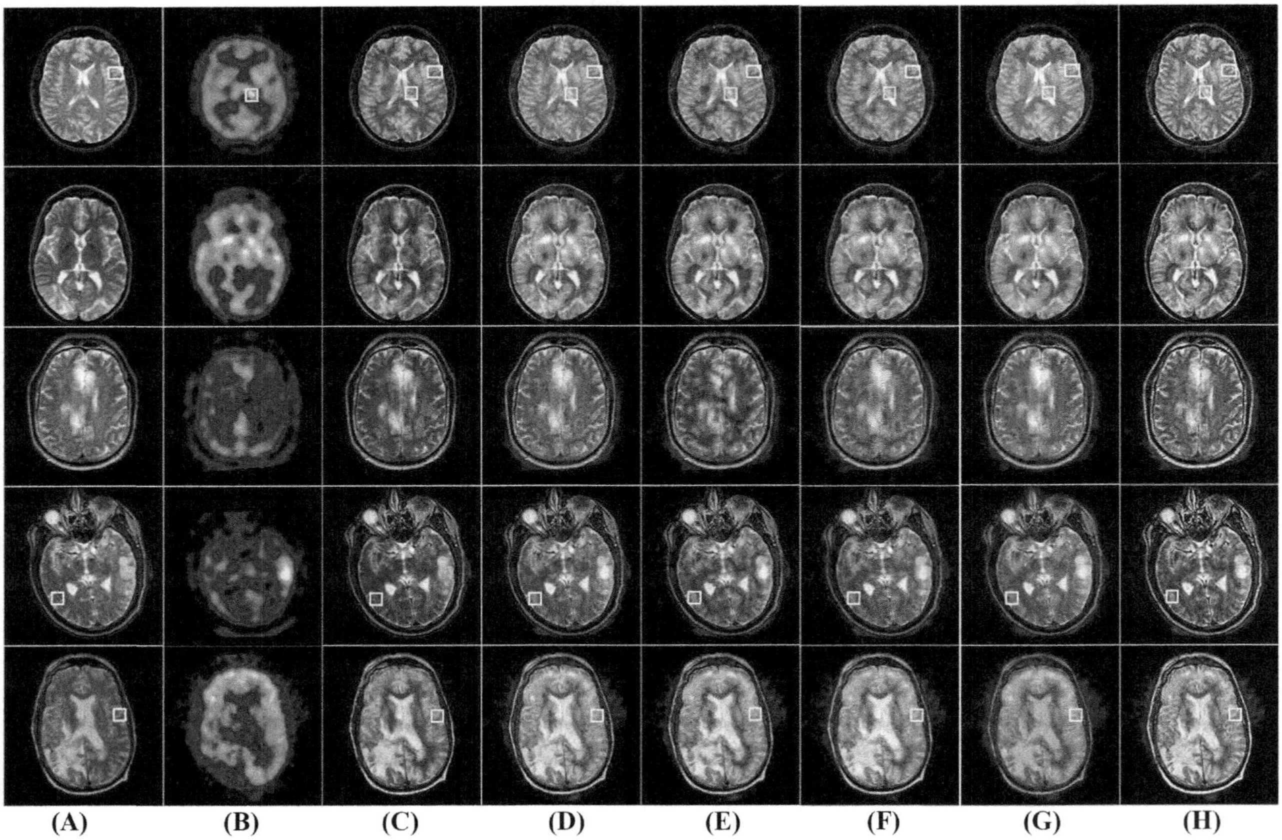

FIGURE 20.3 Fused images of datasets 1–5. (A) MRI images, (B) SPECT images, fused images by (C) IHS, (D) energy, (E) texture, (F) PAPCNN, (G) LRD, and (H) adaptive fusion method.

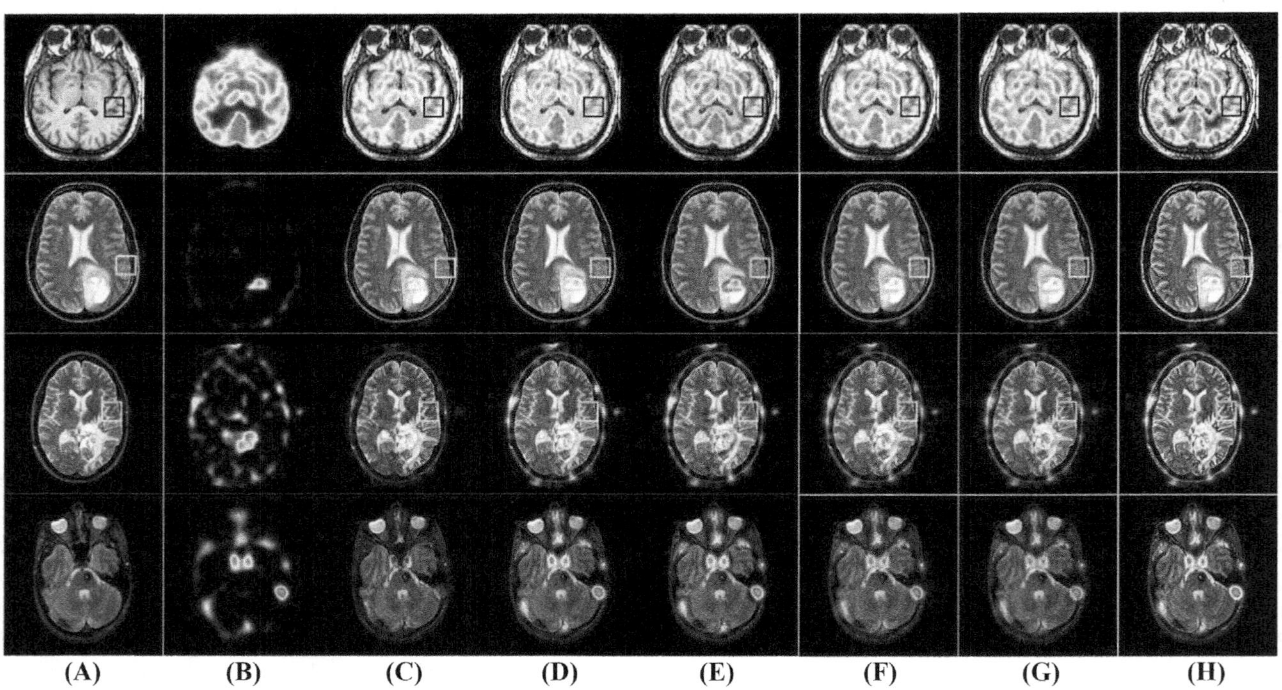

FIGURE 20.4 Fused images of datasets 6–9. (A) MRI, (B) SPECT (C) IHS, (D) energy, (E) texture, (F) PAPCNN, (G) LRD, and (H) adaptive fusion method.

method, the LF fusion rule is texture features-based fusion rule and this technique is mentioned as texture-based method. The fourth method uses parameter adaptive pulse coupled neural network (PAPCNN)-centered strategy for LF and HF subband fusion, and it is referred as PAPCNN method [27]. The fifth method is Laplacian re-decomposition based method [15] and is referred as LRD method. The proposed method is mentioned as adaptive fusion rule-based method. The fused images of datasets 1–5 by diverse approaches are presented in Fig. 20.3 and those of datasets 6–9 are presented in Fig. 20.4. All simulations are done using MATLAB R2020a software on a system with Intel core i5 processor and 8GB RAM.

IHS technique output is shown in Figs. 20.3C and 20.4C. These images suffer from spectral distortion, especially in the high activity regions of the SPECT image. High-activity regions in the SPECT image are white in color. However, they are not presented in similar way in fused images by IHS as shown in Figs. 20.3C. The output by energy-based technique is presented in Figs. 20.3D and 20.4D. These images have good spectral characteristics, that is, the color of these fused images is same as that of the SPECT images. The fused images by texture-based method are shown in Figs. 20.3E and 20.4E. The fused images shown in Fig. 20.4E suffer from spatial distortion in the regions of abnormality. The fused images by PAPCNN methods are shown in Figs. 20.3F and 20.4F. The output by LRD technique [15] is presented in Figs. 20.3G and 20.4G. The chromatic content of these fused images is different from that of the SPECT images. The output by the proposed technique (adaptive fusion method) is presented in Figs. 20.3H and 20.4H. These images have good spatial resolution than all other fused images and no spectral distortion is perceived in these images.

As the visual analysis of the fused images is independent in nature, quantifiable assessment of the experimental results is an essential task. In this work, the fused images by different methods are evaluated with the following image fusion quality metrics: [28] standard deviation (SD), visual information fidelity (VIF), edge index (EI), and spatial frequency (SF).

SD metric quantifies the clarity of the output images. The images having high SD value would have better clarity. Values of SD of the output images of nine datasets by different methods are listed in Fig. 20.5. Fused images by IHS method have less SD value compared to others for most of the datasets. This suggests that the clarity of these output images is poor compared to the other method. SD of the fused images by the proposed method (adaptive) is highest for all the nine datasets. This indicates that the output by the proposed technique has better clarity.

Another important metric used for image fusion quality assessment is VIF-based metric. VIF model the quality assessment problem as an information fidelity problem [29]. The range of VIF is [0 1]. The VIF calculated between an image and its copy is one. High VIF value infers that the more amount of information about source images is conveyed by the fused image [30]. VIF for different datasets is presented in Fig. 20.6. Energy-based fusion method has reported less value of VIF value for all the datasets except dataset 4 and 7. For these datasets, LRD method [15] is giving least value of VIF. The texture-based method and the proposed method are performing better with respect to VIF.

EI values of the fused images by different methods are depicted in Fig. 20.7. Mean of the magnitude of Sobel gradient calculated at each pixel is referred to as the EI. High EI value indicates that the strong edges are present in the fused image. Among all the methods compared, LRD method has reported a least value of EI for all the datasets. EI of energy- and texture-based methods is almost same. The proposed adaptive method recorded a highest value of EI for all datasets. This suggests that the output by the proposed technique is having strong edges and good contrast.

Spatial frequency (SF) of fused images is depicted in Fig. 20.8. SF of the output image is calculated by seeing the square of the difference between adjacent pixels along rows and columns. This metric indicates the total activity of the

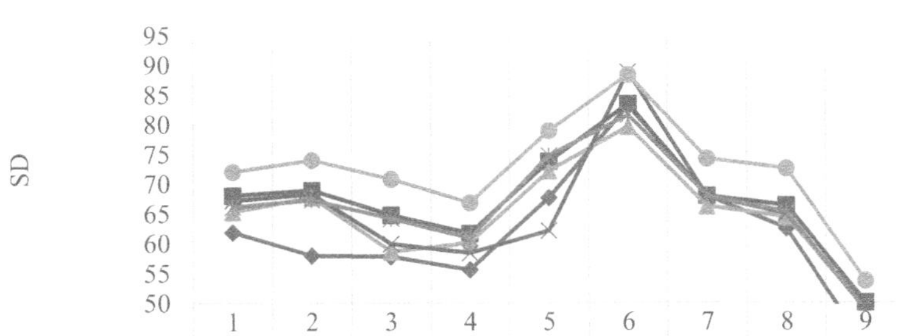

	1	2	3	4	5	6	7	8	9
IHS	61.88	57.97	57.77	55.56	67.61	83.11	67.88	62.47	42.48
Energy	68.05	68.93	64.75	61.65	73.71	83.43	67.88	66.35	50.07
Texture	65.30	67.75	58.58	60.09	72.15	79.59	66.15	64.36	48.98
LRD	67.18	68.28	59.91	58.39	62.10	88.79	67.93	64.92	49.10
PAPCNN	66.09	67.30	64.18	60.82	74.75	81.54	67.93	65.21	48.61
Adaptive	72.06	74.01	70.79	66.74	78.89	88.33	74.18	72.53	53.67

FIGURE 20.5 Standard deviation [26] values of nine fused images by different methods.

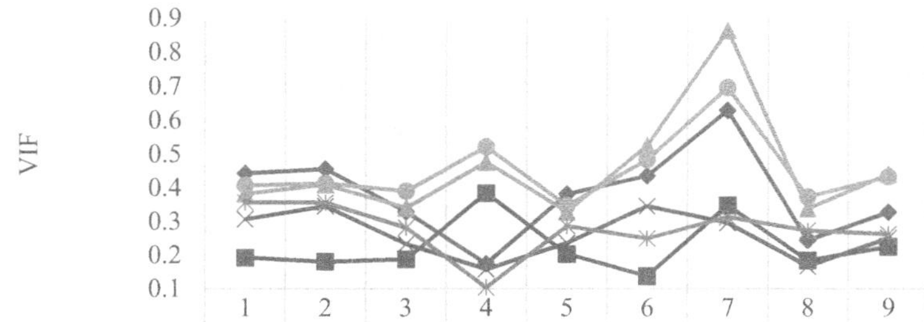

	1	2	3	4	5	6	7	8	9
IHS	0.39	0.41	0.28	0.12	0.33	0.39	0.58	0.19	0.28
Energy	0.14	0.13	0.14	0.33	0.15	0.09	0.30	0.13	0.17
Texture	0.33	0.36	0.29	0.43	0.28	0.48	0.81	0.29	0.39
PAPCNN	0.26	0.30	0.18	0.11	0.19	0.30	0.25	0.12	0.20
LRD	0.31	0.31	0.23	0.05	0.24	0.20	0.26	0.22	0.21
Adaptive	0.36	0.36	0.34	0.47	0.30	0.43	0.64	0.32	0.38

FIGURE 20.6 Visual information fidelity value of nine fused and source image pairs by different methods.

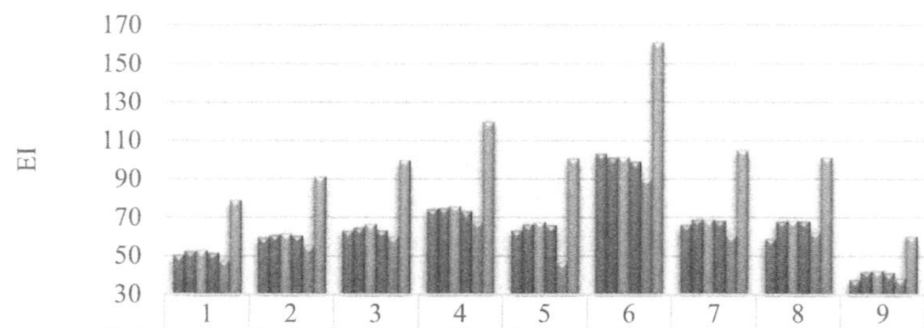

	1	2	3	4	5	6	7	8	9
HIS	50.50	59.75	63.20	74.50	63.37	103.33	66.39	59.09	37.80
Energy	52.41	61.04	64.88	75.16	66.69	101.09	69.11	68.38	42.08
Texture	53.08	61.71	66.83	76.04	67.77	101.42	69.01	68.72	42.42
PAPCNN	51.67	60.62	63.42	73.38	66.31	99.26	68.46	68.17	41.43
LRD	47.85	55.21	60.06	67.85	46.81	90.43	60.76	62.44	38.46
Adaptive	79.26	91.52	99.96	119.97	100.81	161.29	105.12	101.36	60.59

FIGURE 20.7 Edge index value of nine fused images by different methods.

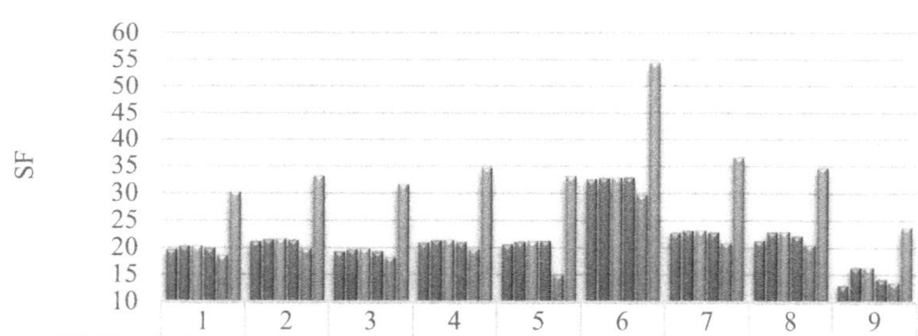

	1	2	3	4	5	6	7	8	9
HIS	19.72	21.20	19.36	21.00	20.68	32.67	22.87	21.32	13.14
Energy	20.30	21.61	19.74	21.41	21.17	32.89	23.22	22.99	16.39
Texture	20.43	21.72	19.90	21.47	21.27	32.95	23.25	23.03	16.44
PAPCNN	19.97	21.46	19.34	21.03	21.28	33.01	22.92	22.26	14.18
LRD	18.69	19.90	18.32	19.69	14.99	29.78	20.95	20.63	13.62
Adaptive	30.31	33.23	31.70	34.74	33.24	54.52	36.69	34.77	23.81

FIGURE 20.8 Spatial frequency of nine fused images by different methods.

output image. The output image having higher value of SF would have additional details. With respect to SF, LRD is performing poor compared to the others. SF values of fused images by the proposed technique are higher than that of the examined approaches.

Further, quantitative comparison of proposed method with two latest image fusion methods is presented in Table 20.1. One of the method is based on joint bilateral filter and gradient energy (BFGE) [22]. It is referred to as

TABLE 20.1 Quantitate comparison of image fusion methods.

Dataset no.	SF			EI			VIFQ			STD		
	BFGE [22]	NPS [21]	Adaptive	BFGE [22]	NPS [21]	Adaptive	BFGE [22]	NPS [21]	Adaptive	BFGE [22]	NPS [21]	Adaptive
1	19.18	19.45	30.31	50.86	50.56	79.26	0.36	0.32	0.36	66.83	66.52	72.06
2	14.10	21.21	33.23	40.65	59.93	91.52	0.26	0.30	0.36	49.43	63.42	74.01
3	19.03	19.27	31.70	63.95	63.62	99.96	0.29	0.23	0.34	63.96	63.96	70.79
4	20.89	21.13	34.74	74.46	73.93	119.97	0.10	0.11	0.47	61.19	60.64	66.74
5	16.00	21.27	33.24	51.85	66.16	100.81	0.26	0.23	0.30	62.73	72.54	78.89
6	31.81	32.38	54.52	94.37	96.72	161.29	0.34	0.32	0.43	80.33	77.49	88.33
7	22.63	22.70	36.69	67.66	67.82	105.12	0.29	0.25	0.64	66.91	66.83	74.18
8	20.96	21.22	34.77	64.36	65.16	101.36	0.21	0.15	0.32	63.98	64.15	72.53
9	14.10	15.02	23.81	40.65	40.89	60.59	0.26	0.16	0.38	49.43	48.61	53.67

TABLE 20.2 Computational time of different methods.

	IHS	Energy	Texture	PAPCNN	LRD	NPS	BFGE	Adaptive
Time (s)	0.035	1.192	1.176	9.812	100.467	7.206	0.92	1.484

BFGE method in Table 20.1. Another method is coupled neural P system-based method. It is referred to as NPS in Table 20.1 [21]. The values SF, EI, VIFQ, and SD for these methods are presented in Table 20.1. The proposed method (adaptive) achieved highest value for all these four image fusion quality metrics compared to these two latest image fusion methods.

Finally, the computational time of different methods is given Table 20.2. IHS method is less complex compared to other methods. The computational complexity of LRD method is high compared to other methods. The parameter adaptive PCNN method is also a relatively high computationally complex method. The quantitative evaluation of fusion results revealed that proposed adaptive fusion method is giving good spatial resolution with better spectral characteristics at relatively lower computational complexity.

20.5 Conclusions

An effective method for SPECT and MRI image fusion is proposed in this work. The proposed method utilizes distinct fusion rules adaptively based on the amount information present in the SPECT image. Morphological features-based postenhancement technique employed in the proposed image fusion framework facilitated the fused images to have high spatial resolution and good contrast. Experiments have been conducted on SPECT and MRI images of various diseases. The proposed method has been comprehensively compared to seven latest image fusion methods. Visual inspection of fusion results demonstrated that abnormalities in the source images are clearly visible in the fused images by the proposed method without any distortion. This demonstrates the clinical use of the proposed method. Further, performance evaluation with visual information fidelity, spatial frequency, edge index, and standard deviation also showed the advantage of the proposed adaptive fusion method over other image fusion methods. In future, suitability of the proposed method for the automatic lesion segmentation, the disease diagnosis, and classification may be tested.

References

[1] K.G. Baum, M. Helguera, A. Krol, Fusion viewer: a new tool for fusion and visualization of multimodal medical data sets, J. Digit. Imaging 21 (Suppl 1) (2008) S59–S68.

[2] T. Li, Y. Wang, Multiscaled combination of MR and SPECT images in neuroimaging: a simplex method based variable-weight fusion, Comput. Methods Prog. Biomed. 105 (2012) 31–39.

[3] T.-M. Tu, S.-C. Su, H.-C. Shyu, P.S. Huang, A new look at IHS-like image fusion methods, Inf. Fusion. 2 (2001) 177–186.

[4] C. He, Q. Liu, H. Li, H. Wang, Multimodal medical image fusion based on IHS and PCA, Procedia Eng. 7 (2010) 280–285.

[5] V. Bhavana, H.K. Krishnappa, Multi-modality medical image fusion using discrete wavelet transform, Procedia Comput. Sci. 70 (2015) 625–631.

[6] N. Amini, E. Fatemizadeh, H. Behnam, MRI-PET image fusion based on NSCT transform using local energy and local variance fusion rules, J. Med. Eng. Technol. 38 (2014) 211–219.

[7] C.T. Kavitha, C. Chellamuthu, Fusion of SPECT and MRI images using integer wavelet transform in combination with curvelet transform, Imaging Sci. J. 63 (2015) 17–23.

[8] G. Easley, D. Labate, W.-Q. Lim, Sparse directional image representations using the discrete shearlet transform, Appl. Comput. Harmon. Anal. 25 (2008) 25–46.

[9] W. Lim, Nonseparable shearlet transform, IEEE Trans. Image Process. 22 (2013) 2056–2065.

[10] Y. Liu, X. Chen, J. Cheng, H. Peng, A medical image fusion method based on convolutional neural networks, in: Proc. 2017 20th International Conference on Information Fusion (Fusion), City.

[11] A. Bengana, M.A. Chikh, I.B. Hacene, Multimodal medical image fusion using multi-resolution transform, Int. J. Biomed. Eng. Technol. 27 (2018) 221–232.

[12] X. Jin, G. Chen, J. Hou, Q. Jiang, D. Zhou, S. Yao, Multimodal sensor medical image fusion based on nonsubsampled shearlet transform and S-PCNNs in HSV space, Signal. Process. 153 (2018) 379–395.

[13] S. Sivasangumani, P.S. Gomathi, B. Kalaavathi, Regional firing characteristic of PCNN-based multimodal medical image fusion in NSCT domain, Int. J. Biomed. Eng. Technol. 18 (2015) 199–209.

[14] C. Panigrahy, A. Seal, N.K. Mahato, MRI and SPECT image fusion using a weighted parameter adaptive dual channel PCNN, IEEE Signal. Process. Lett. 27 (2020) 690–694.

[15] X. Li, X. Guo, P. Han, X. Wang, H. Li, T. Luo, Laplacian redecomposition for multimodal medical image fusion, IEEE Trans. Instrum. Meas. 69 (2020) 6880–6890.

[16] J. Fu, W. Li, J. Du, B. Xiao, Multimodal medical image fusion via laplacian pyramid and convolutional neural network reconstruction with local gradient energy strategy, Comput. Biol. Med. 126 (2020) 104048.

[17] A. Seal, D. Bhattacharjee, M. Nasipuri, D. Rodríguez-Esparragón, E. Menasalvas, C. Gonzalo-Martin, PET-CT image fusion using random forest and à-trous wavelet transform, Int. J. Numer. Methods Biomed. Eng. 34 (2018) e2933.

[18] P.-H. Dinh, A novel approach based on Grasshopper optimization algorithm for medical image fusion, Expert. Syst. Appl. 171 (2021) 114576.

[19] J. Jose, et al., An image quality enhancement scheme employing adolescent identity search algorithm in the NSST domain for multimodal medical image fusion, Biomed. Signal. Process. Control. 66 (2021) 102480.

[20] P. Ganasala, V. Kumar, A.D. Prasad, Performance evaluation of color models in the fusion of functional and anatomical images, J. Med. Syst. 40 (2016) 122.

[21] B. Li, et al., Medical image fusion method based on coupled neural P systems in nonsubsampled shearlet transform domain, Int. J. Neural Syst. 31 (2021) 2050050.

[22] X. Li, F. Zhou, H. Tan, W. Zhang, C. Zhao, Multimodal medical image fusion based on joint bilateral filter and local gradient energy, Inf. Sci. 569 (2021) 302–325.

[23] D.S. Shibu, S.S. Priyadharsini, Multi scale decomposition based medical image fusion using convolutional neural network and sparse representation, Biomed. Signal. Process. Control., 69, 2021, p. 102789.

[24] W. Li, Y. Zhang, G. Wang, Y. Huang, R. Li, DFENet: a dual-branch feature enhanced network integrating transformers and convolutional feature learning for multimodal medical image fusion, Biomed. Signal. Process. Control. 80 (2023) 104402.

[25] B. Li, J.-N. Hwang, Z. Liu, C. Li, Z. Wang, PET and MRI image fusion based on a dense convolutional network with dual attention, Comput. Biol. Med. 151 (2022) 106339.

[26] E.D. Vidoni, The whole brain atlas, Available from: http://www.med.harvard.edu/aanlibJ. Neurol. Phys. Ther. 36 (2012).

[27] M. Yin, X. Liu, Y. Liu, X. Chen, Medical image fusion with parameter-adaptive pulse coupled neural network in nonsubsampled shearlet transform domain, IEEE Trans. Instrum. Meas. 68 (2018) 49–64.

[28] A. Sengupta, A. Seal, C. Panigrahy, O. Krejcar, A. Yazidi, Edge information based image fusion metrics using fractional order differentiation and sigmoidal functions, IEEE Access. 8 (2020) 88385–88398.

[29] H.R. Sheikh, A.C. Bovik, Image information and visual quality, IEEE Trans. Image Process. 15 (2006) 430–444.

[30] Y. Han, Y. Cai, Y. Cao, X. Xu, A new image fusion performance metric based on visual information fidelity, Inf. Fusion. 14 (2013) 127–135.

Index

Note: Page numbers followed by "*f*" and "*t*" refer to figures and tables, respectively.

D

E

F

G

H

Q

R

S

T

U

V

W

X

Y

Z

9780443159992